International Summer School for Advanced Studies 2012 (PREDEAL12)

Dynamics of Open Nuclear Systems

Journal of Physics: Conference Series Volume 413

Predeal, Romania
9-20 July 2012

ISBN: 978-1-62276-991-9
ISSN: 1742-6588

Printed from e-media with permission by:

Curran Associates, Inc.
57 Morehouse Lane
Red Hook, NY 12571

Some format issues inherent in the e-media version may also appear in this print version.

Printed by Curran Associates, Inc. (2013)

For permission requests, please contact the Institute of Physics at the address below.

Institute of Physics
Dirac House, Temple Back
Bristol BS1 6BE UK

Phone: 44 1 17 929 7481
Fax: 44 1 17 920 0979

techtracking@iop.org

Additional copies of this publication are available from:

Curran Associates, Inc.
57 Morehouse Lane
Red Hook, NY 12571 USA
Phone: 845-758-0400
Fax: 845-758-2634
Email: curran@proceedings.com
Web: www.proceedings.com

TABLE OF CONTENTS

I. NUCLEAR STRUCTURE

II. DECAY PROCESSES

III. NUCLEAR REACTIONS AND ASTROPHYSICS

IV. CONTRIBUTIONS

Foreword

This proceedings volume contains the invited lectures and contributions presented at the International Summer School on Nuclear Physics held at Trei Brazi, a summer resort of the Bioterra University, near the city of Predeal, Romania, on 9–20 July 2012.

The long tradition of International Summer Schools on Nuclear Physics in Romania dates as far back as 1964, with the event being scheduled every two years. During this period of almost 50 years, many outstanding nuclear scientists have lectured on various topics related to nuclear physics and particle physics. This year we celebrate the 80th birthday of Aureliu Sandulescu, one of the founders of the Romanian school of theoretical nuclear physics. He was Serban Titeica's PhD student, one of Werner Heisenberg's PhD students, and he organized the first edition of this event. Aureliu Sandulescu's major contributions to the field of theoretical nuclear physics are related in particular to the prediction of cluster radioactivity, the physics of open quantum systems and the innovative technique of detecting superheavy nuclei using the double magic projectile ^{48}Ca (Calcium), nowadays a widely used method at the JINR—Dubna and GSI—Darmstadt laboratories. The title of the event, 'Dynamics of Open Nuclear Systems', is in recognition of Aureliu Sandulescu's great personality. The lectures were attended by Romanian and foreign Master and PhD students and young researchers in nuclear physics. About 25 reputable professors and researchers in nuclear physics delivered lectures during this period. According to a well-established tradition, an interval of two hours was allotted for each lecture (including discussions). Therefore we kept a balance between the school and conference format. Two lectures were held during the morning and afternoon sessions. After lecture sessions, three or four oral contributions were given by young scientists. This was a good opportunity for them to present the results of their research in front of renowned professors and researchers in nuclear physics. This proceedings volume is organized into four chapters, which reflects the traditional chapter structure of nuclear physics textbooks, but seen from the perspective of open quantum systems:

I Nuclear structure
II Decay processes
III Nuclear reactions and astrophysics
IV Contributions

The lectures and contributions are listed alphabetically by author within each chapter. The volume contains many comprehensive reviews related to the topics of the School. The first week of the School was focused on nuclear structure and decay phenomena, considering the nucleus as an open system. Experts in these fields lectured on cluster radioactivity, the stability of superheavy nuclei, alpha-decay fine structure, fission versus fusion, beta and double beta decay and pairing versus alpha-clustering. New experimental results related to the nuclear stability of low-lying and high spin states were also presented. Recent developments at JINR—Dubna and GSI—Darmstadt international laboratories were also reported by their current or former directors. The second week of the event was dedicated to the physics of exotic nuclei, heavy ion reactions and multi-fragmentation, symmetries and phase transitions of open quantum systems. The stability of the atomic nucleus is an important and always interesting discussion point, especially in the context of newly discovered nuclear systems close to the stability line, such as proton/neutron rich or superheavy nuclei. Several lectures and contributions were focused on nuclear structure models describing low-lying states. This includes the status of density functional theory, new developments in Bohr–Mottelsohn Hamiltonian and shell-model theory, proton–neutron correlations, shape coexistence, back-bending phenomena and the thermodynamics of open quantum systems. Open systems in astrophysics, such as supernovae and neutron stars, were presented in detail by several lecturers. Important topics connected to the status of the equation of state, hyperonic and quark matter and neutrino physics, as well as the applications of nuclear structure in astrophysics, were also on the School's agenda.

There were many discussions and questions both during and after presentations. An open and friendly atmosphere characterized our School, although different opinions quite often divided the participants. Many discussions continued during coffee breaks and excursions organized in the beautiful surroundings. We hope that this proceedings volume will be useful for future reference to both young scientists and senior researchers working in various fields of nuclear physics.

We cannot end without expressing our many thanks to the National Authority for Scientific Research and the Romanian Academy (Elias Foundation) for their financial support. We acknowledge the Horia Hulubei National Institute of Physics and Nuclear Engineering and Bioterra University for their important contribution in organizing the School.

Guest Editors

D S Delion, N V Zamfir, A R Raduta and F Gulminelli

International Summer School on Nuclear Physics: First Week

International Summer School on Nuclear Physics: Second Week

Sponsors

IFIN-HH

ANCS
Autoritatea Nationala pentru Cercetare Stiintifica

Int. Summer School for Advanced Studies 'Dynamics of open nuclear systems' (PREDEAL12) IOP Publishing
Journal of Physics: Conference Series **413** (2013) 011002 doi:10.1088/1742-6596/413/1/011002

Aurel Sandulescu—a life dedicated to nuclear physics

I was surprised to receive an invitation letter from Andrei Dorobantu, whom I did not know, to give a talk about pairing excitations at the Predeal International School of 1978, which I accepted. This was the first time that I would visit Romania, and I knew very little about the research that was performed in the country.

In my talk I was showing the role of the pairing mode in a rather popular theory at that time which was called Nuclear Field Theory. Suddenly I was interrupted in a rather brusque fashion by a man with an acute and loud voice, telling me that it has been shown by somebody that the Nuclear Field Theory does not converge. I was very upset by this interruption, particularly because he was right. I told him this and asked him to wait up at the end of my talk. During the coffee break the man came to me and presented himself. It was the first time I came into contact with the extraordinary person that is Aurel Sandulescu.

During the coffee break we spoke about his research, especially in alpha decay. I was interested in this subject because just then I had started to perform calculations in relation to experiments carried out at my Institution in Stockholm, which at that time was called the Research Institute of Physics.

We continued with our discussions during the whole duration of the School, often with his wife, Violeta, present. I became amazed by the extensive work he had done since the late 1950s on microscopic treatments of alpha decay. He had studied the decay of both normal and superfluid as well as spherical and deformed nuclei, all within the framework of the shell model or its BCS equivalent. I was very enthusiastic about this, since I was convinced that one should, in principle, be able to describe the decay process by using a shell model representation. I was disappointed to realize that he did not agree with me. I insisted that from a purely theoretical point of view the shell model was, rather more than a model, an excellent representation to describe nuclear processes. However, with the perspective that the passing years provides, I have since had to admit that one needs cluster components in the shell model wave functions in order to properly describe the clustering and subsequent decay. For Aurel, this realization led him to consider the decay as a fission process. It was through this that he found, just by simply looking through the penetrability, that cluster radioactivity is a valid form of decay, as was indeed confirmed experimentally several years afterwards. This new form of radioactive decay has had a profound influence on nuclear physics. It opened the way to the investigation of the structure and exotic forms that superheavy nuclei may acquire, since one expects that heavy clusters are likely to be emitted from these nuclei. The importance of cluster decay, and the theory behind its discovery, is of great relevance even today. This can be attested, for example, by the recent prediction of Poenaru, Gherghescu and Greiner (2011 *Phys. Rev. Lett.* **107** 062503) that in some superheavy nuclei this form of decay is dominant even over alpha emission, thus making cluster decay a powerful tool in the detection of superheavy nuclei.

The description of cluster decay as a fragmentation process and its relation to heavy and superheavy elements was a natural extension of the research that Aurel Sandulescu was conducting at that time. Just before the cluster decay paper appeared he addressed the important question of the production of superheavy elements. The straightforward method of reaching this is by heavy ions collisions, but the question was: which nuclei should one use as reaction partners? Sandulescu proposed choosing those

Journal of Physics: Conference Series **413** (2013) 011002 doi:10.1088/1742-6596/413/1/011002

nuclei that provide a minimum of the potential energy, which was an important step in the synthesis of superheavy nuclei.

Another subject which is of great importance in nuclear physics and in which Sandulescu played a crucial role was his prediction of neutronless spontaneous binary and ternary fission. He was so elated and eager about this that he even contributed to the experimental discovery of this rare phenomenon. Since then, there have been many theoretical as well as experimental studies performed in this and similar subjects related to cold fission processes. Particularly relevant in the framework of this conference are his studies on the damping of collective modes in deep inelastic collisions. He treated this difficult subject by extending a formalism derived for open systems to nuclear collective motions. The master equation that he thus obtained describes well the dynamics of degrees of freedom not only in the damping of excited nuclear states, but also particularly in heavy ion collisions and in the decaying of the most collective states in nuclear physics—giant resonances.

He has also analyzed the complicated mechanisms that induce the clustering of nucleons, which led to his realization that the cluster structure can be viewed as solitons moving on the nuclear surface.

This intense scientific work was followed by other related activities. In Romania he became a full professor in 1970, a corresponding member of the Romanian Academy of Sciences in 1991, an ordinary member in 1992 and Vice President in the period 1994–1998. Most remarkable is that he tried to influence the development of Science in the country by himself becoming a politician. Thus, he was a member of the Romanian Parliament in the period 1996–2000. Outside of Romania he has also undertaken remarkable activities. He has been Invited Scientist and Invited Professor to many institutions around the world. One should mention in particular his position as Vice Director of the Joint Institute of Nuclear Research, Dubna, Russia, in the period 1983–1986. He has proceeded through all these endeavors by attracting many foreign scientists to Romania, including myself. But especially important was his influence upon students, many of whom are today recognized physicists and leaders of Romanian physics.

It is in recognition of his contribution to physics, especially nuclear physics, as well as his work for Romanian science that this School is organized in his honour, on the occasion of his 80th birthday.

I cannot finish this short account without mentioning the person that accompanied Aurel Sandulescu through all his life, his wife of more than fifty years, Violeta. She also was an exceptional person. In her youth she was a very dedicated sporty girl, very athletic, very courageous, very fair and, above all, very generous. These qualities remained with her throughout her life. With deep sorrow I learnt that she passed away three months ago, still youthful and enthusiastic.

R J Liotta
KTH, Alba Nova University Center, SE-10691 Stockholm, Sweden

List of participants

Claudia Anghel	University of Bucharest	anghel_clau@yahoo.com
Ismahane Ami	Université M'hamed Bougara de Boumerdès	amiismahane@yahoo.fr
Nicusor Arsene	University of Bucharest	nicusorarsene@spacescience.ro
Claudiu Badea	University of Bucharest	claudiu.cornel@yahoo.com
Virgil Baran	University of Bucharest	virgilbaran@yahoo.com
Dorel Bucurescu	IFIN-HH, Bucharest	bucurescu@nipne.ro
Andreea Budaca	IFIN-HH, Bucharest	abudaca@theory.nipne.ro
Radu Budaca	IFIN-HH, Bucharest	budacaradu@gmail.com
Petrica Buganu	IFIN-HH, Bucharest	buganu_p@yahoo.com
Stefana Calinescu	IFIN-HH, Bucharest	stefana.calinescu@tandem.nipne.ro
Vasile Candea	Academy of Romanian Scientists	aosromania@yahoo.com
Andreea Croitoru	University of Bucharest	c.m.andreea@gmail.com
Cristian Costache	IFIN-HH, Bucharest	cristian.costache@nipne.ro
Dan Cozma	IFIN-HH, Bucharest	cozma@niham.nipne.ro
Maria Colonna	LNS-Catania	colonna@lns.infn.it
Dana Danila	University Bioterra	godja_daniela_ileana@yahoo.com
Doru Delion	IFIN-HH, Bucharest	delion@theory.nipne.ro
Alexandru Dumitrescu	University of Bucharest	a.dumitrescu.fizica@gmail.com
Mihai Forascu	University Bioterra	
Ioana Gheorghe	IFIN-HH, Bucharest	ioana.gheorghe@nipne.ro
Georgiana Giubeca	University of Bucharest	giubega_georgiana@yahoo.com
Lavinia Giubeca	University of Bucharest	giubega_elena@yahoo.com
Nicoleta Giurgea	University of Bucharest	giurgeanico@yahoo.com
Walter Greiner	Frankfurt Institute for Advanced Studies (FIAS)	greiner@fias.uni-frankfurt.de
Francesca Gulminelli	LPC-Caen and University of Lower Normandy	gulminelli@lpccaen.in2p3.fr
Mihai Horoi	Central Michigan University	horoi@phy.cmich.edu
Mihail Itkis	JINR-Dubna	itkis@jinr.ru
Liviu Ixaru	IFIN-HH, Bucharest	ixaru@ifin.nipne.ro
Markus Kortelainen	Univesity of Jyväskylä	markus.kortelainen@jyu.fi

Bao-An Li	Texas A&M, Commerce	Bao-An.Li@tamuc.edu
Razvan Lica	IFIN-HH	razvan.lica@nipne.ro
Roberto J Liotta	Royal Institute of Technology, Stockholm	liotta@kth.se
Roland Lombard	IPN-Orsay	roland.lombard@laposte.net
Catalin Manu	University of Bucharest	catalin.manu@infim.ro
Paul Mereuta	University of Bucharest	pmereuta@yahoo.com
Mihai Mirea	IFIN-HH, Bucharest	mirea@theory.nipne.ro
Serban Misicu	IFIN-HH, Bucharest	misicu@theory.nipne.ro
Herve Molique	University of Strasbourg and IN2P3/CNRS	Herve.Molique@iphc.cnrs.fr
Gottfried Münzenberg	GSI-Darmstadt	G.Muenzenberg@gsi.de
Andrei Neacsu	IFIN-HH	nandrei@theory.nipne.ro
Adina Olacel	University of Bucharest	adinaolacel@gmail.com
Dragos Palade	University of Bucharest	paladedrgs@yahoo.com
Margarit Pavelescu	Academy of Romanian Scientists	mpavelescu2002@yahoo.com
Cristina Petrone	IFIN-HH	cristina.petrone@tandem.nipne.ro
Alexandrina Petrovici	IFIN-HH	apetro@nipne.ro
Mihai Petrovici	IFIN-HH	mpetro@nipne.ro
Octavian Postavaru	IFIN-HH	opostavaru@niham.nipne.ro
Tudor Prisacaru	ANCS	
Constança Providência	Universidade de Coimbra	cp@teor.fis.uc.pt
Chong Qi	Royal Institute of Technology, Stockholm	chongq@kth.se
Adriana R Raduta	IFIN-HH	araduta@nipne.ro
Apolodor A Raduta	IFIN-HH	raduta@nipne.ro
Cosmin Romanitan	University of Bucharest	romanitan.cosmin@yahoo.com
Zhongzhou Ren	Nanjing University	zren@nju.edu.cn
Alexandra Saftoiu	IFIN-HH	allixme@gmail.com
Hiroyuki Sagawa	University of Aizu	sagawa@u-aizu.ac.jp
Aurel Sandulescu	Romanian Academy	
Peter Schuck	IPN-Orsay	schuck@ipno.in2p3.fr

Nafiseh Shayan Shakib	Birjand University, Iran	nafiseh.shakib@gmail.com
Armen Sedrakian	ITP, Frankfurt am Main	sedrakian@th.physik.uni-frankfurt.de
Ion Silisteanu	IFIN-HH silist@ifin.nipne.ro	silist@theory.nipne.ro
Eliade Stefanescu	Center for Advanced Studies in Physics	eliadestefanescu@yahoo.fr
Kumar Suresh	University of Delhi	sursvmk123@gmail.com
Jouni Suhonen	University of Jyväskylä	jouni.suhonen@phys.jyu.fi
Livius Trache	Texas A&M University	trache@comp.tamu.edu
Dana Tudor	University of Bucharest	tudordana29@yahoo.com
Lucian Tudor	University of Bucharest	luciantudor89@yahoo.com
Stefan Typel	GSI-Darmstadt	s.typel@gsi.de
Iuliana Visan	RAAN-Pitesti	iuliana.visan@yahoo.com
Michal Warda	Maria Curie-Sklodowska University, Lublin	warda@kft.umcs.lublin.pl
Furong Xu	Peking University	frxu@pku.edu.cn
Roxana Zus	University of Bucharest	roxanazus@gmail.com

Peer review statement

All papers published in this volume of *Journal of Physics: Conference Series* have been peer reviewed through processes administered by the proceedings Editors. Reviews were conducted by expert referees to the professional and scientific standards expected of a proceedings journal published by IOP Publishing.

Journal of Physics: Conference Series **413** (2013) 012001 doi:10.1088/1742-6596/413/1/012001

Geometrical symmetries in atomic nuclei: From theory predictions to experimental verifications

J Dudek[1], A Góźdź[2], H Molique[1] and D Curien[1]

[1]IPHC, Department of Subatomic Research and University of Strasbourg, 23 rue du Loess, F 67-037 Strasbourg Cedex 2, France

[2]Institute of Physics, University of Maria Curie-Skłodowska, pl. Marii Curie-Skłodowskiej 1, PL-20-031 Lublin, Poland

E-mail: Jerzy.Dudek@unistra.fr, Andrzej.Gozdz@umcs.lublin.pl, Herve.Molique@unistra.fr, Dominique.Curien@iphc.cnrs.fr

Abstract. In the lectures delivered at the 2012 Predeal School an overview has been presented of the contemporary theory of the nuclear geometrical (shape) symmetries. The formalism combines two most powerful theory tools applicable in the context: The group- and group-representation theory together with the modern realistic mean-field theory. We suggest that all point-groups of symmetry of the mean-field Hamiltonian, sufficiently rich in symmetry elements (as discussed in the text) may lead to the magic numbers that characterise such a group in analogy with the spherical magic gaps characterising nuclear sphericity. We discuss in simple terms the mathematical and physical arguments for the presence of such symmetries in nuclei. In our opinion: It is not so much the question of *Whether?* - but rather: *Where in the Nuclear Chart several of the point group-symmetries will be seen?* We focus our presentation on the tetrahedral symmetry with the magic numbers calculated to be 32, 40, 56, 64, 70, 90 and 136, and discuss qualitatively the problem of the formulation of the experimental criteria which would allow for the final discovery of the tetrahedral symmetry in subatomic physics.

1. Introduction: Nuclear interactions and nuclear shapes

The notion of geometrical symmetries in physics is usually associated with the notion of shapes. In the present context these will be the *nuclear shapes*. However, already at this point an attentive reader may become alert. Indeed, the nuclei are quantum systems, whereas *shapes* are usually identified with 3-dimensional surfaces, say Σ, with a generic equation of the form: $f(x, y, z) = 0$, the latter usually referred to as 'classical'. This is because, strictly speaking, there is no difference in terms of the mathematical structure between Σ' referring to, e.g., the surface of the rugby-ball and Σ - to a nuclear form. And yet: Some physicists talk about shapes of nuclei as if the latter were metal toys... But even if it should make sense to talk about surfaces describing 'shapes of quantum systems': To what extent should such imaginary surfaces be useful to inform us about nuclear interactions?

To approach the solution of our puzzle recall, that according to the results of numerous experiments: a. The volume of a nucleus is approximately equal to the sum of the volumes of the constituent nucleons, what implies that nuclei can be considered as 'tightly packed' with nuclear-matter balls (read: nucleons) of $d \approx 2.5$ Fm diameter; b. Independently, the strong nuclear interactions in the vacuum are of a short range, comparable with the nucleonic sizes. Because

Journal of Physics: Conference Series **413** (2013) 012001 doi:10.1088/1742-6596/413/1/012001

of the short-range of the nuclear interactions, nucleonic density in a nucleus, $\rho(x, y, z)$, falls off to zero within a very narrow layer, sometimes referred to as nuclear skin. The existence of such a thin layer in which nucleonic density decreases from, say 90% to, say, 10% of the maximum value suggests that the associated notion of an auxiliary surface Σ, defined by the equation: $\rho(x, y, z) = \frac{1}{2}\rho_{max}$, can be used to depict conveniently the area of space in which bulk of the nucleons can be found i.e. inside of such a surface.

Let us consider Hamiltonian of an A-body system, $\hat{H}$, represented with the help of the two-body nucleon-nucleon interactions: $\hat{H} = \hat{t} + \frac{1}{2}\sum_{i\neq j}\hat{v}_{i\leftrightarrow j}$. Such a hypothesis may seem simplistic these days since there exists an increasing evidence for the presence of the nuclear three-body forces - but it will be amply sufficient in the present context. This interaction Hamiltonian is by construction Hermitian, symmetric under the exchange of particles i and j, translation and rotation invariant (thus scalar) etc. This last condition, however, is challenging our imagination: How come, that the Hamiltonian whose orientation with respect to the reference frame cannot be defined ($\hat{H}$ is scalar) can produce non-spherical surfaces allowing to talk about nuclear shapes, their orientation and symmetries in our 3D space? Understanding of this part of the puzzle leads us to the concept of the spontaneous symmetry breaking according to which the symmetry of the lowest energy solution may turn out to be different from that of the underlying Hamiltonian. Indeed, according to numerous experiments, atomic nuclei are in a great majority non-spherical (deformed) and we are lead to conclude that, in reality, the short-range nuclear forces must be present in the non-spherical 'cavities of space' - inside deformed surfaces delimiting the nuclear density distributions introduced earlier.

We arrive in this way at the basic concept of one of the most successful nuclear theories, the nuclear mean-field theory, according to which the nucleonic motion in a nucleus is governed, to an approximation, by a generally non-spherical one-body potential whose action is limited essentially to the area of the nucleonic presence: the area of the nucleonic spatial density - thus defined by the surface Σ. To be more specific, let us denote the observables needed to describe the quantum motion of a nucleon by $\hat{x} \equiv \{\hat{r}, \hat{p}, \hat{s}, \hat{t}\}$, where we have introduced the operators of position, momentum, spin and isospin, respectively. According to the mean-field approximation, our many-body Hamiltonian takes the form ($\hat{t}_i$ - kinetic energy of the ith nucleon):

$$\hat{H}(\hat{x}_1, \hat{x}_2, \ldots\, \hat{x}_A) \approx \sum_{i=1}^{A} \hat{h}(\hat{x}_i) \;\; \text{where} \;\; \hat{h}(\hat{x}_i) = \hat{t}_i + \hat{v}(\hat{x}_i), \tag{1}$$

and where the nuclear mean-field operator, $\hat{v}$, remains common for al the nucleons. One of the most successful phenomenological a realisation of such an interaction is based on the so-called Woods-Saxon deformed potential[1] composed of the nuclear central- and the spin-orbit terms $\hat{v} \equiv V_{cent}(\hat{r}; V_o, r_o, a_o) + V_{so}(\hat{r}, \hat{p}, \hat{s}; V_{so}, r_{so}, a_{so})$ together with the Coulomb term for protons.

2. Nuclear mean-field Hamiltonian and groups of symmetry

It becomes clear from the expressions in footnote[1], that if the potential-generating surface-Σ is invariant under a certain symmetry operation, say $\hat{g}$, so is the generated mean-field potential $\hat{v}$. It then follows that if a group G is a group of symmetry of Σ [read: $\forall \hat{g} \in G$, acting with $\hat{g}$ on Σ leaves the latter unchanged] then G is also a group of symmetry of the mean-field Hamiltonian.

To examine the symmetry properties of the nuclear mean-field problem we actually need to be able to solve an inverse problem: Given group $\mathcal{G} = \{\hat{\mathcal{O}}_1, \hat{\mathcal{O}}_2, \ldots\, \hat{\mathcal{O}}_f\}$ whose impact on the nucleonic motion in the nucleus we wish to examine. How to construct a *realistic* Hamiltonian

[1] The expression for the central Woods-Saxon potential is $V_{cent} \equiv V_c/\{1+\exp[dist_{\Sigma_{(r_c)}}(\vec{r})/a_c]\}$, whereas the one for the spin-orbit term has the form $V_{so} \equiv \left\{\vec{\nabla}\left\{V_{so}/\{1+\exp[dist_{\Sigma_{(r_{so})}}(\vec{r})/a_{so}]\}\right\} \wedge \hat{p}\right\} \cdot \hat{s}$. In both expressions $dist_{\Sigma}(\vec{r})$ denotes the distance from the actual nucleonic point-position $\vec{r}$ to the nuclear surface Σ.

invariant under all transformations in $\mathcal{G}$? To solve this problem let us introduce the basis of spherical harmonics, $\{Y_{\lambda\mu}(\vartheta,\varphi)\}$, and the corresponding expansion expressing Σ under the condition that the volume contained within Σ does not depend on the actual deformation. The latter condition is taken care of by introducing the constant-volume condition-function $c(\{\alpha\})$

$$\Sigma: \quad R(\vartheta,\varphi) = R_o c(\{\alpha\})[1 + \sum_{\lambda=2}^{\lambda_{max}} \sum_{\mu=-\lambda}^{\lambda} \alpha^{\star}_{\lambda\mu} Y_{\lambda\mu}(\vartheta,\varphi)] \text{ with } R_o = r_o A^{1/3}. \tag{2}$$

The condition of invariance reads now: $\forall\, \hat{\mathcal{O}}: \Sigma \xrightarrow{\hat{\mathcal{O}}} \Sigma' \equiv \Sigma$. The latter can be written down as

$$\sum_{\lambda=2}^{\lambda_{max}} \sum_{\mu=-\lambda}^{\lambda} \alpha^{*}_{\lambda\mu} [\hat{\mathcal{O}} Y_{\lambda\mu}(\vartheta,\varphi)] = \sum_{\lambda=2}^{\lambda_{max}} \sum_{\mu'=-\lambda}^{\lambda} \alpha^{*}_{\lambda\mu'} Y_{\lambda\mu'}(\vartheta,\varphi). \tag{3}$$

In what follows we will need a representation of the operators $\hat{\mathcal{O}} \in \mathcal{G}$ adapted to the action on the spherical harmonics. The operators of proper rotations can be written down as

$$\hat{\mathcal{O}} \to \hat{R}(\Omega) \equiv \exp(i\alpha\hat{\jmath}_z)\exp(i\beta\hat{\jmath}_y)\exp(i\gamma\hat{\jmath}_{z'}) \text{ where } \Omega \equiv \{\alpha,\beta,\gamma\} - \text{Euler angles}, \tag{4}$$

and where the matrix elements of operators $\hat{R}(\Omega)$ within the basis of states $\{|\lambda\mu\rangle\}$ are called Wigner functions. We have: $\langle\lambda\mu'|\hat{R}(\Omega)|\lambda\mu\rangle \equiv D^{\lambda}_{\mu'\mu}(\Omega)$, according to the usual notation. Within this notation the invariance condition for the considered surface takes the form (cf. [1])

$$\sum_{\lambda=2}^{\lambda_{max}} \sum_{\mu=-\lambda}^{\lambda} \alpha^{*}_{\lambda\mu} [\hat{\mathcal{O}}\, Y_{\lambda\mu}(\vartheta,\varphi)] = \sum_{\lambda=2}^{\lambda_{max}} \sum_{\mu=-\lambda}^{\lambda} \alpha^{*}_{\lambda\mu} \sum_{\mu'=-\lambda}^{\lambda} D^{\lambda}_{\mu'\mu}(\Omega)\, Y_{\lambda\mu'}(\vartheta,\varphi). \tag{5}$$

The latter can be written down in the form of an identity as

$$\sum_{\mu'=-\lambda}^{\lambda} \sum_{\lambda=2}^{\lambda_{max}} [\sum_{\mu=-\lambda}^{\lambda} \alpha^{*}_{\lambda\mu} D^{\lambda}_{\mu'\mu}(\Omega) - \alpha^{*}_{\lambda\mu'}]\, Y_{\lambda\mu'}(\vartheta,\varphi) = 0, \quad \forall\, \vartheta,\varphi. \tag{6}$$

Spherical harmonics are linearly independent from where it follows the system of equations

$$\sum_{\mu=-\lambda}^{\lambda} [D^{\lambda}_{\mu'\mu}(\Omega_k) - \delta_{\mu'\mu}]\, \alpha^{\star}_{\lambda\mu} = 0; \;\; k = 1, 2, \ldots f. \tag{7}$$

Above, Ω_k are fixed sets of Euler angles corresponding to $\mathcal{O}_k$. For instance for a four-fold $\mathcal{O}_z$-axis we find $\Omega = \{\pi/2, 0, 0\}$ etc. Solutions can be taken as eigen-vectors of the $(2\lambda+1) \times (2\lambda+1)$ matrix $D^{\lambda}_{\mu'\mu}(\Omega_k)$ with the eigen-value equal $+1$.

The following observations can be made. Firstly, for a fixed representation of a given group $\mathcal{G}$ with the help of the triplets of Euler angles $\{\Omega_k;\, k = 1, 2, \ldots f\}$, relation (7) represents f systems of algebraic equations for the ensembles of coefficients $\{\alpha_{\lambda\mu}\}$ for each $\lambda = 2, 3, \ldots \infty$. Such systems are often strongly over-defined i.e. they have more equations than unknowns and generally may only have some non-trivial solutions precisely because of the imposed symmetry (see below for particular examples). Secondly, let us notice that the above system of equations is uniform. This allows to select, e.g. $\bar{\alpha}_{\lambda\mu=0}$ as an independent parameter, which uniquely fixes all the other non-zero components. Therefore, when solutions (i.e. families of surfaces represented at each given order λ by the set of parameters $\{\alpha_{\lambda\mu}\}$) exist, they can be parametrized at each order λ with the help of $(2\lambda+1)$ real parameters[2] out of which only one is independent!

In this way we may consider having prepared the 'workshop' allowing to construct the mean-field Hamiltonian at any predefined point-group symmetry.

[2] Observe that for nuclear surfaces Σ to be real, one needs the condition $\alpha^{*}_{\lambda\mu} = (-1)^{\mu}\alpha_{\lambda-\mu}$ what limits the freedom of originally $(4\lambda+1)$ real parameters to $(2\lambda+1)$ real parameters only.

3. Groups of symmetry and symmetry-implied conditions for nuclear stability

So far we have addressed important fundamental issues in the context of our lectures such as: How come that the rotation-insensitive many-body Hamiltonians may simulate the behaviour of quantum systems whose orientation in space can be defined since the system is non-spherical? Or: How to construct the mean-field Hamiltonian with an arbitrarily predefined symmetry so that we may analyse symmetry's impact on the system of interest through realistic calculations?

In what follows we will discuss the link between the properties of the nucleonic single-particle spectra directly influenced by the shapes of the nuclear surface and in particular the link between the level spacing and the total nuclear energy - thus the overall nuclear stability.

3.1. Single-particle energy-gaps and total energy minima: The mean-field view-point

In this section we wish to discuss yet another fundamental aspect: In order that symmetries of quantum systems become of 'practical' interest it is necessary that *stationary states* of the system manifest such symmetries so that the associated phenomena can be studied in laboratory. Stationary states arise when the total energy of the system possesses a local minimum at the configurations manifesting a geometrical symmetry and it will be now our goal to explain how such a link can be obtained within the mean-field nuclear theories.

Contemporary realistic calculations of nuclear energies can be performed within the nuclear mean-field theory framework, either using the self-consistent Hartree-Fock and/or constrained Hartree-Fock type approaches or phenomenological Hamiltonians like the ones based on the Woods-Saxon potential described earlier. Our discussion leading to Eq. (7) suggests that the phenomenological models in which the symmetry of the generating surface can be predefined in advance are the best suited for the studies of the nuclear symmetry, the argument valid in addition to the argument of the predictive power, superior in the case of the phenomenological as compared to the present-day self-consistent models.

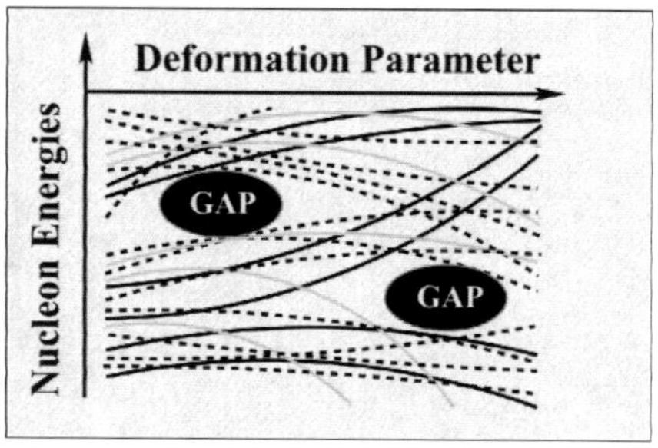

Figure 1. Schematic illustration: The nucleonic single particle energy levels in function of an ensemble of the deformation parameters, the latter represented with the help of a single axis, as calculated within the mean-field theories. Gaps in the spectra represent configurations of increased stability.

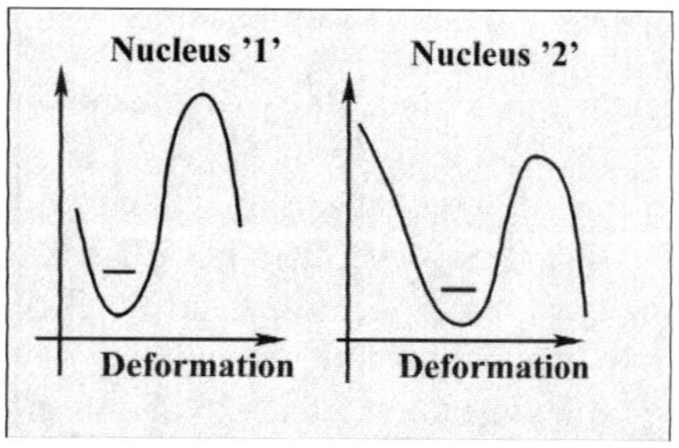

Figure 2. Schematic illustration: Total energy calculations, either Hartree-Fock or phenomenological, give the local minima, when the numbers of nucleons correspond to occupation of all the levels below the gap, figure 1. For neighbouring particle numbers the minima are usually less pronounced.

Whereas self-consistent models give directly an approximation of the total nuclear energies, the phenomenological models use the so-called Strutinsky method allowing to combine the macroscopic liquid drop model energy, E_{ld}, with the so-called shell-energies, E_{shell}, to approximate the final result: $E_{total} = E_{ld} + E_{shell}$. Whichever of the two techniques used, the final outcome is similar: The gaps in the single-particle spectra, figure 1, imply the presence of the total energy minima. In figure 2 they correspond to a larger (left) and a smaller (right) particle numbers, particles occupying the lowest single particle levels below the gaps. It then follows that, semi-qualitatively, the search for the stable nuclear configurations represented by the local minima on the total energy surfaces is strongly related to the search of the big gaps in the single-nucleon energy spectra.

3.2. Search for maximum energy-gaps with group-representation theory as a guideline

Given a mean-field theory Hamiltonian $\hat{H}$ and a group: $\mathcal{G} = \{\mathcal{O}_1, \mathcal{O}_2, \ldots \mathcal{O}_f\}$. Assume, that $\mathcal{G}$ is a symmetry group of $\hat{H}$, i.e. $[\hat{H}, \mathcal{O}_k] = 0$ with $k = 1, 2, \ldots f$. Let irreducible representations of $\mathcal{G}$ be $\{\mathcal{R}_1, \mathcal{R}_2, \ldots \mathcal{R}_r\}$ and denote their dimensions $\{d_1, d_2, \ldots d_r\}$, respectively. One can show that the eigenvalues $\{\varepsilon_\nu\}$ of the problem can be split into families of non-interacting levels

$$\hat{H}\psi_\nu = \varepsilon_\nu \psi_\nu : \quad \{e_\nu\} \to \{e^{(1)}_{\nu_1}\} \oplus \{e^{(2)}_{\nu_2}\} \oplus \ldots \oplus \{e^{(r)}_{\nu_r}\} \tag{8}$$

where $\{e_{\nu_\kappa}\}$ denotes the ensemble of all the levels belonging to the irreducible representation $\mathcal{R}_\kappa$ and thus characterised by the same degeneracy d_κ. When deformation changes while respecting the symmetry (see below) - the levels belonging to the same irreducible representation vary 'repelling each other', the mechanism known in quantum theories as Landau-Zener effect. Thus they fill the whole available energy interval defined by the depth of the potential well, cf. figure 3. Following this repulsion mechanism, the average level spacing increases, very roughly, by the factor of "r". Consequently, a relatively big level-spacing is created in all the r families with the increased chance that by superposing all the levels in one common energy window, some big gaps will 'survive' (right-hand side of figure 3). Let us note in passing that levels belonging to different irreducible representations cross each other in function of deformation without what we call 'level-interaction' - where from the term 'non-interacting' in the text above equation (8).

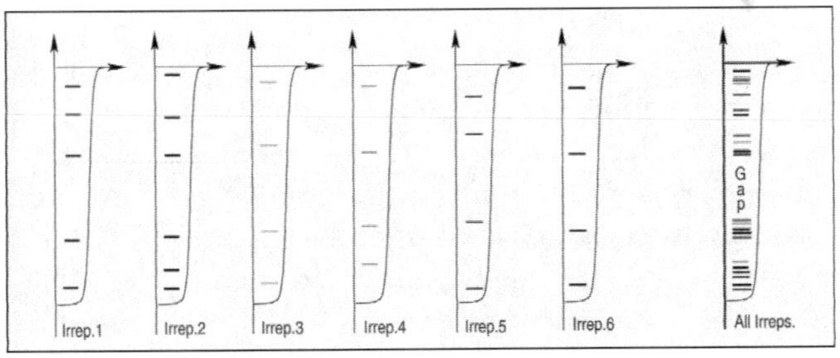

Figure 3. Schematic illustration: Global structure of the mean-field single-particle spectrum of a Hamiltonian invariant under transformations of the symmetry group $\mathcal{G}$ with $r = 6$ irreducible representations. Degeneracies of levels within each of the potential wells are equal to d_κ, for $\kappa = 1, 2, \ldots, 6$.

From what has been said, it becomes clear that the key-message from the group theory considerations is this:

> *Respecting certain geometrical symmetries by a nucleus will increase its chances for the biggest gaps in the single-particle spectra and thus for a possible existence of the energy minima at stable group-symmetric configurations if the underlying symmetry group has:*
> *A. Possibly largest number of irreducible representations, and*
> *B. The irreducible representations in question are of the highest possible dimensions.*

Indeed, since the dimension of a given irreducible representation is equal to the degeneracy of the corresponding levels, it follows that the second of the above conditions may play a helpful role in increasing even more the average level-spacing. It then becomes clear that the next logical step to follow will be to determine which known symmetry groups fulfil the above conditions in the most satisfactory manner?

The groups of symmetry of finite bodies in the 3D space are very well known in particular from the molecular physics considerations. On can show that all the symmetry elements of all these groups must leave at least one point of the body unchanged - and so the symmetry groups in question are refereed to as (molecular) *point-groups*. There exist a well know hierarchy referred to as "32 point-group symmetry chain" illustrated, in figure 4.

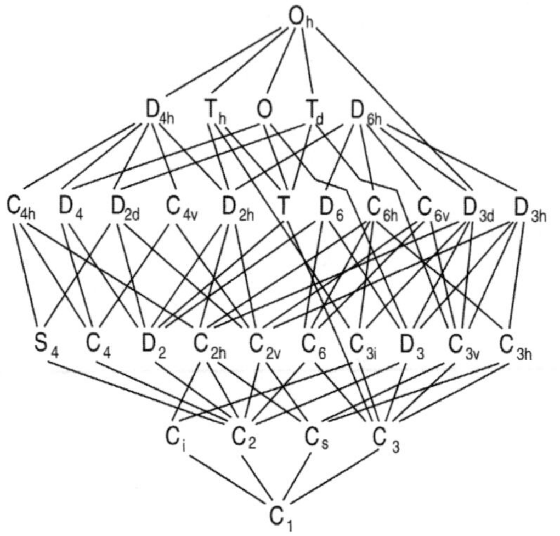

Figure 4. The highest in hierarchy of the 32 symmetry point-groups is O_h, the octahedral group, which is composed of 48 symmetry elements. Several groups in this diagram are of potential interest for the theory of nuclear structure; of particular interest for us (see text) is the tetrahedral group, T_d, next in hierarchy, composed of 24 symmetry elements. The latter does not contain inversion and can serve for modelling certain important properties of the negative parity rotational bands in nuclei. The triaxial quadrupole-deformed nuclei are described by D_{2h}, lower in the hierarchy. [Based on Koster et al., ref. [2]]

The scheme shown in figure 4 contains the so-called classical (or single) as opposed to double point groups. Without entering into mathematical details let us mention that to describe the properties of motion of Fermions whose wave-function have a spinor character we have to assure that, in accordance with the general and well known rules of quantum mechanics, a rotation of a Fermion wave-function through the angle of 2π must change the sign of such a wave function. To ensure such a property one adds, after H. Bethe, a special new element to each of the single groups what, again in agreement with the well know group properties, actually doubles the number of the groups elements, where from the name *double point-groups*. The double groups have very different mathematical properties, as compared to the prototype single groups, and differ in particular in numbers and dimensions of the corresponding irreducible representations.

One finds that among 32 *double* point groups all but octahedral and tetrahedral ones have two-dimensional irreducible representations. The exception from this rule are four-dimensional representations in the groups O_h^D and T_d^D, which will therefore receive a particular attention in what follows. Moreover, the octahedral double point group has 6 irreducible representations whereas the octahedral one has 3 (cf. also overlays 71 and 72 of ref. [3]).

Let us mention in passing the particular and fundamental role in quantum physics of the two-dimensional irreducible representations leading to the two-fold (the so-called Kramers, or spin-down - spin-up) degeneracies, which take place for all systems of Fermions whose Hamiltonians are *not* explicitly time-dependent - thus time-reversal invariant - in particular in all non-spherical nuclei whose the so-called Nilsson orbitals carry always the Kramers degeneracy.

Inspection of figure 4 strongly suggests that there may be several point groups totally unexplored so far in the nuclear structure considerations and our preliminary calculations indicate that some of them are probably present in certain super-deformed nuclei, groups such as, e.g., C_{4h}, D_{3d} or D_{3h}. In what follows we focus on tetrahedral (also octahedral) groups[4].

Journal of Physics: Conference Series **413** (2013) 012001 doi:10.1088/1742-6596/413/1/012001

4. Focus on nuclear tetrahedral and octahedral symmetries

To be able to examine, through realistic calculations, the symmetry implications of the two symmetry-groups for nuclei, we have applied a simple formalism leading to equation (7) in order to arrive at the computer programmable expressions for the tetrahedral-symmetric surfaces. As already mentioned, only *special combinations* of spherical harmonics, $Y_{\lambda\mu}$, may form a basis for surfaces with tetrahedral symmetry and *only* for λ-odd. The three lowest-order solutions are[1]:

$$\left.\begin{array}{ll} \lambda = 3: & \alpha_{3,\pm 2} \equiv t_3 \\ \lambda = 5: & \text{no solution possible} \\ \lambda = 7: & \alpha_{7,\pm 2} \equiv t_7; \quad \alpha_{7,\pm 6} \equiv -\sqrt{\frac{11}{13}} \cdot t_7 \\ \lambda = 9: & \alpha_{9,\pm 2} \equiv t_9; \quad \alpha_{9,\pm 6} \equiv +\sqrt{\frac{28}{198}} \cdot t_9 \end{array}\right\}. \tag{9}$$

Above, symbols t_3, t_7 and t_9 represent the independent tetrahedral symmetry deformations of order 3, 7 and 9. The equation for the tetrahedral-symmetric nuclear surface in the third order will therefore be proportional to $\{1 + t_3[Y_{3,+2}(\vartheta,\varphi) + Y_{3,-2}(\vartheta,\varphi)]\}$, whereas the one in the seventh order, to $\{1 + t_7[Y_{7,+2}(\vartheta,\varphi) + Y_{7,-2}(\vartheta,\varphi) - \sqrt{11/13}Y_{7,+6}(\vartheta,\varphi) + \sqrt{11/13}Y_{7,-6}(\vartheta,\varphi)]\}$, etc. Analogous expressions for the octahedral symmetry are illustrated in the overlays 129-138 of Ref. [3]. Let us note that the classical prototypes of the Platonic figures with the symmetries

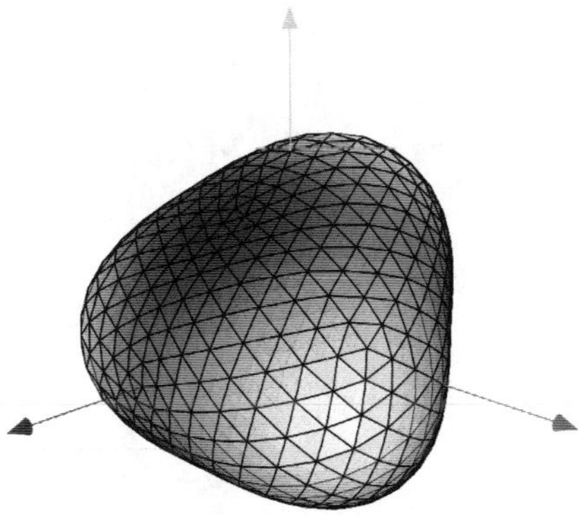

Figure 5. An example of a tetrahedral-symmetric surface at $t_3 = 0.2$; for the corresponding single nucleon levels see overlays 113-114 of ref. [3].

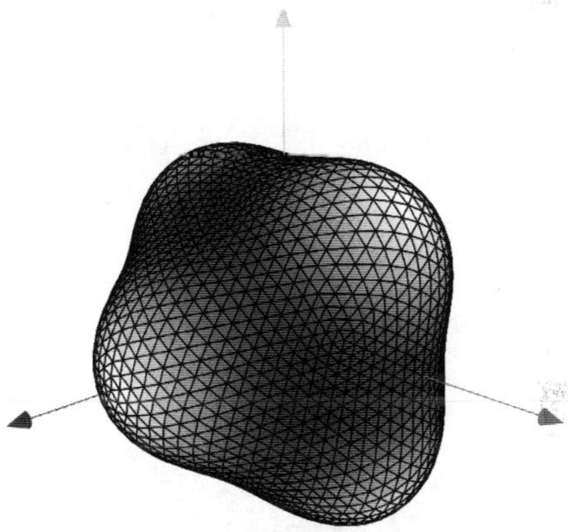

Figure 6. An example of an octahedral-symmetric surface at $o_4 = 0.2$; for the corresponding single nucleon levels see overlays 139-140 of ref. [3].

discussed here are: A 'pyramid' with a triangular basis (tetrahedron) and a 'diamond shape' (octahedron). The mathematical properties of the solutions illustrated guarantee that the surfaces remain symmetric with respect to all the symmetry elements of their respective groups for all values of parameters t_3, t_7, o_4 etc.; although in this context it does not make sense to say that one surface is 'more tetrahedral than the other', nevertheless it makes sense to say that a given tetrahedral symmetric surface is more deformed than the other.

5. Towards experimental verification

Theoretical analysis of the problem of symmetries must aim, among others, at establishing their possible measurable signs. We will arbitrarily introduce two lines of reasoning more related to pedagogy of the discussion rather than the mathematical rigour: Certain criteria can be viewed upon as *necessary vs. sufficient conditions* whereas some others as *static vs. dynamic conditions.* Often the necessary conditions are simpler to be presented. They have the form of statements:

Suppose a nuclear quantum state is generated by the tetrahedral-symmetric Hamiltonian - What are the implied characteristic features of such states? Sufficient conditions follow an opposite logic: By establishing a certain property we will be sure that the studied system possesses a given symmetry. The latter conditions (if exist) are usually more difficult to establish experimentally.

We will begin with the historically first and pedagogically simplest (although somewhat naive) a selection of necessary *static* criteria. Later we will argue that although the related physical properties are worthwhile discussing, they will need to be super-seeded with the *dynamical* (as opposed to static) ones for reasons that will also be discussed.

5.1. Possible experimental signals of tetrahedral symmetry in nuclei: Static, necessary criteria

Given the fact that the tetrahedral symmetric surfaces are expressed with the help of the harmonics of the order $\lambda = 3$, viz. $Y_{3\pm 2}$, the signals of the corresponding shapes can be sought in the negative parity rotational bands in nuclei. This is in analogy to one-phonon negative-parity octupole bands (vibrational or static, see below) associated with the axially symmetric pear-shape nuclei, phonon solutions constructed with the help of the Y_{30} spherical harmonic.

Consider an academic case of four families of octupole deformed nuclear surfaces, defined using $\lambda = 3$ deformations: α_{30}, α_{31}, α_{32} and α_{33}. Introduce next the uniform charge-density distribution, $\rho_\Sigma(\vec{r})$, and express the multipole moments as usual by

$$\rho_\Sigma(\vec{r}) = \begin{cases} \rho_0 : & \vec{r} \in \Sigma \\ 0 : & \vec{r} \notin \Sigma \end{cases} \quad \rightarrow \quad Q_{\lambda\mu} = \int \rho_\Sigma(\vec{r})\, r^\lambda\, Y_{\lambda\mu}\, d^3\vec{r}, \tag{10}$$

to finally calculate one by one the associated *quadrupole* moments as functions of $\alpha_{3\mu}$, using the Taylor expansion valid approximately for small deformations. Somewhat schematically:

$$Q_{\lambda\mu}(\alpha) \approx Q_{\lambda\mu}\Big|_{\alpha=0} + Q'_{\lambda\mu}\Big|_{\alpha=0}\Delta\alpha + \frac{1}{2}Q''_{\lambda\mu}\Big|_{\alpha=0}\Delta\alpha\Delta\alpha, \tag{11}$$

and after setting in equation (10) $\{\lambda = 2$ and $\mu = 0\}$ we obtain from (11):

$$\alpha_{30}: \quad Q_{20} = 15/(2\sqrt{5\pi}) \cdot \alpha_{30}^2 \cdot \rho_0\, R_0^5, \tag{12}$$

$$\alpha_{31}: \quad Q_{20} = 15/(4\sqrt{5\pi}) \cdot \alpha_{3+1}\alpha_{3-1} \cdot \rho_0\, R_0^5, \tag{13}$$

$$\alpha_{32}: \quad Q_{20} = 0, \tag{14}$$

$$\alpha_{33}: \quad Q_{20} = 125/(12\sqrt{5\pi}) \cdot \alpha_{3+3}\alpha_{3-3} \cdot \rho_0\, R_0^5. \tag{15}$$

Among $\lambda = 3$ deformations, only the tetrahedral one, $\alpha_{32} \equiv t_3$, leads to $Q_2 = 0$ - all other octupole deformations induce some quadrupole moments! This observation brings us to the historically first a suggestion that there must exist rotational bands with the usual energy sequences, $E_I \sim I(I+1)$, since the nuclei in question are non-spherical, however, with none or weak E2-transitions. Although there is nothing wrong with this suggestion, a deeper founded analysis discussed below indicates the existence of alternative manifestations of the tetrahedral symmetry (cf. class-1 solutions in figure 11 below).

Let us summarise: In the discussed *purely static scenario* a nucleus with the pure tetrahedral deformation (all the others assumed vanishing) is expected to have the E2 transitions vanishing or being rather weak, whereas the presence of the Coriolis interactions is expected to increase the quadrupole polarisation with increasing spin. These mechanisms are schematically illustrated in figure 7. An example of the corresponding experimental data is presented in figure 8. Why are these observations *not* a direct proof of the presence of the tetrahedral symmetry in the compared nuclei? In experiment, the missing signals from the E2 transitions may be a result of either very weak E2 or, alternatively, very strong, competing E1 towards the ground-state band.

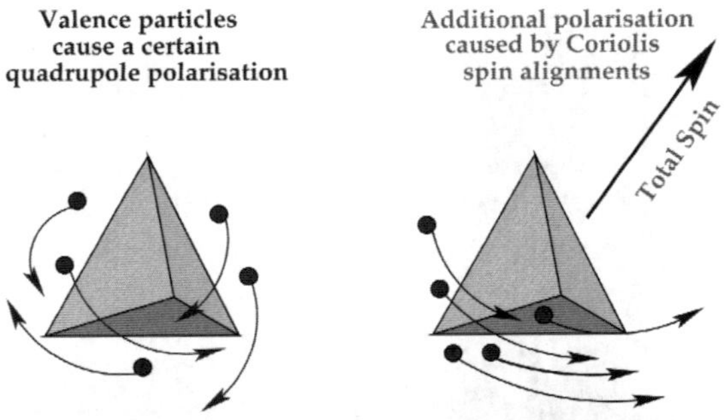

Figure 7. Qualitative illustration: There are only few nuclei with strictly doubly-magic particle occupations, cf. eq. (16). The valence particles or holes in the neighbourhood of the magic numbers will cause necessarily a quadrupole polarisation (left) and thus, possibly, weak E2 transitions. The Coriolis spin alignment is expected to induce an extra quadrupole polarisation at increasing spin (right).

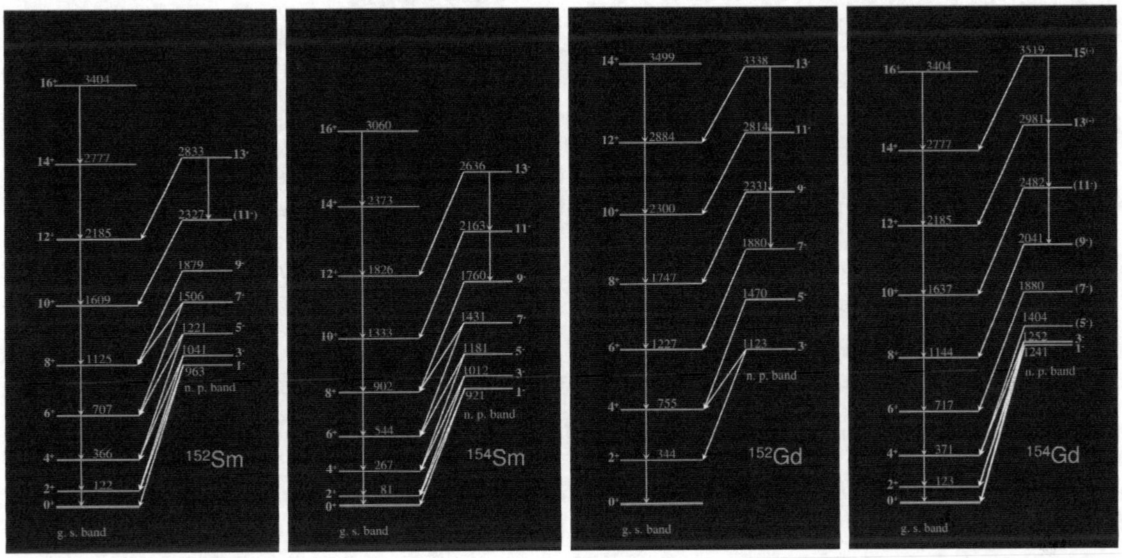

Figure 8. The ground-state and the lowest negative parity bands in selected Samarium (Z=62) and Gadolinium (Z=64) nuclei with the particle numbers close or equal to the tetrahedral magic numbers Z=64 and N=90. The E2 transitions are very weak and at the bottoms of the bands they were not seen experimentally. Only at the higher spins the E2 transitions were detectable; experimental data from ref. [5].

The negative parity bands are traditionally interpreted in terms of either statical octupole minima - cf. one of the best known situations of this type illustrated in figure 9 - or in terms of the octupole vibrations. In this latter case (not shown) it is assumed that the octupole α_{30} deformation is zero but the total energy landscape is flat and the nucleus undergoes a relatively large-amplitude octupole (pear-shape) vibrational motion.

The nuclear rotational structure in this case is represented by two bands of alternating parity with the common moment of inertia what causes an energetic inter-twingling of the alternating parity states. Since the pear-shape deformation implies unequal shifts of the proton and the neutron centres of mass a certain dipole moment is generated. It follows that the competition between the stretched quadrupole and dipole transitions causes a characteristic zig-zag decay pattern as seen in the figure. The measured ratios of the reduced transition probabilities happen to be constant and nearly independent of spin as shown in the accompanying table, whereas in the case of a static tetrahedral deformation an alternative scenario can be envisaged. First of all the negative parity bands, if associated with the tetrahedral minima which are placed higher

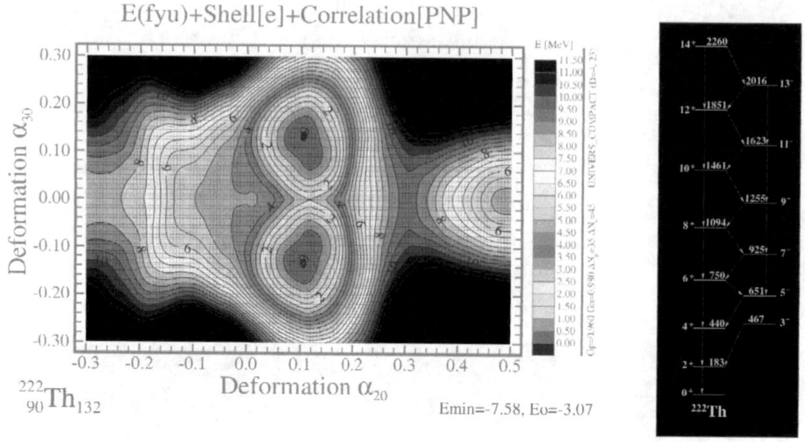

Spin	$(E2)_{in}/(E1)_{out}$
19^-	+0.3(?)
17^-	+0.4(2)
15^-	+0.4(2)
13^-	+0.3(2)
11^-	+0.4(2)
9^-	+0.4(2)
7^-	+0.4(3)
5^-	——
3^-	——

Figure 9. Microscopic Strutinsky-type calculations with the universal Woods-Saxon potential calculations for ^{222}Th in function of the quadrupole and pear-shape (standard octupole) deformations [left]. In the middle - The experimental results of the octupole band associated with the static octupole minima. Observe the characteristic 'zig-zag' E1-transition pattern associated with the static octupole minima. The measured ratios of the $B(E2)_{in}/B(E1)_{out}$ in $10^6\, e \cdot Fm$ show spin-independence of this quantity. [Experimental data from Ref. [5].]

in energy, are expected to be shifted upwards in the energy scale and as a consequence the zig-zag pattern of the E1 transitions becomes energetically impossible. Taking into account the mechanism of the zero-point (vibrational) motion in the quadrupole direction the most probable quadrupole dynamical deformation defined as $\langle|\alpha|\rangle \sim \int |\alpha|\, \varphi_n^2(\alpha)\, d\alpha \neq 0$ is calculated to be different from zero implying certain, possibly weak quadrupole moments, thus E2 transitions. With a possibly weak E1 moments and increasing Coriolis-induced E2 moments a possible

Table 1. The experimental values of the reduced transition ratios $B(E2)_{in}/B(E1)_{out}$ in $10^6\, Fm^2$ for the lowest negative parity bands in the Rare Earth nuclei close to the tetrahedral magic numbers $Z = 64, 70$ and $N = 90$ - compared to the same quantity in the static-octupole case of ^{222}Th illustrated in figure 9. [Experimental data from ref. [5].]

Spin	^{152}Gd	^{156}Gd	^{154}Dy	^{160}Er	^{164}Er	^{162}Yb	^{164}Yb	^{222}Th
19^-	-	50	-	-	-	-	-	+0.3
17^-	-	16	-	-	-	-	-	+0.4
15^-	-	6	-	60	24	-	-	+0.4
13^-	14	7	15	18	23	-	17	+0.3
11^-	4	15	5	10	0	11	11	+0.4
9^-	4	0	-	0	-	11	10	+0.4
7^-	0	0	0	-	-	0	0	+0.4

scenario will be that of an increasing $B(E2)/B(E1)$ ratio as indeed observed, cf. table 1. Observe that the results for the static octupole case and all the other ones differ on the average by an order to two orders of magnitude. We conclude that the underlying quantum mechanisms must

Journal of Physics: Conference Series **413** (2013) 012001 doi:10.1088/1742-6596/413/1/012001

be different - without being able to affirm having seen the tetrahedral symmetry signal; for that the absolute reduced transition probabilities might be necessary - see below.

5.2. Towards convincing signals of tetrahedral symmetry in nuclei: Sufficient dynamic criteria
Convincing criteria for identification of the tetrahedral symmetry in subatomic systems must be constructed in formal analogy to the corresponding criteria developed in molecular physics. For that purposes the Schrödinger equation for the collective nuclear motion (rotational and vibrational) must be solved and the corresponding transition intensity ratios which happen to depend sensitively on the symmetry of the system should be calculated. This work has not been accomplished yet but several steps have already been undertaken. Without entering into

Figure 10. Ground-state solution; invisible α_{20}-axis directed south-east, orthogonal α_{32}-axis - north-east.

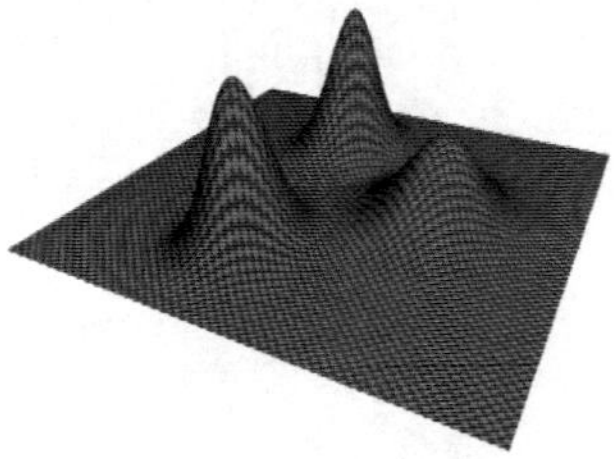

Figure 11. Class-1 solution combining *dominating* tetrahedral and quadrupole symmetries.

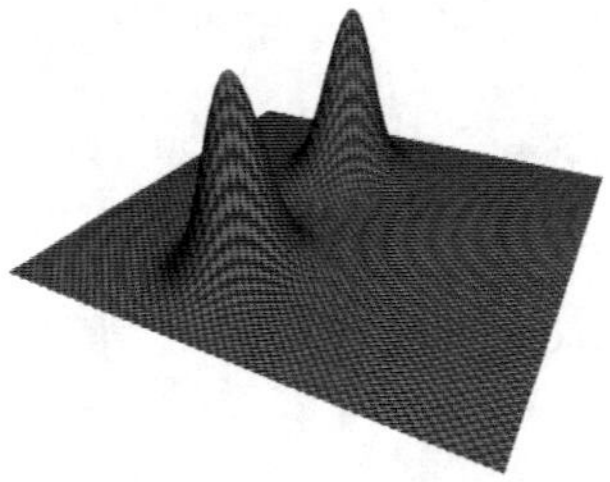

Figure 12. An analogous Class-2 solution representing 'pure' tetrahedral symmetry.

any details let us refer to [6] whose main results are qualitatively illustrated in figures 10-12 showing that in addition to purely quadrupole-symmetric solution (left) and purely tetrahedral solution (right) there exist a new, so-called class-1 solutions which combine the simultaneous presence of two very different-symmetry solutions in one single quantum state (middle). Using the wave functions like the ones whose absolute-square values are presented in the figure the electromagnetic transition probabilities can be calculated. [Interested reader is referred to a recent review [7] of the newest progress for details.]

6. Summary and Conclusions

In these lecture series we have discussed and illustrated the consequences of our strategic suggestion about symmetry-groups, nuclei and their mean fields. Although we have focussed on the tetrahedral symmetry the underlying idea is more general and can be formulated as follows:

> *Each symmetry group sufficiently rich*[3] *in symmetry elements may generate magic numbers that characterise the underlying symmetry in full analogy to the spherical magic numbers 8, 20, 28, 50, 82 and 126 that characterise the nuclear sphericity. In particular, the tetrahedral stability can be associated with the following*

$$\boxed{\textit{Tetrahedral Magic Numbers: 32, 40, 56, 64, 70, 90 and 136}} \tag{16}$$

> *and, more precisely, with the following tetrahedral doubly magic nuclei:*

[3] This phrasing, which cannot be made more precise at this point, alludes to the fact that several point-group symmetries listed in the diagram in figure 4, especially in the parts of the chain close to the 'mother group' O_h, when applied to the realistic mean-field Hamiltonians generate big single-particle gaps - 'big' and 'sufficiently strong' referring to the fact that on the total energy surfaces stable minima are generated.

$$^{64}_{32}\mathrm{Ge}_{32},\ ^{72}_{32}\mathrm{Ge}_{40},\ ^{88}_{32}\mathrm{Ge}_{56},\ ^{80}_{40}\mathrm{Zr}_{40},\ ^{96}_{40}\mathrm{Zr}_{56},\ ^{104}_{40}\mathrm{Zr}_{64},\ ^{110}_{40}\mathrm{Zr}_{70},$$
$$^{112}_{56}\mathrm{Ba}_{56},\ ^{126}_{56}\mathrm{Ba}_{70},\ ^{146}_{56}\mathrm{Ba}_{90},\ ^{134}_{64}\mathrm{Gd}_{70},\ ^{154}_{64}\mathrm{Gd}_{90},\ ^{160}_{70}\mathrm{Yb}_{90},\ ^{226}_{90}\mathrm{Th}_{136}.$$

Characteristically, whereas the spherical symmetry corresponds to a fixed geometry (no sphere is more spherical than the other) the point-group symmetry-driven magic-numbers may correspond to various areas on the (e.g. tetrahedral) deformation axis. In other words: Point-groups allow for 'variable deformation' of nuclear objects, whereas perfectly preserving the symmetry properties.

Tetrahedral and octahedral symmetries, if identified in nature, will very likely cause switching certain accents of importance in nuclear structure research, particularly in the experimental objectives of various measurements. Today, as it seems, the nuclear physics community has adopted thinking about shapes, symmetry and geometry, through the perspective of the so-called prolate-oblate shape coexistence (quadrupole axial shapes), the so-called tri-axial shapes (mainly quadrupole) and occasionally the so-called octupole (axial pear-shape) vibrations and rotations. Because of the rather frequent appearance of those keywords in the literature, some colleagues, mainly from the outside of the field, refer to those quadrupole/octupole shapes as 'trivial'. Limiting our thinking to the quadrupole/octupole shapes and trying to force all interpretations of the rotational motion with the help of those is (probably) historically justified but counterproductive today. Since literally thousands of projects have been focussed on the rotational band properties and the associated effective moments of inertia - a very poor tool for testing the geometry and shapes - the electromagnetic transition probes, technically more difficult, have been used much more seldom and the corresponding literature is scarce.

To approach the issue of the experimental verification of the presence of the hypothetical 'non-trivial' symmetries such as those listed in the upper half of the diagram in figure 4 requires precise measurements of the transition intensities, branching rations and life-times. Those and possibly other measurements would allow for establishing as precisely as possible, first of all, the E1, E2 and M1 reduced transition probabilities. Such an information seems indispensable in view of the fact that various symmetries with various and usually distinct character of the irreducible representations generate various types of forbidden transitions and of selection rules in the form of very characteristic branching ratios which can serve as the 'smoking gun signal' for the possible presence of those exotic symmetries.

On the theory side we will need to implement an already rather well known formalism for constructing the physical solutions (wave-functions) which transform according to the irreducible representations of the group studied. These techniques, although formally rather well known, are not simple in the applications to e.g. determinantal microscopic wave functions which will be needed to evaluate the reduced transition probabilities and the theoretical values of the characteristic branching ratios using various projection techniques. The discovery of the possible presence of what we called class-1 solutions (figure 11) mixing the dominating tetrahedral symmetry with the *totally different* quadrupole one is an extremely intriguing new structure in low-energy subatomic physics - fascinating, but possibly complicating the interpretations.

References

[1] Dudek J *et al* 2007 *Int. J. Mod. Phys.* E 16, 516
[2] Koster G F, Dimmock J O, Weeler R G and Statz H 1963 *Properties of Thirsty-Two Point Groups* MIT Press Cambridge Mass.
[3] Given the volume limitation of the present publication, the illustrative material had be proportionally limited. For reader's convenience we refer here to the overlay numbers in the Predeal Presentation available at: http://www.theory.nipne.ro/predeal12/pdf/SymmetriesPredeal12.pdf
[4] Li X and Dudek J 1994 *Phys. Rev.* C 49 R1250 [First reference to 4-fold degenerate nucleonic levels]; Dudek J *et al* 2002 *Phys. Rev. Lett.* 88 252503; *ibid.* 2006 *Phys. Rev. Lett.* 97 072502
[5] Evaluated Nuclear Structure Data File (ENSDF): http://www.nndc.bnl.gov/ensdf/
[6] Dobrowolski A *et al* 2011 *Int. J. Mod. Phys.* E 20 500
[7] Dudek J *et al* 2012 *Zakopane Conference in Nuclear Physics* - submitted to Acta Physica Polonica

Int. Summer School for Advanced Studies 'Dynamics of open nuclear systems' (PREDEAL12) IOP Publishing
Journal of Physics: Conference Series **413** (2013) 012002 doi:10.1088/1742-6596/413/1/012002

Nuclei: superheavy–superneutronic–strange–and of antimatter[1]

Walter Greiner

Frankfurt Institute for Advanced Studies, J. W. Goethe–Universität, D-60438, Frankfurt a.M., Germany

E-mail: greiner@fias.uni-frankfurt.de

Abstract. The extension of the periodic system into various new areas is investigated. Experiments for the synthesis of superheavy elements and the predictions of magic numbers are reviewed. Further on, investigations on hypernuclei and the possible production of antimatter-clusters in heavy-ion collisions are reported. Various versions of the meson field theory serve as effective field theories at the basis of modern nuclear structure and suggest structure in the vacuum which might be important for the production of hyper- and antimatter.

1. Introduction

I am happy to be here and congratulate my friend and collaborator for more than 30 years, Prof. Aurel Sandulescu, on occasion of his 80-th birthday. After developing the fragmentation theory with Werner Scheid based on the two–center shell model we predicted together with Dorin Poenaru **cluster radioactivity**. As a term it entered the **Encyclopedia Britannica**. Indeed, it is one of the most outstanding discoveries in Nuclear Physics. Looking through Rumanian history of Physics I compare it with Proca's work. I will summarize here the work on the extension of the Periodic System, into which my work with Aurel Sandulescu is embedded.

There are fundamental questions in science, like e. g. "How did life emerge?" or "How does our brain work?" and others. However, the most fundamental of those questions is "How did the world originate?". The material world has to exist before life and thinking can develop. Of particular importance are the substances themselves, i. e. the particles the elements are made of (baryons, mesons, quarks, gluons), i. e. elementary matter. The vacuum and its structure is closely related to that. We want to report on these questions, beginning with the discussion of modern issues in nuclear physics.

The elements existing in nature are ordered according to their atomic (chemical) properties in the **periodic system**, which was developed by Dmitry Mendeleev and Lothar Meyer. The heaviest element of natural origin is uranium. Its nucleus is composed of Z = 92 protons and a certain number of neutrons (N = 128–150). They are called the different uranium isotopes. The transuranium elements reach from neptunium (Z = 93) via californium (Z = 98) and fermium (Z = 100) up to lawrencium (Z = 103). The heavier the elements are, the larger are their radii and their number of protons. Thus, the Coulomb repulsion in their interior increases, and they undergo fission. In other words: the transuranium elements become more unstable as they get

[1] A similar talk I have given and published in The European Physical Journal D (2012).

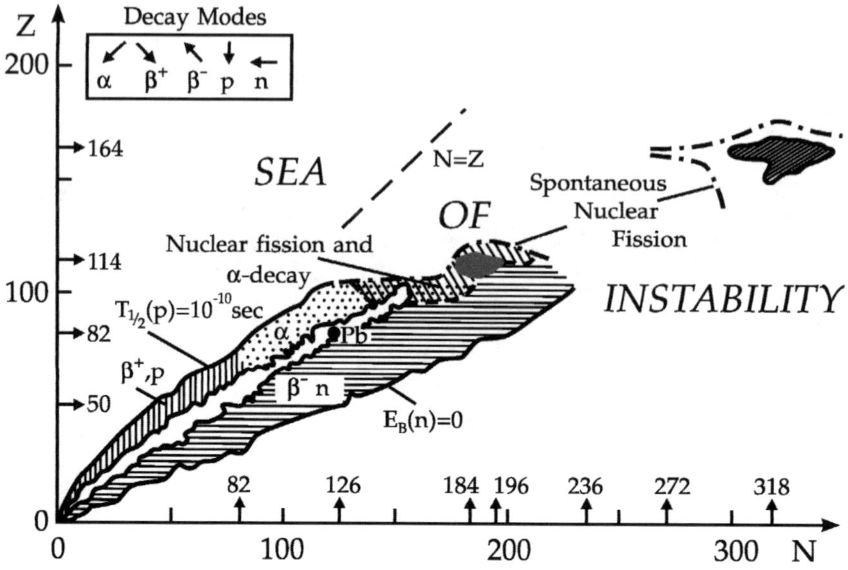

Figure 1. The periodic system of elements as conceived by the Frankfurt school in the late sixties. The islands of superheavy elements (Z = 114, N = 184, 196 and Z = 164, N = 318) are shown as dark hatched areas.

bigger. In the late sixties, the dream of the superheavy elements arose. Theoretical nuclear physicists around S. G. Nilsson (Lund) and from the Frankfurt school [1, 2] predicted that so-called closed proton and neutron shells should counteract the repelling Coulomb forces. Atomic nuclei with these special **"magic" proton and neutron numbers** and their neighbours could again be rather stable. These magic proton (Z) and neutron (N) numbers were thought to be Z = 114 and N = 184 or 196. Typical predictions of their life-times varied between seconds and many thousand years. Figure 1 summarizes the expectations at the time. One can see the islands of superheavy elements around Z = 114, N = 184 and 196, respectively, and the one around Z = 164, N = 318.

2. Cold Valleys in the Potential

The important question was how to produce these superheavy nuclei. There were many attempts, but only little progress was made. It was not until the middle of the seventies that the Frankfurt school of theoretical physics together with foreign guests (R. K. Gupta (India), A. Sandulescu (Romania), D. Poenaru (Romania)) [3] theoretically understood and substantiated the concept of bombarding double magic lead nuclei with suitable projectiles, which had been proposed intuitively by the Russian nuclear physicist Y. Oganessian [4]. The two-center shell model, which is essential for the description of fission, fusion and nuclear molecules, was developed in 1969–1972 by W. Greiner and his students U. Mosel [1] and J. Maruhn [5]. It showed that the shell structure of the two final fragments was visible far beyond the barrier, where the nucleus undergoes fusion. The collective potential energy surfaces of heavy nuclei, which were calculated in the framework of the two-center shell model, exhibit pronounced valleys [6].

These valleys provide promising doorways to the fusion of superheavy nuclei for certain projectile-target combinations (Fig. 2). If projectile and target approach each other through those **"cold" valleys** [3, 7], they get only minimally excited and the barrier, which has to be overcome (fusion barrier) is lowest (as compared to the neighbouring projectile-target

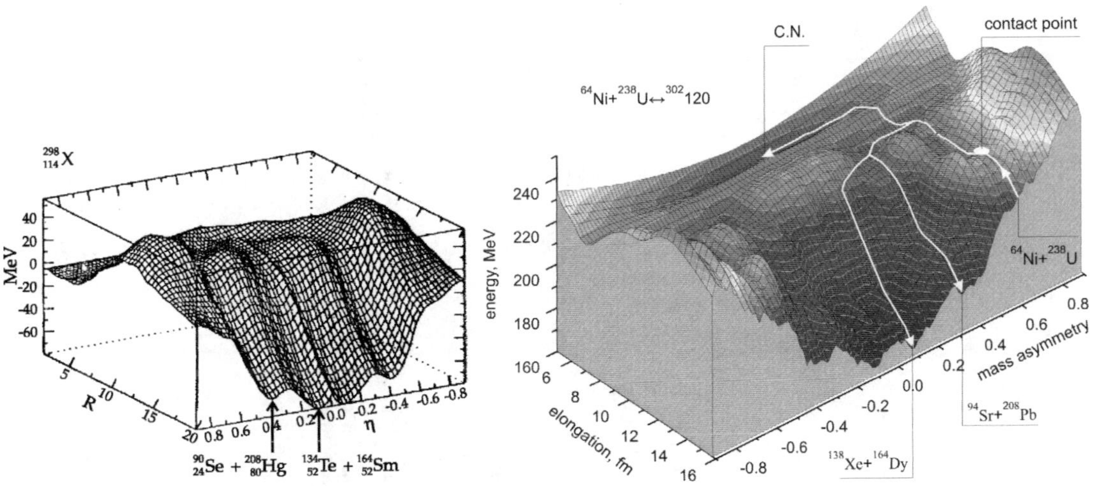

Figure 2. Left: The collective potential energy surface of $^{184}114$ calculated within the two center shell model by J. Maruhn *et al.* shows clearly the cold valleys, which reach up to the barrier and beyond. R denotes the distance between the centers for fragments containing A_1 and A_2 nucleons, while $\eta = (A_1 - A_2)/(A_1 + A_2)$ is the mass asymmetry parameter. Right: Collective potential energy surface of the element $^{302}120$. (C. N. stands for compound nucleus).

combinations). In this way, the correct projectile- and target-combinations for fusion were predicted. Indeed, Gottfried Münzenberg and Sigurd Hofmann and their group at GSI [8] have followed this approach. With the help of the SHIP mass-separator and the position sensitive detectors, which were especially developed by them, they produced the pre-superheavy elements Z = 106, 107, . . . 112, each of them with the theoretically predicted projectile-target combinations, and only with these.

Everything else failed. This is an impressive success, which crowned the laborious construction work of many years. The last but one example of this success, was the discovery of element 112 and its long α–decay chain. The Dubna group produced the six isotopes of Z = 113–118 by bombarding ^{244}Pu–^{248}Cf with ^{48}Ca [9]. This is also a cold valley reaction (in this case due to the combination of a spherical and a deformed nucleus), as predicted by Gupta, Sandulescu and Greiner in 1977 [3]. There exist also cold valleys for which both fragments are deformed [7], or have non-axial orientations [10], but these have not been verified experimentally. The cold valleys also play an important role in nuclear fission giving rise to asymmetric and superasymmetric [11, 12] fission and to cluster radioactivity [13].

3. Shell structure in the superheavy region

Decay properties and stability of heaviest nuclei with $Z \leq 132$ were recently studied within the macro-microscopical approach for nuclear ground state masses and phenomenological relations for the half-lives with respect to α-decay, β-decay and spontaneous fission [14]. It was found (see Fig. 3) that the β-stable isotopes ^{291}Cn and ^{293}Cn with a half-life of about 100 years are the longest-living superheavy nuclei located on the first island of stability. Because of their short half-lives the search in nature of superheavy nuclei may be performed only in cosmic rays. Under terrestrial conditions a measurable amount of superheavies is unlikely to exist. Note, that fusion reactions lead to the proton-rich nuclei. The heaviest synthesized nuclei with $Z = 118$ is situated already quite close to the border of 1 μs half-life. It means that the synthesis and detection of nuclei with $Z > 120$ produced in fusion reactions may be difficult at existing experimental

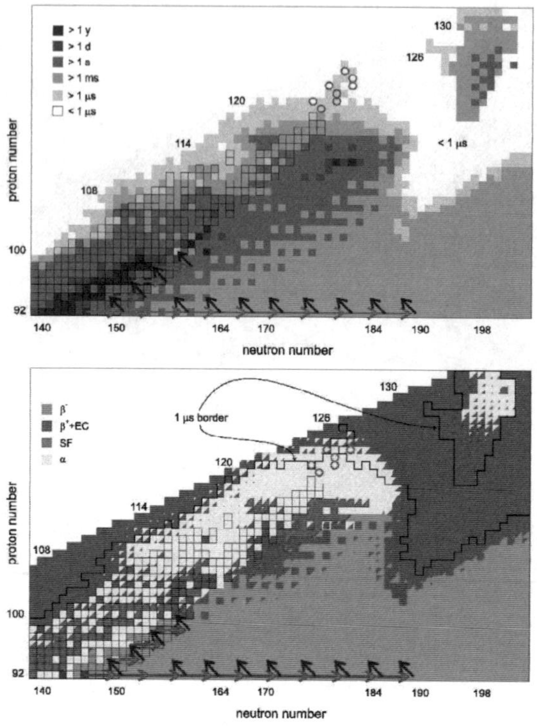

Figure 3. Half-lives (top) and decay modes (bottom) of nuclei in the upper part of the nuclear map. The circles show the nuclei with $Z = 119 - 124$, which may be synthesized in $3n$ channel of fusion reactions $^{50}\mathrm{Ti} + ^{249}\mathrm{Bk}, ^{249}\mathrm{Cf}$ and $^{54}\mathrm{Cr}, ^{58}\mathrm{Fr} + ^{248}\mathrm{Cm}, ^{249}\mathrm{Bk}, ^{249}\mathrm{Cf}$. Outlined squares on the lower panel correspond to the experimentally known nuclei. The most stable Copernicium isotopes are ^{291}Cn and ^{293}Cn. Schematic view of slow (terminated at the short-lived fission Fermium isotopes) and fast neutron capture processes with subsequent β^- decays are shown by the arrows.

facilities due to their short half-lives (shorter than 1 μs). This prediction should be taken into account while planning new experiments and experimental setups. One may see as well that the nearest neutron-rich isotopes of superheavy elements with $111 \leq Z \leq 115$ to those synthesized recently in Dubna in ^{48}Ca-induced fusion reactions are found to be β^+-decaying. This fact may significantly complicate their experimental identification. However, existence of this area of β^+-decaying nuclei gives us a possibility to reach the center of the island of stability. One of the way to produce ^{291}Cn is the triple β^+ (or EC) decay of ^{291}115 which in turn could be, for example, synthesized after α-decay of ^{295}117 in the reaction $^{48}\mathrm{Ca} + ^{249}\mathrm{Bk} \rightarrow ^{295} 117 + 2\mathrm{n}$. The proposed method of reaching the island of stability hopefully may be realized in future with the progress in experimental techniques. We found as well the second area of stability of superheavy nuclei (still with shorter half-lives) situated in the region of $Z \sim 124$ and $N \sim 198$. It is separated from the "continent" by the "gulf" of short-living nuclei with half-lives shorted than 1 μs.

Studies of the shell structure of superheavy elements in the framework of the meson field theory and the Skyrme-Hartree-Fock approach have recently shown that the magic shells in the superheavy region are very isotope dependent [3]. Additionally, there is a strong dependence

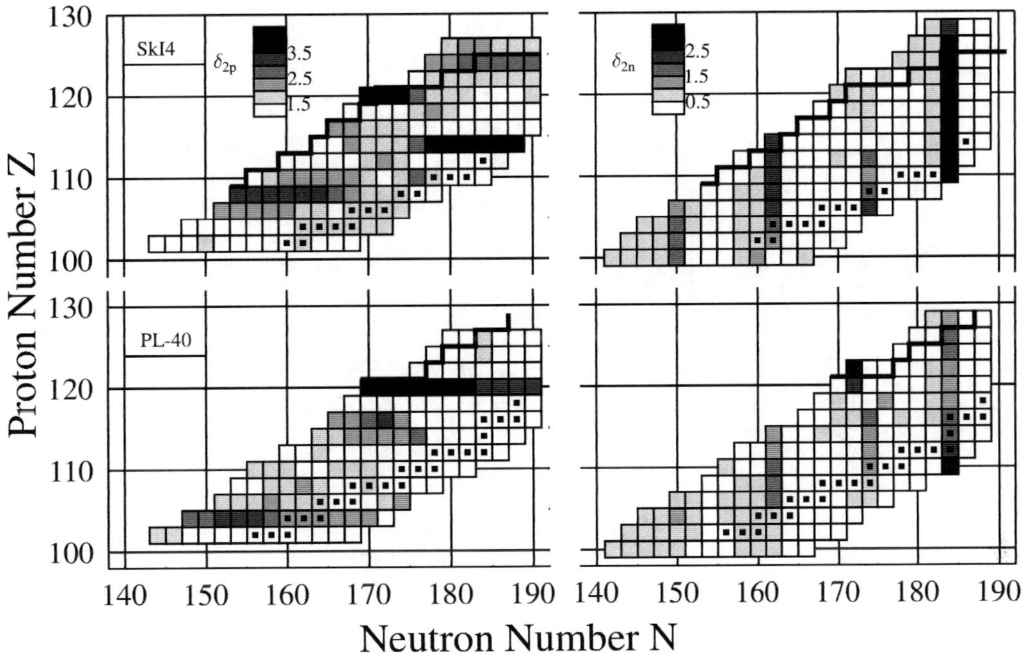

Figure 4. Grey scale plots of proton gaps (left column) and neutron gaps (right column) in the N-Z plane for deformed calculations with the forces SkI4 and PL- 40. Besides the spherical shell closures one can see the deformed shell closures for protons at Z = 104 (PL-40) and Z = 108 (SkI4) and the ones for neutrons at N = 162 for both forces.

on the parameter set and the model. Some forces hardly show any shell structure, while other predict the magic numbers Z = 114, 120 and 126. Using the heaviest known even-even nucleus Hassium $^{264}_{156}108$ as a criterium to find the best parameter sets in each model, it turns out that PL-40 and SkI4 produce best its binding energy. However, these two forces make conflicting predictions for the magic number in the superheavy region: SkI4 predicts Z = 114, 120 and PL-40 Z = 120. Most interesting, Z = 120 **as magic proton number seems to be as probable as** Z = 114. Calculations of deformed systems within the two models [15] reveal again different predictions: Though both parametrizations predict N = 162 as the deformed neutron-shell closure, the deformed proton-shell closures are Z = 108 (SkI4) and Z = 104 (PL-40) (see Fig. 4). Calculations of the potential energy surfaces [15] show single humped barriers; their heights and widths strongly depend on the predicted magic number. Furthermore, recent investigations in a chirally symmetric mean-field theory (see also below) result also in the prediction of these two magic numbers [16, 17]. The corresponding magic neutron numbers are predicted to be N = 172 and to a lesser extend N = 184. Thus, this region provides an open field of research. The charge distribution of the Z = 120,N = 184 nucleus indicates a hollow inside. This may suggest that it might be essentially a fullerene consisting of 60 α–particles and one binding neutron per alpha. The cold valleys in the collective potential energy surface are basic for understanding this exciting area of nuclear physics! It is a master example for understanding the **structure of elementary matter**, which is so important for other fields, especially astrophysics, but even more so for enriching our "Weltbild", i.e. the status of our understanding of the world around us.

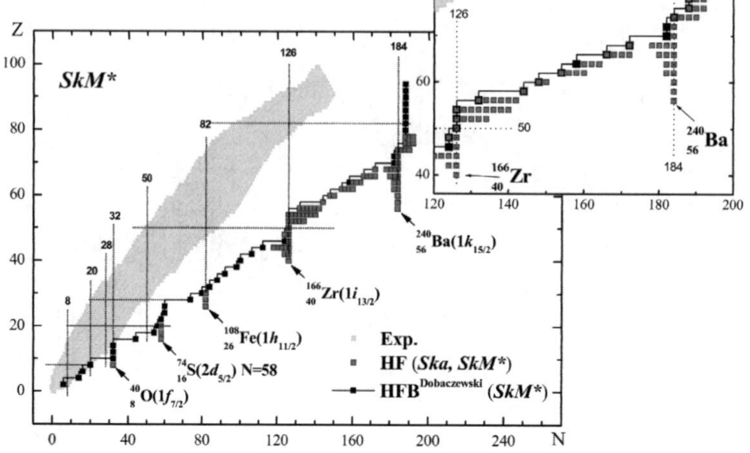

Figure 5. Fragment of the neutron drip line and elements (red squares) that are stable against one neutron emission [18]. One can see the formation of stability peninsulas along neutron magic numbers.

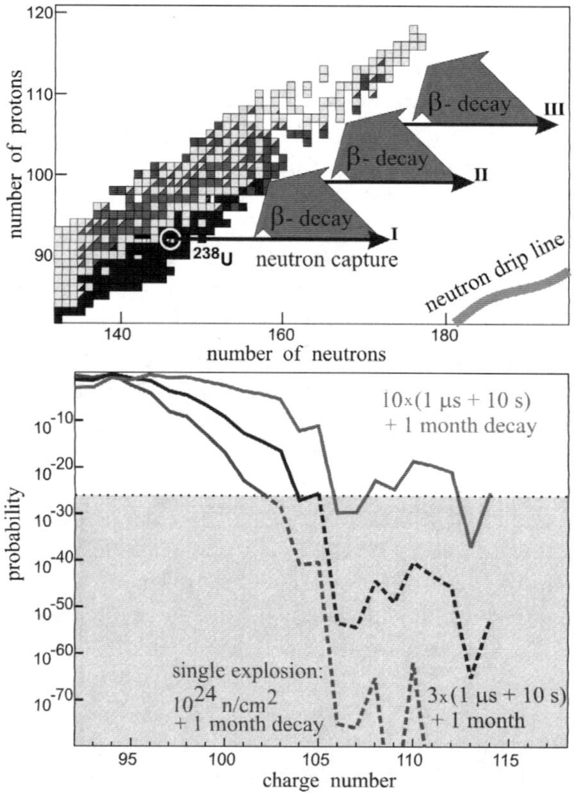

Figure 6. Schematic picture for multiple neutron irradiation of initial ^{238}U material (top) and probability for formation of heavy nuclei (bottom) in such process (one, three and ten subsequent explosions). Dotted line denotes the level of few atoms.

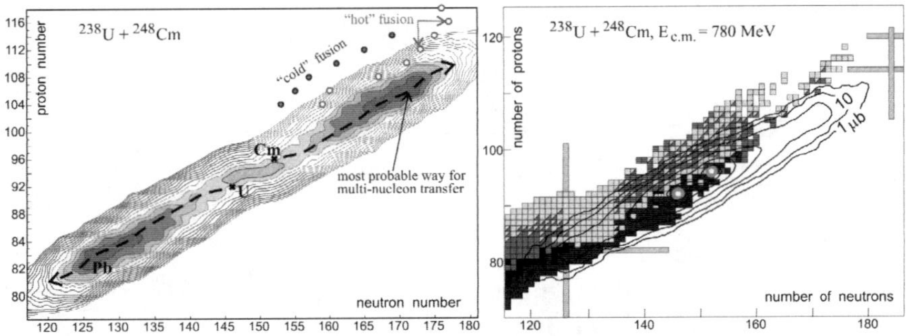

Figure 7. Landscape of potential energy surface (on the left) and the cross sections (microbarns, logarithmic scale) for production of primary fragments (on the right) in collision of ^{238}U with ^{248}Cm.

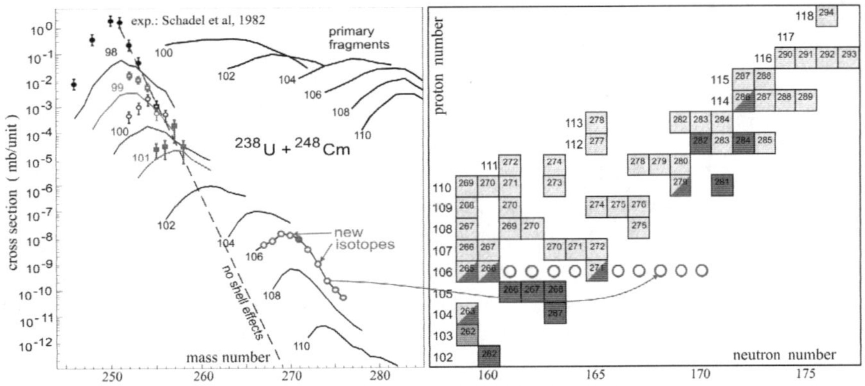

Figure 8. Yield of primary and survived isotopes of SH nuclei produced in collisions of ^{238}U with ^{248}Cm at 800 MeV center-of-mass energy. Dashed line shows the expected locus of transfer reaction cross sections without the shell effects.

The investigation of the neutron drip line by extended Hartree-Fock+BCS calculations led to a surprise: extremely neutron rich nuclei along the magic neutron numbers become stable against one-and two-neutron separation [18], see Fig. 5. The standard production of superheavy nuclei by fusing two smaller stable nuclei leads automatically to neutron poor isotopes near the proton drip line (therefore the lifetime of the produced superheavies is so small). Only a few superheavy atoms are produced this way. This leads us directly to the question of how superheavies with larger neutron numbers (and therefore having larger lifetimes: up to thousands of years) can be produced. One also wants to produce such long–living superheavies in macroscopic quantities (milligrams, grams,...) so that they can eventually be used technically. This can be done either by double (or multiple) underground atomic bomb explosions or by pulsed reactors with very high neutron flux ($\simeq 10^{21}$ neutrons/ sec cm^2) see [19, 20] and Figs. 3 and 6.

The idea to take advantage of the shell effects for the production of SH nuclei in the multi-nucleon transfer processes of low-energy heavy ion collisions was proposed in [21]. The shell effects are known to play an important role in fusion of heavy ions with actinide targets driving the nuclear system to the quasi-fission channels into the deep lead and tin valleys and, thus, decreasing the fusion probability. On the contrary, in the transfer reactions the same effects may lead to enhanced yield of SH nuclei. It may occur if one of heavy colliding nuclei, say ^{238}U, gives

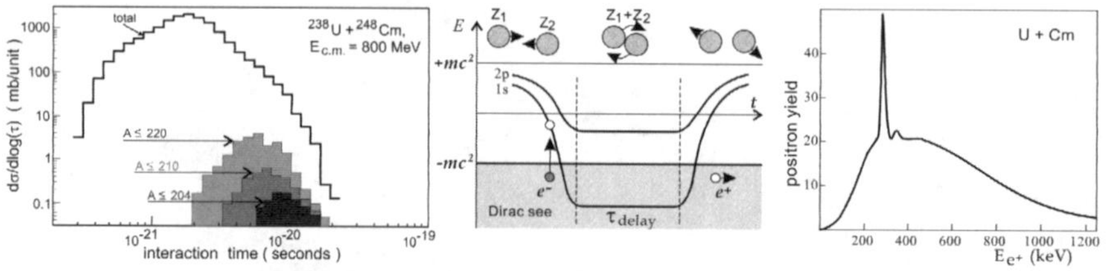

Figure 9. Reaction time distribution (left panel) and spontaneous positron formation (middle panel) in supercritical electric field of long-lived giant quasi-atom formed in collision of U+Cm; the right panel shows the energy distribution of the positron yield.

away nucleons approaching to double magic ^{208}Pb nucleus, whereas another one, say ^{248}Cm, accepts these nucleons becoming superheavy in the exit channel the so called "inverse" (anti-symmetrizing) quasi-fission process. The potential energy surface of the giant nuclear system formed in collision of ^{238}U and ^{248}Cm nuclei is shown in Fig. 7. In low-energy damped collisions of heavy ions just the potential energy surface regulates to a great extent the evolution of the nuclear system driving it to the minimal values of potential energy in the multidimensional space of collective variables. In the course of nucleon exchange the most probable path of the nuclear system formed by ^{238}U and ^{248}Cm lies along the line of stability with formation of SH nuclei which have many more neutrons as compared with those produced in the "cold" and "hot" fusion reactions. Due to fluctuations even more neutron rich isotopes of SH nuclei may be formed in such transfer reactions. The calculated cross sections for formation of primary fragments in low-energy collisions of ^{238}U with ^{248}Cm target are shown in Fig. 7 by the counter lines in logarithmic scale. As can be seen, the superheavy nuclei located very close to the center of the island of stability may be produced in this reaction with rather high cross section of one microbarn. This region of the nuclear map cannot be reached in any fusion reaction with stable projectiles and long-lived targets. Of course, the question arises whether these excited superheavy primary fragments may survive. The calculated cross sections for formation of neutron-rich SH nuclei in low-energy collisions of ^{238}U with ^{248}Cm target are shown in Fig. 8 for final survived fragments. These SH nuclei are located very close to the center of the island of stability and cannot be produced in any fusion reactions with stable projectiles and long-lived targets. These are the shell effects which give us a significant gain as compared to a monotonous exponential decrease of the cross sections with increasing number of transferred nucleons.

We found that the nuclear system consisting of two very heavy nuclei may hold in contact rather long in some cases [22]. During this time the giant nuclear system moves over the multidimensional potential energy surface with almost zero kinetic energy (result of large nuclear viscosity). The reaction time distribution is shown in Fig. 9 for the ^{238}U+^{248}Cm collision. With increase of the energy loss and mass transfer the reaction time becomes longer and its distribution becomes more narrow. The lifetime of a giant composite system more than 10^{-20} s is quite enough to expect positron line structure emerging on top of the dynamical positron spectrum due to spontaneous e^+e^- production from the supercritical electric fields as a fundamental QED process ("decay of the vacuum", Fig. 9) [23]. Formation of the background positrons in these reactions forces one to find some additional trigger for the longest events. For the considered case of ^{238}U+^{248}Cm collision at 800 MeV center-of-mass energy, the detection of the surviving nuclei in the lead region at the laboratory angles of about 25° and at the low-energy border of their spectrum (around 1000 MeV for Pb) could be a real trigger for longest reaction time.

It was recently found that low-energy collision of actinides may lead to quite an exotic process of three-body clusterization, the so-called true ternary quasifission, leading to formation of two

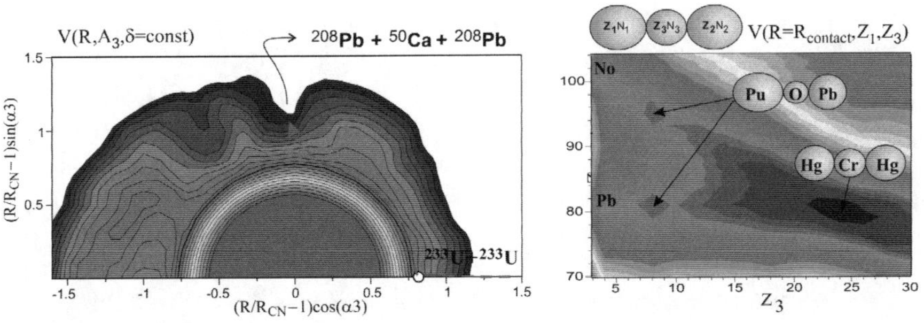

Figure 10. (Left panel) Potential energy (macroscopic plus shell corrections) for ternary quasifission of giant nuclear system formed in ^{233}U+^{233}U collision, depending on elongation and mass of third fragment ($\alpha_3 = \pi \cdot A_3/100$, where A_3 is the mass number of the third fragment). (Right panel) Landscape of potential energy of three-body contact configurations of giant nuclear system formed in collision of ^{238}U+^{238}U.

lead-like fragments and some heavy third particle in between [24]. This type of processes is quite possible because the shell effects significantly reduce the potential energy of the three-cluster configurations with two strongly bound lead-like fragments. In Fig. 10 the landscape of the potential energy surface is shown for a three-body clusterization of the nuclear system formed in collision of U+U. It is seen (left panel) that the shell correction at contact configurations makes a very deep minimum for the "lead-calcium-lead" ($A_3 = 50$) clusterization. In the right panel the potential energy is shown as a function of three variables, Z_1, Z_3 (charges of the first and third fragments) and system elongation R (minimized over the neutron numbers) at fixed (equal) deformations of the fragments being in contact. As can be seen, the giant nuclear system, consisting of two touching uranium nuclei, may split into the two-body exit channel with formation of lead-like fragment and complementary superheavy nucleus (the so-called anti-symmetrizing quasifission process which may lead to an enhanced yield of SH nuclei in multi-nucleon transfer reactions [21]). Beside the two-body Pb–No clusterization and the shallow local three-body minimum with formation of light intermediate oxygen-like cluster, the potential energy has the very deep minimum corresponding to the Pb-Ca-Pb–like configuration (or Hg-Cr-Hg) caused by the $N = 126$ and $Z = 82$ nuclear shells. The extreme clustering process of formation of two lead-like doubly magic fragments in collisions of actinide nuclei is a very interesting subject for experimental study. Such measurements, in our opinion, are not too difficult. It is sufficient to detect two coincident lead-like ejectiles (or one lead-like and one calcium-like fragments) in U+U collisions to conclude unambiguously about the ternary fission of the giant nuclear system.

4. Extension of the periodic system into the field of hyper- and antimatter

Nuclei that are found in nature consist of nucleons (protons and neutrons) which themselves are made of u (up) and d (down) quarks. However, there also exist s (strange) quarks and even heavier flavours, called charm, bottom, top. The latter has just recently been discovered. Let us stick to the s quarks. They are found in the "strange" relatives of the nucleons, the so-called hyperons ($\Lambda, \Sigma, \Theta, \Omega$). The Λ–particle, e. g., consists of one u, d and s quark, the Θ–particle even of an u and two s quarks, while the Ω (sss) contains strange quarks only.

If such a hyperon is taken up by a nucleus, a **hyper-nucleus** is created. Hypernuclei with one hyperon have been known for 20 years [25]. Several years ago, Carsten Greiner, Jürgen Schaffner and Horst Stöcker [26] theoretically investigated nuclei with many hyperons, **hypermatter**, and found that the binding energy per baryon of strange matter is in many cases even higher than that of ordinary matter (composed only of u and d quarks). This leads to the idea of extending the periodic system of elements in the direction of strangeness.

One can also ask for the possibility of building atomic nuclei out of **antimatter**, that means searching e. g. for anti-helium [27], anti-carbon, anti-oxygen. Figure 11 depicts this idea. Due to charge conjugation symmetry, antinuclei should have the same magic numbers and the same spectra as ordinary nuclei. However, as soon as they get in touch with ordinary matter, they annihilate with it and the system explodes. Now the important question arises, how these strange matter and antimatter clusters can be produced. First, one thinks of collisions of heavy nuclei, e. g. lead on lead, at high energies (energy per nucleon $\geq$200 GeV). Calculations with the URQMD-model of the Frankfurt school show that through **nuclear shock waves** [28] nuclear matter gets compressed to 510 times of its usual value, $\rho_0 \simeq 0.17$ fm^{-3}, and heated up to temperatures of $kT \simeq 200$ MeV. As a consequence, about 10 000 pions, 100 Λ's, 40 Σ's and Θ's and about as many antiprotons and many other particles are created in a single collision. It seems conceivable that it is possible in such a scenario for some Λ's to get captured in a nuclear cluster. This happens indeed rather frequently for one or two Λ–particles; however, more of them get built into nuclei with rapidly decreasing probability only. This is due to the low probability for finding the right conditions for such a capture in the phase space of the particles: the numerous particles travel with all possible momenta (velocities) in all directions. The chances for hyperons and antibaryons to meet get rapidly worse with increasing number. In order to produce multi–Λ–nuclei and antimatter nuclei, one has to look for a different source.

In the framework of the meson field theory, the energy spectrum of baryons has a peculiar structure, depicted in upper part of Fig. 12. It consists of an upper and a lower continuum, as it is known for electrons (see, e. g. Ref.[29]). Of special interest in the case of the baryon spectrum is the potential well, built of the scalar and the vector potential, which rises from the lower continuum. Naftali Auerbach and collaborators noticed this first [30]. It is known since P. A. M. Dirac (1930) that the negative energy states of the lower continuum have to be occupied by particles (electrons or, in our case, baryons). Otherwise our world would be unstable, because the "ordinary" particles are found in the upper states which can decay through the emission of photons into lower lying states. However, if the "underworld" is occupied, the Pauli-principle will prevent this decay. Holes in the occupied "underworld" (Dirac sea) are antiparticles.

The occupied states of this underworld, including up to 40 000 occupied bound states of the lower potential well, represent the **vacuum**. The peculiarity of this strongly correlated vacuum structure in the region of atomic nuclei is that — depending on the size of the nucleus — more than 20 000 up to 40 000 (occupied) bound nucleon states contribute to this polarization effect. Obviously, we are dealing here with a **highly correlated vacuum**. Pronounced shell structure can be recognized [31]. Holes in these states have to be interpreted as bound antinucleons (antiprotons, antineutrons). If the primary nuclear density rises due to compression, the lower well increases while the upper decreases and soon is converted into a repulsive barrier. This compression of nuclear matter can only be carried out in relativistic nucleus-nucleus collision with the help of shock waves, which have been proposed by the Frankfurt school (see W. Scheid et al., Ref. [32]) and which have since then been confirmed extensively (see, e. g. Ref. [33]). These **nuclear shock waves** are accompanied by heating of the nuclear matter. Indeed, density and temperature are intimately coupled in terms of the hydrodynamic Rankine–Hugoniot equations. Heating as well as the violent dynamics cause the creation of many holes in the very deep (measured from $-M_B c^2$) vacuum well. These numerous bound holes resemble antimatter clusters which are bound in the medium; their wave functions have large overlap with antimatter clusters. When the primary matter density decreases during the expansion stage of the heavy–ion collision, the potential wells, in particular the lower one, disappear.

The bound antinucleons are then pulled down into the (lower) continuum. In this way

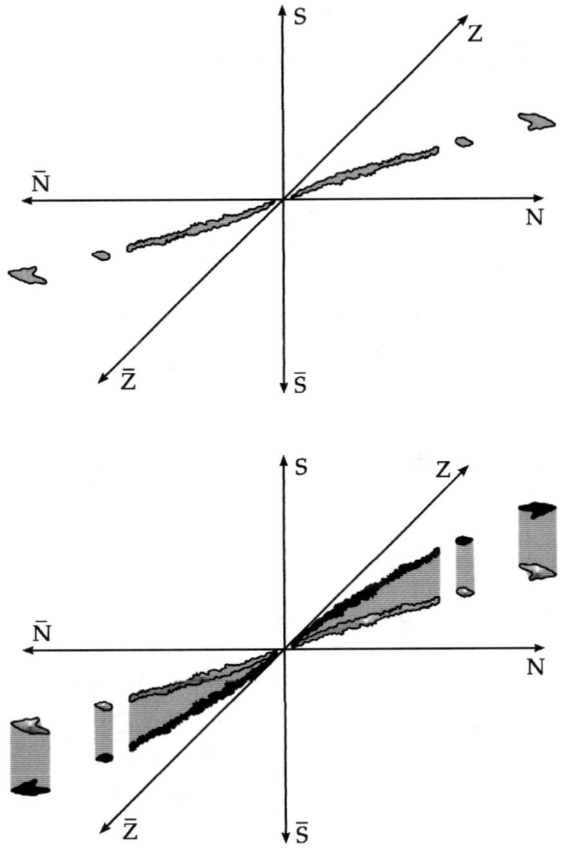

Figure 11. The extension of the periodic system into the sectors of strangeness $S, \overline{S}$ and antimatter $\overline{ZN}$. In the upper part of the figure only the stable valley in the usual proton (Z) and neutron (N) plane is plotted, however, extended into the sector of antiprotons and antineutrons. In the second part of the figure it has been indicated, how the stable valley winds out of the Z-N-plane into the strangeness sector. The same can be observed for the antimatter sector.

antimatter clusters may be set free. Of course, a large part of the antimatter will annihilate on ordinary matter present in the course of the expansion. However, it is important that this mechanism for the production of antimatter clusters out of the highly correlated vacuum does not proceed via the phase space. The required coalescence of many particles in phase space suppresses the production of clusters, while it is favoured by the direct production out of the highly correlated vacuum. In a certain sense, the highly correlated vacuum is a kind of cluster vacuum (vacuum with cluster structure). The shell structure of the vacuum levels (see Fig. 12) supports this latter suggestion. Fig. 13 illustrates this idea. Recently the STAR Collaboration at RHIC observed Anti–^{4}He with production rate in excess of coalescent nucleosynthesis production [27].

The mechanism is similar for the production of multi-hyper nuclei $(\Lambda, \Sigma, \Theta, \Omega)$. Meson field theory predicts also for the Λ energy spectrum at finite primary nucleon density the existence of upper and lower wells. The lower well belongs to the vacuum and is fully occupied by Λ's.

Dynamics and temperature then induce transitions ($\Lambda\overline{\Lambda}$ creation) and deposit many Λ's in

Journal of Physics: Conference Series **413** (2013) 012002 doi:10.1088/1742-6596/413/1/012002

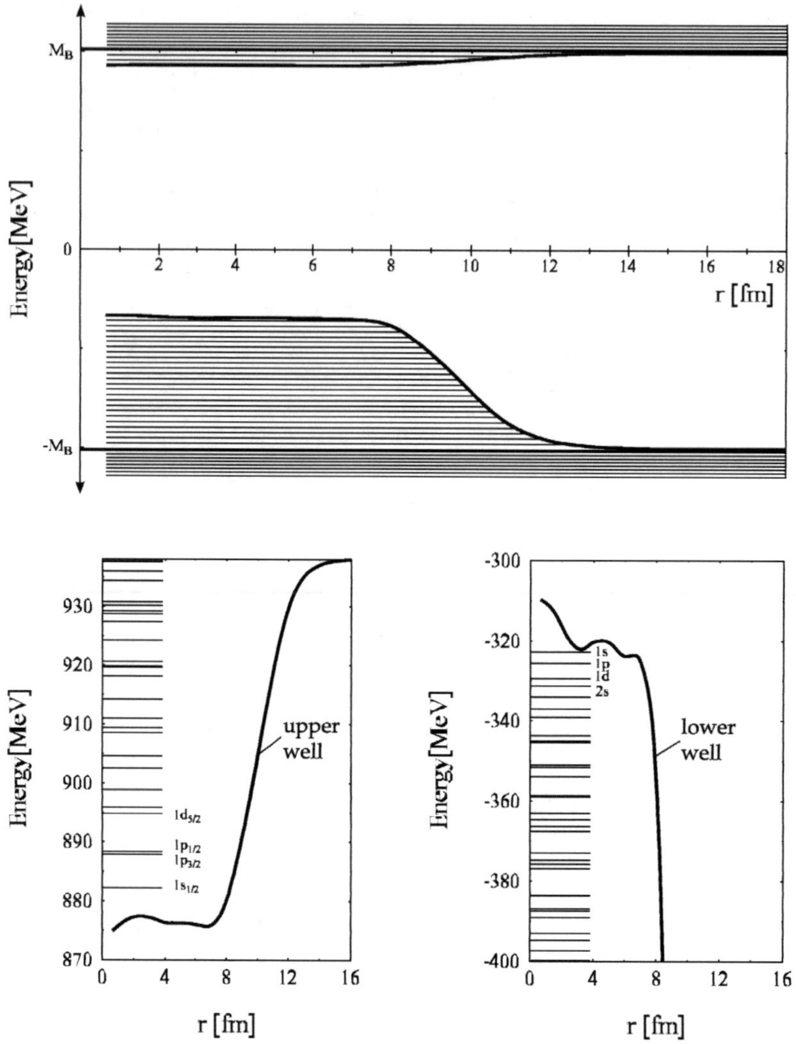

Figure 12. Baryon spectrum in a nucleus (upper panel). Below the positive energy continuum exists the potential well of real nucleons. It has a depth of 50-60 MeV and shows the correct shell structure. The shell model of nuclei is realized here. However, from the negative continuum another potential well arises, in which about 40 000 bound particles are found, belonging to the vacuum. A part of the shell structure of the upper well and the lower (vacuum) well is depicted in the lower figures.

the upper well. These numerous bound Λ's are sitting close to the primary baryons: in a certain sense a giant multi–Λ hypernucleus has been created. When the system disintegrates (expansion stage) the Λ's distribute over the nucleon clusters (which are most abundant in peripheral collisions). In this way multi–Λ hypernuclei can be formed. Of course this vision has to be worked out and probably refined in many respects. This requires a much more and thorough investigation in the future. It is particularly important to gain more experimental information on the properties of the lower well by (e, p) or (e, p) and also $(\overline{p}_c p_b,\, p_c \overline{p}_b)$ reactions at high energy ($\overline{p}_c$ denotes an incident antiproton from the continuum, p_b is a proton in a bound

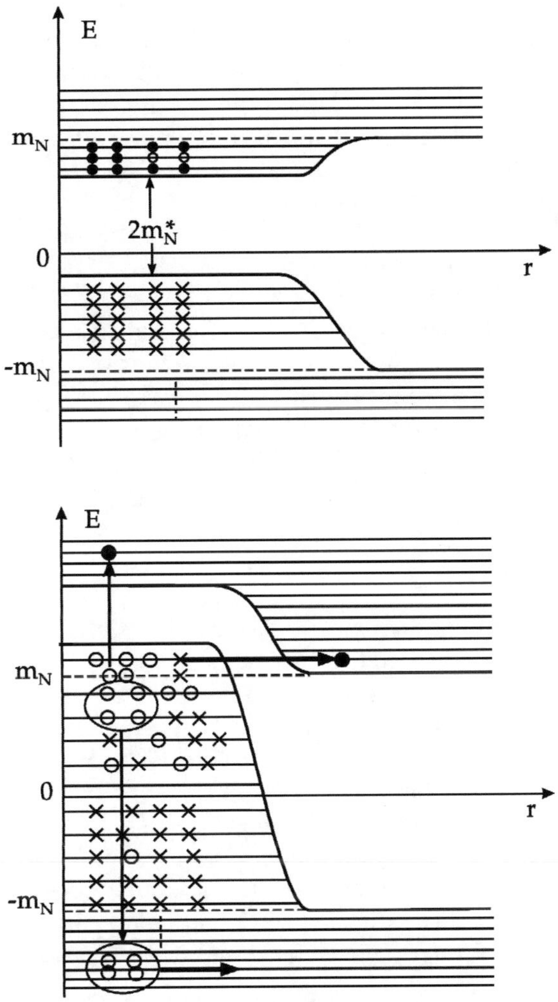

Figure 13. Due to the high temperature and the violent dynamics, many bound holes (antinucleon clusters) are created in the highly correlated vacuum, which can be set free during the expansion stage into the lower continuum (upper panel). In this way, antimatter clusters can be produced directly from the vacuum. The horizontal arrow in the lower part of the figure denotes the spontaneous creation of baryon-antibaryon pairs, while the antibaryons occupy bound states in the lower potential well. Such a situation, where the lower potential well reaches into the upper continuum, is called supercritical. Four of the bound holes states (bound antinucleons) are encircled to illustrate a "quasi-antihelium" formed. It may be set free (driven into the lower continuum) by the violent nuclear dynamics.

state; for the reaction products the situation is just the opposite). Also the reaction (p, p' d), (p, p' ^{3}He), (p, p' ^{4}He) and others of similar type need to be investigated in this context. The systematic studies of antiproton scattering on nuclei can contribute to clarify these questions. Various effective theories, e. g. of the Walecka-type on the one side and theories with chiral invariance on the other side, have been constructed to describe dense strongly interacting matter [4]. It is important to note that they seem to give different strengths of the potential wells and also different dependence on the baryon density.

According to chirally symmetric meson field theories, the antimatter–cluster production and multi–hypermatter–cluster production out of the highly correlated vacuum takes place at approximately the same heavy-ion energies as compared to the predictions of the Dürr-Teller-Walecka type meson field theories. This in itself is a most interesting, quasi-fundamental question to be clarified. In the future, the question of the nucleonic substructure (form factors, quarks, gluons) and its influence on the highly correlated vacuum structure has to be studied. The nucleons are possibly strongly modified in the correlated vacuum: the Δ resonance correlations are probably important. Is this highly correlated vacuum state, especially during the compression, a preliminary stage to the quark-gluon cluster plasma? To which extent is it similar or perhaps even identical with it?

5. Concluding remarks – outlook

Study of nuclear properties and mechanisms of nuclear reactions in the region of superheavy nuclei remains one of the most challenging problems. In particular, the center of island of stability waits for its discovery, that requires new methods (reactions), which may be used for synthesis of neutron-rich superheavy nuclei. The extension of the periodic system into the sectors of hypermatter (strangeness) and antimatter is of general and astrophysical importance. Indeed, microseconds after the big bang, the new dimensions of the periodic system we have touched upon, certainly have been populated in the course of the baryo- and nucleo-genesis. In the early history of the universe, even higher dimensional extensions (charm, bottom, top) may have played a role, which we did not pursue here. It is an open question, how the depopulation (the decay) of these sectors influences the structure and composition of our world today. Our conception of the world will certainly gain a lot through the clarification of these questions.

References

[1] Mosel U and Greiner W 1968, *Z. Phys.* **217** 256; Mosel U and Greiner W 1968, *Z. Phys.* **222** (1969) 261; see also *Memorandum zur Errichtung eines gemeinsamen Ausbildungszentrums für Kernphysik Hessischer Hochschulen*, this was the initiative for building GSI (1966)

[2] Nilsson S G , Nix J R, Sobiczewski A, Szymaimageski Z, Wycech S, Gustafson C and Möller P 1968, *Nucl. Phys.* **A 115** 545; Nilsson S G, Thompson S G and Tsang C F 1969, *Phys. Lett.* **B 28** 458

[3] Săndulescu A, Gupta R K, Scheid W and Greiner W 1976, *Phys. Lett.* **B 60**, 225; Gupta R K, Săndulescu A and Greiner W 1976, *Phys. Lett.* **B 67** 257

[4] Oganessian Yu Ts, Demin A G, Iljinov A S, Tretyakova S P, Pleve A A, Penionzhkevich Yu E, Ivanov M P and Tretyakov Yu P 1975, *Nucl. Phys.* **A 239** 157

[5] Maruhn J A and Greiner W 1974, *Phys. Rev. Lett.* **32** 548; Maruhn J A 1973, Ph. D. Thesis, J. W. Goethe Universität, Frankfurt an Main

[6] Zagrebaev V and Greiner W 2010, *Giant Nuclear Systems of Molecular Type*, in Clusters in Nuclei - Vol. 1, C. Beck (ed.), *Lecture Notes in Physics* **818** 267

[7] Gupta R K, Münzenberg G and Greiner W 1997, *J. Phys.* **G 23** L13

[8] Hofmann S and Münzenberg G 2000, *Rev. Mod. Phys.* **72** 733

[9] Dubna–group (Oganesian Yu Ts, Utyonkov V K and Moody K J) 2000, *Sci. Am.* **282**; Oganesian Yu Ts 2007, *J. Phys* **G 34** R165

[10] Mişicu Ş and Greiner W 2002, *Phys. Rev.* **C 66** 044606

[11] Itkis M G, Luk'yanov S M, Okolovich V N, Penionzkevich Yu E, Rusanov A Ya, Salamatin V S, Smirenkin S N and Chubaryan N N 1990, *Yad. Fiz.* **52**, 23 [Sov. J. Nucl. Phys. **52**, 15]

[12] Gönnenwein F, Gagarski A, Petrov G, Guseva I, Zavarukhina T, Mutterer M, von Kalben J, Kopatch Yu, Tiourine G, Trzaska W, Sillanpäa M, Soldner T, Nesvizhevsky V 2010, *AIP Conference Proceedings* **1224** 338

[13] Poenaru D N and Greiner W 2010, *Cluster Radioactivity* in Clusters in Nuclei 1, C. Beck, ed., vol. **818** of Lecture Notes in Physics, pp. 1–56, Springer

[14] Karpov A V, Zagrebaev V I, Martinez Palenzuela Y, Felipe Ruiz L and Greiner W, *Int. J. Mod. Phys.* **E 21** (2012) 1250013

[15] Rutz K, Bender M, Bürvenich T, Schilling T, Reinhard P G, Maruhn J A and Greiner W 1997, *Phys. Rev.* **C 56** 238; Bürvenich T, Rutz K, Bender M, Reinhard P G, Maruhn J A and Greiner W 1998, *Eu. Phys. J.* **A 3** 139; Bender M, Rutz K, Reinhard P G, Maruhn J A and Greiner W 1998, *Phys. Rev.* **C 58** 2126

[16] Serot B D and Walecka J D 1986, *Advances in Nuclear Physics* **16**, Plenum Press, New York; Theis J, Graebner G, Buchwald G, Maruhn J, Greiner W, Stöcker H and Polonyi J 1983, *Phys. Rev.* **D 28** 2286; Klimt S, Lutz M and Weise W 1990, *Phys. Lett.* **B 249** 386; Mishustin I N 1996, *Proc. Int. Conf. Structure of Vacuum and Elementary Matter*, South Africa, Wilderness, World Scientific, Singapore p. 499; Papazoglou P, Zschiesche D, Schramm S, Schaffner–Bielich J, Stöcker H and Greiner W 1999, *Phys. Rev.* **C 59** 411; Mishustin I N, Satarov L M, Stöcker H and Greiner W 2000, *Phys. Rev.* **C 62** 034901

[17] Papazoglou P 1998, Ph. D. Thesis, University of Frankfurt; Beckmann Ch, Papazoglou P, Zschiesche D, Schramm S, Stöcker H and Greiner W 2002, Phys. Rev. **C 65** 024301

[18] Gridnev K A, Tarasov V N, Tarasov D V, Gridnev D K, Pilipenko V V and Greiner W 2010, *Int. Jour. Mod. Phys.* **E19** 449-457; Tarasov V N, Gridnev K A, Gridnev D K, Tarasov D V, Schramm S and Greiner W, *Stability Peninsulas near Neutron Drip Line*, arXiv:1106.5910v2 [nucl-th], submitted to *Int. Jour. Mod. Phys.*

[19] Botvina A, Mishustin I, Zagrebaev V, and Greiner W 2010, *Int. J. Mod. Phys.* **E19** 2063

[20] Zagrebaev V I, Karpov A V, Mishustin I N and Greiner W 2011, *Phys. Rev.* **C 84** 044617; arXiv:1105.1659 [nucl-th]

[21] Zagrebaev V I, Oganessian Yu Ts, Itkis M, Greiner W 2006, *Phys. Rev.* **C 73** 031602

[22] Zagrebaev V and Greiner W 2007, *J. Phys. G* **34** 1

[23] Reinhardt J, Müller U, Müller B, Greiner W 1981, *Z. Phys.* **A 303** 173

[24] Zagrebaev V I, Karpov A V and Greiner W 2010 *Phys. Rev.* **C 81** 044608

[25] Povh B 1976, *Rep. Progr. Phys.* **39** 823

[26] Schaffner J, Greiner C and Stöcker H 1992, *Phys. Rev.* **C 46** 322

[27] The STAR Collaboration 2011, *Nature* **473** 353

[28] Stöcker H, Greiner W and Scheid W 1978, Z. Phys. **A 286** 121

[29] Greiner W, Müller B and Rafelski J 1985, *QED of Strong Fields*, Springer Verlag, Heidelberg. For a more recent review see Greiner W and Reinhardt J 1998, *Supercritical Fields in Heavy-Ion Physics*, Proc. 15th Advanced ICFA Beam Dynamics Workshop on Quantum Aspects of Beam Physics, World Scientific, Singapore

[30] Auerbach N, Goldhaber A S, Johnson M B, Miller L D and Picklesimer A 1986, *Phys. Lett.* **B 182** 221

[31] Mishustin I N, Satarov L M, Schaffner J, Stöcker H and Greiner W 1993, *J. Phys.* **G 19** 1303; Panda P K , Patra S K, Reinhardt J, Maruhn J, Stöcker H and Greiner W 1997, *Int. J. Mod. Phys.* **E 6** 307

[32] Scheid W, Hoffmann I and Greiner W 1974, in Proc. Symp. Physics with Relativistic Heavy Ions, Lawrence Berkeley Lab., Berkeley

[33] Stöcker H and Greiner W 1986, *Phys. Rep.* **137** 279

Superheavy elements created at the Joint Institute for Nuclear Research

Mikhail Itkis

Joint Institute for Nuclear Research, Dubna 141980, Russia

E-mail: itkis@jinr.ru

Abstract. This study is concerned with the search for the entrance channels suitable to synthesize of superheavy elements. The differences between "cold" and "warm" fusion reactions successfully used for superheavy element productions are discussed. A special attention is made on the quasifission process which competes with fusion process in the reactions with heavy ions and becomes dominant in the reactions leading to the formation of superheavy nuclei.

1. Introduction

The existence of the superheavyheavy nuclei, as well as their decay properties, are mainly determined by fission barrier B_f. According to the Liquid Drop Model (LDM) the limit of existence of nuclei is located immediately after $Z \approx 100$ [1], but the presence of shell effects can lead to a considerable increase of stability of nuclei with respect to various decay modes: α decay or spontaneous fission. The island of stability in the region of nuclei with $Z = 114$ and $N = 184$ predicted theoretically [2] has induced an extensive experimental investigation in the field of superheavy element (SHE) synthesis. At present the SHE are synthesized in the complete fusion reactions of two heavy nuclei. More than 30 SHEs with $Z = 108 - 118$ have been synthesized in the reactions of "cold" fusion (reactions with ^{208}Pb or ^{209}Bi targets) and "warm" fusion (reactions with actinide targets).

In the "cold" fusion reactions the excitation energy of compound nucleus (CN) is 10-20 MeV near the reaction threshold. The excitation energy of a formed CN is about 30-40 MeV in "warm" fusion reactions and de-excitation of CN to the ground state is due to the emission of three or four neutrons and several γ-rays. Nevertheless in the "warm" fusion reactions the formed CN is more neutron rich and nearer to the closed neutron shell at $N = 184$ than in the case of "cold" fusion reactions. Consequently, the "warm" fusion reactions are more preferable for synthesis of the SHE. A big success was achieved in the reactions of actinides with double magic ^{48}Ca ions at FLNR [3]. The production cross sections of SHE in these reactions virtually do not change with increasing atomic number of CN and maintain the level of a few picobarn.

Due to the low production rate for SHE the experiments on synthesis of new elements lasted at least several months. The study of fusion-fission process also allows to obtain information about the fusion probability. Note that the most part of SHEs are strongly fissile nuclei and the counting rate for fusion-fission process is sufficiently larger. In heavy ion-induced reactions the CN formation cross section is suppressed strongly by competing quasifission process (QF) [4, 5]. Therefore, the measurement of binary reaction channel gives the information about the competition between fusion-fission and QF processes and allows to estimate the probability of the CN formation.

2. Influence of entrance channel on the competition between fusion-fission and quasifission

The contribution of QF increases with increasing entrance channel mass asymmetry. As an example, to demonstrate this dependence the mass-energy distributions of binary fragments obtained in the reactions ^{22}Ne+^{249}Cf, ^{26}Mg+^{248}Cm, ^{36}S+^{238}U and ^{58}Fe+^{208}Pb leading to the formation of Hs* ($Z = 108$) [6] are presented in Fig. 1.

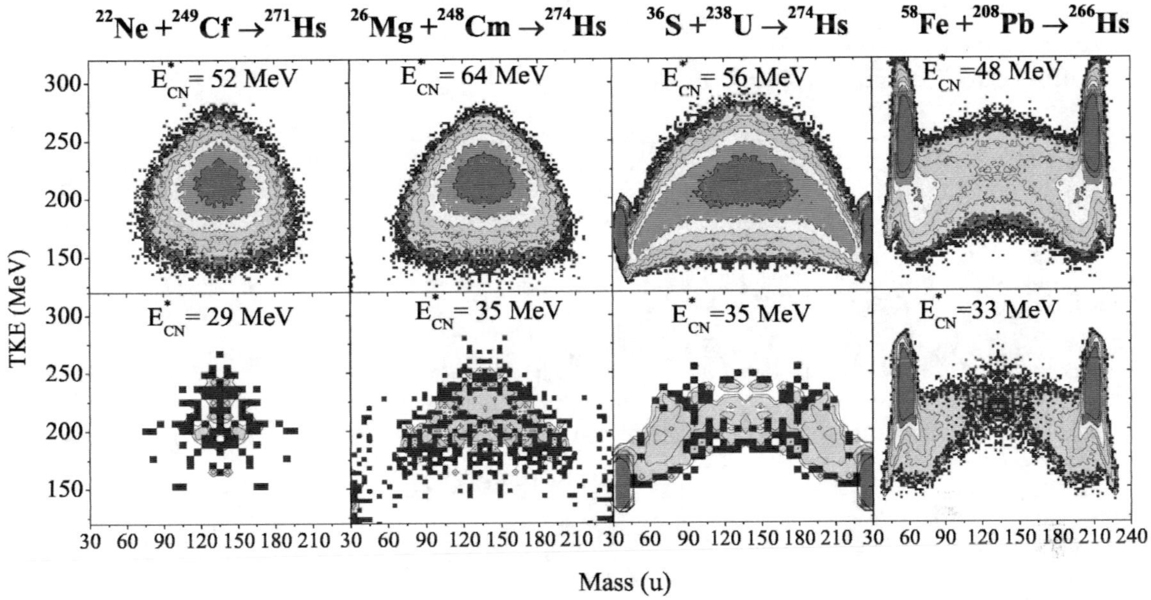

Figure 1. Mass-energy distributions of binary reaction fragments for the reactions ^{22}Ne+^{249}Cf, ^{26}Mg+^{248}Cm, ^{36}S+^{238}U, ^{58}Fe+^{208}Pb leading to the formation of elements with Z=108 at energies above (top panel) and below (bottom panel) the Coulomb barrier.

In the chosen reactions the parameters of mass-asymmetry α and the Coulomb factor Z_1Z_2 in the entrance channel varies strongly: for the reaction ^{58}Fe+^{208}Pb: $\alpha = 0.571$ and $Z_1Z_2 = 980$, for ^{36}S+^{238}U: $\alpha = 0.737$ and $Z_1Z_2 = 1152$, for ^{26}Mg+^{248}Cm,: $\alpha = 0.810$ and $Z_1Z_2 = 1472$, and for ^{22}Ne+^{249}Cf: $\alpha = 0.838$ and $Z_1Z_2 = 2132$. As demonstrated in Fig. 1 the TKE-mass distributions change with decreasing asymmetry in the entrance channel from symmetric for incoming ^{22}Ne-ions to strongly asymmetric for incoming ^{58}Fe-ions. These changes are understood as reflecting the relative contributions of CN-fission and QF to the fission process of excited Hs* isotopes as depending on the reaction studied.

It is clearly seen that even at similar CN excitation energies the TKE-mass distributions are vastly different for these reactions. In the case of the reactions ^{22}Ne+^{249}Cf and ^{26}Mg+^{248}Cm the mass distributions have a near Gaussian shape with no evidence for asymmetric fission. The reactions are considered to be mainly attributed to the CN-fission process. For the ^{36}S+^{238}U reaction the mass distributions of the fission-like fragments change markedly. This difference in mass distribution for the ^{26}Mg+^{248}Cm and ^{36}S+^{238}U reactions is connected with an increasing contribution of the QF process for the ^{36}S-induced reaction. At low excitation energies QF is the dominant process for the reaction ^{36}S+^{238}U. At higher excitation energies the mass distribution becomes symmetric and similar to the reaction $^{26}Mg+^{248}Cm$ though - due to a remaining trace of QF - slightly wider.

The TKE-mass distribution for the $^{58}Fe+^{208}Pb$ reaction has a wide two-humped shape even at excitation energy of 48 MeV (well above the Bass barrier). For this reaction the QF process dominates at energies below and above the Bass barrier. At the last decade the large efforts have been done in the investigation of suppressing of fusion process by QF for different "ion-target" combinations [4, 5, 6, 7, 8, 9, 10, 11]. It was found that the balance between fusion and QF strongly depends on entrance channel properties, such as mass asymmetry, deformation of interacting nuclei, collision energy, and the Coulomb factor Z_1Z_2. The presence of QF has to be taken into account for experiments devoted to the production of SHEs in order to choose the optimal reaction and collision energy.

3. "Cold" and "warm" fusion reactions

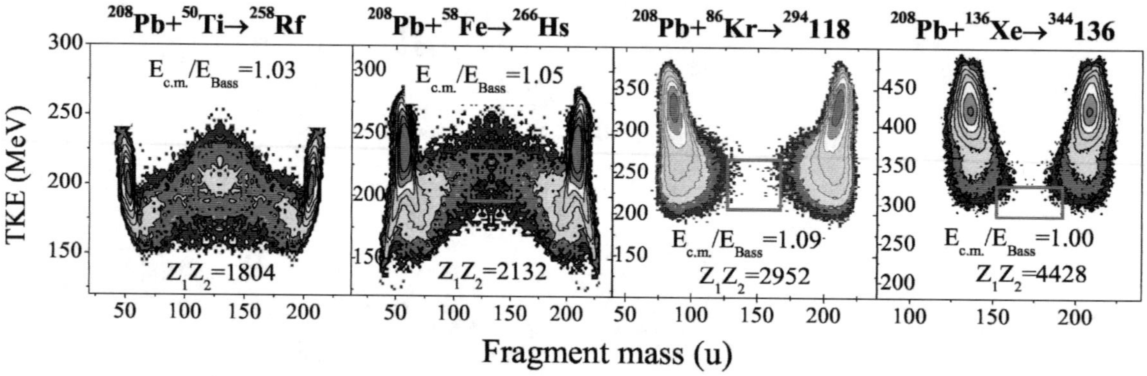

Figure 2. Mass-energy distributions of fragments obtained in the "cold" fusion reactions $^{208}Pb+^{50}Ti,^{58}Fe,^{86}Kr,^{136}Xe$ at energies close to the Bass barrier.

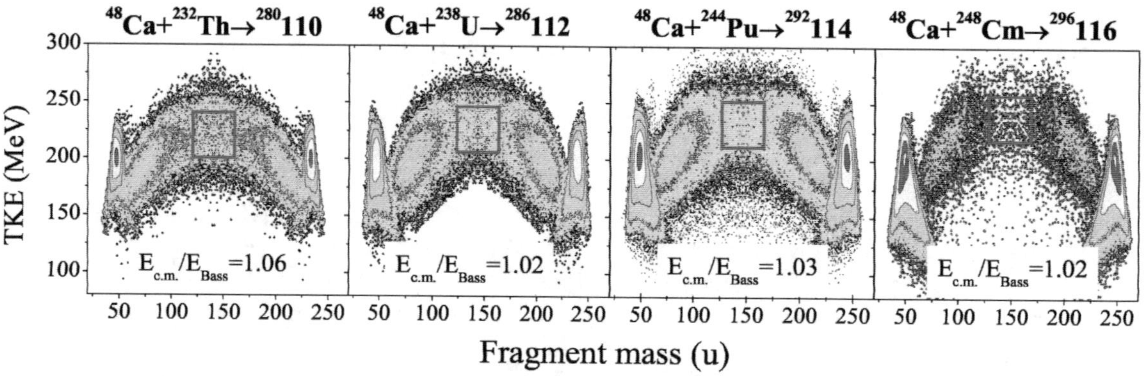

Figure 3. Mass-energy distributions of fragments obtained in the "warm" fusion reactions $^{48}Ca+^{232}Th,^{238}U,^{244}Pu,^{248}Cm$ at energies close to the Bass barrier.

Coming back to "cold" and "warm" fusion reactions one should note that the entrance channel is strongly different. For SHEs with $Z \geq 108$ at the former reaction type the values of the Coulomb factor $Z_1Z_2 \geq 2000$, while at the latter reaction type the Coulomb factor $Z_1Z_2 = 1960$ even in the case of reaction $^{48}Ca+^{249}Cf$ leading to the SHE with $Z = 118$. In Figs. 2 and 3 the mass energy distributions of binary fragments obtained in the reactions of "cold" and "warm" fusion are shown.

It is clearly seen that in the "cold" fusion reactions at the transition from ^{50}Ti to ^{86}Kr the shape of mass-energy distributions of binary reaction fragments changes strongly due to the QF process. For the^{50}Ti ion the contribution of symmetric fragments into the capture cross-section is around 60%; in the case of ^{86}Kr the QF and deep-inelastic scattering are dominant processes. For the heaviest system ^{136}Xe+^{208}Pb the main process is deep-inelastic scattering [14]. For the reactions with ^{48}Ca-ions ("warm" fusion) the mass-energy distributions are similar for all the reactions: the clearly pronounced asymmetric QF with heavy fragments near the double magic lead is observed, the contribution of symmetric fragments into the capture cross section is about 10% .

Today the properties of fission of nuclei up to element with $Z = 106$ are well studied. It is known that at low excitation energy and spontaneous fission the spherical nuclear shells with $Z = 50$ and $N = 82$ and deformed neutron shell at $N = 88$ play a dominant role in the formation of fission fragments. The fission of nuclei with excitation larger than 40 MeV (when shell effects practically disappear) is well described by the LDM and characterized by symmetric mass distribution with the Gaussian shape with parameters depending on the temperature at saddle (or scission) point. The most probable kinetic energy of the fission fragments depends on the parameter $Z^2/A_{CN}^{1/3}$ and can be estimated by the Viola systematics based on experimental data [15]. Therefore as the first approximation the symmetric fragment region with mass $A_{CN}/2 \pm 20$ u may be attributed to CN-fission. The red contours in Figs. 2 and 3 show the regions of masses and TKE's where the main part of CN fission fragments is expected.

However, this estimation corresponds to the upper limit for fusion-fission process due to the population of the mass region $A_{CN}/2 \pm 20$ u by both the CN-fission and QF fragments. The QF is characterized by large nucleon exchange and energy dissipation as well as CN-fission process and separation between CN-fission and QF fragments is a complicated problem, especially for symmetric mass split. A realistic description of deep inelastic scattering, QF and CN-fission processes in low energy heavy ion collisions shows [12, 13] that the potential energy surface for these systems is strongly modulated by shell effects and leads to the appearance of deep valleys corresponding to the formation of well bound magic nuclei. In accordance with these calculations, at least three paths leading to the formation of fission-like fragments can be distinguished: asymmetric QF caused by the influence of proton shells with $Z = 28, 82$ and neutron shells with $N = 50, 126$; symmetric QF determined by the shells with $Z = 50$ and $N = 82$; fusion-fission leading to the formation of symmetric fragments.

It is important to note that in the case of "warm" fusion reactions all target nuclei are well-deformed nuclei. In the reactions with deformed nuclei the potential energy surface strongly depends on the relative orientation of the reaction partners, which changes the Coulomb barrier and the distance between the centers of the colliding nuclei. When two interacting nuclei touch each other by their lateral surfaces (near-side collisions), a high probability of formation of a spherical CN is expected, whereas in the elongated configuration, when nuclei touch each other by their poles (near-tip collision), a high QF probability is expected. One should note that all the events observed in the production of isotopes of elements 112, 114, 116, 117 and 118 with the fusion reactions ^{238}U,^{242,244}Pu, ^{248}Cm, ^{249}Bk, ^{249}Cf+^{48}Ca were detected at energies above the Coulomb barrier [16, 17, 18]. In contrast, in "cold" fusion reactions of the target nuclei of ^{208}Pb and ^{209}Bi with more massive projectiles, the maximum yield of evaporation residues is observed at subbarrier projectile energies [19, 20].

The maximal cross sections for evaporation residues obtained in the "cold" and "warm" fusion reactions are presented in Fig. 4. The evaporation residue cross section drops dramatically with increasing atom number of CN in the case of "cold" fusion reaction. In the "warm" fusion reactions their cross sections virtually do not change. This behavior may be explained by the

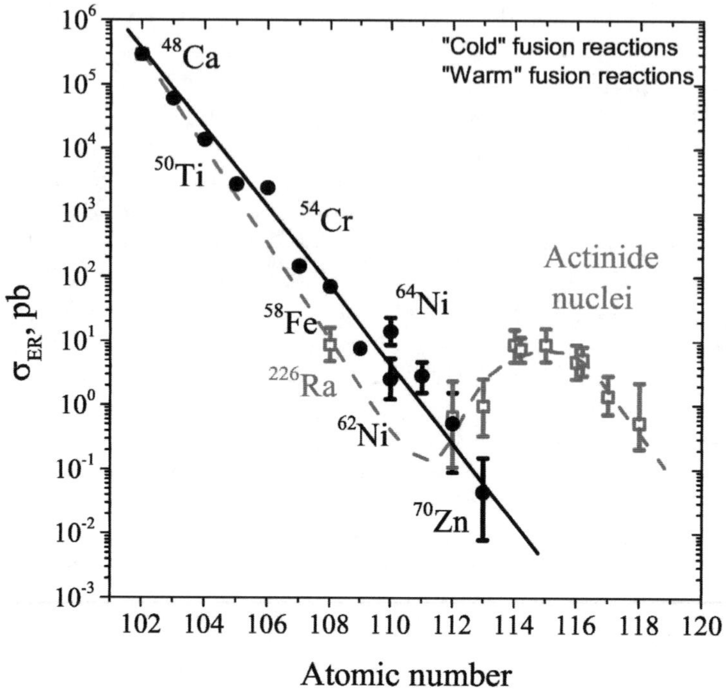

Figure 4. Maximal cross sections of 1n- evaporation channel in the "cold" fusion reactions of ^{208}Pb and ^{209}Bi target nuclei with different projectiles (indicated in the figure) and cross sections of 3n and 4n channels in the "warm" fusion reactions of ^{48}Ca as a function of compound nucleus atomic number. The lines are drawn as guide the eye. Data were taken from [3].

fact that the contribution of QF increases strongly with increasing atom number of CN in the case of "cold" fusion reaction. For the "warm" fusion reaction the contribution of QF process nearly the same due to small difference in the reaction entrance channel. While the survival probability against fission is higher in the "cold" fusion reactions due to lower excitation energy of formed CN, the evaporation residues are some 10-15 u shifted from β stability line, that leads to a considerable decrease in their half-lives. The isotopes of SHE formed in the ^{48}Ca induced reactions cannot reach the neutron closed shell with $N = 184$ due to the lack of 7-9 neutrons, but nevertheless the additional neutrons increase strongly the half-lives of these nuclei up to several seconds.

The analysis of the available experimental data on the fusion and fission of nuclei of ^{286}Cn, ^{292}Fl, and ^{296}Lv, produced in the reactions ^{48}Ca+^{238}U, ^{48}Ca+^{244}Pu, and ^{48}Ca+^{248}Cm, as well as experimental data on the survival probability of those nuclei in evaporation channels of three- and four-neutron emission, enables us to reach the quite reliable conclusion that the fission barriers of those nuclei are really quite high [21], which results in their relatively high stability. The lower limits that we obtained for the fission barriers of nuclei of $^{283-286}$112, $^{288-292}$114, and $^{292-296}$116 are 5.5, 6.7, and 6.4 MeV, respectively.

4. Alternative ways for SHE

While the relative contribution of QF to the capture cross section mainly depends on the reaction entrance channel properties, the features of asymmetric QF are determined essentially by the driving potential of composite system. Generally, in heavy-ion-induced reactions the formation of QF fragments is connected with the strong influence of the nuclear shell at $Z = 82$ and $N = 126$ (doubly magic lead) and $Z = 28$ and $N = 50$ (double magic nickel). The experimental mass distribution of the fission-like fragments formed in the reaction ^{48}Ca+^{248}Cm at energy close to the Coulomb barrier is presented in the top panel of Fig. 5. The symmetric fragment mass distribution (the open circles in Fig. 5) has been extracted from the experimental mass distribution of the all fission-like fragments using the Gaussian fitting procedure of asymmetric peaks. The driving potential for this composite system at scission point calculated in the frame of the proximity model is also shown in the same figure. As it was mentioned above the maximum fragment yields correspond to the positions of the local minima of the driving potential as in the case of formation of asymmetric fragments, as well as symmetric ones.

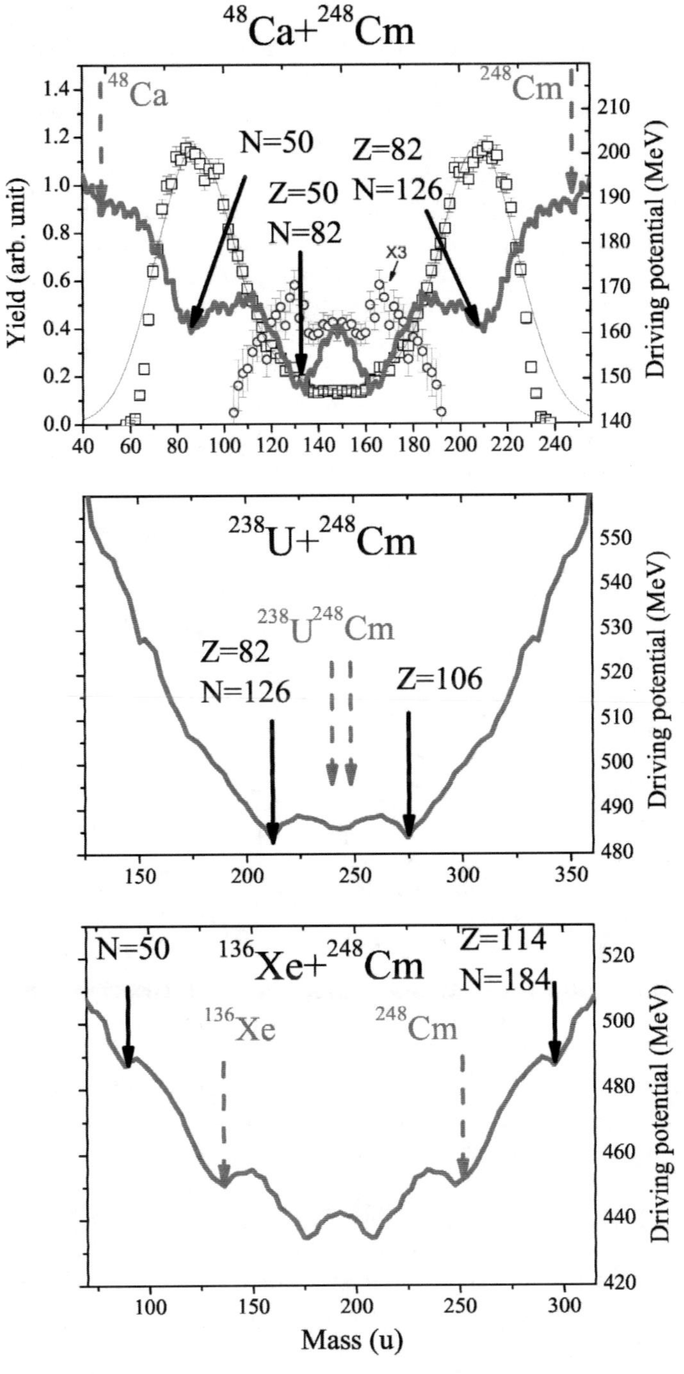

Figure 5. The driving potentials as a function of reaction fragment mass at scission point for the reactions ^{48}Ca+^{248}Cm, ^{238}U+^{248}Cm and ^{136}Xe+^{248}Cm.

In the middle and bottom panels of Fig. 5 the driving potentials for the composite systems formed in the reactions ^{238}U+^{248}Cm and ^{136}Xe+^{248}Cm calculated in the same approximation are presented, respectively. The dashed arrows indicate the position of the entrance channels, while the solid arrows show the position of the proton and neutron closed shells. It is clearly seen that the local minima in the driving potential exist for all reactions, though in the latest two reactions these minima are located from the outside the entrance channel. Thus, we may expect an increase of the fragment yields in the mass region around these minima. W. Greiner and V. Zagrebaev in [22] proposed to call this process inverse QF. Notice, that in the case of the reaction ^{238}U+^{248}Cm one of the minima corresponds to doubly magic lead valley and the complementary fragment is a SHE around Z=106. In the reaction ^{136}Xe+^{248}Cm the both fragments have closed shells: the light fragment is near N=50, the heavy one is close to Z=114 and N=184 (predicted "island of the stability").

The idea of the production of SHE in the multi-nucleon transfer reactions in the collision of U+U nuclei (or similar reactions) was already proposed in [23]. In this work it was found that at an incident energy of 7.42 MeV/u (about 22% above the Coulomb barrier) a direct search for α-decay or fission of SHE being produced in a deep inelastic collision resulted in an upper cross section limit of 2nb. Although the stronger penetration of nuclei leads to enhanced mass transfer, the higher excitation energies involved drasti-cally reduce the survival probability of the nuclei produced. The decrease in collision energy to the Coulomb barrier energy leads to the lower total excitation and consequently to larger cross section of survived SHE. According to the calculation of the cross section of survived SHE formed in the reaction ^{232}Th+^{250}Cf at 800 MeV center-of-mass energy (near the Coulomb barrier) from [22] there is a real chance for production of the long-lived neutron-rich SHE in such reactions.

Even though the "warm" fusion reaction leads to the formation of more neutron-rich nuclei than in the case of cold fusion even after the de-excitation process, the isotopes of SHE formed in these ^{48}Ca induced reactions cannot reach the neutron closed shells with $N = 184$ due to the lack of 7-9 neutrons. From the investigation of the mass-energy distributions of binary reaction fragments obtained in the reactions ^{48}Ca+^{238}U, ^{58}Fe+^{244}Pu and ^{64}Ni+^{238}U [9] it was found that the cross section drops three order of magnitude for the formation of the compound nucleus with $Z = 120$ obtained in the reaction ^{64}Ni+^{238}U compared to the formation of the compound nucleus with $Z = 112$ obtained in the reaction ^{48}Ca+^{238}U at an excitation energy of the compound nucleus of about 45 MeV. This is unfortunately a limiting factor. Furthermore, the relative contribution of the CN-fission fromed in ^{64}Ni+^{238}U is much lower than in the case of ^{58}Fe+^{244}Pu, leading to the formation of the same composite system.

Recently the experiments aimed at the synthesis of isotopes of element $Z = 120$ have been performed using the ^{244}Pu (^{58}Fe, xn)$^{302-x}$120 reaction [24] and ^{238}U(^{64}Ni, xn)$^{302-x}$120 reaction [25]. A cross section limit of 0.4 pb at E* = 44.7 MeV for the former reaction and 0.09 pb at E* = 36.4 MeV for the latter reaction were obtained. In the case of ^{48}Ca induced reactions the evaporation residue cross section for 3n, 4n channels is about a few picobarns even for the heaviest nucleus with $Z = 118$.

A possible alternative pathway is represented by the "inverse" QF or deep-inelastic reactions in the collision of ^{136}Xe, ^{232}Th and ^{238}U with actinide targets. According to the theoretical expectations the cross section for the survived nuclei formed in such processes is higher than in the reaction of complete fusion. The reaction ^{136}Xe+^{248}Cm can be more interesting due to the fact that the heavy valley of the driving potential for this system corresponds to the nuclei close to the "island of stability" while in the case of the collision of Th and U with actinide targets the heavy valley lies outside the stability line. To estimate the formation probabilities of superheavy elements in these reactions the additional investigations are needed.

References

[1] Bohr N and Wheeler J A 1939 *Phys. Rev.* **56** 426
[2] Sobiczewsky, A Gareev, F A Kalinkin B N 1966 *Phys. Lett.* **22** 500

[3] Oganessian Yu Ts 2007 *J. Phys. G: Nucl. Part. Phys.* **34** R165
[4] Bock R et al. 1982 *Nucl. Phys.* **A388** 334
[5] Töke et al. 1985 *Nucl. Phys.* **A440** 327
[6] Itkis Yu M et al. 2011 *Phys. Rev. C* **83** 064613
[7] Simenel C et al. 2012 *Phys. Lett.* **B 710** 607
[8] Knyazheva G N et al. 2007 *Phys. Rev. C* **75** 064602
[9] Kozulin E M et al. 2010 *Phys. Lett.* **B 686** 227
[10] Nishio K et al. 2008 *Phys. Rev. C* **77** 064607
[11] Hinde D J et al. 1996 *Phys. Rev. C* **53** 1290
[12] Zagrebaev V I and Greiner W 2005 *J. Phys. G* **31** 825
[13] Aritomo Y, Hagino K, Nishio K and Chiba S 2012 *Phys. Rev. C* **85** 044614
[14] Kozulin E M et al. 2012 *Phys. Rev.* **C 86** 044611
[15] Viola V E Kwiatkowski K and Walker M 1985 *Phys. Rev. C* **31** 1550
[16] Oganessian Yu Ts et al. 2004 *Phys. Rev. C* **70** 064609
[17] Oganessian Yu Ts et al. 2006 *Phys. Rev. C* **74** 044602
[18] Oganessian Yu Ts et al. 2010 *Phys. Rev. Lett.* **104** 142502
[19] Hofmann S and Münzenberg, 2000 *Rev. Mod. Phys* **72** 733
[20] Hofmann S et al., 2002 *Eur. Phys. J. A* **14** 147
[21] Itkis M G, Zagrebaev V I, Oganessian Yu Ts 2002 *Phys. Rev. C* **65** 044602
[22] Greiner W and Zagrebaev V I 2006 , *in Proceedings of International Symposium on Exotic Nuclei (EXON 2006), edited by Yu.E.Penionzhkevich and E.A. Cherepanov, AIP Conf. Proc.* **912** 221
[23] Freiesleben H et al. 1979 *Z. Phys.* **A 292** 171
[24] Oganessian Yu Ts et al. 2009 *Phys. Rev. C* **79** 024603
[25] Hofmann S et al. 2008 *GSI Scientific Reports* 131

Int. Summer School for Advanced Studies 'Dynamics of open nuclear systems' (PREDEAL12) IOP Publishing
Journal of Physics: Conference Series **413** (2013) 012004 doi:10.1088/1742-6596/413/1/012004

The Density functional method for nuclear systems. History and perspectives

R J Lombard

Groupe de Physique Théorique, Institut de Physique Nucléaire,
91406 Orsay Cedex, France

Abstract. The earlier attempts to construct a density functional for nuclear systems are shortly reviewed. Some results obtained with the functional derived by Beiner, Mas and myself are then recalled, as well as the extension to excited states. The universality of the functional is discussed. A way to improve the efficiency of the functional is proposed, in which the parameters of the functional are allowed to vary according to the underlying shell structure of each nucleus.

PACS number(s): 03.65.Ge ; 31.15.ac ; 31,15.B-

1. Foreword.

I should apologise to colleagues actually working in this field : I haven't followed the recent developments, and thus references to relevant papers published in the last 10-15 years will be missing.

What will be presented here is merely the history of the earlier attempts to put the density functional method for nuclear systems on a valuable track, without any attempt to deliver a semi-exhaustive review on the subject. This paper is mostly based on personal experience, thus many references to similar or related works are missing. The basic motivation is to suggest a way to go beyond the present status of the density functional.

This paper is dedicated to my friend and colleague Marcel Beiner, who passed away some years ago after a long illness.

My first meetings with physicists from the Eastern countries goes back to the Summer school in the Tatra mountains in 1962 "Selected Topics in Nuclear Theory". A. Sandulescu, who is honoured during this present meeting, was among the participants.

2. Introduction

Since the earlier days of nuclear physics, the problem of modelising the atomic nucleus as an ensemble of bound particles has retained the attention of theorists. The first attempt to calculate the binding energy per particle in the homogeneous medium (infinite nuclear matter) was achieved by Euler [1]. He used plane wave, Gaussian forces, and worked up to second order in the perturbation theory. Details of his calculations are quoted in the book of G. Brown on the nuclear many-body problem [2].

Short after second world war, descriptions of finite systems started, first using the Thomas-Fermi approximation for the kinetic energy and the concept of local density. In this respect, we shall quote the work of P. Gombas, with two main papers entitled "Statistische Theorie der Atomkern" [3]. The basic idea is to count the number of available states in a box, or in a sphere, assuming plane wave as eigenfunctions. This leads immediately to an estimate of the density in function of the Fermi momentum k_f, taking into account the degeneracy due to the spin and isospin degrees of freedom :

$$\rho \propto k_f^3. \tag{1}$$

By using this relation, the total energy of a nucleus can be calculated by using plane wave in first order, as a function of k_f, then transformed into a function of $\rho(r)$ and integrated over the nuclear volume. Various techniques can be applied to get the ground state energy by minimisation. Gombas used Gaussians to parametrised $\rho(r)$ and looked for the minimum in the parameter space.

Several authors have followed this method to either to directly estimate binding energies of finite nuclei, or to derive the proton and neutron numbers dependence of a mass formula. Let us quote here the earlier papers of Skyrme [4], Seyler and Blanchard [5], Bodmer [6], Hara [7] and the work by Kumar, Lecouteur and Roy [8].

However, the first person to formulate the total binding energy of a nucleus under the form of a density functional was Wilets. His model appeared in an article with Berg studying the nuclear surface [9]. Their functional reads

$$E[\rho] = \int \left[k_{tf}\rho^{5/3} + \zeta(\rho)\frac{(\nabla\rho(r))^2}{\rho(r)} + \epsilon(\rho) \right] dv \tag{2}$$

The first term is the kinetic energy at the Thomas-Fermi approximation. The second contribution contains the Weiszäcker correction to the Thomas-Fermi expression and surface effect, and the third is the potential energy in nuclear matter. The Coulomb energy is neglected ; proton and neutron densities are assumed to be proportional. The variational principle was applied together with the constraint on the particle number A

$$\int \rho dv = A\ . \tag{3}$$

Few cases were solved in the one-dimensional space, which corresponds to semi infinite nuclear matter. The effect of Coulomb energy and non proportional proton and neutron densities were considered later by Wilets.

3. The La Jolla time.

I became involved in the density functional industry while working as a post-doc by Keith Brueckner at La Jolla. He had recently achieved nuclear matter calculations for various ratio of proton to neutron numbers with Coon and Dabrowsky [10]. At the same time, Walter Kohn was also professor at La Jolla. He had established his theorem with

Hohenberg proving that the ground state of any fermion systems could be described by a functional of the one-body density [11]. The theorem proves the existence of the functional but does not say how to derive it. Brueckner gave the nuclear matter results to S. Jorna, J.-R. Buchler and myself with the aim of constructing a density functional based on the Thomas-Fermi approximation to describe finite nuclei.

Our first goal was to obtain the nuclear matter potential energy as a function of ρ from the numbers provided us by Brueckner. On very simple and natural arguments, based on dimensionality, we found that the most appropriate expression was an expansion in the Fermi momentum k_f. It means an expansion in $\rho^{1/3}(r)$, which has the dimension of a length and corresponds to the local density approximation. The potential energy per particle took the following form

$$\epsilon/A = a_1[1+s_1\alpha^2(r)]\rho(r) + a_2[1+s_2\alpha^2(r)]\rho^{4/3} + a_3[1+s_3\alpha^2(r)]\rho^{5/3}(r)\ . \quad (4)$$

The neutron-proton asymmetry was efficiently taken into account by the factor α^2 defined by

$$\alpha(r) = \frac{\rho_n(r) - \rho_p(r)}{\rho(r)} \quad ; \quad \rho(r) = \rho_n(r) + \rho_p(r), \quad (5)$$

where $\rho_n(r)$ and $\rho_p(r)$ are the neutron and proton densities, respectively. The fit to the Brueckner, Coon and Dabrowsky results was quite good, as it can be judged from the original drawing displayed in fig. 1.

Surface effects were included very much in the same way as proposed by Wilets. They originate from the first order corrections to the Thomas-Fermi approximation due to Weiszäcker [12], as well as the finite range of nuclear forces. In both cases, next order would involve the density Laplacians. Such contributions are more delicate to handle numerically, giving rise to fourth order differential equations and generating instabilities. Moreover the convergence of the series expansion is not guaranteed. Consequently, it is safer to take the first order terms as effective components of the functional.

Finally, the constructed functional took the following form

$$E[\rho(r)] = a_k\rho^{5/3}(r) + \eta_w\frac{|\nabla\rho(r)|^2}{\rho(r)} + \eta|\nabla\rho(r)|^2 + \epsilon\,(\rho(r)) + E_{coul}\ . \quad (6)$$

The last component is the Coulomb energy, which contains the direct term and an approximate exchange term. The field variables can be $\rho(r)$ and $\alpha(r)$ or $\rho_n(r)$ and $\rho_p(r)$. The total energy is obtained by integration over the whole space

$$E_{tot} = \int E[\rho_n(r), \rho_p(r)]d^3r \quad (7)$$

with the constraints on the particle numbers

$$\int \rho_n(r)d^3r = N\ ;\ \int \rho_p(r)d^3r = Z\ . \quad (8)$$

Applying the Euler-Lagrange minimisation procedure with respect to $\rho_n(r)$ and $\rho_p(r)$ generates a system of two coupled second order differential equations. Simultaneously, the Poisson equation for the charge density is solved to get the electric field seen by the protons.

As a very first attempt, these equations were solved for proportional proton and neutron densities, taking $\alpha = 0$ and assuming spherical symmetry. The functional is left with the single free parameter η [13]. The binding energies of the $N = Z$ nuclei were reproduced correctly up to $A \propto 100$, with the main characteristics : the increase of the binding energies with A up to a saturation point followed by a slow decrease. The original drawing is displayed in fig 2.

Separated proton and neutron densities were considered in a subsequent work [14]. Departure from spherical symmetry was achieved with R. Clark [15]. Ellipsoidal shapes were found to lower the ground state energy with respect to a spherical shape for $Z \geq 106$. The mean field experienced by a single neutron or a single proton was obtained by taking the first derivative of the functional with respect to $\rho_n(r)$ or $\rho_p(r)$, respectively. Solving the Schrödinger equation with this mean field as potential yields the single particle energies. This was the subject of the Lin thesis, and applied to the isotope shift of nuclear charge distributions, in particular to the ^{40}Ca and ^{48}Ca shift [16].

Finally, the density functional were also used to calculate the interaction potential between to colliding nuclei in the sudden approximation :

$$V(\vec{r_1} - \vec{r_2}) = E[\rho_1 + \rho_2] - E[\rho_1] - E[\rho_2] \ . \tag{9}$$

When confronted to experiments, the calculated binding energies and charge radii were reproduced at a semi-quantitative level. The discrepancies arise from three main points : the properties of nuclear matter entering through $\epsilon(\rho)$, the shell effects and the fact that, at very low density, the usual model of nuclear matter has to be corrected for clustering of nucleons. The easiest part is to play with the parameters of $\epsilon(\rho)$. They can be fixed by imposing the energy per particle $E_0(\rho_0)/A$ at the equilibrium density ρ_0 for $\alpha = 0$ and the "compressibility', namely the value of the ϵ second derivative at ρ_0. The asymmetry coefficient has also to be considered. Such adjustments result in a satisfactory global description of ground state properties [17]. Further improvements require the introduction of shell effects.

4. Shell structure.

The shell structure observed in nuclei arises from the Pauli principle. It has important implication at the level of the one-body density. In fact, the Thomas-Fermi approach yields averaged nuclear densities and cannot reproduce the density fluctuations due to the Pauli correlations. The simplest way to introduce the shell structure is to expand the one-body density on a basis of orthonormal functions :

$$\rho_q(r) = \sum_j |\psi_{q,j}(r)|^2 \tag{10}$$

Here q is the isospin state and j the set of quantum numbers attached to each single particle state. This procedure is known as the Kohn-Sham approximation [18]. The total energy is thus a functional of the single particle wave functions. Applying the

Journal of Physics: Conference Series **413** (2013) 012004 doi:10.1088/1742-6596/413/1/012004

Euler-Lagrange principle leads to a set of second order differential equation similar to the Hartree-Fock system.

Two supplementary contributions have to be added in order to built a functional able to come close to the observed binding energies and charge densities. The first one is the pairing effect, the second is a spin-orbit component. In our work, the pairing was treated at the BCS approximation, matrix elements being calculated by using the Hamada-Johnson potential [19]. We greatly benefited from the careful study that Marcel Beiner had achieved in this topics for his PhD thesis. The pairing correlations have the important property of giving the probability occupation $v_j^2 \leq 1$ of the $\{j\}$ state.

The origin of the spin-orbit component is still debated, though its presence is well demonstrated experimentally. Our choice was more or less dictated by various models. We verified that the precise radial shape of this term was not decisive. The adopted expression was simple but sufficient to get the correct energy splitting among spin-orbit partners.

Finally the total energy was given by

$$E_{tot} = \int [\frac{\hbar^2}{2m} \sum_{q,j} |\vec{\nabla}\psi_{q,j}(\vec{r})|^2 + \epsilon(\rho(\vec{r})) + \eta_0 |\vec{\nabla}\rho(\vec{r})|^2 + \eta_1 |\vec{\nabla}(\rho_n(\vec{r}) - \rho_p(\vec{r})|^2$$

$$+ \epsilon_{coul} - \frac{\lambda}{2}(\rho_n^\gamma + \rho_p^\gamma)\vec{\nabla} \cdot \vec{J}\ ]\ d^3r + E_{pairing}\ . \qquad (11)$$

Here, the current $\vec{J}$ is defined by

$$\vec{J} = -i \sum_{q,j} \psi_{q,j}^* (\vec{\sigma} \wedge \vec{\nabla}) \psi_{q,j}\ , \qquad (12)$$

with $\vec{\sigma}$ being the Pauli matrices. This functional contains 12 free parameters. They were adjusted on the properties of 4 nuclei, ^{40}Ca, ^{48}Ca, ^{56}Ni and ^{208}Pb, reproducing the binding energies, charge radii and separation energies. As far as the symmetry energy is concerned, the functional was fitting the infinite neutron matter of Buchler and Ingber [20].

The functional has been applied to a large ensemble of nuclei [21]. A first set of about 250 nuclei situated along the line of stability has been extended to proton and neutron rich nuclei, up to the drip line for light element [22] . Binding energies have been published in the 1975 Nuclear Data table [23].

The differences between calculated and experimental binding energies show a structure. The agreement is excellent for magic nuclei. The discrepancies reach about 1-3 Mev for other spherical nuclei. Being maximum in the middle of the major shells they climb up to 20 Mev for deformed nuclei.

The situation is similar with the charge radii, the discrepancies exhibiting the same behaviour with respect to the shell structure. Besides the charge radii, the charge densities were also well reproduced. The sample displayed in fig 3 underlines the remarkable agreement in the nuclear surface [21]. However, the shell effects affecting the interior part is only qualitatively reproduced. This is an important point to stress : according to the Kohn theorem, the correct functional must produce the correct binding energy for the correct ground state density. Thus, the fit of the charge distributions is of key importance. The discrepancies are a clear indication that the way the shell effects have been incorporated is not sufficient. We conjecture that other correlations have be included besides the pairing contribution.

5. Predictability

Reproducing measured quantities is a first step in establishing the validity of a model. However, its predictive power is more interesting. The high quality of the fits to magic and specially doubly magic nuclei was a confidence criterion to predict properties of similar nuclei. This was the case with ^{132}Sn, for which a binding energy of 1102.5 Mev and a charge radius of 4.71 fm, were predicted, and later confirmed experimentally : 1102.8 Mev and 4.709 fm, respectively [24, 25].

The position of the neutron drip line was well reproduced for light nuclei. It allows us to predict ^{26}O to be unbound, which was confirmed. Furthermore, because the model solved Hatree-Fock BCS equations, it gives a natural criterion to fix magic numbers. It relies on the fact that for magic numbers the occupation probability $v_{q,j}$ of a $\{q, j\}$ state is 0 or 1. Consequently, a magic N or Z corresponds to

$$\sigma_q = \sum_j v_{q,j} u_{q,j} = 0 \tag{13}$$

with $u_{q,j} = 1 - v_{q,j}$ and $v_{q,j}^2 + u_{q,j}^2 = 1$. In other words, the magic character of N or Z is defined as the stability of the energy gap between occupied and empty shells with respect to the pairing correlations. Conversely, the maximum of σ_q corresponds to the middle of major shells, a place were deformed nuclei occur. Two examples are displayed in fig. 4.

This criterion was applied to neutron rich nuclei. N = 16 was predicted to be magic, which was confirmed later on. The next neutron magic number, close to the neutron drip line is N = 34. On the side of the superheavy elements, N = 182 and N = 228 are found to be magic. Less attention has been paid to proton rich nuclei, for which the magic character of proton number is lost. For stable or almost stable nuclei, we could not find any magic proton number beyond Z = 82. This situation is due to the Coulomb force, which increases with Z^2 while the strong interaction increases with A. The Coulomb force being repulsive, there is a necessity for the proton orbitals to be confined over the nuclear volume to exist as bound or discrete states. In other words the proton orbitals are sensitive to the Coulomb and centrifugal barriers. This phenomenon increases the density of proton states close to the Fermi surface and thus diminishes the gap between orbitals.

Note that our calculations assume spherical symmetry, which is not necessarily a safe ansatz beyond approximately Z = 106. However, as long as the neutrons are spherically distributed, this hypothesis is a sound one. This is the case for N= 182 and 228. A point which merits to be further investigated.

To the extend that the symmetry energy is well taken into account, proton and neutron properties are treated at the same level. This is reflected in the binding energies as well as in the one- and two-particle separation energies. Both quantities are actually reproduced in a similar way. In such a case, the neutron densities are predicted with the same degree of accuracy as the charge densities.

Another firm prediction : bubble nuclei, namely nuclei with a large particle depression in the centre, do not exist as ground states.

6. Improving the functional.

For years, the density functional method was considered less reliable than the Hatree-Fock-Bogoliubov approach using Skyrme or Gogny interactions. The pioneering work in this field is due Brink and Vautherin [26]. They were followed by many others. Attempts to use forces as realistic as possible were achieved by Negele [27], as well as Campi and Sprung [28].

The two approaches have some part in common, since Skyrme of Gogny forces are effective two-body density dependent interactions based on nuclear matter properties [29]. The great advantage of the functional is that it does not postulate a specific 2-body effective force. The different components can be derived independently and thus the functional is submitted to less constraints. This aspect is particularly efficient for the spin-orbit contribution and the pairing matrix elements.

The functional is expected to be universal, in other words it must describe the ensemble of fermion systems belonging to the same class. In particular, this universality is expressed by the form of $\epsilon(\rho)$, that is its expansion in powers of k_f. The dimensionality argument that was advocated from the nuclear matter calculations, get recently support from effective field theory [30].

Consequently, the universality of the functional is to be found in the potential energy form $\epsilon(\rho)$ and the gradient terms besides contributions which have a conventional character.

The parameters are determined from the properties of a specific class of nuclei. Here use was made of double magic nuclei, and their one-body densities were expanded on the self-consistent basis obtained from the Hartree-Fock BCS calculations. This appeals to the two following remarks.

1. If indeed the charge densities of the doubly magic nuclei are reproduced accurately, it means we have touched the "true functional". As we have seen from the ^{208}Pb result, this goal is not totally achieved.

2. The parameters determined from a given ensemble of nuclei are not necessary valid for another ensemble. The parameter values reflect the underlying complicated many-body structure. This last is sensitive to various types of correlations, which vary with the density of states near the Fermi surface as exemplified by the pairing correlations.

Consequently, the parameters should be allowed to vary with the nucleus according to its substructures. Obviously, the variation should be smooth and under control. An ad hoc adjustment case by case would be unacceptable. The behaviour of the discrepancies recorded for the binding energies suggests to take advantage of the amount of pairing correlations through the σ_q to drive these changes.

To be explicit, the parameters we used were adjusted to magic nuclei. A similar procedure for deformed nuclei would lead to a different set of parameters. We conjecture that both sets could be linked by simple functions of the σ_q. To examples bring support for this conjecture.

6.1. A perturbative approach to the binding energies.

A perturbative estimate of the four quasi-particle contribution was achieved with Behrman [31]. Since the variational principle leads to Hartree-Fock-BCS equations, the ground state is stable against one-particle one-hole excitations. We therefore assume the dominant correction to come from two-phonon configurations. We take each phonon to have spin and parity 2^+, and couple them together to give zero angular momentum. For the two-body residual interaction, we use a simple ad hoc parametrisation of the two-body matrix elements. Considering A, B, a, b single-particle state in the spherical representation and denoting $\bar{a}, \bar{b}$ the time reversal states of a, b, we ended up with a correction of the form

$$\Delta E = - \sum_{A,B,\bar{a},\bar{b}} | < BCS|w|BCS >_{A,B,a,b} |^2 \cdot \frac{(1-\delta_{Aa})(1-\delta_{Ab})(1-\delta_{Ba})(1-\delta_{Bb})}{E_A + E_B + E_a + E_b} . \tag{14}$$

The matrix element was given by

$$\begin{aligned} < BCS|W|BCS >_{A,B,\bar{a},\bar{b}} &= -G\sqrt{2J+1}[u_A u_B v_a v_b][u_a u_b v_A v_B]\{2\delta_{t_{zA}+t_{zB},t_{za}+t_{zb}} \\ &- \eta[1+(-)^{j_A+j_B+j_a+j_b}]\delta_{t_{za}+t_{zB},t_{zA}+t_{zb}} \\ &+ \eta[(-)^{j_A+j_B-J} + (-)^{j_a+j_b-J}]\delta_{t_{za}+t_{zA},t_{zb}+t_{zB}}\} \end{aligned} \tag{15}$$

Here,

$$\eta = [(1+\delta_{AB})(1+\delta_{ab})]^{-1} , \tag{16}$$

and G is an adjustable constant. The Kronecker symbol δ of t_z ensures charge conservation with $t_z = \pm 1/2$ for protons and neutrons, respectively. The v_j are the occupation probability and the E_j the single particle energies with respect to the BCS ground state. The summation runs over all single particle levels which conform to the constraints on angular momentum, parity and charge conservation which we have imposed. In practice the calculation was restricted to single particle state with v_j^2 or u_j^2 between 0.99 and 0.01. We have verified that more than 90 % of the contribution comes from configurations within this truncated space.

Results are displayed in fig 5 for $G = 5.144$ MeV. It shows the clear improvement brought by taking into account the two-phonon correlations on the binding energies.

6.2. A mass formula with coefficients depending on valence nucleons.

Let us consider the Weiszäcker-Bethe mass formula and allow its parameters to depend on the number of valence nucleons [32]. Assume the magic numbers N_m to be known, we choose the simplest probabilistic way of counting the valence nucleons. It is nothing but the product of the number of particle pairs by the number of hole pairs between two magic numbers $N_{m1,q} < N_{m2,q}$ [33]. With

$$n_q = N_q - N_{m1,q} \tag{17}$$

the number of particles above $N_{m1,q}$, we define

$$f_q = n_q \frac{\Delta_q - n_q}{\Delta_q^2} \quad ; \quad \Delta_q = N_{m2,q} - N_{m1,q} \ . \tag{18}$$

The function f_q is zero at each end of a major shell and maximum in the middle. It more or less simulates the σ_q of the Hartree-Fock-BCS calculations, ignoring subshell structures. The mass formula reads

$$E(N,Z) = -a_v(f_p, f_n)A + a_\tau \frac{(N-Z)^2}{A} + a_s R^2(f_p, f_n) + a_c \frac{Z^2}{R(f_p, f_n)} \ . \tag{19}$$

For the radius, we found the following form to yield a very good fit to experimental charge radii (to 1 % or better)

$$R(f_p, f_n) = r_0(1 + 0.0069[f_p + f_n + f_p f_n]) \tag{20}$$

with

$$r_0 = 1.291[0.9428 - 0.3382\frac{(N-z)^2}{A^2} + 33.55A^{-2}]a^{1/3} \ . \tag{21}$$

The four parameters of r_0 have been fitted on the charge radii of ^{16}O, ^{40}Ca, ^{48}Ca, ^{132}Sn and ^{208}Pb. The binding energies of this set of magic nuclei has been used to fix the four parameters of the mass formula. Proton and neutron radii were supposed to be proportional, which is a crude approximation valid only for stable nuclei.

The following parameters have been determined :

$$\begin{aligned} a_v &= 15.5935 \text{ MeV}; a_s = 11.29 \text{ MeVfm}^{-2} \\ a_\tau &= 23.59 \text{ MeV} \quad ; a_c = 0.9428 \text{ MeVfm} \ . \end{aligned} \tag{22}$$

The mass formula has been applied to stable nuclei with $50 \leq Z \leq 82$. In this case, it fits the binding energies of ^{132}Sn and ^{208}Pb. If the f_q are set to zero, the binding energies of other nuclei are overestimated, up to 15 MeV in the middle of the major shell. The discrepancies are reduced by a factor 2 or more if the f_q are introduced in the radii R. Another factor of 2 is gained if a correction of the form

$$E_{corr} = [1. + \frac{1.1}{1. + f_p + f_n + f_p f_n}](f_p + f_n) \text{ MeV} \tag{23}$$

is added to the volume term. This correction is independent on the number of particles except through the f_q factors. It thus becomes negligible as $A \to \infty$. The difference between experimental and theoretical binding energies are displayed in fig 6. Most of

Journal of Physics: Conference Series **413** (2013) 012004 doi:10.1088/1742-6596/413/1/012004

the results are situated within a band 3.0 MeV wide, which is remarkable, considering the crudeness of the occupation number approximation which has been used, and the simplicity of the model. It gives us a hint that a more sophisticated method based on the σ_q determined self-consistently should be quite efficient.

The exercise shows also that one of the major improvement came from a better description of the radii. It suggests that, at the level of $\epsilon(\rho)$, the most important change with σ_q would be to lower the equilibrium density with increasing σ_q.

7. Excited states.

The density functional describes essentially the ground states. This is the gist of the Kohn theorem . Thus the question of the excited states generated by a functional sounds irrelevant. However, to the extent that the functional represent a semi-classical field theory of a many-particle ground state, the Landau theory of quasi-particle excitation should apply, at least in first approximation.

In this framework, the quasi-particle interaction is defined as the second derivative of the functional with respect to the basic field variables, i.e. the proton and neutron densities in each spin state. For the particle-hole interaction, it reads [34]

$$W_{ph}(\vec{r_1}-\vec{r_2}) = \delta(\vec{r_1}-\vec{r_2}) \sum_{s,t,s',t'} \frac{1}{16}\left(1+(-1)^{s-s'}\vec{\sigma_1}\cdot\vec{\sigma_2}\right) \tag{24}$$

$$x\left(1+(-1)^{t-t'}\right)\vec{\tau_1}\cdot\vec{\tau_2}\left[\frac{\partial^2\epsilon(\rho_{,t},\rho_{s',t'})}{\partial\rho_{st}\partial\rho_{s't'}}\right] .$$

Attempts to calculate selected excited levels have shown qualitative agreements with experimental results, specially for the collective states. Use was made of an approximate RPA technique. The 2^+ and 3^- levels of spherical nuclei were found between 0.5 and 1.0 Mev too high. Beyond excitation energies, transition densities were also calculated, and transition probabilities were in good agreement with experiments [35].

For instance, for the lowest 3^- state of ^{208}Pb the excitation energy was found to be 3.2 MeV compared to the experimental value of 2.6 MeV. The corresponding transition density was used to calculate the proton- ^{208}Pb differential cross section with excitation of this 3^-. The calculations were made at 1 Gev incident energy by means of the Glauber model and found in good agreement with experimental data [36].

Although the description of excited states with the density functional method cannot compete with more specialised approaches designed to fit nuclear spectra, it should give a way to test functionals beyond ground state properties. For instance, we have verified that the functionals used with Beiner and Mas yield ground state stable against spin excitations. This is not automatically the case, and some Skyrme forces were known to have this deficiency.

On the other hand, if a quasi particle description of low lying quadrupole and octupole states is not able to reproduce the bulk experimental behaviours, it suggests that some kind of correlations should be added explicitly in the functional. This point has not retained sufficient attention till now, as far as I know. It merits more investigations.

8. Conclusions.

The density functional method has proven to be a sound approach to the nuclear many-body problem. The universality of the functional is contained in the functional expression through the $\rho^{1/3}$ expansion of the potential energy and the gradient terms. The effective field theory has brought support to the potential energy expansion in $\rho^{1/3}$, obtained first from nuclear matter calculations. The coefficients of the successive terms are not accurately predicted by a fundamental theory. They are fitted on a sample of selected nuclei. Thus they are supposed to vary with nuclei according to criterions based on the concept of magic nucleons numbers, valence nucleons, the density of single particle levels near the Fermi surface and the lowest excitation energies. We conjecture that such criterions are well represented by the particle number dispersions of the BCS states.

Finally, the universality of the functional should incite its application to other systems of particles in strong interaction, like the hypernuclei.

Playing with the common means of access to information, asking for papers concerning the nuclear applications of the density functional method produces an enormous amount of references. There is no point here to propose a list of relevant articles. However, I would like to mention the works of the Bulgarian group with Antonov, Petkov and Stoitsov. They proposed a original approach to the problem in the momentum space [37].

References

[1] Euler H Z, Z. Physik **105** (1937) 553.
[2] Brown G E, *Unified Theory of Nuclear Models and Forces*, North Holland Publishing Co, Amsterdam-London (1971).
[3] Gombas P, Ann Physik **10** (1952) 253.
[4] Skyrme T H R, Phil. Mag. **1** (1956) 1093.
[5] Seyler R G and Blanchard C H, Phys. Rev. **131** (1956) 355.
[6] Bodmer A R, Nucl. Phys. **17** (1960) 1179.
[7] Hara Y, Progr. Thoeret. Phys. (Kyoto) **24** (1960) 1179.
[8] Kumar K, Lecouteur K J and Roy M K, Nucl. Phys. **42** (1963) 524.
[9] Berg R A and Wilets L, Phys. Rev. **101** (1956) 201 ; Wilets L, Phys. Rev. **101** (1956) 1805 ; Rev. Mod. Phys. **30** (1958) 542.
[10] Brueckner K A, Coon S A and Dabrowski J, Phys. Rev. **168** (1968) 1184.
[11] Hohenberg P and Kohn W, Phys. Rev, **136** (1964) B864.
[12] von Weiszäcker C F, Z. Physik **96** (1935) 431.
[13] Brueckner K A, Buchler J R, Jorna S and Lombard R J, Phys. Rev. **171** (1968) 1188.
[14] Brueckner K A, Buchler J R, Jorna S and Lombard R J, Phys. Rev. **173** (1968) 944.
[15] Brueckner K A, Clark R C, Wing-Fai Lin and Lombard R J, Phys. Rev C **1** (1970) 249.
[16] Brueckner K A, Wing-Fai Lin and Lombard R J, Phys. Rev. **181** (1969) 1506.
[17] Lombard R J, Ann. Phys. (N.Y.) **77** (1973) 380.

[18] Sham L J and Kohn W, Phys. Rev. **145** (1966) 561.
[19] Hamada T and Johnson T D, Nucl. Phys. **34** (1962) 383.
[20] Buchler J R and Ingber L, Nucl. Phys. A **170** (1971) 1.
[21] Beiner M and Lombard R J, Ann. of Phys. (N.Y.) **86** (1974) 262.
[22] Beiner M, Lombard R J and Mas D, Nucl. Phys. A **249** (1975) 1.
[23] Beiner M, Lombard R J and Mas D, *At. and Nucl. Data Tables* **17** (1976) 450.
[24] Audi G, Wapstra A H and Thibault C, *The AME2003 atomic mass evaluation*, Nucl. Phys. A **729** (2003) 337.
[25] Leblanc F *et al.*, Phys. Rev. C **72** (2005) 034305.
[26] Vautherin D and Brink D M, Phys. Rev. C **5** (1972) 626.
[27] Negele J W, Phys. Rev. C **1** (1970) 1260.
[28] Campi X and Sprung D W, Nucl. Phys. A **194** (1972) 401 ; Campi X *Nuclear self-consistent fields, Ed Ripka G and Porneuf M*, North Holland Pub Cie, Amsterdam (1975) 271.
[29] Gogny D, *Nuclear self-consistent fields, Ed Ripka G and Porneuf M*, North Holland Pub Cie, Amsterdam (1975) 333.
[30] see for instance Girlanda L, Rusetsky A and Weise W, Ann. Phys. **312** (2004) 92.
[31] Behrman R H and Lombard R J, Phys. Lett. **72B** (1977) 37.
[32] Abualrob H and Lombard R J, unpublished work.
[33] Deloncle I and Roussière B, arXiv : nucl-th/0309050v1 (2003) and arXiv : nucl-th/0405037v2 (2004).
[34] Lombard R J and Mas D, Nucl. Phys. A **336** (1980) 185.
[35] Lombard R J and Mas D, Ann. Phys, (NY) **167** (1986) 2.
[36] Lombard R J, *Particle distribution in the atomic nucleus*, in honor of Glauber R J, In : Coherence, Cooperation and fluctuations, Ed. Haake, Narducci and Walls, Cambridge university press (1986).
[37] see for instance Petkov I Zh and Stoitsov M V, *Nuclear Density Functional Theory*, Clarendon Press, Oxford, 1991.

Int. Summer School for Advanced Studies 'Dynamics of open nuclear systems' (PREDEAL12) IOP Publishing
Journal of Physics: Conference Series **413** (2013) 012005 doi:10.1088/1742-6596/413/1/012005

Covariant density functional theory for exotic nuclei near the neutron drip-line

Y Chen[1], L L Li[2], Z P Li[3], H Z Liang[1], W H Long[4], P Ring[1,5], N Van Giai[6], J M Yao[3], S Q Zhang[1], Y Zhang[1], P W Zhao[1], S G Zhou[7,8] and J Meng[1,9,10]

[1]State Key Laboratory of Nuclear Physics and Technology, School of Physics, Peking University, Beijing 100871, China
[2]Beijing Institute of Applied Physics and Computational Mathematics, Beijing 100088, China
[3]School of Physics Science and Technology, Southwest University, Chongqing, 400715, China
[4]School of Nuclear Science and Technology, Lanzhou University, Lanzhou 730000, China
[5]Physik Department, Technische Universitä München, D-85748 Garching, Germany
[6]Institut de Physique Nucléaire, IN2P3-CNRS and Université Paris-Sud, 91406 Orsay, France
[7]State Key Laboratory of Theoretical Physics, Institute of Theoretical Physics, Chinese Academy of Sciences, Beijing 100190, China
[8]Center of Theoretical Nuclear Physics, National Laboratory of Heavy Ion Accelerator, Lanzhou 730000, China
[9]School of Physics and Nuclear Energy Engineering, Beihang University, Beijing 100191, China
[10]Department of Physics, University of Stellenbosch, Stellenbosch 7602, South Africa

E-mail: mengj@pku.edu.cn

Abstract. The covariant density functional theory with a few number of parameters allows a very successful description of ground-state and excited-state properties for nuclei all over the nuclear chart. The recent progress on the covariant density functional theory and its applications for nuclear structure and astrophysics as well as extensions by the Beijing group and collaborators are summarized. In particular, the halos in exotic nuclei are discussed in details.

1. Introduction

The nuclear physics knowledge we have known today is basically built on our understanding of nuclei along the β-stability line. One of the current nuclear physics frontiers is the understanding of the nuclear structure far from the β-stability line and the origin of the elements. The tool we use for such task is the covariant density functional theory (CDFT). Naturally, the first question to be answered is why the CDFT is chosen.

The density functional theory (DFT) with a small number of parameters allows a very successful description of ground-state and excited-state properties of nuclei all over the nuclear chart. In particular, the covariant version of DFT takes the Lorentz symmetry into account in a self-consistent way, and has received wide attention during the past years due to its successful description of lots of nuclear phenomena [1, 2, 3, 4, 5]. The representations with large scalar and vector fields in nuclei, of the order of a few hundred MeV, provide more efficient descriptions than non-relativistic approaches that hide these scales. Moreover, it is of particular importance that the CDFT includes nuclear magnetism [6], i.e., a consistent description of currents and

Journal of Physics: Conference Series **413** (2013) 012005 doi:10.1088/1742-6596/413/1/012005

time-odd fields, which plays an important role in the nuclear rotations [7, 8, 9, 10] but is usually difficult to adjust to experimental data in non-relativistic functionals.

The relativistic Bruekner Hartree Fock theory and the CDFT with density-dependent meson-nucleon couplings can reproduce well the nuclear saturation properties in infinite nuclear matter [11, 12, 13]. Furthermore, the CDFT can reproduce well the measurements of the isotopic shifts in the Pb region [14], give naturally the spin-orbit potential, explain the origin of the pseudospin symmetry [15, 16] as a relativistic symmetry [17, 18, 19, 20, 21, 22] and the spin symmetry in the anti-nucleon spectrum [23, 24], and is reliable for nuclei far away from the β-stability line, etc. Obviously, the CDFT is one of the best candidates for the description of exotic nuclei.

Recent applications of the CDFT and its extensions by the Beijing group include

- a new point-coupling energy functional PC-PK1, which improves the description for isospin dependence of binding energy along either the isotopic or the isotonic chains [25];
- halo nuclei in the relativistic Hartree-Bogoliubov (RHB) and relativistic Hartree-Fock-Bogoliubov (RHFB) theory in continuum as well as continuum Skyrme Hartree-Fock-Bogoliubov (HFB) theory with Green's function method [26, 27, 28, 29, 30, 31];
- deformed halo nuclei in a deformed RHB model in continuum based on the Woods-Saxon basis (DRHBWS) [32, 33, 34];
- the first study of r-process nucleosynthesis using relativistic mean-field mass model and its application in the nucleocosmochronology [35, 36, 37];

In this contribution, only the applications to halos in spherical and deformed nuclei will be discussed in details due to the limit in length.

2. Covariant Density Functional Theory

The details of CDFT can be found in Refs. [1, 2, 3, 4, 5]. Taking the CDFT with the point-coupling interaction as an example, the basic building blocks are densities and currents

$$(\bar{\psi}\mathcal{O}\Gamma\psi), \qquad \mathcal{O}\in\{1,\vec{\tau}\}, \qquad \Gamma\in\{1,\gamma_\mu,\gamma_5,\gamma_5\gamma_\mu,\sigma_{\mu\nu}\}, \tag{1}$$

where ψ is Dirac spinor field of nucleon, $\vec{\tau}$ is the isospin Pauli matrix, and Γ generally denotes the 4×4 Dirac matrices. There are ten such building blocks characterized by their transformation characteristics in isospin and Minkowski space [38]. In this proceeding, the isospin vectors and space ones are denoted by arrows and bold types, respectively.

The starting point of the CDFT with the point-coupling interaction is an effective Lagrangian of the point-coupling form which can be written as a power series in $\bar{\psi}\mathcal{O}\Gamma\psi$ and their derivatives,

$$\begin{aligned}\mathcal{L} &= \bar{\psi}(i\gamma_\mu\partial^\mu - m)\psi - \frac{1}{4}F^{\mu\nu}F_{\mu\nu} - e\frac{1-\tau_3}{2}\bar{\psi}\gamma^\mu\psi A_\mu \\ &\quad -\frac{1}{2}\alpha_S(\bar{\psi}\psi)(\bar{\psi}\psi) - \frac{1}{2}\alpha_V(\bar{\psi}\gamma_\mu\psi)(\bar{\psi}\gamma^\mu\psi) - \frac{1}{2}\alpha_{TV}(\bar{\psi}\vec{\tau}\gamma_\mu\psi)(\bar{\psi}\vec{\tau}\gamma^\mu\psi) \\ &\quad -\frac{1}{3}\beta_S(\bar{\psi}\psi)^3 - \frac{1}{4}\gamma_S(\bar{\psi}\psi)^4 - \frac{1}{4}\gamma_V[(\bar{\psi}\gamma_\mu\psi)(\bar{\psi}\gamma^\mu\psi)]^2 - \frac{1}{2}\delta_S\partial_\nu(\bar{\psi}\psi)\partial^\nu(\bar{\psi}\psi) \\ &\quad -\frac{1}{2}\delta_V\partial_\nu(\bar{\psi}\gamma_\mu\psi)\partial^\nu(\bar{\psi}\gamma^\mu\psi) - \frac{1}{2}\delta_{TV}\partial_\nu(\bar{\psi}\vec{\tau}\gamma_\mu\psi)\partial^\nu(\bar{\psi}\vec{\tau}\gamma_\mu\psi),\end{aligned} \tag{2}$$

where all the symbols have their usual meaning as in Ref. [5].

The mean-field approximation leads to the replacement of the operators $\bar{\psi}(\hat{\mathcal{O}}\Gamma)_i\psi$ in Eq. (2) by their expectation values which become bilinear forms of the nucleon Dirac spinor ψ_k,

$$\bar{\psi}(\hat{\mathcal{O}}\Gamma)_i\psi \to \langle\Phi|\bar{\psi}(\hat{\mathcal{O}}\Gamma)_i\psi|\Phi\rangle = \sum_k v_k^2\bar{\psi}_k(\hat{\mathcal{O}}\Gamma)_i\psi_k, \tag{3}$$

where i indicates S, V, and TV. The sum $\sum$ runs over only positive energy states with the occupation probabilities v_k^2, i.e., the "no-sea" approximation. Based on these approximations, one finds the energy density functional for a nuclear system

$$E_{\rm DF}[\boldsymbol{\tau}, \rho_S, j_i^\mu, A_\mu] = \int d^3r\ \mathcal{E}(\boldsymbol{r}), \tag{4}$$

with the energy density

$$\mathcal{E}(\boldsymbol{r}) = \mathcal{E}^{\rm kin}(\boldsymbol{r}) + \mathcal{E}^{\rm int}(\boldsymbol{r}) + \mathcal{E}^{\rm em}(\boldsymbol{r}), \tag{5}$$

which is composed of a kinetic part

$$\mathcal{E}^{\rm kin}(\boldsymbol{r}) = \sum_k v_k^2\ \psi_k^\dagger(\boldsymbol{r})\,(\boldsymbol{\alpha}\cdot\boldsymbol{p} + \beta m)\,\psi_k(\boldsymbol{r}), \tag{6}$$

and an interaction part

$$\begin{aligned}\mathcal{E}^{\rm int}(\boldsymbol{r}) &= \frac{\alpha_S}{2}\rho_S^2 + \frac{\beta_S}{3}\rho_S^3 + \frac{\gamma_S}{4}\rho_S^4 + \frac{\delta_S}{2}\rho_S\triangle\rho_S + \frac{\alpha_V}{2}j_\mu j^\mu + \frac{\gamma_V}{4}(j_\mu j^\mu)^2 + \frac{\delta_V}{2}j_\mu\triangle j^\mu \\ &\quad + \frac{\alpha_{TV}}{2}\vec{j}^{\,\mu}_{TV}\cdot(\vec{j}_{TV})_\mu + \frac{\delta_{TV}}{2}\vec{j}^{\,\mu}_{TV}\cdot\triangle(\vec{j}_{TV})_\mu \end{aligned} \tag{7}$$

with the local densities and currents

$$\rho_S(\boldsymbol{r}) = \sum_k v_k^2\bar{\psi}_k(\boldsymbol{r})\psi_k(\boldsymbol{r}), \tag{8a}$$

$$j_V^\mu(\boldsymbol{r}) = \sum_k v_k^2\bar{\psi}_k(\boldsymbol{r})\gamma^\mu\psi_k(\boldsymbol{r}), \tag{8b}$$

$$\vec{j}^{\,\mu}_{TV}(\boldsymbol{r}) = \sum_k v_k^2\bar{\psi}_k(\boldsymbol{r})\vec{\tau}\gamma^\mu\psi_k(\boldsymbol{r}), \tag{8c}$$

and an electromagnetic part

$$\mathcal{E}^{\rm em}(\boldsymbol{r}) = \frac{1}{4}F_{\mu\nu}F^{\mu\nu} - F^{0\mu}\partial_0 A_\mu + eA_\mu j_p^\mu. \tag{9}$$

Minimizing the energy density functional Eq. (4) with respect to $\bar{\psi}_k$, one obtains the Dirac equation for the single nucleon

$$[\gamma_\mu(i\partial^\mu - V^\mu) - (m+S)]\psi_k = 0. \tag{10}$$

The single-particle effective Hamiltonian contains local scalar $S(\boldsymbol{r})$ and vector $V^\mu(\boldsymbol{r})$ potentials,

$$S(\boldsymbol{r}) = \Sigma_S, \quad V^\mu(\boldsymbol{r}) = \Sigma^\mu + \vec{\tau}\cdot\vec{\Sigma}^\mu_{TV}, \tag{11}$$

where

$$\Sigma_S = \alpha_S\rho_S + \beta_S\rho_S^2 + \gamma_S\rho_S^3 + \delta_S\triangle\rho_S, \tag{12a}$$

$$\Sigma^\mu = \alpha_V j_V^\mu + \gamma_V(j_V^\mu)^3 + \delta_V\triangle j_V^\mu + eA^\mu, \tag{12b}$$

$$\vec{\Sigma}^\mu_{TV} = \alpha_{TV}\vec{j}^{\,\mu}_{TV} + \delta_{TV}\triangle\vec{j}^{\,\mu}_{TV}. \tag{12c}$$

The Coulomb field A_0 is determined by Poisson's equation.

As the translation symmetry is broken in the mean-field approximation, proper treatment of center-of-mass (c.m.) motion is very important, and here the c.m. correction energy is taken into account in a microscopic way,

$$E^{\mathrm{mic}}_{\mathrm{c.m.}} = -\frac{1}{2mA}\langle \hat{\boldsymbol{P}}^2_{\mathrm{c.m.}} \rangle \tag{13}$$

with A mass number and $\hat{\boldsymbol{P}}_{\mathrm{c.m.}} = \sum_i^A \hat{\boldsymbol{p}}_i$ the total momentum in the c.m. frame [39, 40, 41].

For the formulas presented above, we mainly focus on the *ph* channel of the CDFT. For the *pp* channel, both the BCS and the Bogoliubov methods are widely used. However, it has been proved that for the exotic nuclei near the neutron drip-line, the Bogoliubov method could provide much more reliable results [53].

In a phenomenological construction of the CDFT, the nucleon-nucleon effective interaction is usually adjusted to the ground-state data of finite nuclei. For instance, a new parametrization PC-PK1 for the CDFT with point-coupling interaction is proposed by fitting to the observables of 60 selected nuclei, including the binding energies, charge radii, and empirical pairing gaps [42].

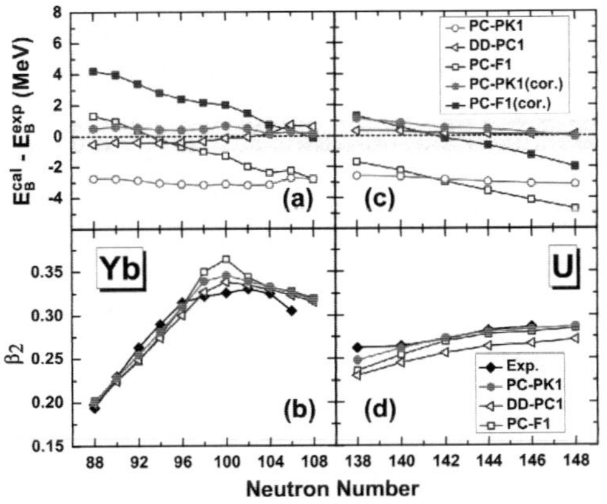

Figure 1. Deviations of the binding energies calculated with PC-PK1, DD-PC1, and PC-F1 from the data [43](upper panels) as well as the corresponding calculated ground state deformations in comparison with data [44] (lower panels) for Yb and U isotopes . Taken from Ref. [42].

As reported in Ref. [42], the binding energies and quadrupole deformations of the ground states for Yb and U isotopes are investigated with PC-PK1. In the upper panels of Figure 1, the deviations of the binding energies calculated with PC-PK1, DD-PC1 [45], and PC-F1 [46] from the data [43] are shown. A systematic underestimation of the binding energies around 3 MeV for both Yb and U isotopes by PC-PK1 can be reduced to within 1 MeV after taking into account the rotational correction energies with the cranking approximation [47]. For PC-F1, the difference between the calculated and the observed binding energy decreases monotonically with the neutron number and the agreements can not be improved by the rotational correction. As almost all the isotopes in Figure 1 are used to adjust DD-PC1, it is not surprising to see its good agreement with the data without taking into account the rotational corrections.

The calculated ground-state deformations are shown in the lower panels of Figure 1 in comparison with the available data [44]. It is found that PC-PK1 can provide also a good description for the deformations as well as their corresponding evolutions with neutron number for both Yb and U isotopes.

Journal of Physics: Conference Series **413** (2013) 012005 doi:10.1088/1742-6596/413/1/012005

3. Halos in spherical nuclei

Since the discovery of a large radius in ^{11}Li [48], the exotic nuclear halo phenomenon becomes one of the most interesting topics close to the nucleon drip-lines. In halo nuclei, the weakly binding leads to many new features, e.g., the coupling between bound states and continuum due to the pairing correlations, and the extended spatial density distributions. For the reliable theoretical description of the halo structures, it is of special significance to have the appropriate asymptotic behavior of nuclear densities at large radial distance and keep the self-consistency in dealing the discrete bound states and the continuum as well as the couplings between them. In fact such goal can be achieve by the relativistic continuum Hartree-Bogoliubov (RCHB) theory [26, 27, 28, 29] which fully takes into account the mean-field effects of the continuum.

In Ref. [29], the properties of even-even Ca, Ni, Zr, Sn, and Pb isotopes from the β-stability line to the neutron drip-line are studied within the RCHB theory with the parametrization NL-SH [49] and a zero range density-dependent pairing force

$$V^{pp}(\mathbf{r},\mathbf{r}') = \frac{V_0}{2}(1-P^\sigma)(1-\frac{\rho(\mathbf{r})}{\rho_{\text{sat}}})\delta(\mathbf{r}-\mathbf{r}'), \tag{14}$$

where all the symbols have their usual meaning as in Ref. [29]. The available experimental two-neutron separation energies $S_{2\text{N}}$ are reproduced very well. After careful analysis on the systematic of two-neutron separation energies, single-particle levels, the orbital occupation and nucleon density distributions, the giant halo phenomenon and the contribution from the continuum are demonstrated to appear in Ca isotopes with $A > 60$.

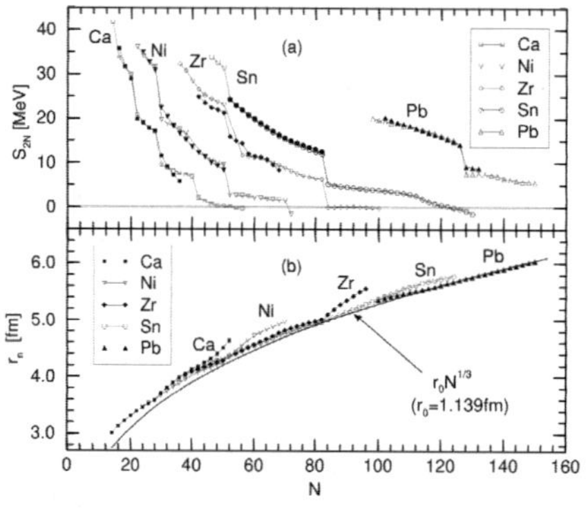

Figure 2. (a) The two-neutron separation energies $S_{2\text{N}}$ calculated by RCHB theory with NL-SH (Open symbols) in comparison with data available (solid ones) for the even Ca, Ni, Zr, Sn, and Pb isotopes against the neutron number N. (b) The root mean square (rms) neutron radii r_n for even Ca, Ni, Zr, Sn, and Pb isotopes by RCHB calculations against the neutron number N. The curve $r_0 N^{1/3}$ with $r_0 = 1.139$ fm has been included to guide the eye. Taken from Ref. [29].

In Figure 2, the two-neutron separation energies $S_{2\text{N}}$ (a) and neutron rms radii r_n (b) in even Ca, Ni, Zr, Sn, and Pb isotopes calculated by the RCHB theory with NL-SH are compared with the experiment data. In Figure 2(a), the good agreement between experiment and calculation is clearly shown. The experimental magic or submagic numbers $N = 20, 28$, and 40 are reproduced. The $S_{2\text{N}}$ values for exotic Ca isotopes are extremely close to zero in several isotopes. As discussed in Ref. [29], if taking ^{60}Ca as a stable core, the valence neutrons will gradually occupy the loosely bound states and the continuum above the sub-shell of $N = 40$, particularly for $^{66-72}$Ca. Similar systematics are also found for the $S_{2\text{N}}$ of Zr isotopes beyond $N = 82$ [27]. In fact such typical behavior of $S_{2\text{N}}$ can be taken as an evidence of the occurrence of "giant halos" in Ca chain, just as that in Zr chain [27]. In Figure 2(b), it is very interesting to see that r_n follows the $N^{1/3}$ systematics well for stable nuclei although their proton numbers are quite different. Near the drip-line, abnormal behaviors appear at $N = 40$ in Ca isotopes and at $N = 82$ in Zr isotopes. As pointed in Ref. [29], the nuclei having the abnormal r_n increase correspond to those having small $S_{2\mathrm{N}}$, which provides another evidence to prove the halo emergence.

In the Hartree level of the CDFT, the important ingredients such as the spin-dependent tensor forces are missing. In the density-dependent relativistic Hartree-Fock-Bogoliubov (DDRHFB) theory [50], the tensor forces owing to π and ρ meson exchanges can be naturally taken into account and the significant improvements have been made in the consistent description of shell evolution and appropriate conservation of pseudospin symmetry. In Ref. [30], the giant halos are found in Ce isotopes close to the neutron drip-line with the DDRHFB theory using PKA1 [51] and Gogny pairing force D1S [52].

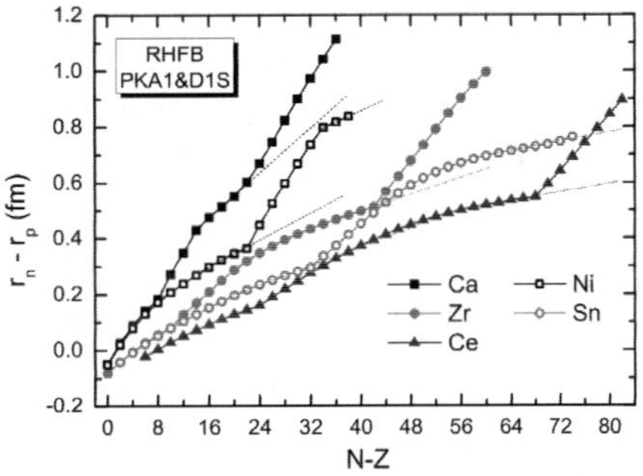

Figure 3. Neutron skin thickness $(r_n - r_p)$ calculated by DDRHFB with PKA1 and Gogny paring force D1S for Ca, Zr, Ni, Sn, and Ce isotopes versus $(N - Z)$. Taken from Ref. [30].

Figure 3 displays the isospin dependence of the neutron skin thickness $(r_n - r_p)$ calculated by the DDRHFB theory with PKA1 for Ca, Ni, Zr, Sn, and Ce isotopes. Against the smooth behavior in the stable region (dotted lines), continuously growing deviations are found in the chains of Ca, Zr, and Ce until the neutron drip-line. As discussed in Ref. [30], this can be considered as an evidence of a halo, which is also found for Ca and Zr isotopes in RCHB calculations [27, 29]. Despite of the existing deviations, Ni and Sn show a similar isospin dependence in both stable and neutron drip-line regions, which may indicate possible formation of the neutron skin.

In order to study further about the possible halos in Ce isotopes, in Fig. 4, the nuclear matter distributions and neutron canonical single-particle configurations for Ce isotopes closed to the drip-line are plotted. As shown in Fig. 4(a), the neutron densities become more and more extended after the isotope ^{186}Ce ($N = 128$), which is evidence for halo. From Fig. 4(b) it is clear that such extremely extended matter distribution, for example, in ^{198}Ce, is mainly caused by the low-l states, namely, the halo orbits $\nu 4s_{1/2}$, $\nu 3d_{5/2}$, and $\nu 3d_{3/2}$. In Fig. 4(c) it is found that the halo orbits $\nu 4s_{1/2}$, $\nu 3d_{5/2}$, and $\nu 3d_{3/2}$ are located around the particle continuum threshold when they are gradually occupied. For the isotopes beyond ^{184}Ce ($N = 126$), the Fermi levels approach the continuum threshold such closely that the stability of these halo isotopes becomes sensitive to pairing effects. Near the low-l states, the high-l states $\nu 2g_{9/2}$ and $\nu 2g_{7/2}$ are found. As discussed in Ref. [30], because of the relatively large centrifugal barrier for g orbits, they do not contribute much to the extended neutron distributions. Nonetheless, the existence of high-l states near halo orbits is of particular significance, because it leads to a rather high level density around the Fermi surface, and evidently the pairing effects are enhanced to stabilize the halo isotopes. Seen from the occupations of the halo orbits N_{halo} in Fig. 4(d), the halos in ^{186}Ce,

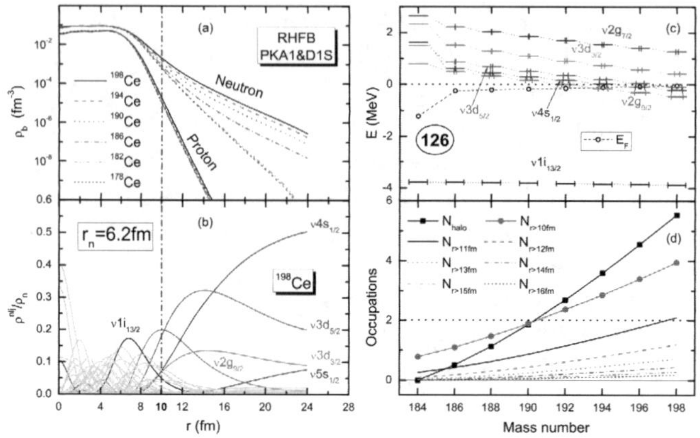

Figure 4. (a) Neutron and proton densities, (b) relative contributions of different orbits to the full neutron density in ^{198}Ce, (c) neutron canonical single-particle energies, occupation probabilities (x-axis error bars), and Fermi energies E_F (open circles), and (d) neutron numbers filling the halo orbits $4s_{1/2}$, $3d_{5/2}$, and $3d_{3/2}$ (N_{halo}) and those lying beyond the spheres with radii $r = 10$, 11, 12, 13, 14, 15, and 16 fm, respectively. Results were calculated by DDRHFB with PKA1 plus the Gogny pairing force D1S. The box radius adopted was $R = 28$ fm. Taken from Ref. [30].

^{188}Ce, and ^{190}Ce are ordinary, while ^{192}Ce, ^{194}Ce, ^{196}Ce, and ^{198}Ce have giant halos as more than two neutrons occupy the halo orbits. Similar conclusions can be drawn from the neutron numbers beyond $r = 10$ fm [$N_r > 10$ fm in Fig. 4(d)], which is large enough for halos.

In the above RHB and RHFB theories, and also most of the non-relativistic HFB approaches, the continuum states are approximated by discretization, thus the proper asymptotic behavior for the continuum state needs a big box, and the information for the width of the quasiparticle resonance is missing. In Ref. [31], the self-consistent continuum Skyrme HFB theory using the Green's function technique in coordinate space is established. In this approach, the densities needed in the energy density functional can be obtained by the contour integral of the Green's function, which is constructed by the wave functions with proper asymptotic behavior for the continuum states. From quasiparticle spectrum of the occupation and pair number densities, one can obtain the widths of the quasiparticle resonances.

Taking neutron-rich nucleus ^{128}Mo as an example, the neutron density $\rho(r)$ and pair density (also called the pairing tensor) $\tilde{\rho}(r)$ calculated by Skyrme HFB theory with Green's functional method and "discretized" method using different box sizes R are shown in Fig. 5. The densities from Green's function method exponentially decay in the asymptotic region, while the densities from "discretized" method drop artificially at the edge of the box due to the box boundary condition, i.e., the wave function equals zero at the edge of the box. Therefore, compared with the "discretized" method, the Green's function approach with proper boundary condition for the continuum states can provide the appropriate asymptotic behavior of the density distribution.

So far, the Green's function method has only been used in the spherical non-relativistic HFB approach, which has already been used to investigate the ground states of the nuclei near the neutron drip-line [31]. One of the future topics will be the development of the CDFT theory based on the Green's function method and also for deformed cases.

4. Deformed halos

It is well known that most of the nuclei observed are deformed. Therefore the existence of the deformed halos is an interesting question. If the deformed halos exist, what new features are

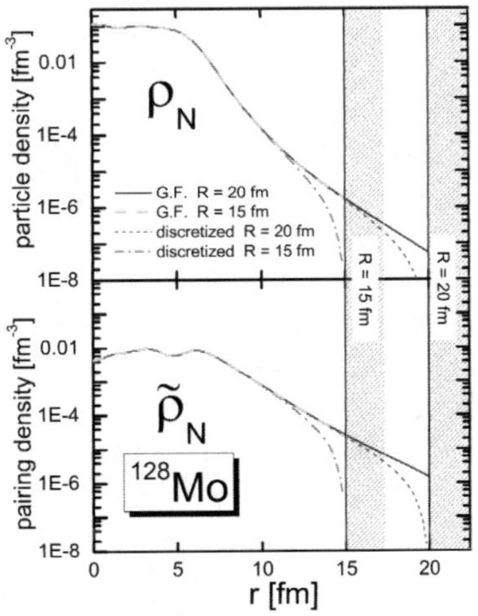

Figure 5. Neutron density $\rho(r)$ and pair density $\tilde{\rho}(r)$ for neutron-rich nucleus ^{128}Mo calculated by Shyrme HFB theory with "discretized" method and Green's function method with different box sizes R. Take from Ref. [54].

expected?

In order to describe the halo phenomena in deformed nuclei, the deformed RHB theory in continuum based on the Woods-Saxon basis (DRHBWS) has been developed recently [56, 32, 33, 57]. In Ref. [32], with the parametrization NL3 [58] and zero range density-dependent pairing force in Eq. (14), the deformed neutron-rich and weakly bound nucleus ^{44}Mg is investigated within the DRHBWS theory and an interesting shape decoupling between the core and halo is found.

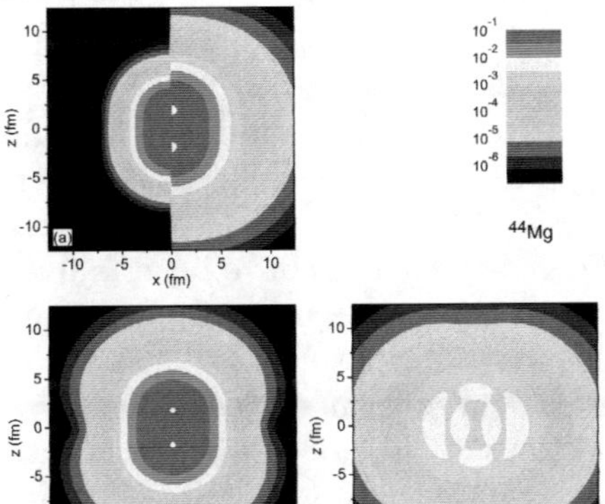

Figure 6. Density distributions of ^{44}Mg with the z axis as the symmetry axis. (a) The proton density ($x < 0$) and the neutron density ($x > 0$), (b) the density of the neutron core, and (c) the density of the neutron halo. Taken from Ref. [32].

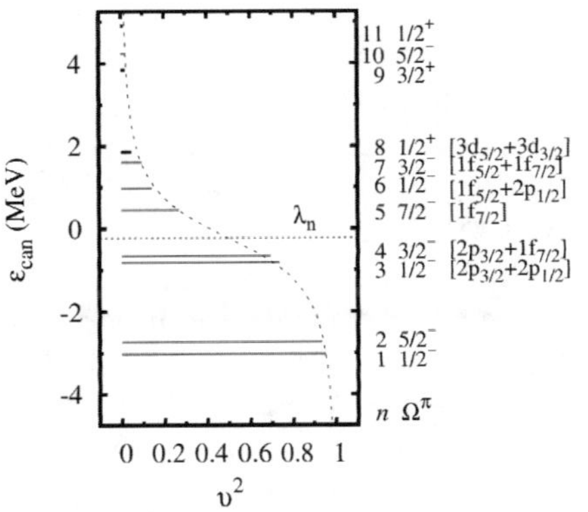

Figure 7. Single-particle spectrum around the Fermi level (dotted line) in the canonical basis for ^{44}Mg as a function of the occupation probability v^2. Taken from Ref. [32].

In Fig. 6(a) the density distributions of all protons and neutrons in ^{44}Mg are shown. Owing to the large neutron excess, the neutron density not only extends much farther in space but also shows a halo structure. The neutron density is decomposed into the contribution of the core in Fig. 6(b) and that of the halo in Fig. 6(c). It is found that the core of ^{44}Mg is prolate and the halo has a slightly oblate deformation. As pointed in Ref. [32], this indicates the decoupling between the deformations of core and halo.

Following the discussion in Ref. [32], the single-particle spectrum around the Fermi level for the ground state of ^{44}Mg is shown in Fig. 7. The orbitals are labeled by the third component of total angular momentum and parity Ω^π. The n numbers characterize the different orbitals appearing in this figure according to their energies. In Fig. 7, there is a considerable gap between the two levels with $n = 2$ and $n = 3$. As discussed in Ref. [32], the levels with $\varepsilon_{\rm can} < -2.5$ MeV contribute to the "core" and the other remaining weakly bound and continuum orbitals with $\varepsilon_{\rm can} > -1$ MeV naturally form the "halo". Therefore, the neutron density $\rho_n(r)$ is decomposed into two parts, one part coming from the orbitals with canonical single-particle energies $\varepsilon_{\rm can} < -2.5$ MeV (called "core") and the other from the remaining weakly bound and continuum orbitals (called "halo"). A further decomposition shows that the two weakly bound orbitals, that is, the third ($\Omega^\pi = 1/2^-$) and the fourth ($\Omega^\pi = 3/2^-$), contribute the most to the halo. If the deformed wave functions of the two weakly bound orbitals are decomposed in the spherical Woods-Saxon basis, it turns out that in both cases the major part comes from p waves, as indicated on the right-hand side of Fig. 7. It is known that the angular distribution of $|Y_{10}(\theta,\phi)|^2 \propto \cos^2\theta$ with a projection of the orbital angular momentum on the symmetry axis $\Lambda = 0$ is prolate and those of $|Y_{1\pm1}(\theta,\phi)|^2 \propto \sin^2\theta$ with $\Lambda = 1$ are oblate. It is found that in the third ($\Omega^\pi = 1/2^-$) orbital, both $\Lambda = 0$ and $\Lambda = 1$ components contribute and the latter dominates. Therefore, this orbital has a slightly oblate shape. For the fourth ($\Omega^\pi = 3/2^-$) state, there is only the $\Lambda = 1$ component from the $p_{3/2}$ wave, and thus an oblate shape is also expected.

5. Conclusion

In summary, the recent progress in the applications of the CDFT as well as its extensions by the Beijing group are summarized. In particular, its applications to halos in spherical and deformed nuclei are discussed in details. For spherical halo nuclei, the properties of even Ca, Ni, Zr, Sn, and Pb isotopes from the β-stability line to the neutron drip-line have been studied by RCHB theory. It has been suggested that the giant halo phenomena appear in Ca isotopes with $A > 60$ apart from Zr isotopes predicted earlier. Considering the tensor forces owing to π and ρ meson exchanges, halo has been studied by DDRHFB and the giant halos have been found in Ce isotopes close to the neutron drip-line. The self-consistent continuum Skyrme HFB theory using the Green's function technique in coordinate space is introduced. It is found that the Green's function approach with proper boundary condition for the continuum states can provide the appropriate asymptotic behaviors of the density distributions, which does not depend on the space size. For deformed halo nuclei, ^{44}Mg has been investigated within DRHBWS theory in continuum and an interesting shape decoupling between the core and halo has been found.

Acknowledgments

This work was partially supported by the Major State 973 Program 2013CB834400, Natural Science Foundation of China under Grants No. 10975008, No. 11175002, No. 11105005, No. 11105006, No. 11105110, No. 11105111, No. 11075066, No. 10979066, No. 11275248, No. 11120101005, the Research Fund for the Doctoral Program of Higher Education under Grant No. 20110001110087, the Fundamental Research Funds for Central Universities under Contracts No. lzujbky-2010-25 and No. lzujbky-2012-k07, and the Program for New Century Excellent Talents in University

References

[1] Serot B and Walecka J 1986 *Adv. Nucl. Phys.* **16** 1
[2] Ring P 1996 *Prog. Part. Nucl. Phys.* **37** 193
[3] Vretenar D, Afanasjev A V, Lalazissis G A and Ring P 2005 *Phys. Rep.* **409** 101
[4] Meng J, Toki H, Zhou S G, Zhang S Q, Long W H and Geng L S 2006 *Prog. Part. Nucl. Phys.* **57** 470
[5] Nikšić T, Vretenar D and Ring P 2011 *Prog. Part. Nucl. Phys.* **66** 519
[6] Koepf W and Ring P 1989 *Nucl. Phys. A* **493** 61
[7] Afanasjev A, Ring P and König J 2000 *Nucl. Phys. A* **676** 196
[8] Zhao P W, Peng J, Liang H Z, Ring P and Meng J 2012 *Phys. Rev. C* **85**(5) 054310
[9] Zhao P W, Peng J, Liang H Z, Ring P and Meng J 2011 *Phys. Rev. Lett.* **107**(12) 122501
[10] Zhao P W, Zhang S Q, Peng J, Liang H Z, Ring P and Meng J 2011 *Phys. Lett. B* **699** 181
[11] Brockmann R and Machleidt R 1990 *Phys. Rev. C* **42** 1965
[12] Brockmann R and Toki H 1992 *Phys. Rev. Lett.* **68** 3408
[13] Ban S F, Li J, Zhang S Q, Jia H Y, Sang J P and Meng J 2004 *Phys. Rev. C* **69** 045805
[14] Sharma M M, Lalazissis G A and Ring P 1993 *Phys. Lett. B* **317** 9
[15] Arima A, Harvey M and Shimizu K 1969 *Phys. Lett. B* **30** 517
[16] Hecht K T and Adler A 1969 *Nucl. Phys. A* **137** 129
[17] Ginocchio J 1997 *Phys. Rev. Lett.* **78** 436
[18] Meng J, Sugawara-Tanabe K, Yamaji S, Ring P and Arima A 1998 *Phys. Rev. C* **58** R628
[19] Meng J, Sugawara-Tanabe K, Yamaji S and Arima A 1999 *Phys. Rev. C* **59** 154
[20] Long W H, Sagawa H, Meng J and Van Giai N 2006 *Phys. Lett. B* **639** 242
[21] Liang H Z, Zhao P W, Zhang Y, Meng J and Van Giai N 2011 *Phys. Rev. C* **83** 041301(R)
[22] Lu B N, Zhao E G and Zhou S G 2012 *Phys. Rev. Lett.* **109** 072501
[23] Zhou S G, Meng J and Ring P 2003 *Phys. Rev. Lett.* **91** 262501
[24] Liang H Z, Long W H, Meng J and Van Giai N 2010 *Eur. Phys. J. A* **44** 119
[25] Zhao P W, Li Z P, Yao J M and Meng J 2010 *Phys. Rev. C* **82** 054319
[26] Meng J and Ring P 1996 *Phys. Rev. Lett.* **77** 3963
[27] Meng J and Ring P 1998 *Phys. Rev. Lett.* **80** 460
[28] Meng J 1998 *Nucl. Phys. A* **635** 3
[29] Meng J, Toki H, Zeng J Y, Zhang S Q and Zhou S G 2002 *Phys. Rev. C* **65** 041302(R)
[30] Long W H, Ring P, Meng J, Van Giai N and Bertulani C A 2010 *Phys. Rev. C* **81** 031302(R)
[31] Zhang Y, Matsuo M and Meng J 2011 *Phys. Rev. C* **83** 054301
[32] Zhou S G, Meng J, Ring P and Zhao E G 2010 *Phys. Rev. C* **82** 011301(R)
[33] Li L L, Meng J, Ring P, Zhao E G and Zhou S G 2012 *Phys. Rev. C* **85** 024312
[34] Chen Y, Li L L, Liang H Z and Meng J 2012 *Phys. Rev. C* **85** 067301
[35] Sun B, Montes F, Geng L S, Geissel H, Litvinov Y A and Meng J 2008 *Phys. Rev. C* **78** 025806
[36] Sun B and Meng J 2008 *Chin. Phys. Lett.* **25** 2429
[37] Niu Z M, Sun B H and Meng J 2009 *Phys. Rev. C* **80** 065806
[38] Liang H Z, Zhao P W, Ring P, Roca-Maza X and Meng J 2012 *Phys. Rev. C* **86** 021302(R)
[39] Bender M, Rutz K, Reinhard P G and Maruhn J A 2000 *Eur. Phys. J. A* **7** 467
[40] Long W H, Meng J, Van Giai N and Zhou S G 2004 *Phys. Rev. C* **69** 034319
[41] Zhao P W, Sun B Y and Meng J 2009 *Chin. Phys. Lett.* **26** 112102
[42] Zhao P W, Li Z P, Yao J M and Meng J 2010 *Phys. Rev. C* **82** 054319
[43] Audi G, Bersillon O, Blachot J and Wapstra A H 2003 *Nucl. Phys. A* **729** 3
[44] Raman S, Nestor C W and Tikkanen P 2001 *At. Data Nucl. Data Tables* **78** 1
[45] Nikšić T, Vretenar D and Ring P 2008 *Phys. Rev. C* **78** 034318
[46] Bürvenich T, Madland D G, Maruhn J A and Reinhard P G 2002 *Phys. Rev. C* **65** 044308
[47] Girod M and Grammaticos B 1979 *Nucl. Phys. A* **330** 40
[48] Tanihata I and et al 1985 *Phys. Rev. Lett.* **55** 2676
[49] Sharma M M, Nagarajan M A and Ring P 1993 *Phys. Lett. B* **312** 377
[50] Long W H, Ring P, Van Giai N and Meng J 2010 *Phys. Rev. C* **81** 024308
[51] Long W H, Sagawa H, Van Giai N and Meng J 2007 *Phys. Rev. C* **76** 034314
[52] Berger J, Girod M and Gogny D 1984 *Nucl. Phys. A* **428** 23
[53] Dobaczewski J, Flocard H and Treiner J 1984 *Nucl. Phys. A* **422** 103
[54] Meng J and et al 2011 *Prog. in Phys.* **31** 119
[55] Zhou S G, Meng J and Ring P 2003 *Phys. Rev. C* **68** 034323
[56] Zhou S G, Meng J and Ring P 2006 *AIP Conf. Proc.* **865** 90
[57] Li L L, Meng J, Ring P, Zhao E G and Zhou S G 2012 *Chin. Phys. Lett.* **29** 42101
[58] Lalazissis G A, König J and Ring P 1997 *Phys. Rev. C* **55** 540

Int. Summer School for Advanced Studies 'Dynamics of open nuclear systems' (PREDEAL12) IOP Publishing
Journal of Physics: Conference Series **413** (2013) 012006 doi:10.1088/1742-6596/413/1/012006

NUSTAR - presence and prospects of nuclear structure research at GSI and FAIR

G Münzenberg[1,2] and H Geissel[1,3]
[1]GSI Helmholtzzentrum für Schwerionenforschung Planckstrasse1, 64291 Darmstadt Germany
[2]Manipal University, Manipal 576104 Karnataka, India
[3]II Physikalisches Institut Justus Liebig Universität Gießen, Heinrich Buff Ring 16 35392 Giessen, Germany

E-mail: g.muenzenberg@gsi.de

Abstract. New experimental methods at rare-isotope facilities including in-flight separation, storage rings, and traps, setups for reaction studies with beams of exotic nuclei in reversed kinematics and large gamma arrays give new insights into the structure of atomic nuclei at the limits of stability. The discoveries of new isotopes and elements, as well as new phenomena including one and two proton decay, nuclear halos and skins as well as coexistence of nuclear shapes will be discussed. Workhorses are the UNILAC and SIS accelerators in combination with the in-flight separators SHIP and FRS. Physics at the ESR storage ring at GSI addresses the interaction of atomic and nuclear physics and large-scale explorations of the landscape of nuclear masses. A brief outlook on the future, NUSTAR at FAIR will be given.

1. Introduction

The principal goal of nuclear structure research at rare-isotope facilities is the investigation of atomic nuclei at the limits of existence of elementary matter: along the proton- and neutron driplines where protons and neutrons become unbound and in the region of heavy elements, the limit in mass and charge. Research topics are: nuclear shells and shapes far-off stability, pairing and the proton-neutron interaction, exotic decay modes such as proton emission and new forms of nuclear matter including nuclear skins and halos. Of special interest is the evolution of nuclear structure along the paths of cosmic nucleosynthesis [1,2,3].

New experimental methods at exotic nuclear beam facilities including storage rings and traps offer unique possibilities for the exploration of the nuclear landscape to search for new phenomena. Modern in-flight facilities like FAIR at GSI combine heavy-ion accelerators for intense ion beams with in-flight separators of high transportation efficiency and selectivity. The heavy-ion accelerators deliver beams of all chemical elements up to uranium - including isotopes occurring with low abundance in nature - with energies from the Coulomb barrier to more than 1 GeV/u. In-flight separators deliver isotopic energetic beams of nuclides far-off stability with lifetimes down to microseconds. Efficient detector systems with 4π geometry for particle and γ-detection allow spectroscopy with the highest sensitivity and the capability of single-particle detection and investigation. The high energy of the in-flight separated beams allows reaction studies with unstable short-lived nuclei in reversed kinematics.

A special feature of NUSTAR is the combination of the FRagment Separator FRS and the Experimental Storage Ring ESR opening-up new experimental possibilities such as large–scale direct

Journal of Physics: Conference Series **413** (2013) 012006 doi:10.1088/1742-6596/413/1/012006

mass and live-time measurements of nuclei with half-lives down to microseconds and experiments which involve both, nuclear and atomic physics. The stored atomic nuclei are bare or carry only few electrons, thus allowing nuclear decay studies of highly ionized ions under variable ionization degrees, a scenario of hot stellar media.

After a brief excursion to the upper end of the nuclei chart, the heavy and superheavy elements, we will discuss physics with relativistic beams of unstable rare nuclei, experimental techniques including recent developments, physics results, and prospects on the example of NUSTAR at GSI and FAIR.

2. Heavy and superheavy elements – the upper limit for the existence of elementary matter

With the discovery of nuclear fission by Hahn and Straßmann and its physical interpretation by Meitner and Frisch it became clear that nuclear fission will terminate the number of chemical elements. They predicted the limit for the existence of chemical elements to around Z = 100, later refined models predicted this limit for Z = 104 to Z = 106. With the discovery of the nuclear shells and their relevance to nuclear stability, the question arose as to whether "super heavy" nuclei, can exist at the next doubly magic shell closure after ^{208}Pb beyond the droplet limit estimated by Meitner and Frisch only due to shell stabilization [4]. Most of the early predictions place the island around Z = 114 and N = 184. Modern calculations with selfconsistent models place the magic proton number at Z = 120 or even Z = 126. At the same time the progress in accelerator development allowed the acceleration of heavy projectiles. Taking advantage of this new development Y. Oganessian proposed the use of the doubly magic ^{208}Pb as a target and irradiated it with appropriate projectile beams such as the most neutron-rich isotopes of titanium or chromium to create cold compound systems in order to enhance the survival probability of these highly fissile elements at the top of the chart of nuclides. There were however two obstacles: Super heavy nuclei were predicted as an *island* in the "sea of instability", beginning somewhere beyond rutherfordium, Z = 104, and the *extra push* which forbids the fusion of massive systems at the Coulomb barrier. Both problems were solved in SHIP experiments [5]. With the synthesis of rutherfordium in an irradiation of ^{208}Pb with ^{50}Ti cold heavy ion fusion was proven, with the discovery of hassium, Z = 108, a new shell region of deformed nuclei was found. It is centred at Z = 108 and N = 162 and forms the bridge from the transuranium nuclei to the predicted SHE region. The old and the new situation are schematically depicted in Fig. 1.

At present two ways to produce elements beyond seaborgium (Z = 106) are known, Fig. 3. The cold heavy ion fusion with ^{208}Pb and ^{209}Bi targets which was successful in the discoveries of elements 107 to 112 at SHIP (Fig. 2) and element 113 at RIKEN, Japan, and the hot fusion with ^{48}Ca beams and appropriate actinide targets bred in the reactor, which was successful in producing the new elements 114 to 118 at JINR Dubna. Both methods were proposed and pioneered by Y. Oganessian. The theoretical basis for the production of heavy elements in cold fusion was laid by Sandulescu et al. [6].

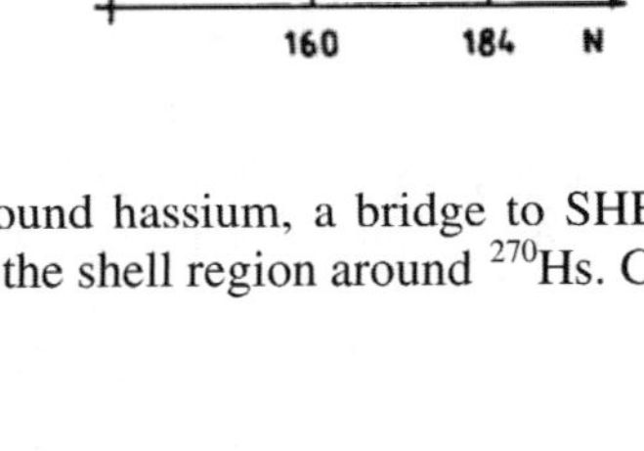

Fig. 1. The shell region around hassium, a bridge to SHE. Upper panel: situation before and lower panel after the discovery of the shell region around ^{270}Hs. Courtesy A. Sobiczewski.

The key to the discoveries of the heaviest elements beyond Z = 106 (seaborgium) is the combination of a powerful heavy-ion accelerator delivering intense projectile beams with separation of reaction products in-flight and single-atom detection and identification. These methods were pioneered at GSI with the UNILAC heavy ion accelerator and the velocity filter SHIP [5]. SHIP, a two stage velocity filter with large acceptance (Fig. 2), separates the heavy fusion products by their kinematic properties from the projectile beam and background from other reactions. The energetic fusion products are implanted into a position sensitive silicon surface barrier detector where their decay history in observed in situ. The heavy and superheavy elements were identified by their long α–decay chains leading to known nuclides on the basis of individual nuclei.

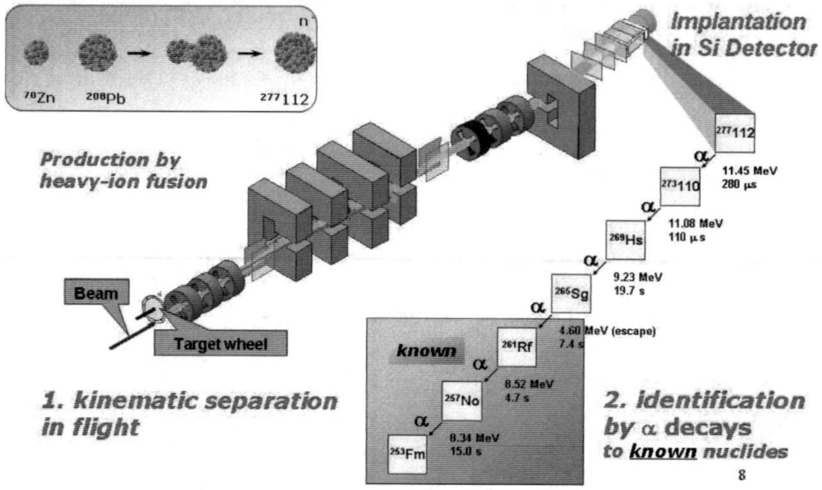

Fig. 2. The velocity filter SHIP. Insets: upper left: the fusion reaction, lower right: the α-decay chain of an single atom of copernicium running down to known nuclides indicated by the box.

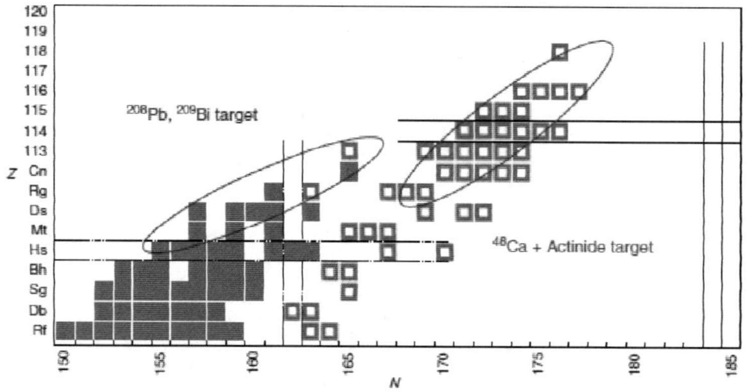

Fig. 3. Production of superheavy elements by cold (blue color) and hot (red color) fusion. The subshell at Z = 108 and N = 162 and the spherical shell closure at Z = 114 and N = 184 are indicated.

What did we learn? The idea of superheavy nuclei has been confirmed: nuclei can exist only by shell stabilization in "the sea of instability" beyond the liquid drop limit. The heaviest element reported up to now is element 118 [7], searches for element 120 are under way at SHIP and TASCA, GSI. Though the magic number 114 has already been passed, there is no indication for spherical superheavy nuclei. The explored region is too far away from the magic neutron number 182 (Fig. 3). Future work will concentrate on SHE chemistry to place the chemical elements in the periodic table, structure research, and, most important, the search for the spherical superheavy nuclei. Intensity upgrades of accelerators and new separators with the capability of mass identification are under way, e.g. the design study in the GSI-Giessen-Manipal collaboration.

3. The GSI accelerators with SHIP, FRS, and ESR

As a typical example of a first-generation rare isotope facility, the UNILAC-SHIP and SIS-FRS-ESR setup at GSI Darmstadt, is displayed in Fig. 4. Heavy-ion beams of all chemical elements from hydrogen to uranium are accelerated by the UNILAC to energies near and above Coulomb barrier for the production of exotic nuclei and heavy elements by complete fusion reactions. SHIP is located in the straight beam line. In parallel operation beam bunches are alternately injected into the Heavy-Ion Synchrotron SIS where they are accelerated to energies of typically 1 GeV/u. They are impinged onto the production target of the FRS for the production of radioactive nuclei at relativistic energies by projectile-fragmentation or -fission. The nuclides of interest are separated, event wise identified in flight, either implanted into a detector system for nuclear spectroscopy, or injected into the Experimental Storage Ring ESR where they are stored and cooled, or directed to the setup for reaction studies in complete kinematics (Fig. 4). Challenges are:

- Single-atom in-flight identification in mass A and nuclear charge Z
- Reactions with short-lived species in complete kinematics
- Physics with stored and electron-cooled beams and atomic nuclei.

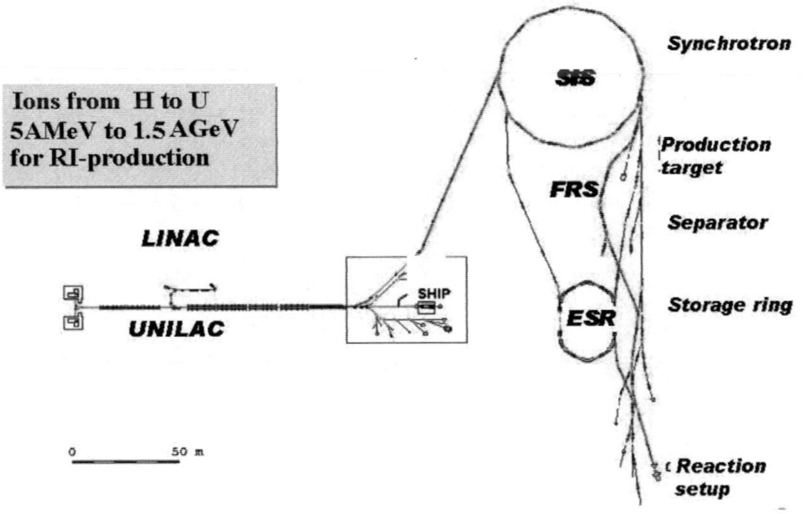

Fig. 4. The GSI accelerators UNLIAC and SIS with the in-flight separators SHIP and FRS, and ESR

The separation scheme of GSI projectile fragment separator FRS is explained in Fig.5. The relativistic heavy ion beam from SIS hits the production target where it is transmuted by fragmentation to a beam of radioactive isotopes preserving the kinematic properties of the beam. In the first part of the separator we have a kinematic separation according to the magnetic rigidity. As all fragments have approximately the same velocity as the projectiles and are fully stripped at the high energy they are separated in A/Z. A is the mass number and Z the element number. The inset in Fig. 5 shows the example of ^{78}Ni produced by an ^{86}Kr beam. All fragments with A/Z = 78/28 are separated. The second separation uses the energy loss $\Delta E \sim Z^2$ in matter (degrader in Fig. 5) which is analyzed in the second separator stage. As a result the isotopic clean energetic beam of ^{78}Ni is obtained at the exit of FRS. It can be implanted in detector systems, directed to the ESR or to a reaction setup for reaction studies with radioactive nuclei. It should be noted here that the number of separated isotopes can be adjusted by the degrader thickness. Thin degraders are used to produce cocktail beams.

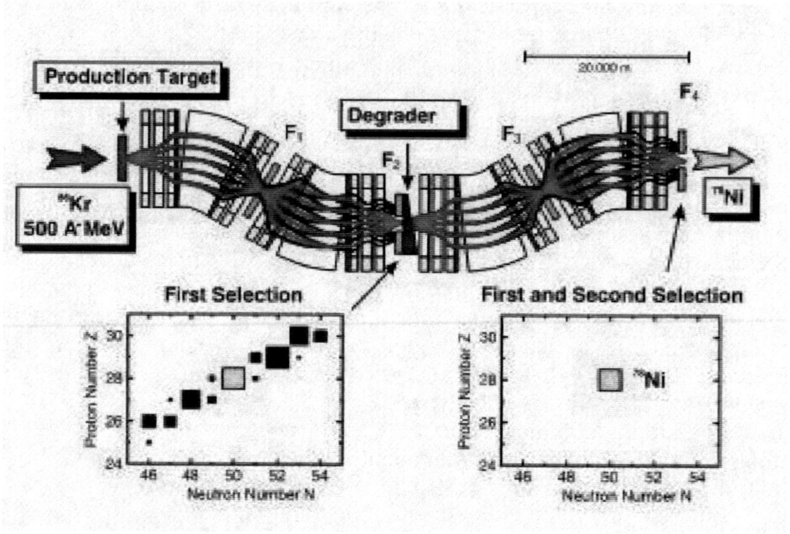

Fig. 5. Isotope separation with FRS

4. Research with nuclear beams at the limits of stability – some highlights from FRS and SHIP

At present 118 chemical elements are known, 80 of them are stable. All elements above the doubly magic lead, Z = 82, are unstable. The 254 stable isotopes form only a small band in the chart of nuclei within the more than 3000 unstable isotopes known at present. The region within which atomic nuclei can exist is limited by the proton- and neutron driplines, beyond they are proton and neutron unbound, the upper end is terminated by prompt fission.

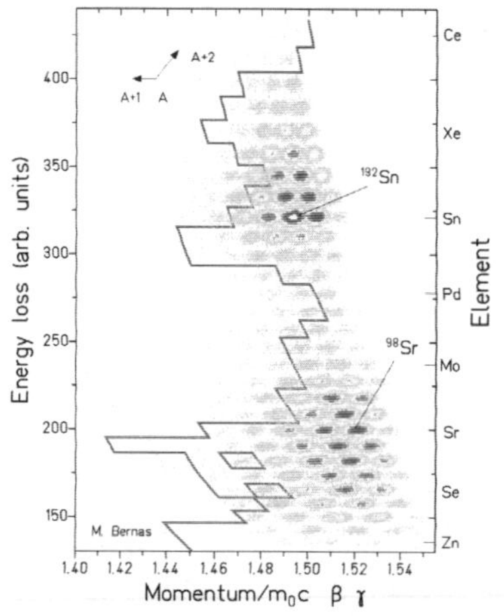

Fig. 6. Projectile fission (including abrasion-fission) of ^{238}U

Fig. 6 displays an example of isotope production, separation, and identification with FRS by projectile fission in-flight of ^{238}U. The heavy and the light fission group are collected in one spectrum. With the FRS such spectrum could be achieved for the first time [9]. Each dot in the spectrum represents an atomic nucleus, each cluster an isotope. Right of the red line we find isotopes discovered in this experiment. Some highlights and discoveries obtained with the in-flight separators at GSI are illustrated in Fig. 7.

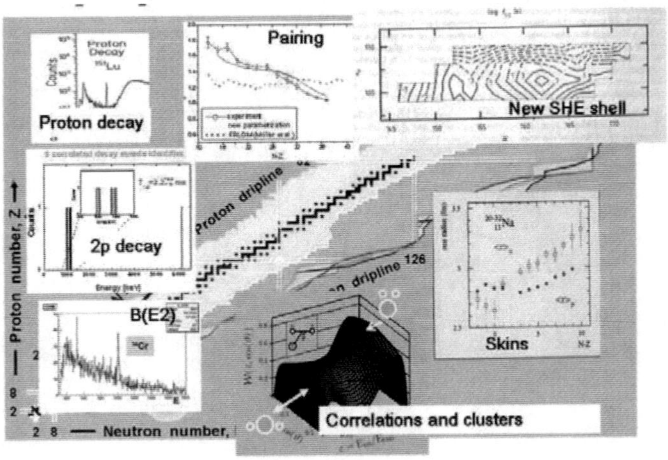

Fig. 7. Highlights and discoveries with SHIP and FRS

4.1 Nuclear masses

Mass measurements led to the discovery and understanding of basic properties of nuclear matter. The present generation of experiments employ time-of flight spectrometers, traps, and storage rings for

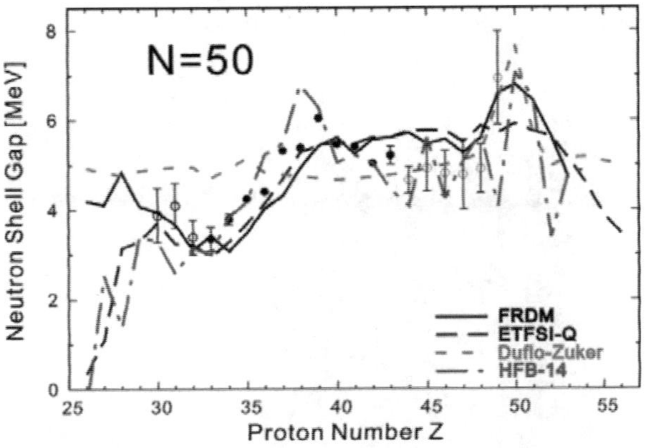

Fig. 8. The neutron shell gap for N = 50 isotones

direct mass measurements of short-lived nuclei at the limits of stability with a precision of better than 10^{-6} and half-lives down to 100 nanoseconds, and the exploration of large areas of the nuclear chart. Large-scale mass measurements as we performed at the ESR allow crucial tests of nuclear models [10]. As a typical example in Fig. 8 displays the experimental and theoretical mass values in the

Journal of Physics: Conference Series **413** (2013) 012006 doi:10.1088/1742-6596/413/1/012006

representation of shall gaps for N = 50 isotones which cross two proton shells: at nickel (Z = 28) and tin (Z = 50). The theories reproduce well the proton-rich species including the doubly magic ^{100}Sn but diverge for the neutron-rich species around the doubly magic ^{78}Ni. This result suggests that the exploration of neutron-rich species will reveal new nuclear properties.

4.2 Nuclear Shapes

A nuclear liquid drop is spherical. Structure effects may create non-spherical nuclei. Nuclides may even co-exist in different shapes. Such an example of shape coexistence is ^{186}Pb [11]. The decay of ^{190}Po feeds three 0+ states: prolate, oblate, and spherical (Fig 9). Such deformations are measured by decay spectroscopy.

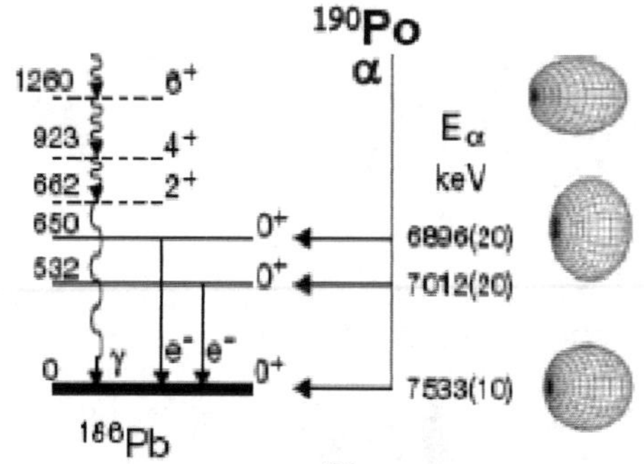

Fig. 9. The coexistence of nuclear shapes in ^{186}Pb

4.3 New Decay Modes

When crossing the proton dripline, protons become unbound and proton emission from the ground state can be observed. This new decay mode was discovered by S. Hofmann et al. in 1981in the decay of ^{151}Lu [12]. Presently about 30 ground state proton emitters are known. Ground state proton emission is an excellent example for quantum tunnelling. The proton energy and the half live of the decay yield information on the initial quantum state. Fine structure in the proton decay spectrum, first observed in ^{131}Eu, yields more details on nuclear structure and provides a stringent test for nuclear models at the limit of stability.

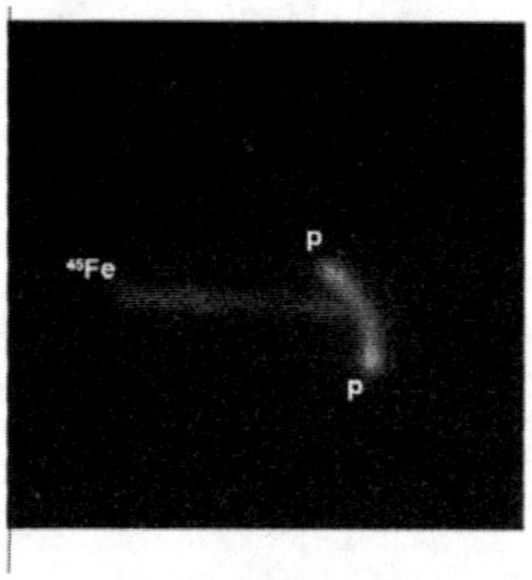

Fig. 10. Image of two proton decay observed with an optical CCD camera

For proton-dripline nuclei with an even number of protons, two-proton decay is observed. The first ground state two-proton emitter ^{45}Fe was discovered in 2002 by M. Pfützner et al. [13] and independently B. Blank et al.. Fig. 10 shows an example of a two-proton decay of ^{45}Fe observed with a CCD camera looking at an optical ionization chamber from a recent experiment. A theoretical analysis of the observed tracks shows that the two protons are emitted neither fully correlated nor uncorrelated. The decay is a three - body decay.

4.4 New Forms of Nuclear Matter

In 1985 Tanihata et al. at the Berkeley BEVALAC made reaction studies with relativistic beams of ^{11}Li. They measured a large interaction cross section [14], which they interpreted as large nuclear radius explained by Hansen and Jonson as a neutron halo [15]. The weakly bound two outer neutrons of ^{11}Li surround the nuclear ^{9}Li core as dilute neutron matter.

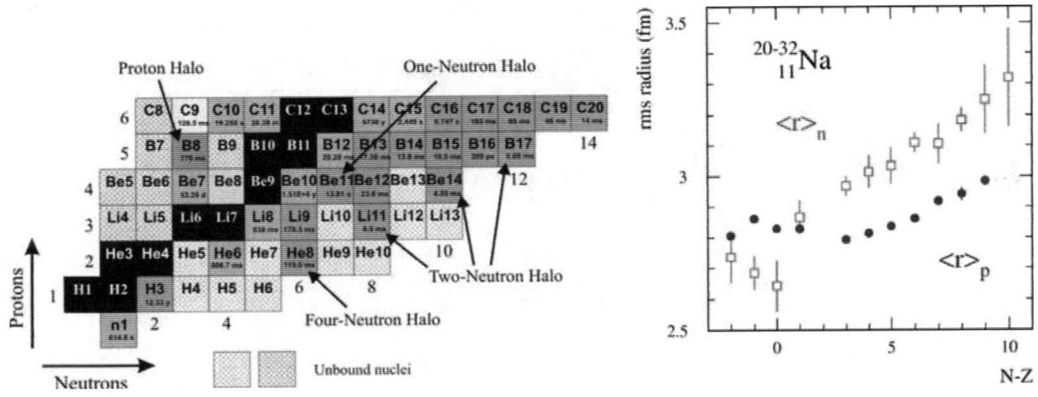

Fig. 11. Chart of neutron and proton halo nuclei (left), neutron skin in sodium isotopes (right)

In the meantime one-neutron halo, four neutron halo, and proton halo nuclei have been observed (Fig. 11). Because of the large reaction cross section the halo neutrons can be individually removed in a nuclear collision and give a frozen picture of their nuclear state.

The tendency of excess-neutrons to separate from the nuclear core is also observed in heavier nuclear systems where they form a neutron skin. Fig. 11 shows the evolution of a neutron skin towards the neutron rich isotopes for sodium [16]. The radii of neutron matter are from experimental interaction cross section values employing Glauber scattering theory (matter radii), the radii of the proton distribution are deduced from the hyperfine structure measured in a laser experiment. Neutron skins are a testing ground for the nuclear equation of state and the neutron crust in neutron stars as the crust of the neutron stars has the same matter density as the neutron skin.

4.5 Interplay of atomic and nuclear effects

The experimental storage ESR ring offers unique conditions to store atomic nuclei as fully stripped or few-electron systems which allows to study radioactive species under conditions as the exist in the universe [10]. Bound-State Beta Decay is a classical example. If the energy of the emitted β is not sufficient to escape to the continuum, the atomic nucleus can not decay, it is stable. However if the electron orbits are empty it can decay into the K shell. This new decay mode, Bound-State Beta Decay, was observed first for the decay of $^{163}Dy^{66+}$ in 1992 by Jung et al. [17]. Atomic nuclei in the cosmos undergoing β decay in a hot stellar environment will make such decays and have a different half-live from that observed on earth which has dramatic consequences if e.g. such nuclides are used as cosmic clocks as for instance $^{187}Re^{75+}$. Decays of bare atomic nuclei are investigated at the GSI storage ring ESR. They are produced at relativistic energies where they are fully stripped, separated in-flight in the FRS, and injected at full energy into the ESR where they are stored and cooled for experimental investigations [10]. In general bound state beta decay decreases the half-lives of heavy beta emitters by about 15-20 %. New features of nuclear electron capture decay have also been discovered for H- and He-like ions.

5. FAIR – the future of GSI and NUSTAR

Fig. 13 shows the new international GSI facility, FAIR, the Facility for Antiproton and Ion Research [18]. A double ring synchrotron system delivers intense beams of all ions up to uranium. For

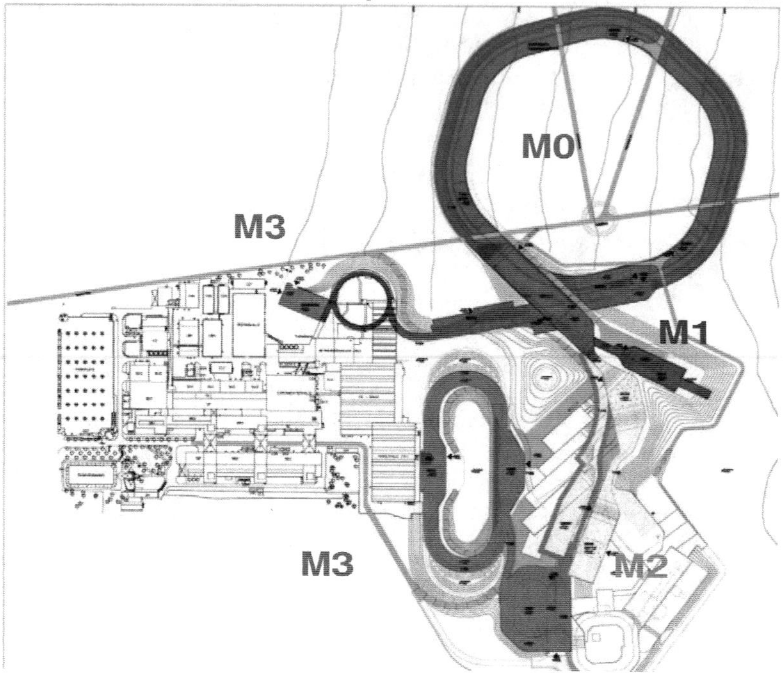

Fig. 13. FAIR-the new GSI project

Energies between 0.5 AGeV to 1.5 AGeV will be used for NUSTRAR experiments at GSI-FAIR. Module 2 (yellow) is NUSTAR, the facility for NUclear Structure Astrophysics, and Reactions. The workhorse of NUSTAR [19] is the Super-FRS, a three-stage fragment separator of large acceptance with superconducting magnets. Increased primary beam intensity and separator transmission will increase the intensity of radioactive beams by up to four orders of magnitude, especially for experiments with stored exotic nuclei. This will give access to new regions of the nuclear landscape, to the borderlines of the existence of elementary matter for structure research, astrophysics, and reactions.

[1] Geissel H, Huyse M, Münzenberg G, van Duppen P 2013, "Encyclopaedia of Applied High-Energy and Particle Physics" Wiley, to be submitted
[2] Mackinstosh R, Al-Khalili J, and Jonson B 2001, *"Nucleus: A trip into the Heart of Matter"*, Canopus Publishing Ltd. UK
[3] Thoenessen M and Sherrill B 2011, *Nature* **473** 25
[4] Mosel U and Greiner W 1969, *Z. Phys.* **222** 261
[5] Hofmann S and Münzenberg G 2000, *Rev. Mod. Phys.* **72** 733.
[6] Sandulescu A et al 1976, *Phys. Lett.* B **60** 225
[7] Oganessian Yu 2007, *J. Phys. G. Nucl. Part. Phys.* **34** R165.
[8] Geissel H et al 1992, *Nucl. Instr. Meth. Phys. Res.* A **70** 286
[9] Bernas M et al 1994, *Phys. Lett.* B **331** 19

[10] Franzke B, Geissel H, and Münzenberg G, 2001, *Mass Spectometry Reviews* **27** 428
[11] Andreyev A N et al. 2000, *Nature* **405** 430
[12] Hofmann S et al 1982, *Z. Phys.* A **305**111
[13] Pfützner M et al. 2002, *Eur. Phys. J.* A **14** 279
[14] Tanihata I et al 1985, *Phys. Lett.* B **160** 380
[15] Hansen G and Jonson B 1987, *Europhys. Lett.* **4** 409
[16] Suzuki T et al 1995, *Phys. Rev. Lett.* **75** 3241
[17] Jung M et al 1992, *Phys. Rev. Lett.* **69** 2164.
[18] http://www.gsi.de/forschung_beschleuniger/fair.htm
[19] http://www.gsi.de/start/forschung/forschungsgebiete_und_experimente/nustarenna/nustarfair.htm

Exotic structure and decay of medium mass nuclei near the drip lines

A Petrovici[1,2], K W Schmid[2] and A Faessler[2]

[1]National Institute for Physics and Nuclear Engineering, R-077125 Bucharest, Romania
[2]Institut für Theoretische Physik, Universität Tübingen, D-72076 Tübingen, Germany

E-mail: spetro@nipne.ro

Abstract. Nuclei at or near the N=Z line are of particular interest as micro-laboratory for fundamental effects. The simulation of many astrophysical objects requires the knowledge of properties and decay rates of nuclei near the drip lines. The structure and dynamics of proton-rich A≃70 nuclei and neutron-rich A≃100 nuclei are influenced by shape coexistence and mixing. A realistic description of shape coexistence phenomena requires beyond-mean-field approaches. Results concerning the self-consistent description of exotic phenomena including anomalies in mirror energy differences of proton-rich nuclei, triple shape coexistence in neutron-rich N=58 Sr and Zr isotopes as well as Gamow-Teller β-decay strength distributions, half-lives, and β-delayed neutron emission probabilities of neutron-rich nuclei in the A≃100 mass region obtained in the frame of the *complex* Excited Vampir model using realistic effective interactions in large model spaces are presented.

1. Introduction

The interest for the investigation of medium mass nuclei near the drip lines goes beyond the frontier of nuclear structure and dynamics. Shape coexistence and mixing, isospin mixing, competition between neutron-proton and like-nucleon pairing correlations are characteristic features of nuclei near the N=Z line in the A≃70 mass region. Nuclei at or near the N=Z line are of particular interest as micro-laboratory for high precision tests of fundamental effects which requires particularly high accuracy of the nuclear structure models [1].

The isotopic chain of neutron-rich zirconium nuclei offers an example of rapid transition from spherical to deformed shape with a possible identification of the sudden onset of quadrupole deformation between N=58 and 60. The rapid neutron-capture process involving neutron-rich nuclei in the A≃100 region remains as one of the most challenging problem in nuclear astrophysics [2]. The β-decay properties of some specific nuclei that are important contributors to the reactor decay heat are also required. Previous theoretical investigations revealed the difficulty of describing in a unitary way the experimental data at low and high spins as well as the Gamow-Teller (GT) β-decay properties of these nuclei. The realistic description of shape coexistence phenomena in proton-rich A≃70 nuclei and neutron-rich A≃100 nuclei requires beyond-mean-field approaches. We shall present results on exotic phenomena in these nuclei obtained within the *complex* Excited Vampir variational approach with symmetry projection before variation. We use a large model space and a realistic effective nucleon-nucleon interaction obtained starting from a nuclear matter G-matrix based on Bonn A potential. In the next section the *complex* Excited Vampir variational procedure will be briefly described and the effective

Journal of Physics: Conference Series **413** (2013) 012007 doi:10.1088/1742-6596/413/1/012007

Hamiltonian will be defined. In Section 3 results concerning shape coexistence phenomena specific for the A≃70 mass region will be presented. Particular features of neutron-rich A≃100 nuclei influenced by shape coexistence and mixing will be discussed in Section 4.

2. Theoretical framework

Shape coexistence phenomena characterizing the structure and dynamics of proton-rich A≃70 nuclei and neutron-rich A≃100 nuclei are investigated within the *complex* Excited Vampir model which succesfully explained the structure and weak decays of the nuclei dominated by shape coexistence in the A≃70 mass region [3-7]. Many predictions that emerged have been experimentally confirmed. The *complex* Excited Vampir model uses Hartree-Fock-Bogoliubov (HFB) vacua as basic building blocks which are only restricted by time-reversal and axial symmetry. The underlying HFB transformations are essentially *complex* and do mix proton- with neutron-states as well as states of different parity and angular momentum. The broken symmetries (nucleon numbers, parity, total angular momentum) are restored by projection techniques and the resulting symmetry-projected configurations are used as test wave functions in chains of successive variational calculations to determine the underlying HFB transformations as well as the configuration mixing. The HFB vacua account for arbitrary two-nucleon correlations and thus simultaneously describe like-nucleon as well as isovector and isoscalar proton-neutron pairing. Unnatural parity correlations are also included. Furthermore, the *complex* Excited Vampir model (EXVAM) allows the use of rather large model spaces and realistic effective interactions.

For nuclei in the $A \simeq 70$ mass region we use a ^{40}Ca core and include the $1p_{1/2}$, $1p_{3/2}$, $0f_{5/2}$, $0f_{7/2}$, $1d_{5/2}$ and $0g_{9/2}$ oscillator orbits for both protons and neutrons in the valence space. We start with an isospin symmetric basis and then introduce the Coulomb shifts for the proton single-particle levels resulting from the ^{40}Ca core by performing spherically symmetric Hartree-Fock calculations using the Gogny-interaction D1S in a 21 major-shell basis [4].

For nuclei in the A $\simeq$ 100 mass region is used a rather large model space above the ^{40}Ca core built out of $1p_{1/2}$, $1p_{3/2}$, $0f_{5/2}$, $0f_{7/2}$, $2s_{1/2}$, $1d_{3/2}$, $1d_{5/2}$, $0g_{7/2}$, $0g_{9/2}$, and $0h_{11/2}$ oscillator orbits for both protons and neutrons in the valence space. The corresponding single-particle energies had been adjusted in *complex* Monster (Vampir) calculations for odd mass nuclei in the A $\simeq$ 100 mass region [8]. This rather large model space allows the realistic description of the Gamow-Teller β-decay properties of neutron-rich nuclei in this region.

The effective two-body interaction is constructed from a nuclear matter G-matrix based on the Bonn A potential. In order to enhance the pairing properties the G-matrix was modified by three short-range (0.707 fm) Gaussians for the isospin T = 1 proton-proton, neutron-neutron, and neutron-proton matrix elements. The isoscalar spin 0 and 1 particle-particle matrix elements are enhanced by an additional Gaussian. In addition, the isoscalar interaction was modified by monopole shifts including all T = 0 matrix elements of the form $\langle 0g_{9/2}0f; IT = 0|\hat{G}|0g_{9/2}0f; IT = 0\rangle$ involving protons and neutrons occupying the $0f_{5/2}$ and the $0f_{7/2}$ orbitals. The Coulomb interaction between the valence protons was added.

3. Coexistence phenomena in proton-rich A≃70 nuclei

Coexistence phenomena at low, intermediate, and high spins have been predicted in the frame of the variational approaches of the Vampir model family [3-7] and later identified experimentally. Low-lying isomeric 0^+ states based on a strong mixing of projected configurations with large or small prolate and oblate deformations in the intrinsic system have been predicted for some Ge, Se, and Kr isotopes. Small spectroscopic quadrupole moments for the lowest 2^+ and 4^+ states of these nuclei and collective B(E2;$\Delta I = 2$) values as well as significant B(E2) strengths for the transitions connecting states of the same spin and parity represent fingerprints of strong oblate-prolate mixing. The rapid variation of the deformation with the number of nucleons and

Journal of Physics: Conference Series **413** (2013) 012007 doi:10.1088/1742-6596/413/1/012007

excitation energy, the shape coexistence and mixing, the competition of T=0 and T=1 pairing correlations may have important effects on the mirror energy differences which are expected to be determined by the isospin mixing induced by the interplay of all these facts.

Anomalies in the mirror energy differences are experimentally identified in this mass region ([9, 10] and references therein) for nuclei supposed to manifest shape mixing at low spins [7]. The

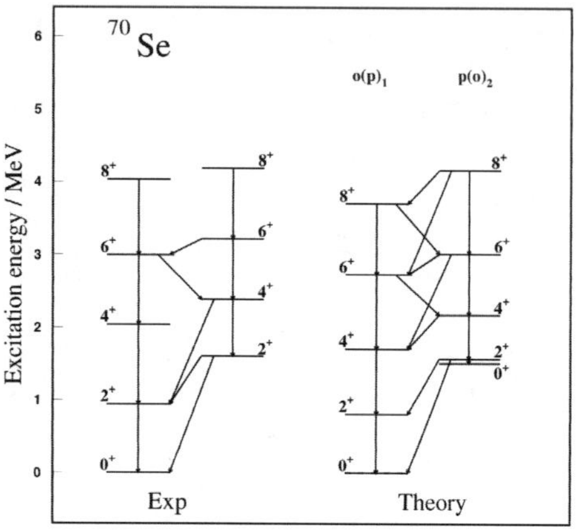

Figure 1. The theoretical EXVAM spectrum for ^{70}Se compared to the available data.

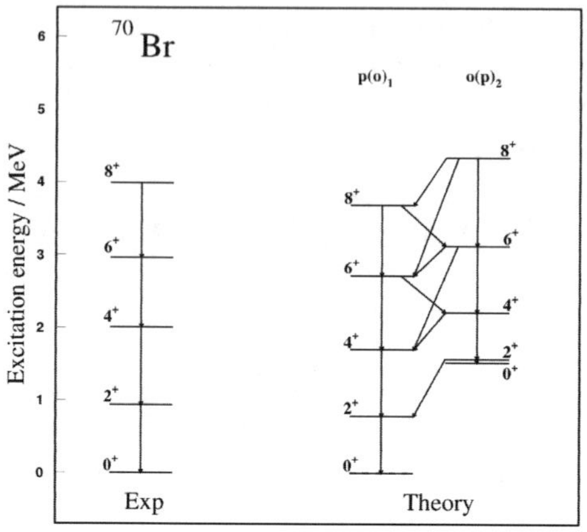

Figure 2. The theoretical EXVAM spectrum for ^{70}Br compared to the available data.

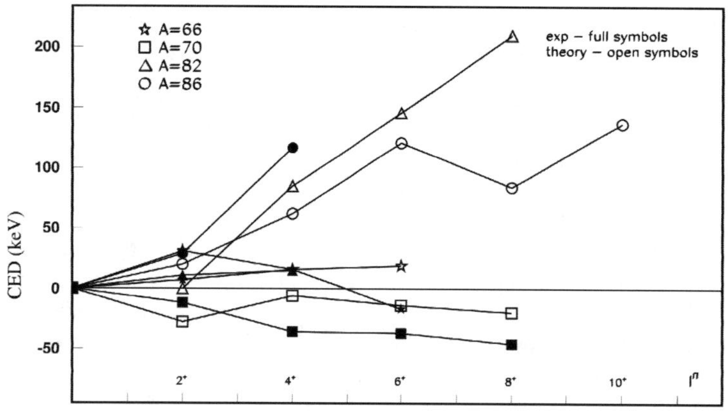

Figure 3. The mirror energy differences obtained within the *complex* Excited Vampir model is compared to data [10].

anomalous negative trend for the A=70 nuclei can be explained based on the strong competition and mixing between configurations of oblate and prolate deformation in the intrinsic system. In ^{70}Br the ground state is dominated by prolate components, while for ^{70}Se the calculations predict a predominantely oblate ground state. In ^{70}Br the oblate (prolate) components represent 36% (64%), 41% (59%), 41% (58%), 20% (80%) of the total amplitude of the wave function for the lowest 0^+, 2^+, 4^+, and 6^+, respectively, while in ^{70}Se the corresponding amplitudes amount to 57% (42%), 59% (41%), 64% (36%), and 39% (61%). In figure 1 and 2 we present the low-spin spectra of ^{70}Se and ^{70}Br, respectively, compared to data. In ^{70}Se the calculated spectroscopic quadrupole moment for the yrast 2^+ state, 4.5 efm^2, is in agreement with recent experimental data [11], whereas for the spins 4^+ and 6^+ we found 11.5 efm^2 and -17.5 efm^2, respectively. In ^{70}Br the corresponding calculated values are: -6.4 efm^2, -9.8 efm^2, and -39.7 efm^2 [10].

The experimental energy differences for the A=66 analogs ^{66}As and ^{66}Ge show a slightly positive trend up to spin 4^+ becoming slightly negative for 6^+ [10]. The predominantly prolate states in the ^{66}As - ^{66}Ge pair show smooth changes of the prolate-oblate mixing for increasing spin. In ^{66}As the oblate (prolate) components amount to 16% (84%), 29% (70%), 18% (81%), and 4% (95%) for the lowest 0^+, 2^+, 4^+, and 6^+ states, whereas for the analog states in ^{66}Ge the corresponding amplitudes are 20% (80%), 38% (61%), 32% (66%), 9% (91%). The EXVAM mirror energy differences manifest a slightly positive trend.

A positive trend is observed for the ^{82}Nb - ^{82}Zr and for ^{86}Tc - ^{86}Mo nuclei, which is reproduced by our calculations. The situation is different for the A=82 and A=86 analogs where it was found that one prolate component strongly dominates the structure of the yrast states representing more than 90% of the total amplitude [5]. In figure 3 the mirror energy differences obtained within the *complex* Excited Vampir using the Bonn A potential and the Coulomb interaction for the mirror nuclei ^{66}Ge - ^{66}As, ^{70}Se - ^{70}Br, ^{82}Zr - ^{82}Nb, and ^{86}Mo - ^{86}Tc are compared to the available data. We conclude that the oblate-prolate shape mixing strongly perturbs the isospin symmetry, altering the excitation energies of the levels in the nuclei of the same isospin multiplet which is reflected in the mirror energy differences behaviour. The overall trend of the mirror energy differences is reasonably well reproduced by the *complex* Excited Vampir model even if we take into account only the isospin mixing determined by the Coulomb interaction [5, 10].

4. Coexistence phenomena in neutron-rich A$\simeq$100 nuclei

The structure of neutron-rich nuclei in the A $\simeq$ 100 mass region relevant for the astrophysical r-process manifests drastic changes in some isotopic chains and often sudden variations of particular nuclear properties have been identified [12-21]. Reliable predictions concerning the properties of neutron-rich nuclei along the r-process path are required from robust nuclear structure models to explain the observational data in the frame of the astrophysical models.

4.1. Sape coexistence and shape evolution in neutron-rich A$\simeq$100 nuclei

For neutron-rich nuclei in the region the N=58 is considered a critical number of neutrons. Since the onset of deformation is supposed to take place for N $>$ 58 we investigated the structural changes and the evolution of deformation with increasing spin aiming at a unitary description of the lowest few 0^+ states and the low, intermediate, and high spins in ^{96}Sr and ^{98}Zr. We studied the properties of the positive parity states up to spin 20^+ in the frame of the *complex* Excited Vampir model including in the many-nucleon bases up to 12 EXVAM configurations [8].

A particular situation is found for the 0^+ states: the lowest projected EXVAM configuration is spherical in both nuclei (the quadrupole deformation parameter amounts to $\beta_2 \simeq 0.03$). In ^{96}Sr the second 0^+ configuration is oblate deformed in the intrinsic system (β_2 = -0.32), the third one is prolate ($\beta_2 = 0.37$) and all three orthogonal configurations are situated in an energy interval of 375 keV. In ^{98}Zr the second configuration is prolate ($\beta_2 = 0.37$), the third is oblate (β_2 = -0.30) and the separation energy between the spherical one and the third one is 323 keV. For each nucleus we constructed the lowest 10 0^+ orthogonal EXVAM configurations and the result of the diagonalization of the residual interaction between them in terms of spherical, prolate, and oblate content in the structure of the lowest four 0^+ states is presented in table 1. The occupation of the $1d_{5/2}$ neutron orbital is essential for the spherical 0^+ EXVAM configuration.

The theoretical lowest bands of ^{96}Sr and ^{98}Zr are compared to the experimental spectra in figure 4 and 5, respectively. The labels of the bands indicate the prolate (p) or oblate (o) intrinsic quadrupole deformation for the dominant projected EXVAM configurations underlying the states of the corresponding bands. The states building the *po*(*p*)-band in each nucleus are characterized by strong prolate-oblate mixing at low spins and variable prolate mixing at intermediate and high spins. The almost pure oblate states belonging to the most right band in the theoretical spectrum are feeding the second 4^+ (2^+) state in ^{96}Sr (^{98}Zr) that manifests maximum o-p mixing as it is illustrated in table 2. The shape evolution with increasing spin and excitation energy is suggested by the amount of mixing in the structure of the wave functions illustrated in table 3 for ^{98}Zr. The occupation of the $0g_{9/2}$ proton orbital is significantly changing from the intrinsically oblate deformed configurations to the prolate deformed ones in both nuclei.

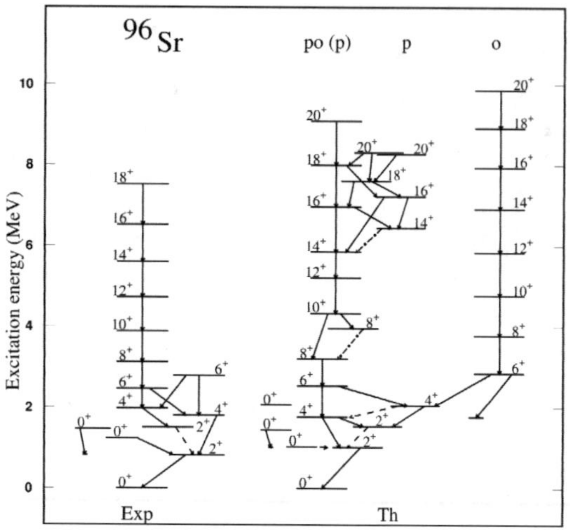

Figure 4. The theoretical EXVAM spectrum for ^{96}Sr is compared to the available data [13, 15, 18].

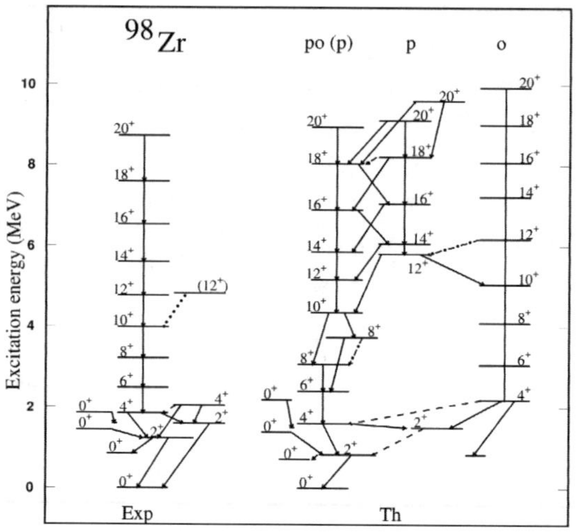

Figure 5. The theoretical EXVAM spectrum for ^{98}Zr is compared to the available data [13, 15, 18].

Table 1. The structure of the wave functions for the lowest 0^+ states.

	^{96}Sr			^{98}Zr		
I[$\hbar$]	spherical	prolate	oblate	spherical	prolate	oblate
0_1^+	36%	20%	44%	12%	43%	45%
0_2^+	57%	18%	25%	84%	12%	4%
0_3^+		69%	31%	1%	57%	42%
0_4^+	4%	6%	90%	2%	10%	88%

Table 2. The o-p mixing of the lowest 2^+ and 4^+ states.

	^{96}Sr		^{98}Zr	
I[$\hbar$]	o-mixing	p-mixing	o-mixing	p-mixing
2_1^+	58(5)%	34(2)%	31%	60(8)%
2_2^+	33(2)%	65%	63(1)%	36%
4_1^+	36(6)%	56(1)%	10%	83(7)%
4_2^+	52(5)%	43%	85(1)%	13(1)%

Table 3. The amount of mixing for the states of the lowest bands of ^{98}Zr.

$I[\hbar]$	o-mixing	p-mixing
6_1^+	2%	90(5)(1)%
6_2^+	80(14)%	4(2)%
8_1^+		89(7)(2)(2)%
8_2^+	3%	86(5)(3)(1)(1)%
8_3^+	84(4)(2)%	4(4)(2)%
10_1^+		43(26)(24)(7)%
10_2^+	74(8)(2)%	13(1)(1)(1)%
12_1^+		82(9)(4)(2)%
12_2^+	7%	75(8)(4)(4)%
12_3^+	67(16)%	11(3)(1)(1)(1)%
14_1^+		55(30)(7)(6)%
14_2^+		59(37)(2)(2)%
14_3^+	65(8)%	19(4)(2)%
16_1^+		56(33)(4)(3)(2)(1)%
16_2^+		51(42)(5)(1)%
16_3^+	89%	9(1)%
18_1^+		49(28)(21)(1)%
18_2^+		68(14)(14)(2)%
18_3^+	98%	1%
20_1^+		87(5)(3)(2)(2)(1)%
20_2^+		91(4)(3)(1)%
20_3^+		98(1)%
20_4^+	99%	

The experimental B(E2) values for the transitions connecting the lowest 0^+ and 2^+ states in ^{96}Sr support the idea of strong shape mixing. The retardation of the $2_1^+ \rightarrow 0_1^+$ transition is induced from the strength B(E2) = 340(209) e^2fm^4 deduced from the measured lifetime of 7(4) ps [12]. Recent preliminary results indicate for this lifetime 4.1 ps [20]. The EXVAM result is B(E2) = 795 e^2fm^4. Lifetimes for the $12^+ \rightarrow 10^+ \rightarrow 8^+$ cascade in ^{96}Sr and $12^+ \rightarrow 10^+ \rightarrow 8^+ \rightarrow 6^+$ cascade in ^{98}Zr have been measured [15]. By a simultaneous fit to several levels, assuming that they form a rotational band characterized by a common quadrupole moment the reported value is Q_0= 220(15) efm^2 in ^{96}Sr and Q_0= 200(10) efm^2 in ^{98}Zr. The theoretical results indicate significant fragmentation of the E2 strength decaying a given state due to the high density of states obtained at intermediate and high spins in both nuclei as it is illustrated in table 4 for ^{98}Zr. At intermediate spins the quadrupole deformation parameter in both nuclei amounts to $\beta_2 \simeq 0.3$ for the $po(p)$ band. For the intermediate spins the states building the o-band in both nuclei indicate smaller deformation for the underlying oblate configurations. The β_2 parameter varies between -0.19 and -0.23 in ^{96}Sr for the spins 8^+, 10^+, 12^+, while in ^{98}Zr even smaller values have been obtained for the corresponding spins: between -0.17 and -0.13. Other preliminary experimental results give support to the theoretical o-p mixing: the spectroscopic quadrupole moment of the 2_1^+ state in ^{96}Sr is equal 0 [20], while the calculated value is $Q^{spec}(2_1^+)$ = 9.5 efm^2 using the effective charges e_p = 1.3 and e_n = 0.3. The shape evolution within the bands is illustrated by the spectroscopic quadrupole moments for ^{98}Zr presented in table 5.

Table 4. $B(E2; I \rightarrow I-2)$ values (in e^2fm^4) for the lowest bands of ^{98}Zr (EXVAM). Strengths for secondary branches are given in parentheses.

$I[\hbar]$	$po(p)$	p	o
2^+	1140 (198)(161)		1305 (28) (18) (15)
4^+	2072 (620)		1593 (56)
6^+	2558 (101)		1662
8^+	1802 (942)(153)		1572 (123)
10^+	1430 (719)		1314 (119) (100)
12^+	2300 (216)	731 (345) (212)	663 (621) (307)
14^+	2428 (123)	1840 (392)	1094 (494)
16^+	1360 (832) (190)	548 (246) (1421)	602 (250)
18^+	863 (1416) (207)	1347 (808) (713)	1115
20^+	409 (1958)	347 (185) (1972)	1313

The angular momentum alignment indicates that maximum contribution in the direction of the total angular momentum is brought by $0h_{11/2}$ neutrons and $0g_{9/2}$ protons. The particularities of the $I^\pi \geq 8^+$ states are reflected by the g-factor values that are larger in ^{96}Sr ($\simeq$ 0.24) with respect to the corresponding values in ^{98}Zr ($\simeq$ 0.14), both corroborated by the faster alignment of the valence neutrons occupying the $0h_{11/2}$ orbital. For the $I^\pi \geq 10^+$ states along the o-band the g-factors are slightly larger than 0.5 (0.44) in ^{96}Sr (^{98}Zr) supporting the role of the $0g_{9/2}$-proton alignment. In figures 4 and 5 are indicated by dashed-dotted lines some M1 transitions connecting states of the same spin. The strongest calculated strengths amount to $B(M1;8_3^+ \rightarrow 8_1^+) = 1.292\ \mu_N^2$ in ^{96}Sr and $B(M1;8_2^+ \rightarrow 8_1^+) = 1.600\ \mu_N^2$ in ^{98}Zr where the states are ordered according to the excitation energy.

Table 5. Spectroscopic quadrupole moments Q_2^{sp} (in efm^2) for the lowest bands of ^{98}Zr.

$I[\hbar]$	$po(p)$	p	o
2^+	-36.6		7.1
4^+	-89.6		54.7
6^+	-115.5		76.7
8^+	-126.7		70.7
10^+	-130.1		58.2
12^+	-129.1	-98.5	55.6
14^+	-126.1	-121.2	30.5
16^+	-126.6	-123.0	60.8
18^+	-124.2	-134.4	74.9
20^+	-125.6	-135.4	68.9

In summary, we found triple shape coexistence specific for the 0^+ states and evolution of the shape coexistence and mixing with increasing spin and excitation energy for the positive parity states up to spin 20^+ in the N = 58 Sr and Zr isotopes within the *complex* Excited Vampir model using a realistic effective interaction in a large model space. The lowest four 0^+ states are described as a variable mixing of spherical, prolate, and oblate deformed Excited Vampir configurations. The oblate-prolate mixing characterizes the low-spin states. At intermediate and high spins prolate and oblate bands coexist in both nuclei. The comparison to the available data indicates rather good agreement.

The situation is changed for the ^{104,106}Zr isotopes where the oblate minima are much higher in energy and the lowest bands are built on prolate strongly deformed configurations [22] in agreement with recent data [17, 21]. Our investigations indicated that the ground states of ^{104}Zr and ^{106}Zr are dominated (99%) by a prolate deformed configuration. In ^{104}Zr beginning with spin 6^+ the yrare states manifest significant mixing of two or three prolate deformed projected configurations whereas for spins 10^+, 12^+, and 14^+ even for the lowest band the states manifest some mixing. In ^{106}Zr significant mixing of different prolate deformed configurations was found starting with spin 10^+.

4.2. Gamow-Teller β-decay of neutron-rich A≃100 nuclei

Aiming at a unitary description of the structure and dynamics of neutron-rich nuclei in the A≃100 mass region we extended our investigations to the Gamow-Teller β-decay of the even-even and odd-odd isotopes studying the strength distributions, β-decay half-lives, and β-delayed neutron emission probabilities. A completely self-consistent description of the Gamow-Teller β-decay properties of ^{104}Zr and ^{106}Zr nuclei involved the calculation of the lowest 50 1^+ states in the odd-odd ^{104,106}Nb nuclei [23]. The calculated 1^+ states are all situated within 6.2 MeV excitation energy in ^{104}Nb and 6.0 MeV in ^{106}Nb. In both nuclei the states manifest strong mixing of differently deformed configurations in the intrinsic system. The Excited Vampir configurations are strongly or moderately deformed and more than 75% manifest prolate deformation. Some states are dominated by strong prolate-oblate mixing. The theoretical results for the GT accumulated strength for the decay of the ground state of ^{104}Zr and ^{106}Zr are presented in figure 6 and 7, respectively. The strong GT β-decay branches for both ^{104}Zr and ^{106}Zr nuclei indicate essential contribution from the $g^{\pi}_{9/2}g^{\nu}_{7/2}$, $d^{\pi}_{5/2}d^{\nu}_{3/2}$, and $d^{\pi}_{5/2}d^{\nu}_{5/2}$ matrix elements. The

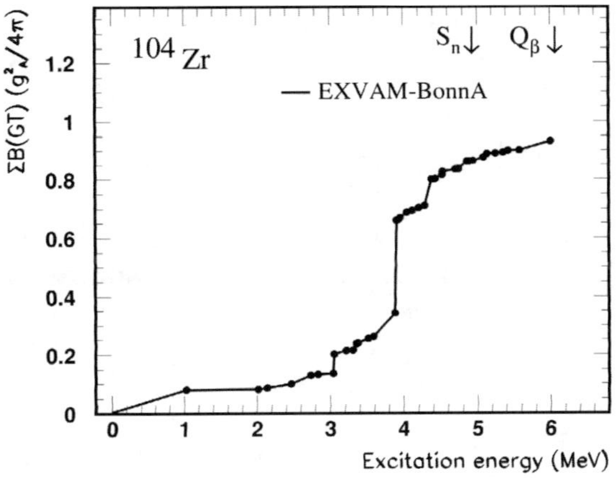

Figure 6. The Gamow-Teller accumulated strength for the decay of ^{104}Zr.

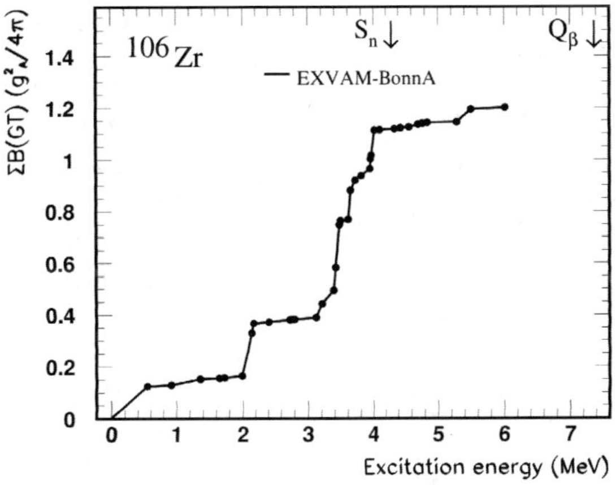

Figure 7. The Gamow-Teller accumulated strength for the decay of ^{106}Zr.

involved β window (Q_β) and the one-neutron separation energy in the daughter nucleus (S_n) are the measured ones in ^{104}Nb, but the predicted ones by mass models in ^{106}Nb [24, 25]. The lowest 1^+ state was considered the ground state according to the experimental assignment [24, 25]. The experimental value of the half-life for the decay of the ground state of ^{104}Zr of 1200(300) ms [26] was changed recently to 870(50)(30) ms [25]. The Excited Vampir result obtained calculating the lowest 50 1^+ in ^{104}Nb is 2040 ms. Recent experimental data [25] for the half-life of ^{106}Zr indicate 260(20)(30) ms, while the Excited Vampir result is 230 ms involving the lowest 50 1^+ in ^{106}Nb. Of course, the theoretical results on the Gamow-Teller strength distributions could be changed, more for the high-excitation energy region, by the higher-lying configurations not included in the *complex* Excited Vampir many-nucleon basis. The probability of β-delayed neutron emission (P_n) was calculated taking into account the GT contributions from the 1^+ states with excitation energy larger than S_n. The experimental results indicate for P_n an upper limit of 1% for the decay of ^{104}Zr and an upper limit of 7% for the decay of ^{106}Zr [25]. The Excited Vampir P_n value based on the lowest 50 1^+ states is smaller than 1% for ^{104}Zr decay and amounts to 2% for ^{106}Zr. Of course, these results are strongly dependent on the 1^+ states higher in energy than S_n and require a much larger EXVAM basis than the presently involved one.

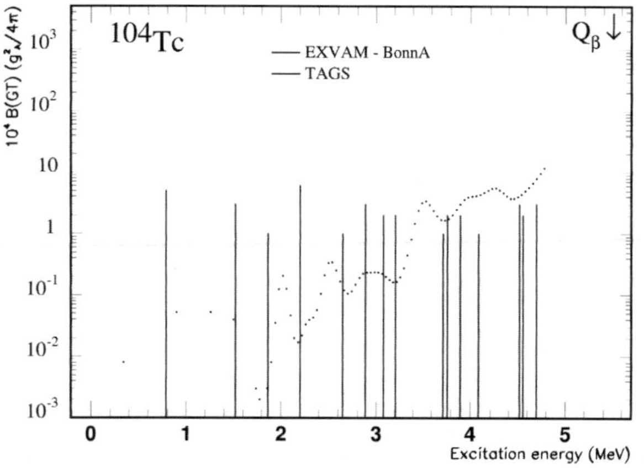

Figure 8. The Gamow-Teller strength distribution for the decay of ^{104}Tc obtained within the *complex* Excited Vampir model compared to TAGS data [27, 28].

Some of the neutron-rich A$\simeq$100 nuclei play a key role in the safe operation of nuclear reactors being important contributors to the decay heat. The estimation and control of the heat emitted by the decay of fission products requires certain still missing useful information, such as a knowledge of the decay properties of specific nuclei that contribute to the heating of the reactor during and after operation [27].In the following we shall present some results on the Gamow-Teller β-decay properties of ^{102}Tc to ^{102}Ru and of ^{104}Tc to ^{104}Ru within the *complex* Excited Vampir model compared to recent results of the total absorption gamma spectrometer (TAGS) measurements [27, 28]. We calculated the lowest 1^+ states in ^{102}Tc and the positive-parity states up to spin 4^+ in ^{102}Ru. For the description of the involved states in ^{102}Ru we included in the Excited Vampir many-nucleon basis up to 26 EXVAM configurations. The dimension of the many-nucleon EXVAM basis for the 1^+ states in ^{102}Tc was 7. The final solutions have been obtained diagonalizing the residual interaction between the considered Excited Vampir configurations for each spin. The obtained results indicate that the structure of the wave function for the lowest 1^+ state of ^{102}Tc manifest a strong mixing of differently deformed prolate and oblate configurations in the intrinsic system. In ^{104}Tc the results indicate completely different structure properties. The 3^+ parent state is dominated by one prolate component which represents more than 99% of the total amplitude. The 2^+ and 4^+ daughter states in ^{104}Ru manifest variable prolate-oblate mixing. For the states with significant GT strength the contribution of the prolate configurations to the structure of the 2^+ states varies from 82% to 9%, while the prolate mixing of the 4^+ states varies from 96% to 8%. The deformation of the states is reflected by the spectroscopic quadrupole moments. The 2^+ states with significant Gamow-Teller strength contribution display quadrupole moments varying from -43.99 efm^2 to 54.44 efm^2 (we used as effective charges $e_p = 1.3$, $e_n = 0.3$). The Gamow-Teller strength distribution for the decay of the 3^+ parent state in ^{104}Tc to the calculated 2^+ and 4^+ daughter states in ^{104}Ru (the 3^+ states do not have a significant contribution) is compared to TAGS results in figure 8. The strong Gamow-Teller β-decay branches indicate essential contribution from the $g^{\pi}_{9/2}g^{\nu}_{7/2}$, $d^{\pi}_{5/2}d^{\nu}_{3/2}$, and $d^{\pi}_{5/2}d^{\nu}_{5/2}$ matrix elements. Smaller contributions are obtained from $p^{\pi}_{1/2}p^{\nu}_{3/2}$ and $p^{\pi}_{3/2}p^{\nu}_{1/2}$ matrix elements. In the case of ^{104}Tc to ^{104}Ru decay all matrix elements are relatively small and the cancellations produce the final small strength for each Gamow-Teller contributing state. The strong mixing of prolate and oblate projected configurations in the parent state as well as in the daughter states is responsible for the significant difference in the GT β-decay properties of ^{102}Tc with respect to ^{104}Tc. Also the deformation of the main configurations in the structure of the wave functions is smaller in the first case where the number of neutrons in the daughter nucleus is the critical N = 58 [8], while in the second case the larger deformation is determined by the N = 60 neutrons in the ^{104}Ru daughter nucleus.

In conclusion, the *complex* Excited Vampir results concerning the influence of shape coexistence and mixing on the structure and dynamics of neutron-rich nuclei in the A$\simeq$100 mass region are in rather good agreement with the available data. Of course the theoretical results could be changed by the higher-lying configurations not included in the *complex* Excited Vampir many-nucleon basis so far. Also changes in the renormalization of the effective interaction could influence the shape mixing in the structure of the wave functions as well as the corresponding electromagnetic and Gamow-Teller β-decay properties.

Acknowledgments
This work was supported by the Romanian National Authority for Scientific Research, CNCS - UEFISCDI - project number PN-II-ID-PCE-2011-3-0153 and the CERN-ISOLDE contract 6/2012, and the DFG (Germany) under the project FA67/36-1.

References
[1] Petrovici A 2010 *J. Phys. G: Nucl. Part. Phys.* **37** 064036
[2] Cowan J J, Thielemann F -K and Truran J W 1991 *Phys. Rep.* **208** 267
[3] Petrovici A, Schmid K W and Faessler A 2000 *Nucl. Phys.* A **665** 333
[4] Petrovici A, Schmid K W, Radu O and Faessler A 2008 *Phys. Rev.* C **78** 044315
[5] Petrovici A, Schmid K W, Radu O and Faessler A 2008 *Phys. Rev.* C **78** 064311
[6] Petrovici A, Schmid K W, Andrei O and Faessler A 2009 *Phys. Rev.* C **80** 044319
[7] Petrovici A 2009 *ISOLDE Workshop and Users meeting 2009*
[8] Petrovici A 2012 *Phys. Rev.* C **85** 034337
[9] Nara Singh B S *et al* 2007 *Phys. Rev.* C **75** 061301
[10] de Angelis G *et al* 2012 *Phys. Rev.* C **85** 034320
[11] Ljungvall J *et al* 2008 *Phys. Rev. Lett.* **100** 102502
[12] Mach H *et al* 1991 *Nucl. Phys.* A **523** 197
[13] Lhersonneau G *et al* 1994 *Phys. Rev.* C **49** 1379
[14] Hamilton J H *et al* 1995 *Prog. Part. Nucl. Phys.* **35** 635
[15] Urban W *et al* 2001 *Nucl. Phys.* A **689** 605
[16] Campbell P *et al* 2002 *Phys. Rev. Lett.* **89** 082501
[17] Hua H *et al* 2004 *Phys. Rev.* C **69** 014317
[18] Wu C Y *et al* 2004 *Phys. Rev.* C **70** 064312
[19] Simpson G S *et al* 2006 *Phys. Rev.* C **74** 064308
[20] Clément E *et al* 2010 *CERN-INTC 2010-009/INTC-P-216-ADD-108/01/2010*
[21] Sumikama T *et al* 2011 *Phys. Rev. Lett.* **106** 202501
[22] Petrovici A, Schmid K W and Faessler A 2011 *J.Phys.:Conf. Series* **312** 092051
[23] Petrovici A, Schmid K W and Faessler A 2011 *Prog. Part. Nucl. Phys.* **66** 287
[24] Rinta-Antila S *et al* 2007 *Eur. Phys. J.* A **31** 1
[25] Pereira J *et al* 2009 *Phys. Rev.* C **79** 035806
[26] Schmitt A, Kaffrell N and Trautmann N 1980 *Jahresbericht* **1979** 34
[27] Algora A *et al* 2010 *Phys. Rev. Lett.* **105** 202501
[28] Jordan D, Tain J L and Algora A *private communication.*

Journal of Physics: Conference Series **413** (2013) 012008 doi:10.1088/1742-6596/413/1/012008

Pairing correlations in exotic nuclei

H. Sagawa[1,2] **and K. Hagino**[3]

[1]Center for Mathematics and Physics, University of Aizu, Aizu-Wakamatsu, 965-8580 Fukushima, Japan
[2]Nishina Center, RIKEN, Wako, 351-0198, Japan
[3]Department of Physics, Tohoku University, Sendai, 980-8578, Japan

E-mail: sagawa@u-aizu.ac.jp

Abstract. The BCS and HFB theories which can accommodate the pairing correlations in the ground states of atomic nuclei are presented. As an application of the pairing theories , we investigate the spatial extension of weakly bound Ne and C isotopes by taking into account the pairing correlation with the Hartree-Fock-Bogoliubov (HFB) method and a 3-body model, respectively. We show that the odd-even staggering in the reaction cross sections of ^{30,31,32}Ne and ^{14,15,16}C are successfully reproduced, and thus the staggering can be attributed to the unique role of pairing correlations in nuclei far from the stability line. A correlation between a one-neutron separation energy and the anti-halo effect is demonstrated for s- and p-waves using the HFB wave functions. We also propose effective density-dependent pairing interactions which reproduce both the neutron-neutron (nn) scattering length at zero density and the neutron pairing gap in uniform matter. Then, we apply these interactions to study pairing gaps in semi-magic finite nuclei, such as Ca, Ni, Sn and Pb isotopic chains.

1. Introduction

It has been known that the pairing correlations play an important role in finite and also infinite nuclear systems. Just after the BCS theory was proposed as a fundamental theory of metallic superconductor, Bohr, Mottelson and Pines proposed a possible analogy of superfluidity in atomic nuclei in 1958[1]. The most prominent evidence for pairing correlation in nuclei is found in the odd-even staggering in binding energies and the energy gap in the excitation spectrum of even-even nuclei in contrast to a compressed quasi-particle spectrum in odd-A nuclei [1, 2, 3]. There are also dynamical effects of pairing correlations seen in the moment of inertia associated with nuclear rotation and large amplitude collective motion [3, 4, 5]. The Hartree-Fock (HF)+BCS method and Hartree-Fock-Bogoliubov (HFB) method have been commonly used to study the ground state properties of superfluid nuclei in a broad mass region [6, 7, 8, 9].

As new phenomena in nuclei near the neutron-drip line, large odd-even staggering (OES) phenomena have been revealed experimentally in reaction cross sections of the isotopes ^{14,15,16}C [10], ^{18,19,20}C [11], ^{28,29,30}Ne[12], ^{30,31,32}Ne [12], and ^{36,37,38}Mg [13]. In Ref. [14], we have argued that the pairing correlations play an essential role in these OES. That is, the OES in reaction cross sections is intimately related to the so called pairing anti-halo effect discussed in Ref. [15]. In this lecture , we discuss the pairing correlations close to the zero energy by a Hartree-Fock Bogoliubov (HFB) method. This problem is also related with the superfluidity of neutron gases in the outer crust of neutron stars [16].

Recently, we proposed new types of density dependent contact pairing interaction, which reproduce pairing gaps in a wide range of nuclear mass table [9]. We discussed also the relation between the proposed paring interactions and the pairing gaps in symmetric and neutron matters obtained by a microscopic treatment based on the nucleon-nucleon interaction. We will show the necessity of the isovector type pairing interaction on top of the isoscalar term to reproduce systematically nuclear empirical pairing gaps.

This lecture note is organized as follows. Section 2 gives basic formulas of BCS and HFB theories. In section 3, we will discuss the odd-even staggering in the reaction cross sections in nuclei near the neutron drip line. Section 4 is devoted to study the isospin dependent pairing interaction for the study of pairing correlations in semi-magic nuclei.

2. BCS and HFB theories

A very important generalization of the HF theory is to accommodate pairing interaction on top of the mean field. The general theoretical framework with the pairing correlations for the single-particle orbitals is called Hartree-Fock-Bogoliubov (HFB) equations, that is also called the Bogoliubov-deGenne equations in condensed matter physics. A simpler version of HFB is called the BCS theory which has been often employed in nuclear physics. In dealing with even-even nuclei, the HF equations are invariant under time reversal. This implies that each orbital i has its time-reversed partner $\bar{i}$, and the two orbits have the same single-particle energy ε_i. The basic BCS Ansatz for the many-particle wave function is

$$|\mathrm{BCS}\rangle = \prod_{i>o}(u_i + v_i a_i^\dagger a_{\bar{i}}^\dagger)|0\rangle \tag{1}$$

where $a_i^\dagger$ is the particle creation operator acting on the HF vacuum $|0\rangle$. The parameters u_i and v_i will be determined by minimizing the expectation value of the Hamiltonian. The normalization of the state requires

$$|u_i|^2 + |v_i|^2 = 1. \tag{2}$$

The BCS state (1) is further rewritten to be

$$|\mathrm{BCS}\rangle \propto \exp\Big(\sum_{i>0}\frac{v_i}{u_i}a_i^\dagger a_{\bar{i}}^\dagger\Big)|0\rangle. \tag{3}$$

The BCS state is not an eigenstate of the particle number. In condensed matter physics, this does not cause any serious problem since the number of particles is close to the Avogadro number $\sim 10^{23}$. On the other hand, in nuclear physics, the number of nucleons is in the order of at most 200 so that we have to take care of the number conservation in some way. One possible way is to introduce a Lagrange multiplier term in the Hamiltonian

$$H' = H - \lambda\hat{N}. \tag{4}$$

Then the particle number expectation value can be fixed to the desired value N

$$\langle\mathrm{BCS}|\hat{N}|\mathrm{BCS}\rangle = 2\sum_{i>0} v_i^2 = N \tag{5}$$

on average. The quasi-particles are introduced by a unitary transformation, so called the Bogoliubov-Valatin transformation

$$\begin{aligned}\alpha_i^\dagger &= u_i a_i^\dagger - v_i a_{\bar{i}} \\ \alpha_{\bar{i}} &= u_i a_{\bar{i}} + v_i a_i^\dagger\end{aligned} \tag{6}$$

where a_i is the physical (or bare) annihilation operator. For the unitary transformation, the quasi-particles preserve the anti-commutation relations

$$\{\alpha_i, \alpha_j^\dagger\} = \delta_{i,j}, \quad \{\alpha_i, \alpha_j\} = \{\alpha_i^\dagger, \alpha_j^\dagger\} = 0. \tag{7}$$

The BCS vacuum is defined by an equation

$$\alpha_i|\mathrm{BCS}\rangle = 0. \tag{8}$$

We can prove that the definition (8) is equivalent to the BCS vacuum (1). Firstly, the definition (8) can read

$$|\mathrm{BCS}\rangle = \prod_{i \in all} \alpha_i|0\rangle = \prod_{i>0} \alpha_i\alpha_{\bar{i}}|0\rangle. \tag{9}$$

By performing the transformation (6) from the quasi-particles to the bare particles in Eq. (9), we obtain

$$\begin{aligned} |\mathrm{BCS}\rangle &= \prod_{i>0}\{u_i^2 a_i a_{\bar{i}} + u_i v_i(1 - a_i^\dagger a_i) - u_i v_i a_{\bar{i}}^\dagger a_{\bar{i}} + v_i^2 a_i^\dagger a_{\bar{i}}^\dagger\}|0\rangle \\ &= \prod_{i>0} v_i\{u_i + v_i a_i^\dagger a_{\bar{i}}^\dagger\}|0\rangle \propto \prod_{i>0}\{u_i + v_i a_i^\dagger a_{\bar{i}}^\dagger\}|0\rangle. \end{aligned} \tag{10}$$

The hamiltonian density can be evaluated for the BCS state as

$$\langle \mathrm{BCS}|H'|\mathrm{BCS}\rangle = \sum_i (\varepsilon_i - \lambda)v_i^2 - \sum_{i,j>0} V_{i\bar{i}j\bar{j}} u_i v_i u_j v_j \tag{11}$$

The variation of the Hamiltonian density with respect to the v_i (or equivalently u_i) gives a BCS gap equation to be solved;

$$\Delta_i = \sum_{j>0} \frac{2j_j + 1}{2} V_{i\bar{i}j\bar{j}} u_j v_j \tag{12}$$

with

$$\begin{aligned} u_i^2 &= \frac{1}{2}\left(1 + \frac{\varepsilon_i - \lambda}{\sqrt{(\varepsilon_i - \lambda)^2 + \Delta_i^2}}\right) \\ v_i^2 &= \frac{1}{2}\left(1 - \frac{\varepsilon_i - \lambda}{\sqrt{(\varepsilon_i - \lambda)^2 + \Delta_i^2}}\right) \end{aligned} \tag{13}$$

The quasi-particle energy E_i is evaluated by a commutator relation

$$[H', \alpha_i^\dagger] = E_i \tag{14}$$

to be

$$E_i = \sqrt{(\varepsilon_i - \lambda)^2 + \Delta_i^2}. \tag{15}$$

The BCS theory is generalized to HFB theory by using the following unitary transformation. The Hartree-Fock-Bogoliubov (HFB) transformation for quasi-particle can read

$$\begin{aligned} \alpha_i^\dagger &= \sum_i u_{ij} a_j^\dagger - v_{ij} a_{\bar{j}} \\ \alpha_{\bar{i}} &= \sum_i u_{ij} a_{\bar{j}} + v_{ij} a_j^\dagger \end{aligned} \tag{16}$$

where u_{ij} and v_{ij} are the upper and the lower components of BCS transformation, respectively. The normalization condition for the u_{ij} and v_{ij} factors is given by

$$\sum_j u_{ij}^2 + v_{ij}^2 = 1. \tag{17}$$

The HFB model can be generalized to the coordinate space representation as [17]

$$\begin{aligned} \alpha_i^\dagger &= \int d\mathbf{r}\{u_i(\mathbf{r})\psi^\dagger(\mathbf{r}) + v_i^*(\mathbf{r})\psi(\mathbf{r})\} \\ \alpha_i &= \int d\mathbf{r}\{u_i^*(\mathbf{r})\psi(\mathbf{r}) + v_i(\mathbf{r})\psi^\dagger(\mathbf{r})\} \end{aligned} \tag{18}$$

where $\psi^\dagger(\mathbf{r})$ and $\psi(\mathbf{r})$ are the creation and the annihilation operators of bare particle in the coordinate space representation and obey the anti-commutator relation

$$\{\psi(\mathbf{r}), \psi^\dagger(\mathbf{r}')\} = \delta(\mathbf{r} - \mathbf{r}'). \tag{19}$$

The orthonormal conditions of $u_i(\mathbf{r})$ and $v_i(\mathbf{r})$ functions are given by

$$\begin{aligned} &\int d\mathbf{r}\{u_i^*(\mathbf{r})u_j(\mathbf{r}) + v_i^*(\mathbf{r})v_j(\mathbf{r})\} = \delta_{i,j} \\ &\sum_i \{u_i^*(\mathbf{r})u_i(\mathbf{r}') + v_i^*(\mathbf{r})v_i(\mathbf{r}')\} = \delta(\mathbf{r} - \mathbf{r}'). \end{aligned} \tag{20}$$

The density and the pair density (abnormal density) are obtained as

$$\rho(\mathbf{r}, \mathbf{r}') = \langle \mathrm{HFB}|\psi^\dagger(\mathbf{r})\psi(\mathbf{r}')|\mathrm{HFB}\rangle = \sum_i v_i(\mathbf{r})v_i^*(\mathbf{r}') \tag{21}$$

$$\kappa(\mathbf{r}, \mathbf{r}') = \langle \mathrm{HFB}|\psi(\mathbf{r})\psi(\mathbf{r}')|\mathrm{HFB}\rangle = \sum_i v_i(\mathbf{r})u_i(\mathbf{r}') \tag{22}$$

where the HFB vacuum is defined by

$$\alpha_i|\mathrm{HFB}\rangle = 0. \tag{23}$$

The hamiltonian density for the constrained Hamiltonian H' (4) is expressed to be

$$\langle \mathrm{HFB}|H'|\mathrm{HFB}\rangle = Tr\{(T - \lambda)\rho\} + \frac{1}{2}Tr\{V\rho\rho\} + \frac{1}{2}\{V\kappa\kappa\} \tag{24}$$

where T and V are the kinetic density and the hamiltonian density due to a two-body interaction, respectively.

The HFB equations are obtained by the variation of the hamiltonian density in Eq. (24) with respect to the $u_i(\mathbf{r})$ and $v_i(\mathbf{r})$ functions to be

$$\begin{pmatrix} T + V_{HF} - \lambda & \Delta(\mathbf{r}) \\ \Delta(\mathbf{r}) & -T - V_{HF} + \lambda \end{pmatrix} \begin{pmatrix} u_i(\mathbf{r}) \\ v_i(\mathbf{r}) \end{pmatrix} = E_i \begin{pmatrix} u_i(\mathbf{r}) \\ v_i(\mathbf{r}) \end{pmatrix}. \tag{25}$$

where V_{HF} is the HF potential,

$$V_{HF} = Tr\{V\rho\} \tag{26}$$

and $\Delta(\mathbf{r})$ is the pairing gap potential,

$$\Delta(\mathbf{r}) = Tr\{V\kappa\}, \tag{27}$$

respectively. In Eq. (25), $u_i(\mathbf{r})$ and $v_i(\mathbf{r})$ are radial wave functions and could have different shapes from the HF single particle wave function, which is the eigenstate of HF hamiltonian $h_{HF} = T - V_{HF}$. In contrast, in the BCS model, the u_i and v_i are factors and simply multiplied to the HF single particle wave function φ_i by the Bogoliubov-Valatin transformation (6).

A question is whether the HFB is necessary for realistic framework to treat the pairing correlations in nuclei. When the active orbitals are well bound as in nuclei along the valley of stability, the single-particle wave functions are not influenced by the pairing field and the BCS model could be entirely adequate. This might not be the case in nuclei near the drip lines, where the pairing correlations would make the difference between bound and unbound orbitals. In a recent paper, we showed that a major part of the effects of HFB could be included by a perturbative treatment of the BCS wave functions even in nuclei far from the valley of stability [18].

3. Halo and Pairing correlations

The pairing correlations play a crucial role to develop the halo structure of loosely bound nuclei. At the same time, the pairing will act to prevent the divergence of very loosely bound nucleons in the mean field. Let us discuss how the single-particle wave function will expand its tail in the mean field potential and how the pairing correlations will prevent the exponential growth of expansion to be infinity. In the mean field model, the single particle wave function

$$\psi_{lm}(\mathbf{r}) = \varphi_{lm}(r)Y_{lm}(\hat{r}) = \frac{u_l(r)}{r}Y_{lm}(\hat{r}) \tag{28}$$

is calculated with a Schrödinger equation

$$\left(-\frac{\hbar^2}{2m}\frac{d^2}{dr^2} + \frac{l(l+1)\hbar^2}{2mr^2} + V_{HF}(r) - \varepsilon_l\right)u_l(r) = 0. \tag{29}$$

The asymptotic behavior in the limit $r \to \infty$ with $V_{HF} \to 0$ is given by the modified spherical Bessel function and behaves

$$u_{l=0}(r) \to \exp(-\kappa r) \;\; \text{in the limit} \;\; r \to \infty \tag{30}$$

for the angular momentum $l = 0$ case. Then the mean square radius for $s-$wave is evaluated to be

$$\langle r^2\rangle_{l=0} = \frac{\int r^2 u_{l=0}(r)^2 dr}{\int u_{l=0}(r)^2 dr} \propto \frac{1}{\kappa^2} = \frac{\hbar^2}{2m|\varepsilon|} \to \infty \tag{31}$$

which diverges in the limit of very loosely bound case $|\varepsilon| \to 0$. It is also shown in the case of $p-$wave case, $l = 1$, that the mean square radius also diverges but slightly slowly proportional to $1/\sqrt{|\varepsilon|}$ [19]. On the other hand, the wave function with $l \geq 2$ does not show any divergence of mean square radius in the limit of $|\varepsilon| \to 0$ [20].

In contrast, the upper component of a HFB wave function, which is relevant to the density distribution, behaves [21]

$$v(r) \propto \exp(-\beta r), \tag{32}$$

where β is proportional to the square root of the quasi-particle energy E,

$$\beta = \sqrt{\frac{2m}{\hbar^2}(E - \lambda)}, \tag{33}$$

λ being the chemical potential. If we evaluate the quasi-particle energy in the BCS approximation or HFB with canonical basis, it is given as

$$E = \sqrt{(\varepsilon - \lambda)^2 + \Delta^2}, \tag{34}$$

where Δ is the pairing gap. For a weakly bound single-particle state with $\varepsilon \sim 0$ and $\lambda \sim 0$, the asymptotic behavior of the wave function $v(r)$ is therefore determined by the gap parameter as,

$$v(r) \propto \exp\left(-\sqrt{\frac{2m}{\hbar^2}\Delta} \cdot r\right). \tag{35}$$

The radius of the HFB wave function will then be given in the limit of small separation energy $|\varepsilon| \to 0$ as

$$\langle r^2 \rangle_{\rm HFB} = \frac{\int r^2 |v(r)|^2 d\mathbf{r}}{\int |v(r)|^2 d\mathbf{r}} \propto \frac{1}{\beta^2} \to \frac{\hbar^2}{2m\Delta}. \tag{36}$$

As we show, the gap parameter Δ stays finite even in the zero energy limit of ε with a density dependent pairing interaction. Thus the extremely large extension of a halo wave function in the HF field will be reduced substantially by the pairing field and the root-mean-square (rms) radius of the HFB wave function will not diverge. This is called the anti-halo effect due to the pairing correlations [15].

We now numerically carry out mean-field calculations with a Woods-Saxon (WS) potential and also HFB calculations using the single-particle wave functions in the WS potential. Most of following materials of this section are taken from refs. [14]. As examples of s-wave and p-wave states, we choose the $2s_{1/2}$ state in ^{23}O and $2p_{3/2}$ state in ^{31}Ne, respectively. Although ^{31}Ne is most likely a deformed nucleus[22, 23], for simplicity we assume a spherical Woods-Saxon mean-field potential. Notice that a Woods-Saxon potential with a large diffuseness parameter a yields the $2p_{3/2}$ state which is lower in energy than the $1f_{7/2}$ state, as was shown in Ref. [24]. We use a similar potential with a=0.75 fm as in Ref. [24] for ^{31}Ne, while that in Ref. [25] for ^{23}O. For the HFB calculations, we use the density-dependent contact pairing interaction of surface type, in which the parameters are adjusted in order to reproduce the empirical neutron pairing gap for ^{30}Ne [26]. While we fix the Woods-Saxon potential for the mean-field part, the pairing potential is obtained self-consistently with the contact interaction.

The left top panel of Fig. 1 shows the mean square radius of the $2s_{1/2}$ state for ^{23}O, while right top panel of Fig. 1 shows the mean square radius of the $2p_{3/2}$ state for ^{31}Ne. In order to investigate the dependence on the single-particle energy, we vary the depth of the Woods-Saxon wells for the $s_{1/2}$ and $p_{3/2}$ states for ^{23}O and ^{31}Ne, respectively. The dashed lines are obtained with the single-particle wave functions, while the solid lines are obtained with the wave function for the canonical basis in the HFB calculations. One can see an extremely large increase of the radius of the WS wave function for both the s-wave and p-wave states in the limit of $\epsilon_{\rm WS} \to 0$. In contrast, the HFB wave functions show only a small increase of radius even in the limit of $\epsilon_{WS} \to 0$. This feature remains the same even when the contribution of the other orbits are taken into account, as shown in the middle panel of Figs. 1. Due to the pairing effect in the continuum, the HFB calculations yield a larger radius than the HF calculations for the cases of $\epsilon_{\rm WS} \leq -1$ MeV. On the other hand, in the case of -1 MeV $< \epsilon_{\rm WS} < 0$ MeV the HF wave function (equivalently one quasi-particle wave function in HFB) extends largely, while the HFB wave function does not get much extension due to the pairing anti-halo effect.

In the bottom panel of Figs. 1, the average pairing gaps are shown as a function of the single particle energy $\epsilon_{\rm WS}$. It is seen that the average pairing gap increases as $\epsilon_{\rm WS}$ approaches zero. This is due to the fact that the paring field couples with the extended wave functions of weakly

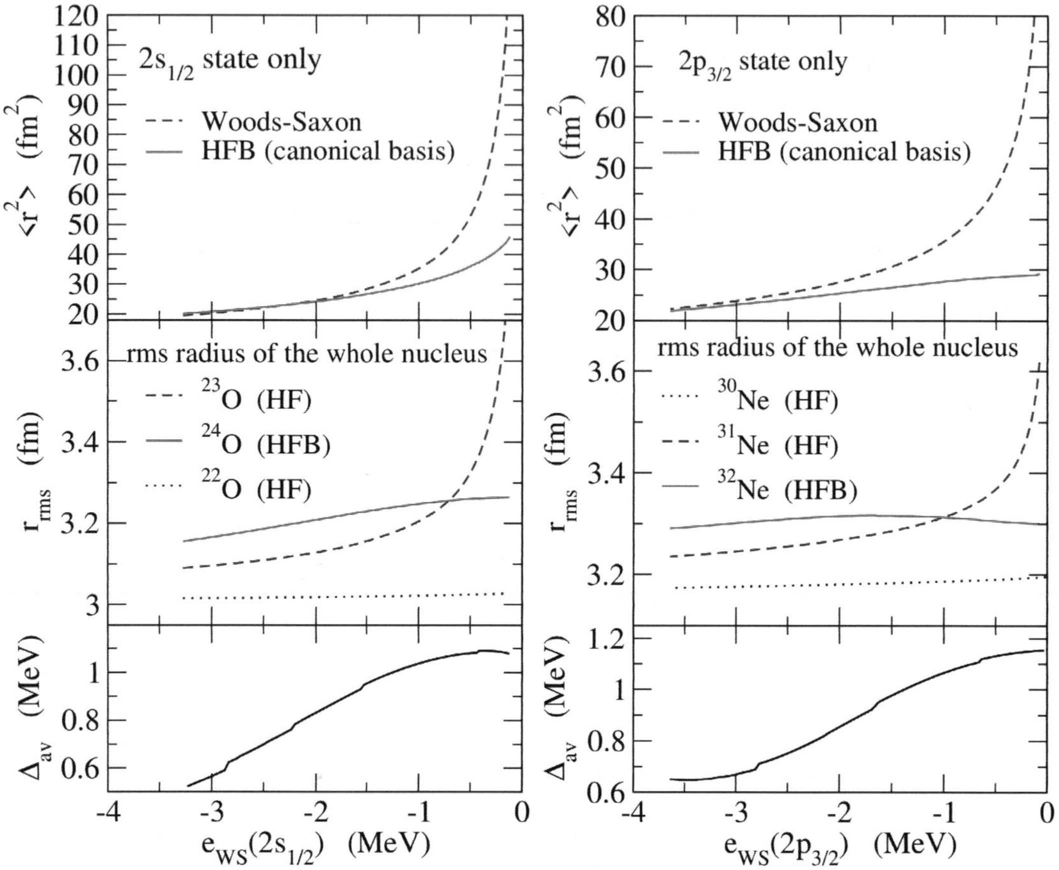

Figure 1. (Color online) (left panel) The mean square radii and the average paring gap as a function of the single particle energy $\varepsilon_{\rm WS}$ in a Woods-Saxon mean-field potential. The top panel shows the mean square radius of the $2s_{1/2}$ wave function with and without the pairing correlation, denoted by HFB and Woods-Saxon, respectively. The middle panel shows the rms radii for ^{22}O (the dotted line), ^{23}O (the dashed line), and ^{24}O (the solid line), obtained with the Hartree-Fock (^{22}O and ^{23}O) and the Hartree-Fock-Bogoliubov (^{24}O) calculations. The bottom panel shows the results of the HFB calculations for the average paring gap of ^{24}O. (right panel) Same as the left panel, but for the $2p_{3/2}$ state and for ^{30,31,32}Ne isotopes.

bound nucleons in the self-consistent calculations. That is, the pairing field is extended as the wave functions do and becomes larger for a loosely bound system.

Let us now calculate the reaction cross sections for the ^{30,31,32}Ne isotopes and discuss the role of pairing anti-halo effect. To this end, we use the Glauber theory, in which we adopt the prescription in Refs. [24, 27] in order to take into account the effect beyond the optical limit approximation. The left panel of Fig. 2 shows the reaction cross sections of the ^{30,31,32}Ne nuclei on a ^{12}C target at 240 MeV/nucleon. We use the target density given in Ref. [28] and the profile function for the nucleon-nucleon scattering given in Ref. [29]. In order to evaluate the phase shift function, we use the two-dimensional Fourier transform technique [30]. The cross sections σ_R shown in Fig. 2 are calculated by using projectile densities constrained to two different separation energies of the $2p_{3/2}$ neutron state. The dashed line with triangles is obtained using the wave functions with the separation energy S_n (^{31}Ne) $=|\epsilon_{\rm WS}|$=1.49 MeV, while the solid line

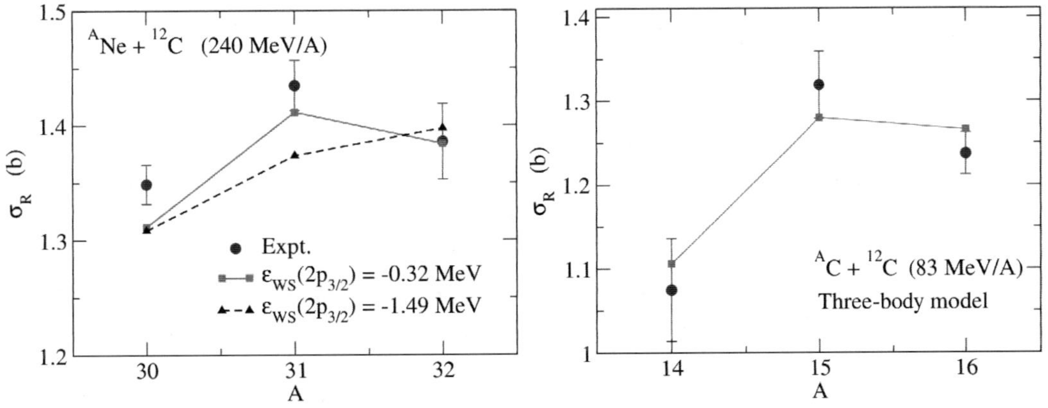

Figure 2. (Color online) (left panel) Reaction cross sections of Ne isotopes on a ^{12}C target at $E_{\rm lab}$=240 MeV/A. The cross sections are calculated with the Glauber theory with HF and HFB densities. The solid line with the filled squares shows the results of $S_n(^{31}\text{Ne}) = |\varepsilon(2p_{3/2})|$=0.32 MeV, while the dashed line with the open triangles is obtained for $S_n(^{31}\text{Ne}) = |\varepsilon(2p_{3/2})|$=1.49 MeV. The experimental data are taken from Ref. [12]. (right panel) Reaction cross sections of C isotopes on a ^{12}C target at $E_{\rm lab}$=83 MeV/A. The cross sections are calculated with the Glauber theory with three-body model densities. The experimental data are taken from Ref. [10].

with squares is calculated with the wave functions of $S_n(^{31}\text{Ne})=|\epsilon_{\rm WS}|$=0.32 MeV. The empirical separation energy of ^{31}Ne has a large ambiguity with S_n=0.29±1.64 MeV[31]. The cross section σ_R of ^{30}Ne is already much larger than the systematic values of Ne isotopes with A<30. On top of that, we can see a clear odd-even staggering in the results with the smaller separation energy, S_n=0.32 MeV, as much as in the experimental data, while almost no staggering is seen in the case of the larger separation energy, S_n=1.49 MeV. This difference is easily understood by looking at the anti-halo effect for $|\epsilon_{\rm WS}| \leq$ 1MeV shown in the middle panel of Fig. 1. Recently, the effect of deformation of neutron-rich Ne isotopes on reaction cross sections was evaluated using a deformed Woods-Saxon model [32, 33]. It was shown that the deformation is large as much as $\beta_2 \sim 0.42$ in ^{31}Ne and enhances the reaction cross section by about 5%. However, the calculated results did not show any significant odd-even staggering in σ_R between ^{28}Ne and ^{32}Ne [32].

We investigate next neutron-rich C isotopes. We particularly study the ^{16}C nucleus using a three-body model given in Ref. [34]. In this case, the valence neutron in ^{15}C occupies the $2s_{1/2}$ level at $\epsilon_{\rm WS} = -1.21$ MeV, while ^{16}C is an admixture of mainly the $(2s_{1/2})^2$ and $(1d_{5/2})^2$ configurations. Assuming the set D given in Ref. [34] for the parameters of the Woods-Saxon and the density distribution for ^{14}C given in Ref. [35], the rms radii are estimated to be 2.53, 2.90, and 2.81 fm for ^{14}C, ^{15}C, and ^{16}C, respectively. The corresponding reaction cross sections σ_R calculated with the Glauber theory are shown in the right panel of Fig. 2. The calculation well reproduces the experimental odd-even staggering of the reaction cross sections, that is a clear manifestation of the pairing anti-halo effect.

In order to quantify the OES of reaction cross sections, we introduce the staggering parameter defined by

$$\gamma_3 = (-)^A \frac{\sigma_R(A+1) - 2\sigma_R(A) + \sigma_R(A-1)}{2}, \tag{37}$$

where $\sigma_R(A)$ is the reaction cross section of a nucleus with mass number A. We can define the same quantity also for rms radii. Notice that this staggering parameter is similar to the one

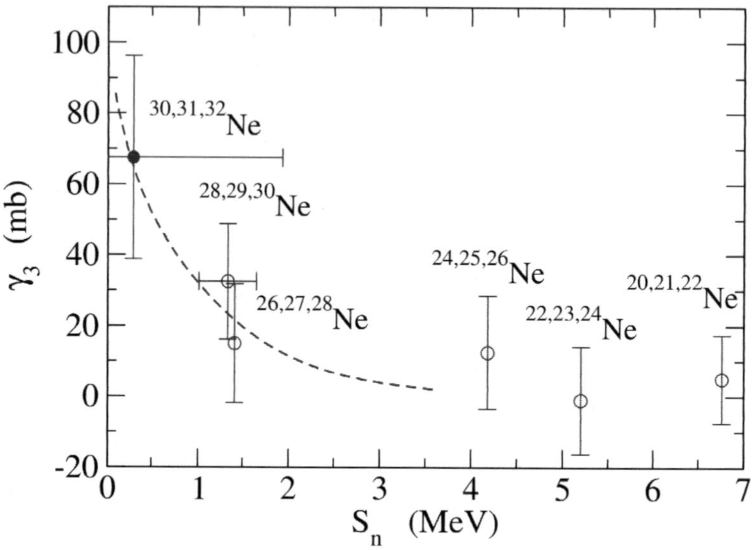

Figure 3. (Color online) The staggering parameter γ_3 of reaction cross sections defined by Eq. (37) for the Ne isotopes with the ^{12}C target at E=240 MeV/nucleon. This is plotted as a function of the neutron separation energy S$_n$ of the odd-A nuclei. The experimental data for the reaction cross sections are taken from Ref. [12], while the empirical separation energies are taken from Refs. [36, 31]. The dashed line is the calculated staggering parameter for the ^{30,31,32}Ne isotopes, assuming that the valence neutron of ^{31}Ne occupies the 2p3/2 orbit.

often used for the OES of binding energy, that is, the pairing gap.

The experimental staggering parameters γ_3 are plotted in Fig. 3 for Ne isotopes as a function of the neutron separation energy for the odd-mass nuclei. We use the experimental reaction cross sections given in Ref. [13] while we evaluate the separation energies with the empirical binding energies listed in Ref. [36]. For the neutron separation energy for the ^{31}Ne nucleus, we use the value in Ref. [31]. The experimental uncertainties of the staggering parameter are obtained as

$$\delta\gamma_3 = \frac{\sqrt{(\delta\sigma_R(A+1))^2 + 4(\delta\sigma_R(A))^2 + (\delta\sigma_R(A-1))^2}}{2}, \tag{38}$$

where $\delta\sigma_R(A)$ is the experimental uncertainty for the reaction cross section of a nucleus with mass number A. The figure also shows by the dashed line the calculated staggering parameter for the ^{30,31,32}Ne nuclei with the 2p$_{3/2}$ orbit. One sees that the experimental staggering parameter agrees with the calculated value for ^{30,31,32}Ne nuclei when one assumes that the valence neutron in ^{31}Ne occupies the 2p$_{3/2}$ orbit. Furthermore, although the structure of lighter odd-A Ne isotopes is not known well, it is interesting to see that the empirical staggering parameters closely follow the calculated values for the 2p$_{3/2}$ orbit. This may indicate that the low-l single-particle orbits are appreciably mixed in these Ne isotopes due to the deformation effects[22, 23].

4. Isospin Dependent Pairing Interaction and Pairing Gaps in Finite Nuclei

There are mainly two different approaches for a calculation of pairing correlations in finite nuclei. The first approach is based on phenomenological pairing interactions whose parameters are determined using some selected data [37], while the second approach starts from a bare nucleon-nucleon interaction and eventually includes the effect of phonon coupling [38]. The latter approach has shown that the medium polarization reduces the pairing gaps in neutron matter while, in symmetric matter, the neutron pairing gaps are much enlarged at low density compared to that of the bare calculation. This enhancement takes place especially for neutron Fermi momenta $k_{\mathrm{Fn}} < 0.7$ fm^{-1}.

We propose effective density-dependent pairing interactions which reproduce both the neutron-neutron (nn) scattering length at zero density and the neutron pairing gap in uniform matter. In order to simultaneously describe the density dependence of the neutron pairing gap for both symmetric and neutron matter, it is necessary to include an isospin dependence in the effective pairing interaction [9]. Depending on whether the medium polarization effects on the pairing gap given in Ref. [39] are taken into account or not, we invent two different density dependent functionals in the pairing interaction. Then, we apply these interactions to study pairing gaps in semi-magic finite nuclei, such as Ca, Ni, Sn and Pb isotopic chains.

The density−dependent pairing interaction can be read as

$$v_{\mathrm{pair}}(1,2) = \frac{1-\mathrm{P}_\sigma}{2} v_0\, \mathrm{g}[\rho, I]\, \delta(\vec{\mathrm{r}}_1 - \vec{\mathrm{r}}_2). \tag{39}$$

where ρ is the nuclear density and I is defined as $I = (\rho_n - \rho_p)/\rho \cdot \tau_z$ [9]. In Ref. [9], an isovector dependence in the density-dependent term g is separated to two parts $\mathrm{g} = \mathrm{g}^1 + \mathrm{g}^2$. The function g^1 is determined to mimic the bare pairing gaps in nuclear matters and the function g^2 takes care of the medium polarization effect. The functional form of g^1 is given by

$$\mathrm{g}^1[\rho, I] = 1 - \mathrm{f_s}(I)\eta_\mathrm{s}\left(\frac{\rho}{\rho_0}\right)^{\alpha_\mathrm{s}} - \mathrm{f_n}(I)\eta_\mathrm{n}\left(\frac{\rho}{\rho_0}\right)^{\alpha_\mathrm{n}}, \tag{40}$$

where ρ_0=0.16 fm^{-3} is the saturation density of symmetric nuclear matter and the functions $\mathrm{f_s}(I)$ and $\mathrm{f_n}(I)$ are $\mathrm{f_s}(I) = 1 - \mathrm{f_n}(I)$ and $\mathrm{f_n}(I) = I = (\rho_\mathrm{n}(r) - \rho_\mathrm{p}(r))/\rho(r) \cdot \tau_z$. The values of parameters $\eta_\mathrm{s}, \eta_\mathrm{n}$ and powers of density dependence $\alpha_\mathrm{s}, \alpha_\mathrm{n}$ are given elsewhere [9].

We perform Hartree-Fock-Bogoliubov (HFB) calculations for semi-magic Calcium, Nickel, Tin and Lead isotopes using these density-dependent pairing interactions in Eq. (39) derived from a microscopic nucleon-nucleon interaction. Our calculations reproduce well the neutron number dependence of experimental data for binding energy, two neutrons separation energy, and odd-even mass staggering of these isotopes [40]. Especially the interaction IS+IV Bare without the medium polarization effect gives satisfactory results of all the isotopes as is seen in Fig. 4. It is clear in the comparison between IS bare and IS+IV bare that the isospin dependence of the pairing interaction plays an important role in the pairing gaps of neutron-rich nuclei. The isospin dependent pairing interaction is further applied to both even-even and even-odd nuclei by using EV8-odd program. The results successfully reproduce the empirical isotope and isotone dependent functionals of the odd-even mass differences of several medium-heavy and heavy nuclei [41]. Yamagami et al., studied also the quadratic isospin dependent terms of pairing interaction [42, 43]. They adopted several Skyrme interactions with different effective mass and showed the effective mass dependence of the isospin dependent terms. It is also shown in ref. [44] that the effect of Coulomb two-body interaction on the proton pairing gaps can be mimicked by the leaner isospin dependent term in the pairing interaction.

5. Summary

We have studied the mass radii of Ne isotopes with the Hartree-Fock (HF) and Hartree-Fock-Bogoliubov (HFB) methods with a Woods-Saxon potential. The reaction cross sections σ_R were calculated using the Glauber theory with these microscopic densities. We have shown that the empirical odd-even staggering in the reaction cross sections of neutron-rich Ne isotopes with

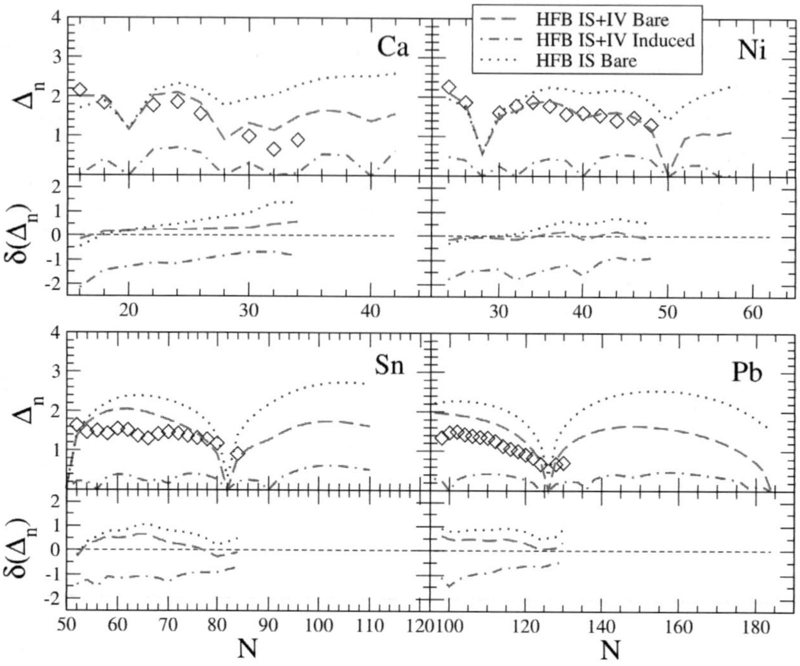

Figure 4. Comparison of the neutron HFB pairing gaps Δ_n with the odd-even mass staggering given by the three-point formula $\Delta^{(3)}$. The dotted line shows the results of the pairing interactions IS+IV Bare, while the short dashed, and long dashed lines are obtained with the pairing interactions IS+IV Screened and IS Bare, respectively. The difference $\delta(\Delta_n)$ is defined as $\delta(\Delta_n) = \Delta_n(\text{th.}) - \Delta_n(\text{exp.})$. All units are given in MeV.

the mass A=30 $\sim$ 32 is well described by the HFB density and can be considered as a clear manifestation of the pairing anti-halo effect associated with a loosely-bound $2p_{3/2}$ wave function. The index of the odd-even staggering is proposed and applied successfully for the study of the reaction cross sections of Ne isotopes to revel the role of the pairing correlations in exotic nuclei.

A new type of density dependent contact pairing interaction was obtained to reproduce the microscopic pairing gaps in symmetric and neutron matter. We performed also HFB calculations for semi-magic Calcium, Nickel, Tin and Lead isotopes using these density-dependent pairing interactions. Our calculations reproduce well the neutron number dependence of experimental data of binding energy, two neutrons separation energy, and odd-even mass staggering of these isotopes with the interaction IS+IV Bare. Recently, the isospin dependent interaction is extended to introduce the quadratic terms of isospin and applied to a wide region of nuclei in the mass table. These result suggests that by introducing the isospin dependent term in the pairing interaction, one can construct a global effective pairing interaction which is applicable to nuclei in a wide range of the nuclear chart.

Acknowledgments

This work was supported by the Japanese Ministry of Education, Culture, Sports, Science and Technology by Grant-in-Aid for Scientific Research under the program numbers (C) 22540262.

References

[1] Bohr A Mottelson B R and Pines D 1958 *Phys. Rev.* **110** 936

[2] Bohr A and Mottelson B R 1969 *Nuclear Structure* (Benjamin, New York) Vol. I.
[3] Brink D M and Broglia R 2005 *"Nuclear superfluidity, pairing in Finite Systems"*, Cambridge monographs on particle physics, nuclear physics and cosmology, vol 24
[4] Bertsch G F 2012 *" 50 Years of Nuclear BCS"*, (World Scientific, edited by Broglia R and Zelevinsky) V, arXiv:1203.5529v1
[5] Matsuyanagi K Hinohara N and Sato K 2012 arXiv:1205.0078v1
[6] Bender M Heenen P -H and Reinhard P -G 2003 **Rev. Mod. Phys. 75** 121
[7] Stoitsov M V Dobaczewski J Nazarewicz W and Borycki P 2006 *Int. J. Mass Spectrum* **251** 243
[8] Duguet T Bonche P Heenen P -H and Meyer J 2001 **Phys. Rev. C 65** 014311
[9] Margueron J Sagawa H and Hagino K 2007 *Phys. Rev. C* **76**, 064316; 2008 *Phys. Rev. C* **77** 054309
[10] Fang D Q *et al* 2004 *Phys. Rev. C* **69** 034613
[11] Ozawa A *et al* 2001 *Nucl. Phys. A* **691** 599
[12] Takechi M *et al* 2010 *Nucl. Phys. A* **834** 412c and private communication
[13] Takechi M *et al* private communication
[14] Hagino K and Sagawa H 2011 *Phys. Rev. C* **84** 011303(R); 2012 *Phys. Rev. C* **85** 014303; 2012 *Phys. Rev. C* **85** 037604
[15] Bennaceur K Dobaczewski J and Ploszajczak M 2000 *Phys. Lett. B* **496** 154
[16] Schuck P and Vinas X 2011 *Phys. Rev. Lett.* **107** 205301
[17] Bulgac A arXiv:nucl-th/9907088
[18] Hagino K and Sagawa H 2005 *Phys. Rev. C* **71** 044302
[19] Riisager K Jensen A S and Moller P 1992 *Nucl. Phys. A* **548** 393
[20] Sagawa H 1992 *Phys. Lett. B* **286** 7
[21] Dobaczewski J *et al* 1996 *Phys. Rev. C* **53** 2809
[22] Hamamoto I 2010 *Phys. Rev. C* **81** 021304(R)
[23] Urata Y Hagino K and Sagawa H 2011 *Phys. Rev. C* **83** 041303(R)
[24] Horiuchi W Suzuki Y Capel P and Baye D 2010 *Phys. Rev. C* **81** 024606
[25] Hagino K and Sagawa H 2005 *Phys. Rev. C* **72** 044321
[26] Yamagami M and Giai Nguyen Van 2004 *Phys. Rev. C* **69** 034301
[27] Abu-Ibrahim B and Suzuki Y 2000 *Phys. Rev. C* **62** 034608
[28] Ogawa Y Yabana K and Suzuki Y 1992 *Nucl. Phys. A* **543** 722
[29] Abu-Ibrahim B *et al* 2008 *Phys. Rev. C* **77** 034607
[30] Bertulani C A and Sagawa H 1995 *Nucl. Phys. A* **588** 667
[31] Jurado B *et al* 2007 *Phys. Lett. B* **649** 43
[32] Minomo K *et al* 2011 *Phys. Rev. C* **84** 034602; 2012 *Phys. Rev. Lett.* **108** 052503
[33] Urata Y Hagino K and Sagawa H 2012 *Phys. Rev. C* **86** 044613
[34] Hagino K and Sagawa H 2007 *Phys. Rev. C* **75** 021301(R)
[35] Parfenova Yu L Zhukov M V and Vaagen J S 2000 *Phys. Rev. C* **62** 044602
[36] Audi G and Wapstra A H 1995 *Nucl. Phys. A* **595** 409
[37] Dobaczewski J Nazarewicz W and Reinhard P -G 2001 *Nucl. Phys. A* **693** 361
[38] Barranco F *et al* 1999 *Phys. Rev. Lett.* **83** 2147
[39] Cao L G Lombardo U and Schuck P 2006 *Phys. Rev. C* **74** 064301
[40] Margueron J Sagawa H and Hagino K 2008 *Phys. Rev. C* **77** 054309
[41] Bertulani C A Lu H F and Sagawa H 2009 *Phys. Rev. C* **80** 027303
[42] Yamagami M Shimizu Y R and Nakatsukasa T 2009 *Phys. Rev. C* **80** 064301
[43] Yamagami M Margueron J Sagawa H and Hagino K 2012 *Phys. Rev. C* **86** 034333
[44] Nakada H and Yamagami M 2011 *Phys. Rev. C* **83** 031302

Journal of Physics: Conference Series **413** (2013) 012009 doi:10.1088/1742-6596/413/1/012009

Alpha-Particle Condensation in Nuclear Systems

P. Schuck[1 2 3], Y. Funaki[4], H. Horiuchi[5], G. Röpke[6], A. Tohsaki[5], T. Yamada[7]

[1]Institut de Physique Nucléaire, CNRS, UMR8608,
Orsay, F-91406, France
[2]Université Paris-Sud, Orsay, F-91505, France
[3]Laboratoire de Physique et de Modélisation des Milieux Condensés, CNRS et, Université Joseph Fourier, 25 Av. des Martyrs, BP 166, F-38042 Grenoble Cedex 9, France
[4]The Institute of Physical and Chemical Research (RIKEN), Wako,
Saitama, 351-0198, Japan
[5]Research Center for Nuclear Physics (RCNP), Osaka University,
Ibaraki, Osaka 567-0047, Japan
[6]Institut für Physik, University of Rostock, Universitätsplatz 1,
18051 Rostock, Germany
[7]Laboratory of Physics, Kanto Gakuin University,
Yokohama 236-8501, Japan

Abstract. The onset of quartetting, i.e. α-particle condensation, in symmetric and asymmetric nuclear matter is studied with the help of an in-medium modified four nucleon equation. It is found that at very low density quartetting wins over pairing, because of the strong binding of the α-particles. The critical temperature can reach values up to around 8 MeV. The disappearance of α-particles with increasing density, i.e. the Mott transition, is investigated. In finite nuclei the Hoyle state, that is the $0_2{}^+$ of ^{12}C is identified as an 'α-particle condensate' state. It is conjectured that such states also exist in heavier nα-nuclei, like ^{16}O, ^{20}Ne, etc. The sixth 0^+ state in ^{16}O is proposed as an analogue to the Hoyle state. The Gross-Pitaevski equation is employed to make an estimate of the maximum number of α particles a condensate state can contain. Possible quartet condensation in other systems is discussed briefly.
Keywords: nuclear matter, α-matter, superfluidity, Bose-Einstein condensation, strongly coupled systems

1. Introduction

One of the most amazing phenomena in quantum many-particle systems is the formation of quantum condensates. At present, the formation of condensates is of particular interest in strongly coupled fermion systems in which the crossover from Bardeen-Cooper-Schrieffer (BCS) pairing to Bose-Einstein condensation (BEC) may be investigated. Among very different quantum systems such as the electron-hole-exciton system in excited semiconductors, atoms in traps at extremely low temperatures, etc., nuclear matter is especially well suited for the study of correlation effects in a quantum liquid.

Neutron matter, nuclear matter, but also finite nuclei are superfluid. However, at low density, nuclear matter will not cluster into pairs, i.e. deuterons but rather into α -particles which are much more stable. Also heavier clusters, starting with Carbon, may be of importance but are presently not considered for condensation phenomena. Therefore, one may ask the question whether there exists quartetting, i.e. α-particle condensation, in nuclei, analogous to nuclear pairing. The only nucleus which in its ground state has a pronounced α -cluster structure is ^{8}Be. In section 4 we will show a figure of ^{8}Be in the laboratory frame and in the intrinsic deformed frame. We will see that ^{8}Be is formed out of two almost free α - particles roughly 4 fm apart and barely touching one another. Actually ^{8}Be is slightly unstable and the two α's only hold together via the Coulomb barrier. Because of the large distance of the two α-particles, the 0^+ ground state of ^{8}Be has, in the laboratory frame, a spherical density distribution whose average is very low: about 1/3 of ordinary saturation density ρ_0. ^{8}Be is, therefore, a very large object with an rms radius of about 3.7 fm to be compared with the nuclear systematics of $R = r_0 A^{1/3}$ = 2.44 fm. Definitely ^{8}Be is a rather unusual and, in its kind, unique nucleus. One may ask the question what happens when one brings a third α-particle alongside of ^{8}Be. We know the answer: the 3-α system collapses to the ground state of ^{12}C which is much denser than ^{8}Be and can not accommodate with its small radius of 2.4 fm three more or less free α-particles barely touching one another. One nevertheless may ask the question whether the dilute three α configuration ^{8}Be- α, or $\alpha - \alpha - \alpha$, may not form an isomeric or excited state of ^{12}C. That such a state indeed exists will be one of the main subjects of our considerations.

Once one accepts the idea of the existence of an α-gas state in ^{12}C, there is no reason why equivalent states at low density should not also exist in heavier nα-nuclei, like ^{16}O, ^{20}Ne, etc. In a mean field picture, i.e. all α's being ideal bosons (in this context remember that the first excited state of an α-particle is at $\sim$ 20 MeV, by factors higher than in all other nuclei), all α's will occupy the lowest 0S-state, i.e. they will condense into this state. This forms, of course, not a macroscopic condensate but it can be understood in the same sense as we know that nuclei are superfluid because of the presence of a finite number of Cooper pairs. On the other hand, for example during the cooling process of compact stars [1], where one predicts the presence of α-particles [2], a real macroscopic phase of condensed α's may be formed. In the present contribution we will mainly concentrate on nuclear systems but we also can think about the possibility of quartetting in other Fermi-systems. One should, however, keep in mind that a pre-requisite for its existence is, as in nuclear physics, that there exists a bound quartet in free space. This is facilitated with the existence of four different types of fermions, all attracting one another. For example to form quartets with cold atoms one could try to trap fermions in four different magnetic substates, a task which eventually seems possible [3].

In the next section we will investigate how the binding energy of various nuclear clusters changes with density. In section 3 we study the critical temperature of α-particle condensation in infinite matter via an in- medium four-nucleon equation (Thouless criterion) and in section 4 we treat α-particle condensation in finite nuclei. Finally in section 5 we conclude with outlook and further discussions.

2. Nuclear clusters in the medium

With increasing density of nuclear matter, medium modifications of single-particle states as well as of few-nucleon states become of importance. The self-energy of an A-particle cluster can in principal be deduced from contributions describing the single-particle self-energies as well as medium modifications of the interaction and the vertices. A guiding principle in incorporating medium effects is the construction of *consistent* ("conserving") approximations, which treat medium corrections in the self-energy and in the interaction vertex at the same level of accuracy. This can be achieved in a systematic way using the Green functions formalism [4] [5]. At the mean-field level, we have only the Hartree-Fock self-energy $\Gamma^{\rm HF} = \sum_2 V(12,12)_{\rm ex} f(2)$ together with the Pauli blocking factors, which modify the interaction from $V(12,1'2')$ to $V(12,1'2')[1-f(1)-f(2)]$ with the $f(i)$ the Fermi-Dirac distribution. In the case of the two-nucleon system ($A=2$), the resulting effective wave equation which includes those corrections reads

$$\left[E^{\rm HF}(1) + E^{\rm HF}(2) - E_{2,n,P}\right] \psi_{2,n,P}(12) + \sum_{1'2'} [1 - f(1) - f(2)]\, V(12, 1'2')\psi_{2,n,P}(1'2') = 0. \quad (1)$$

This *effective wave equation* describes bound states as well as scattering states. The onset of pair condensation is achieved when the binding energy $E_{d,P=0}$ coincides with 2μ.

Similar equations have been derived from the Green function approach for the case $A = 3$ and $A = 4$, describing triton/helion (^{3}He) nuclei as well as α-particles in nuclear matter. The effective wave equation contains in mean field approximation the Hartree-Fock self-energy shift of the single-particle energies as well as the Pauli blocking of the interaction. We give the effective wave equation for $A = 4$,

$$\begin{aligned} &\left[E^{\rm HF}(1) + E^{\rm HF}(2) + E^{\rm HF}(3) + E^{\rm HF}(4) - E_{4,n,P}\right] \psi_{4,n,P}(1234) \\ &+ \sum_{1'2'} [1 - f(1) - f(2)] V(12, 1'2')\psi_{4,n,P}(1'2'34) + \text{permutations} = 0. \end{aligned} \quad (2)$$

A similar equation is obtained for $A = 3$.

The effective wave equation has been solved using separable potentials for $A = 2$ by integration. For $A = 3, 4$ we can use a *Faddeev approach* [6]. The shifts of binding energy can also be calculated approximately via perturbation theory. In Fig. 1 we show the shift of the binding energy of the light clusters ($d, t/h$ and α) in symmetric nuclear matter as a function of density for temperature $T = 10$ MeV.

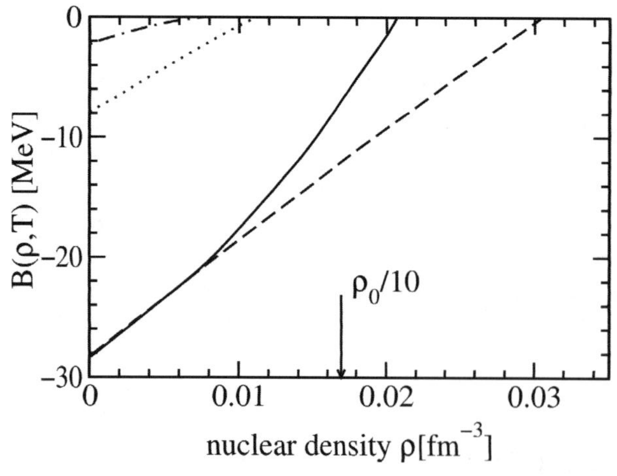

Figure 1. Shift of binding energies of the light clusters (d - dash dotted, t/h - dotted, and α - dashed: perturbation theory, full line : non-perturbative Faddeev-Yakubovski equation) in symmetric nuclear matter as a function of density for given temperature $T = 10$ MeV.

It is found that the cluster binding energy decreases with increasing density. Finally, at the *Mott density* $\rho^{\rm Mott}_{A,n,P}(T)$ the bound state is dissolved. The clusters are not present at higher densities, merging into the nucleonic medium. For a given cluster type characterized by A, n, we can also introduce the Mott momentum $P^{\rm Mott}_{A,n}(\rho, T)$ in terms of the ambient temperature T and nucleon density ρ, such that the bound states exist only for $P \geq P^{\rm Mott}_{A,n}(\rho, T)$.

3. Four-particle condensates and quartetting in nuclear matter

In general, it is necessary to take account of *all bosonic clusters* to gain a complete picture of the onset of superfluidity. As is well known, the deuteron is weakly bound compared to other nuclei. Higher A-clusters can arise that are more stable. In this section, we will consider the formation of α-particles, which are of special importance because of their large binding energy per nucleon (7 MeV). We will not include tritons or helions, which are fermions and not so tightly bound. Moreover, we will not consider nuclei in the iron region, which have even larger binding energy per nucleon than the α-particle and thus constitute, in principle, the dominant component at low temperatures and densities. However, the latter are complex structures of many particles and are strongly affected by the medium as the density increases, since their excitation spectrum starts at much lower energy than the one of the α particle, for instance. So they are assumed not to be of relevance in the density region considered here.

The in-medium wave equation for the four-nucleon problem has been solved using the Faddeev-Yakubovski technique, with the inclusion of Pauli blocking. The binding energy of an α-like cluster with zero c.o.m. momentum vanishes at around $\rho_0/10$, where $\rho_0 \simeq 0.16$ nucleons/fm^3 denotes the saturation density of isospin-symmetric nuclear matter, see Fig. 1. Thus, the four-body bound states make no significant contribution to the composition of the system above this density. Given the medium-modified bound-state energy $E_{4,P}$, the bound-state contribution to the EOS is

$$\rho_4(\beta,\mu) = \sum_P \left[e^{\beta(E_{4,P}-2\mu_p-2\mu_n)} - 1 \right]^{-1} . \tag{3}$$

We will not include the contribution of the excited states or that of scattering states. Because of the large specific binding energy of the α particle, low-density nuclear matter is predominantly composed of α particles. This observation underlies the concept of α matter and its relevance to diverse nuclear phenomena.

As exemplified by Eq. (2), the effect of the medium on the properties of an α particle in mean-field approximation (i.e., for an uncorrelated medium) is produced by the Hartree-Fock self-energy shift and Pauli blocking. The shift of the α-like bound state has been calculated using perturbation theory [7] as well as by solution of the Faddeev-Yakubovski equation [6]. It is found that this bound state merges with the continuum of scattering states at a Mott density $\rho_\alpha^{\mathrm{Mott}} \approx \rho_0/10$, see Fig. 1. The bound states of clusters d, t, and h with $A < 4$ are already dissolved at the density $\rho_\alpha^{\mathrm{Mott}}$. Consequently, if we neglect the contribution of the four-particle scattering phase shifts in the different channels, we can now construct an equation of state $\rho(T,\mu) = \rho^{\mathrm{free}}(T,\mu) + \rho^{\mathrm{bound},d}(T,\mu) + \rho^{\mathrm{bound},\alpha}(T,\mu)$ such that α-particles determine the behavior of symmetric nuclear matter at densities below $\rho_\alpha^{\mathrm{Mott}}$ and temperatures below the binding energy per nucleon of the α-particle. The formation of deuteron clusters alone gives an incorrect description because the deuteron binding energy is small, and the abundance of d-clusters is small compared with that of α-clusters. In the low density region of the phase diagram, α-matter emerges as an adequate model for describing the nuclear-matter equation of state.

With increasing density, the medium modifications – especially Pauli blocking – will lead to a deviation of the critical temperature $T_c(\rho)$ from that of an ideal Bose gas of α-particles (the analogous situation holds for deuteron clusters, i.e., in the isospin-singlet channel).

Symmetric nuclear matter is characterized by the equality of the proton and neutron chemical potentials, i.e., $\mu_p = \mu_n = \mu$. Then an extended Thouless condition based on the relation for the four-body T-matrix, equivalent to Eq. (2) at eigenvalue 4μ serves to determine the onset of Bose condensation of α-like clusters, noting that existence of a solution of this relation signals a divergence of the four-particle correlation function.

An approximate solution has been obtained by a variational approach, in which the wave function is taken of the following form [8] (see also [9] for another variational form):

$$\psi(1234) = \delta(\mathbf{k}_1 + \mathbf{k}_2 + \mathbf{k}_3 + \mathbf{k}_4)\varphi(\mathbf{k}_1)\varphi(\mathbf{k}_2)\varphi(\mathbf{k}_3)\varphi(\mathbf{k}_4)\chi(1234) \tag{4}$$

where the φ's are single particle wave functions in momentum space to be determined variationally out of the in medium four body wave equation and $\chi(1234)$ is the scalar spin-isospin function. The delta function is a projector on total momentum zero.

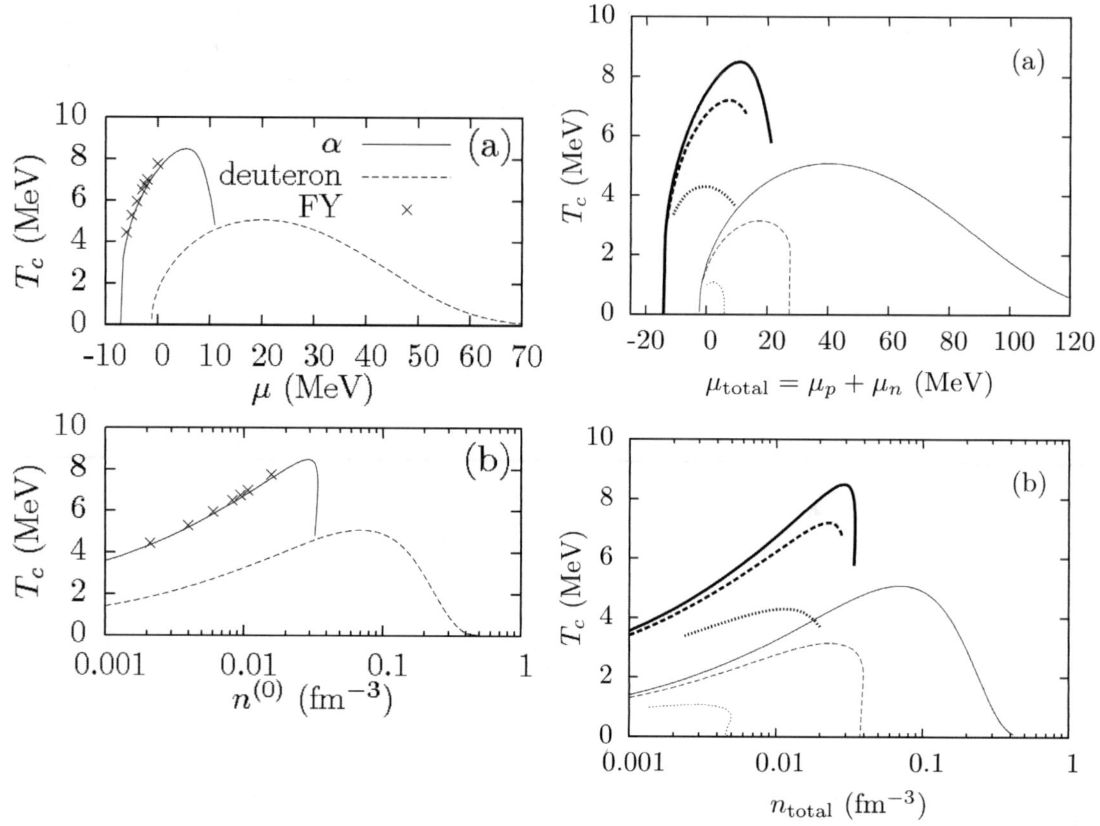

Figure 2. (Left) Critical temperatures for α particle and deuteron condensation in symmetric nuclear matter as a function of μ (a) and density $n^{(0)}$ (b). (Right) Same as in Fig.2 but for asymmetric nuclear matter for various asymmetry parameters $\delta = (n_n - n_p)/(n_n + n_p)$. $\delta = 0$ (full line); $\delta = 0.5$ (broken line); $\delta = 0.9$ (dotted line).

The results for symmetric and asymmetric nuclear matter with a separable force which reproduces binding energy and radius of a free α particle are presented in Figs. 3 and 3. The crosses on the figure indicate an exact solution of the in medium four body equation with a realistic nucleon-nucleon force [8] [10]. The excellent agreement with the approximate solution justifies the choice of our variational wave function. An important feature is that at the lowest temperatures, Bose-Einstein condensation occurs for α particles rather than for deuterons. As the density increases within the low-temperature regime, the chemical potential μ first reaches -7 MeV, where the α's Bose-condense. By contrast, Bose condensation of deuterons would not occur until μ rises to -1.1 MeV. In the asymmetric case, it is interesting to note that at strong asymmetry α particle condensation wins over the deuteron condensation because of the strong binding of the α's.

The *"quartetting"* transition temperature sharply drops as the rising density approaches the critical Mott value (μ 0) at which the four-body bound states disappear. At that point, pair formation in the isospin-singlet deuteron-like channel comes into play, and a deuteron condensate will exist below the critical temperature for BCS pairing up to densities above the nuclear-matter

Journal of Physics: Conference Series **413** (2013) 012009 doi:10.1088/1742-6596/413/1/012009

saturation density ρ_0, as described in the previous Section. The critical density at which the α condensate disappears is estimated to be $\rho_0/5$. Therefore, α- particle condensation primarily only exists in the Bose-Einstein-Condensed (BEC) phase and there does not seem to exist a phase where the quartets acquire a large extension as Cooper pairs do in the weak coupling regime. However, our variational approach on which this conclusion is based represents only a first attempt at the description of the transition from quartetting to pairing. The detailed nature of this fascinating transition remains to be clarified. Further discussions on this phenomenon can be found in [11].

A very intriguing question in relation with non existence of an α condensate at higher densities can be asked: is it possible that in heavy nuclei an α particle condensate exists in the nuclear surface, at least in its fluctuating form? The preformation assumption of α particles in the surface to explain α decay may give a hint to this.

4. Alpha-Particle Condensate States in Self-Conjugate 4n Nuclei

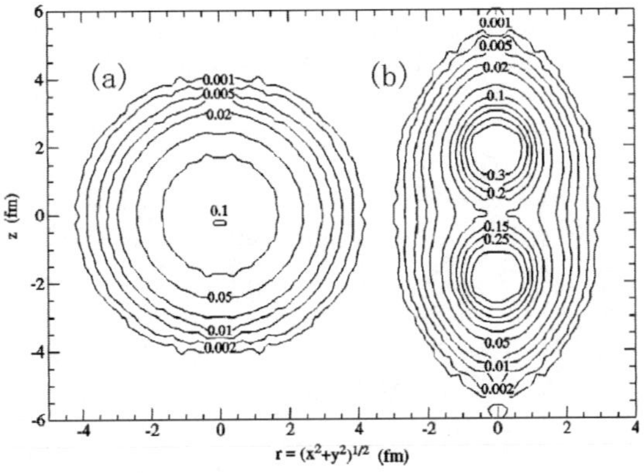

Figure 3. Contours of constant density (taken from Ref. [12]), plotted in cylindrical coordinates, for ^{8}Be(0^+). The left side (a) is in the 'laboratory" frame while the right side (b) is in the intrinsic frame.

Let us discuss the possibility of quartetting in nuclei. The only nucleus having a pronounced α-cluster structure in its ground state is ^{8}Be. In Fig. 3(a), we show the result of an exact calculation of the density distribution of ^{8}Be in the laboratory frame. In Fig. 3(b) we show, for comparison, the result of the same calculation in the intrinsic, deformed frame. We see a pronounced two α-cluster structure where the two α's are $\sim$ 4 fm apart, giving rise to a very low average density $\rho \sim \rho_0/3$ as seen in Fig. 3(a). As already discussed in the introduction, ^{8}Be is a rather unusual and unique nucleus. One may be intrigued by the question, already raised earlier, whether loosely bound α-particle configurations may not also exist in heavier nα-nuclei, at least in excited states, naturally close to the nα disintegration threshold. Since α-particles are rather inert bosons (first excited state at $\sim$ 20 MeV), these α-particles then would all condense in the lowest S-wavefunction, very much in the same way as do bosonic atoms in magneto-optical traps [13]. This question and exploring related issues of quartetting in finite nuclei will consume most of the rest of the present outline on α particle condensation. In fact, we will be able to offer strong arguments that the 0_2^+ state of ^{12}C at 7.654 MeV is a state of α-condensate nature.

First, it should be understood that the 0_2^+ state in ^{12}C is in fact hadronically unstable (as ^{8}Be), being situated about 300 keV above the three α-break up threshold. This state is stabilized only by the Coulomb barrier. It has a width of 8.7 eV and a corresponding lifetime of 7.6×10^{-17} s. As well known, this state is of paramount astrophysical (and biological!) importance due to its role in the creation of ^{12}C in stellar nucleosynthesis. Its existence was predicted in 1953 by the astrophysicist Fred Hoyle [14]; his prediction was confirmed experimentally a few years later by Willy Fowler and coworkers at Caltech [15]. It is also well known that this *Hoyle state*, as it is called, is a notoriously difficult state for any nuclear theory to explain. For example, the most

Journal of Physics: Conference Series **413** (2013) 012009 doi:10.1088/1742-6596/413/1/012009

modern no-core shell-model calculations predict the 0_2^+ state in ^{12}C to lie at around 17 MeV above the ground state – more than twice the actual value [16]. This fact alone tells us that the Hoyle state must have a very unusual structure. It is easy to understand that, should it indeed have the proposed loosely bound three α-particle structure, a shell-model type of calculation would have great difficulties in reproducing its properties.

An important development bearing on this issue took place some thirty years ago. Two Japanese physicists, M. Kamimura [17] and K. Uegaki [18], along with their collaborators, almost simultaneously reproduced the Hoyle state from a microscopic theory. They employed a twelve-nucleon wave function together with a Hamiltonian containing an effective nucleon-nucleon interaction. At that time, their work did not attract the attention it deserved; the true importance of their achievement has been appreciated only recently. The two groups started from practically the same ansatz for the ^{12}C wave function, which has the following three α-cluster structure:

$$\langle \mathbf{r}_1 ... \mathbf{r}_{12} | ^{12}\mathrm{C} \rangle = \mathcal{A} \left[\chi(\mathbf{R}, \mathbf{s}) \phi_1 \phi_2 \phi_3 \right] . \tag{5}$$

In this expression, the operator $\mathcal{A}$ imposes antisymmetry in the nucleonic degrees of freedom and ϕ_i, with $i = 1, 2, 3$ for the three α's, is an intrinsic α-particle wave function of prescribed Gaussian form,

$$\phi(\mathbf{r}_1, \mathbf{r}_2, \mathbf{r}_3, \mathbf{r}_4) = \exp\left\{ - \left[(\mathbf{r}_1 - \mathbf{r}_2)^2 + (\mathbf{r}_1 - \mathbf{r}_3)^2 + ... \right] / 2b^2 \right\} , \tag{6}$$

where the size parameter b is adjusted to fit the rms of the free α-particle, and $\chi(\mathbf{R}, \mathbf{s})$ is a yet-to-be determined three-body wave function for the c.o.m. motion of the three α's, their corresponding Jakobi coordinates being denoted by $\mathbf{R}$ and $\mathbf{s}$. The unknown function χ was determined via calculations based on the Generator Coordinate Method [18] (GCM) and the Resonating Group Method [17] (RGM) calculations using the Volkov I and Volkov II nucleon-nucleon forces, which fit α-α phase shifts. The precise solution of this complicated three body problem, carried out three decades ago, was truly a pioneering achievement, with results fulfilling expectations. The position of the Hoyle state, as well as other properties including the inelastic form factor and transition probability, successfully reproduced the experimental data. Other states of ^{12}C below and around the energy of the Hoyle state were also successfully described. Moreover, it was already recognized that the three α's in the Hoyle state form sort of a gas-like state. In fact, this feature had previously been noted by H. Horiuchi [19] prior to the appearance of Refs. [17, 18], based on results from the orthogonality condition model (OCM) [20]. All three Japanese research groups concluded from their studies that the linear-chain state of three α-particles, postulated by Morinaga many years earlier [21], had to be rejected.

Although the evidence for interpreting the Hoyle state in terms of an α gas was stressed in the cited papers from the late 1970's, two important aspects of the situation were missed at that time. First, because the three α's move in identical S-wave orbits, one is dealing with an α-condensate state what may, in fact, be a general feature in $n\alpha$ nuclei. The second and most important point is that the complicated three-body wave function $\chi(\mathbf{R}, \mathbf{s})$ for the c.o.m. motion of the three α's can be replaced by a structurally and conceptually very simple microscopic three-α wave function of the condensate type, which has practically 100 percent overlap with the previously constructed ones [22] [23] (see also Ref. [24]). We now describe this condensate wave function.

We start by examining the BCS wave function of ordinary fermion pairing, obtained by projecting the familiar BCS ground-state ansatz onto an N-particle subspace of Fock space. In the position representation, this wave function is

$$\langle \mathbf{r}_1 ... \mathbf{r}_N | \mathrm{BCS} \rangle = \mathcal{A} \left[\phi(\mathbf{r}_1, \mathbf{r}_2) \phi(\mathbf{r}_3, \mathbf{r}_4) ... \phi(\mathbf{r}_{N-1} \mathbf{r}_N) \right] , \tag{7}$$

where $\phi(\mathbf{r}_1, \mathbf{r}_2)$ is the Cooper-pair wave function (including spin and isospin), which is to be determined variationally through the familiar BCS equations. The condensate character of the BCS ansatz is born out by the fact that within the antisymmetrizer $\mathcal{A}$, one has a product of $N/2$ times the same pair wave function ϕ, with one such function for each distinct pair in the reference partition of $\{1, 2, \ldots, N-1, N\}$. Formally, it is now a simple matter to generalize (7) to quartet or α-particle condensation. We write

$$\langle \mathbf{r}_1, \ldots, \mathbf{r}_N | \Phi_{n\alpha} \rangle = \mathcal{A}\left[\phi_\alpha(\mathbf{r}_1, \mathbf{r}_2, \mathbf{r}_3, \mathbf{r}_4)\phi_\alpha(\mathbf{r}_5, \ldots, \mathbf{r}_8) \cdots \phi_\alpha(\mathbf{r}_{N-3}, \ldots, \mathbf{r}_N)\right], \quad (8)$$

where ϕ_α is the wave function common to all condensed α-particles. Of course, finding the variational solution for this function is, in general, extraordinarily more complicated than finding the Cooper pair-wave function ϕ of Eq. (7). Even so, in the present case that the α-particle is the four-body cluster involved, and for applications to relatively light nuclei, the complexity of the problem can be reduced dramatically. This possibility stems from the fact that an excellent variational ansatz for the intrinsic wave function of the α-particle is provided [as in Eq. (6)], by a Gaussian form with only the size parameter b to be determined. The new aspect in [22] is that in addition even the center-of-mass motion of the system of α-particles can be described very well by a Gaussian wave function with, this time, a size parameter $B \gg b$ to account for the motion over the nuclear space. This is a strong technical simplification and, at the same time, underlines the boson condensate character of the wave function. We therefore write

$$\phi_\alpha(\mathbf{r}_1, \mathbf{r}_2, \mathbf{r}_3, \mathbf{r}_4) = e^{-2\mathbf{R}^2/B^2} \phi(\mathbf{r}_1 - \mathbf{r}_2, \mathbf{r}_1 - \mathbf{r}_3, \cdots), \quad (9)$$

where $\mathbf{R} = (\mathbf{r}_1 + \mathbf{r}_2 + \mathbf{r}_3 + \mathbf{r}_4)/4$ is the c.o.m. coordinate of one α-particle and $\phi(\mathbf{r}_1 - \mathbf{r}_2, ...)$ is the same intrinsic α-particle wave function of Gaussian form as already used in Refs. [17, 18] and given explicitly in Eq. (6). Naturally, in Eq. (8) the center of mass $\mathbf{X}_{\rm cm}$ of the three α's, i.e., of the whole nucleus, should be eliminated; this is easily achieved by replacing $\mathbf{R}$ by $\mathbf{R} - \mathbf{X}_{\rm cm}$ in each of the α wave functions in Eq. (8). The α-particle condensate wave function specified by Eqs. (8) and (9), proposed in Ref. [22] and henceforth called the THSR wave function, now depends on only two parameters, B and b. The expectation value of an assumed microscopic Hamiltonian H,

$$\mathcal{H}(B, b) = \frac{\langle \Phi_{n\alpha}(B, b) | H | \Phi_{n\alpha}(B, b) \rangle}{\langle \Phi_{n\alpha} | \Phi_{n\alpha} \rangle}, \quad (10)$$

can be evaluated, and the corresponding two-dimensional energy surface can be quantized using the two parameters B and b as Hill-Wheeler coordinates.

Before presenting the results, let us discuss the THSR wave function in somewhat more detail. This innocuous-looking variational ansatz, namely Eq. (8) together with Eq. (9), is actually more subtle than it might at first appear. One should realize that two limits are incorporated exactly. One is obtained by choosing $B = b$, for which Eq. (8) reduces to a standard Slater determinant with harmonic-oscillator single-nucleon wave functions, leaving the oscillator length b as the single adjustable parameter. This holds because the right-hand-side of expression (9), with $B = b$, becomes a product of four identical Gaussians, and the antisymmetrization creates all the necessary P, D, etc. harmonic oscillator wave functions automatically [22]. On the other hand, when $B \gg b$, the density of α-particles is very low, and in the limit $B \to \infty$, the average distance between α's is so large that the antisymmetrisation between them can be neglected, i.e., the operator $\mathcal{A}$ in front of Eq. (8) becomes irrelevant and can be removed. In this limiting case, our wave function then describes an ideal gas of independent, condensed α-particles – it is a pure product state of α's! An elucidating study on this aspect is given in Ref. [25].
Evidently, however, in realistic cases the antisymmetrizer $\mathcal{A}$ cannot be neglected, and evaluation of the expectation value (10) becomes a nontrivial analytical task.

		condensate w.f. (Hill-Wheeler)	RGM [17]	Exp.
E(MeV)	0_1^+	−89.52	−89.4	−92.2
	0_2^+	−81.79	−81.7	−84.6
$R_{\text{r.m.s.}}$(fm)	0_1^+	2.40	2.40	2.44
	0_2^+	3.83	3.47	
$M(0_2^+ \to 0_1^+)$(fm^2)		6.45	6.7	5.4

Table 1. Comparison of the binding energies, rms radii ($R_{\text{r.m.s.}}$), and monopole matrix elements ($M(0_2^+ \to 0_1^+)$) for ^{12}C given by solving Hill-Wheeler equation based on Eq. (8) and by Ref. [17]. The effective two-nucleon force Volkov No. 2 was adopted in the two cases for which the 3α threshold energy is calculated to be −82.04 MeV.

The Hamiltonian in Eq. (10) was taken to be the one used in Ref. [26], which features an effective nucleon-nucleon force of the Gogny type, with parameters fitted to α-α scattering phase shifts as available about fifteen years ago. This force also leads to very reasonable properties of ordinary nuclear matter. Our theory is therefore free of any adjustable parameters. The energy landscapes $\mathcal{H}(B, b)$ for various $n\,\alpha$ nuclei are interesting in themselves [27], but for the sake of brevity they are not shown here.

As we discussed already, the variational wave function constructed from the Hill-Wheeler equation based on Eqs. (8), (9), and (10) has practically 100 percent overlap with the RGM and GCM wave functions constructed in Refs. [17] and [18], once the same Volkov force is used [23]. However, one can take fixed optimised values for b and B parameters and, then, the corresponding single THSR wave function still has 98 percent squared overlap with the RGM or GCM solutions. It is, thus, not astonishing that our results are very similar to theirs. For ^{12}C we obtain two eigenvalues: the ground state and the Hoyle state. Theoretical values for positions, rms values, and transition probabilities are given in Table 1 and compared to the data. Inspecting the rms radii, we see that the Hoyle state has a volume 3 to 4 larger than that of the ground state of ^{12}C. This is the primary aspect of the dilute-gas state we highlighted above. Constructing a pure-state α-particle density matrix $\rho(\mathbf{R}, \mathbf{R}')$ from our wave function, integrating out of the total density matrix all intrinsic α-particle coordinates, and diagonalizing this reduced density matrix, we find that the corresponding $0S$ α-particle orbit is occupied to 70 percent by the three α-particles [25, 28] whereas the occupation of all other states is down by at least a factor of ten. This is a huge percentage, giving vivid support to the view that the Hoyle state is an almost ideal α-particle condensate. By way of contrast, we should also mention that in the ground state of ^{12}C, the α-particle occupation is about equally shared between the $0S$, $0D$, and $0G$ orbits, clearly invalidating a condensate picture of the ground state (it is important to note that the ground-state energy of ^{12}C is also reasonably reproduced by our theory).

Let us now discuss what to our mind is the most convincing evidence that our description of the Hoyle state is the correct one. Like the authors of Ref. [17], we reproduce very accurately the inelastic form factor $0_1^+ \to 0_2^+$ of ^{12}C, as shown in Fig. 4. As such, the agreement with experiment is already quite impressive in view of the fact that we did not use a single adjustable parameter. Additionally, however, the following study was made, results from which are presented in Fig. 5. We artificially varied the extension of the Hoyle state and examined the influence on the form factor. It was found that the overall shape of the form factor shows little variation, for example in the position of the minimum. On the other hand, we found a strong dependence of the absolute magnitude of the form factor; Fig. 5 illustrates this behavior with a plot showing the

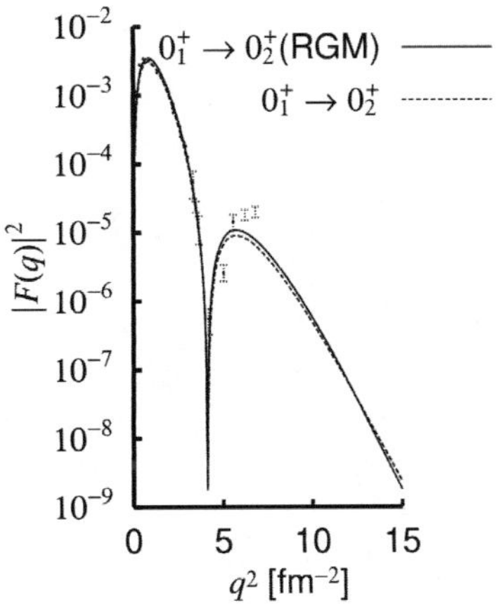

Figure 4. Experimental values of inelastic form factor in ^{12}C to the Hoyle state are compared with our values and those given by Kamimura et al. in Ref. [17] (RGM). In our result, the Hoyle-state wave function is obtained by solving the Hill-Wheeler equation based on Eq. (8).

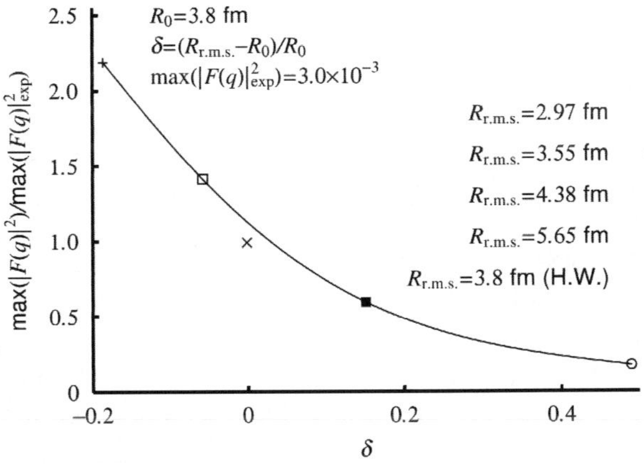

Figure 5. The ratio of the value of the maximum height, theory versus experiment, of the inelastic form factor, i.e. $\max|F(q)|^2/\max|F(q)|^2_{\rm exp}$, is plotted as a function of $\delta = (R_{\rm r.m.s.} - R_0)/R_0$. Here $R_{\rm r.m.s.}$ and R_0 are the rms radii corresponding, respectively, to the wave function of Eq. (8) and that obtained by solving the Hill-Wheeler equation based on Eq. (8).

variation of the height of the first maximum of the inelastic form factor as a function of the percentage change of the rms radius of the Hoyle state [23]. It can be seen that a 20 percent *increase of the rms radius* produces a remarkable decrease of the maximum – by a factor of two! This strong sensitivity of the magnitude of the form factor to the size of the Hoyle state enhances our firm belief that the agreement with the actual measurement is tantamount to a proof that the calculated wide extension of the Hoyle state corresponds to reality.

We thus advocate and support the view that the Hoyle state can be regarded as the ground state of an α-particle condensate [29]. We should, however, be aware of the fact that it is not an ideal condensate and that the α's occupy the lowest S-state only with 70 percent, as already discussed. The major part of the correlations comes from th Pauli principle and intermediate 8Be formation. It is by the way not clear whether a gas of α particles condenses as such or as 'molecules' of 8Be. The latter are also bosons, of course. We furthermore performed a deformed calculation to investigate the structure of the 2_2^+ state in ^{12}C. The state came at the right energy. Our analysis showed that this state essentially corresponds to exciting one α-particle out of the condensate and putting it into the $0D$ orbit. Without going into details, we also affirm that the width of this state is correctly reproduced [30]. It should also be mentioned that this 2_2^+ state has in our calculations an enormous extension with an rms radius of 4.3 fm what corresponds approximately to eight times the ground state volume of ^{12}C or also to the size of ^{40}Ca. As a matter of fact the properties of this state have been subject of a vivid debate among the experimentalists in the recent past. The situation seems clarified now [31][32][33]. It is for instance, the very nice experiment by Moshe Gai [33] which seems to confirm the properties of the 2_2^+ state beyond any doubt. In that reference it also is given an estimate of the radius which agrees with our value. One can talk about an α halo state.

It is tempting to imagine that the 0_3^+ state which – experimentally – is almost degenerate with the 2_2^+ state, is obtained by lifting one α-particle into the $1S$ orbit. Initial theoretical studies [34] indicate that this scenario might indeed apply. However, the width of the 0_3^+ state ($\sim$ 3 MeV) is very broad, rendering a theoretical treatment rather delicate. Further investigations are necessary to validate or reject this picture. At any rate, it would be quite satisfying if the triplet of states ($0_2{}^+$, $2_2{}^+$, $0_3{}^+$) could all be explained from the α-particle perspective, since those three states are *precisely* the ones which cannot be reproduced within a (no core) shell-model approach [16].

Summarizing our inquiry into the possible role of α clustering in ^{12}C, we have accumulated enough facts to be convinced that the Hoyle state is, indeed, what one may call to first approximation an α-particle condensate state. At the same time, we acknowledge that referring to only three particles as a "condensate" constitutes a certain abuse of the word. However, in this regard it should be remembered that also in the case of nuclear Cooper pairing, only a few pairs are sufficient to obtain clear signatures of superfluidity in nuclei!

What about α-particle condensation in heavier nuclei? Once one accepts the idea that the Hoyle state is essentially a state of three free α-particles held together only by the Coulomb barrier, it is hard to see why analogous states would not also exist in heavier $n\,\alpha$ nuclei like ^{16}O, ^{20}Ne, ^{24}Mg, etc. In fact, our calculations on such nuclei systematically yield a 0^+-state close to the α-particle disintegration threshold. For example in ^{16}O we obtain three 0^+-states [22]: the ground state at $E_0 = -124.8$ MeV (experimental value: 127.62 MeV), a second state at excitation energy $E_{0_2{}^+} = 8.8$ MeV, and a third one at $E_{0_3{}^+} = 14.1$ MeV. The threshold in ^{16}O is at 14.4 MeV. Unfortunately, the relevant experimental information in ^{16}O is not nearly so complete as in ^{12}C. In particular, no measurements are available for transition probabilities of 0^+-states near the threshold or for inelastic form factors.

In contrast to the situation for ^{12}C, the THSR wave function is certainly not able to describe the structure of all 0^+-states in ^{16}O lying below the disintegration threshold. In ^{12}C knocking loose one α particle, the other two are also loosely bound, since what remains is ^{8}8Be. However, in ^{16}O this is not the case. Before reaching a four α particle gas state, there will appear configurations where one α particle orbits around a ^{12}C core in its ground state or in excited states of particle- hole type. A case in point is the first excited state in ^{16}O, i.e., the $0_2{}^+$-state at 6.06 MeV, which is believed to have a structure corresponding to an α-particle orbiting in an S wave around a ^{12}C core in its ground state. Such a configuration is clearly missing from our wave function (8). As a matter of fact a calculation for the first six 0^+ states has been performed in the mean while employing a somewhat different, not completely microscopic approach [35]. This method is the so-called 'orthogonality condition model (OCM)' which generally works also quite well for the description of cluster states. The novelty with respect to former calculations of ^{16}O states with this method was that the configuration space was strongly enlarged. In Fig. 6, we show the comparison of the calculated with the experimental sprectrum. In view of the fact that the states are complicated cluster states, the agreement between experiment and theory is very good. The interpretation goes as follows: the first four excited 0^+ states have a 12C + α structure where the α orbits in 0S, 0D, 1S waves around the ground state core of 12C and also in a 0P wave orbiting around the first 1^- of 12C. It is only the first state above the four α particle threshold at 15.1 MeV which is interpreted as an α gas state or a condensate. This state has, indeed, some analogies with the Hoyle state: it is just some hundreds of keV above threshold, it is strongly excited by inelastic electron scattering what means that monopole transition is quite large. Unfortunately the inelastic form factor has not been measured so far.

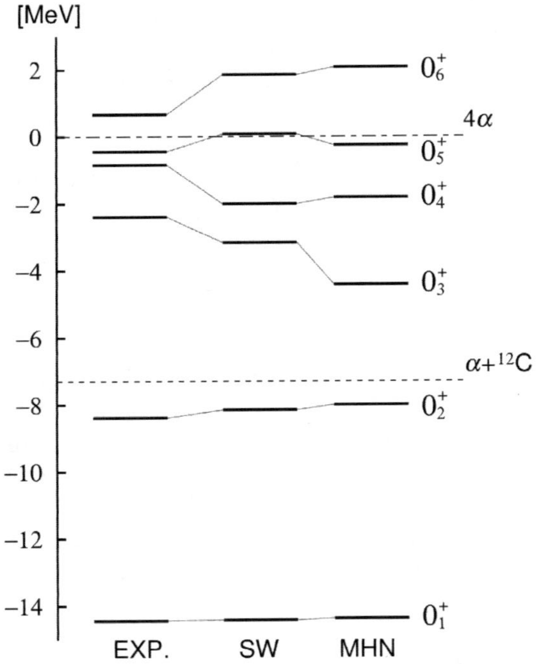

Figure 6. Spectrum of first six 0^+ states in ^{16}O. Left:experiment; middle: theory with SW force; right: theory with MHN force; see [35] for more details.

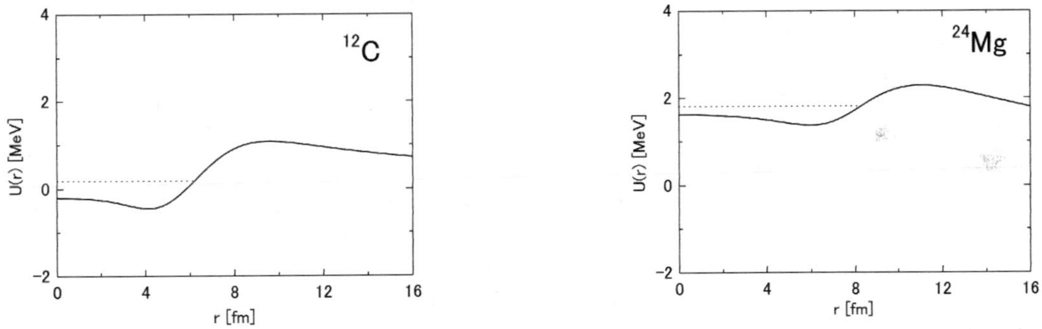

Figure 7. Alpha-particle mean-field potential for three α's in ^{12}C and six α's in ^{24}Mg. Note the lower Coulomb barrier for ^{24}Mg (from Ref. [38]).

One interesting question that can be asked at this point is: How many α's can maximally exist in a self-bound α-gas state? Seeking an answer, we performed a schematic investigation using an effective α-α interaction of the Ali-Bodmer form [36] within an α-gas mean-field calculation of the Gross-Pitaevskii type [37]. The parameters of the force were slightly adjusted to reproduce our microscopic results for ^{12}C. The corresponding α mean-field potential is shown in Fig. 4. One sees the $0S$-state lying slightly above threshold but below the Coulomb barrier. As more α-particles are added, the Coulomb repulsion drives the loosely bound system of α-particles farther and farther apart, so that the Coulomb barrier fades away. According to our estimate [38], a maximum of eight to ten α-particles can be held together in a condensate. However, there may be ways to lend additional stability to such systems. We know that in the case of ^{8}Be, adding one or two neutrons produces extra binding without seriously disturbing the pronounced α-cluster structure. Therefore, one has reason to speculate that adding a few of neutrons to a many-α state may stabilize the condensate. But again, state-of-the-art microscopic investigations are necessary before anything definite can be said about how extra neutrons will influence an α-particle condensate.

Another interesting idea concerning α-particle condensates was put forward by von Oertzen and collaborators [39, 40]. Adding more and more α-particles to the, e.g., ^{40}Ca core, one will arrive sooner or later at the point of α-particle drip. Therefore minimal further excitation may be sufficient to shake loose some α-particles, so that an $n\,\alpha$-condensate could be created on top of an inert ^{40}Ca core. Similar ideas also have been advanced by Ogloblin [41], who envisions a three-α-particle condensate on top of ^{100}Sn, and earlier by Brenner and Gridnev, who have presented evidence of experimental detection of gaseous α-particles in ^{28}Si and ^{32}S on top of an inert ^{16}O core [42].

5. Conclusions, Discussion, Outlook

We have investigated the role that pairing and multiparticle correlations may play in nuclear matter existing in dense astrophysical objects and in finite nuclei. A complete and quantitative description of nuclear matter must allow for the presence of clusters of nucleons, bound or metastable, possibly forming a quantum condensate. In particular, quartetting correlations, responsible for the emergence of α-like clusters, are identified as uniquely important in determining the behavior of nuclear matter in the limiting regime of low density and low temperature. We have calculated the transition temperature for the onset of quantum condensates made up of α-like and deuteron-like bosonic clusters, and considered in considerable detail the intriguing example of Bose-Einstein condensation of α particles. It turns out that contrary to pairing, quartet condensation primarily exists in the BEC phase at low density. In which way quartet condensation is lost by increasing the density is still an open question. It may be similar to a liquid-gas phase transition. Anyway, it is clear that there can not exist a condensate of quartets with a long coherence length for arbitrarily small attraction as this is the case for pairing in the BCS phase. It is inevitable that under increasing density or pressure, the bound α, d, or other nuclidic clusters present at low density experience significant modification due to the background medium (and eventually merge with it). We have shown how self-energy corrections and Pauli blocking alter the properties of cluster states, and we have formulated a cluster mean-field approximation to provide an initial description of this process. One result of special interest is the suppression of the α-like condensate, which is dominant at lower densities, as the density reaches and exceeds the Mott value, allowing the pairing transition to occur. Even at lower densities α-particle condensation may be influenced by neutron excess, i.e. in the case of asymmetric nuclear matter, see Fig. 3. A genuine theory for the quartet order parameter in homogeneous infinite matter, similar to BCS theory, is demanded. First results in this direction have been published [11].

A truly remarkable manifestation of α-particle condensation seems to be present in finite nuclei. Indeed, the so-called Hoyle state ($0_2{}^+$) in ^{12}C at 7.654 MeV is very likely a dilute gas of three α-particles, held together only by the Coulomb barrier. This view is encouraged by the fact that we can explain all the experimental data in terms of a conceptually simple wave function of the quartet-condensate type. Within the same model, we also systematically predict such states in heavier $n\,\alpha$ nuclei, and the search is on for their experimental identification. With the more phenomenological OCM method, we found that in ^{16}O the sixth 0^+ state at 15.1 MeV should be the candidate for an α condensed state. It is quite natural that such states should exist up to some maximum number of α particles. We estimate that the phenomenon will terminate at about eight to ten α's as the confining Coulomb barrier fades away. However, there is the possibility that larger condensates could be stabilized by addition of a few neutrons. Indeed, consider ^{9}Be, which, contrary to ^{8}Be, is bound by ~ 1.5 MeV, still showing a pronounced two α-structure similar to the one of Fig. 3 (b). One could imagine ten α's or more, stabilised by two or four extra neutrons in a low density phase. However, even without being stabilised, if a compressed hot nuclear blob as e.g. produced in a central Heavy Ion collision expands and cools, it may turn on its way out, at a certain low density, into an expanding α condensed state where all α's are in relative S-waves. One may also, in one way or the other (photons?) excite, e.g., ^{40}Ca to the 10 α threshold at about 60 MeV where a slow Coulomb explosion of an α gas would then take place. This would be an analogous situation to an expanding Bose condensate of atoms, once the trapping potential has been switched off. Future dedicated experiments with high resolution multiparticle detectors will tell whether such scenarios can be realised. Intriguing news in this respect come from GANIL where one may have achieved the disintegration of ^{56}Ni into 14 α's (seven α's have been detected with high yield, the other seven may not have been seen because of the detectors were not sensitive to very low energy α particles [43]). Other possibilities of loose α-gas states may exist on top of particularly stable cores, like ^{16}O or ^{40}Ca. Indeed in adding α's to e.g. ^{40}Ca, one will reach the α-particle drip line. Compound states of heavy N = Z nuclei of this kind may be produced in heavy ion reactions and an enhanced α-decay rate may reveal the existence of an α-particle condensate. Ideas of this type are presently promoted by von Oertzen [40], and also M. Brenner [42], and A. Ogloblin [41]. However, coincidence measurements of multiple α's of decaying lighter nuclei like ^{16}O may also be very useful [44][45] [46] to detect at least one additional α- condensate state beyond the only one that has been identified so far, namely the $0_2{}^+$-state in ^{12}C.

Another issue which may be raised in the context of α-particle condensation is the question, also discussed in condensed matter physics [47], whether α's condense as singles or as doubles, i.e. as ^{8}Be. In microscopic studies of ^{12}C one, indeed, can see that in the $0_2{}^+$-state two of the three α's are slightly more closer to one another than to the third one [48]. The question is definitely very interesting and desserves future investigation. However, quantitatively, it constitutes probably only a slight modification over the present formulation of α-condensation.

What about 'ab initio' calculations of the Hoyle state? Presently several groups are on the track [49][50]. The Los Alamos group has achieved to calculate the density of the Hoyle state, though it is not yet converged in the far tail [50]. It agrees very well with the density from the THSR wave function besides in the far tail.

In conclusion, we see that the idea of α-particle condensation in nuclei and nuclear systems has already triggered many new ideas ,calculations, and experiments, in spite of the fact that, so far, a compelling case for such a state has only been made in ^{12}C. Even so, the possible existence of a completely new nuclear phase in which α-particles play the role of quasi-elementary constituents is surely fascinating. Hopefully, many more α-particle states of nuclei will be detected in the near future, bringing deeper insights into the role of clustering and quantum condensates in systems of strongly interacting fermions.

Let us mention in the end that a more elaborate report on α particle condensation can be found in [51]

6. Acknowledgements

We thank T. Sogo for discussions and contributions.

References

[1] Shapiro L S and Teukolsky S A 1983 *Black holes, white Dwarfs and Neutron Stars: The Physics of Compact Objects* (Wiley,NY); Pines D Tamagaki R and Tsuruta S (eds.) 1992 *Neutron Stars* (Addison-Wesley,NY)
[2] Shen H, Toki H, Oyamatsu K and Sumiyoshi K 1998 *Progr. Theor. Phys.* **100** 1013
[3] Salomon Ch Zwierlein M W 2012 private communications.
[4] Kraeft W D, Kremp D, Ebeling W and Röpke G 1986 *Quantum Statistics of Charged Particle Systems* (Berlin, Akademie-Verlag); Dukelsky J, Röpke G and Schuck P 1998 *Nucl. Phys.* A **628** 17
[5] Dukelsky J, Röpke G and Schuck P 1998 *Nucl. Phys.* A **628** 17
[6] Beyer M, Sofianos S A, Kurths C, Röpke G and Schuck P 2000 *Phys. Lett.* B **478** 86
[7] Röpke G, Münchow L and Schulz H 1982 *Nucl. Phys.* A **379** 536
[8] Sogo T, Lazauskas R, Röpke G and Schuck P 2009 *Phys. Rev.* C **79** 051301

[9] Röpke G, Schnell A, Schuck P and Nozières P 1998 *Phys. Rev. Lett.* **80** 3177
[10] Sogo T, Röpke G and Schuck P 2010 *Phys. Rev.* C **82** 034322
[11] Sogo T, Röpke and G Schuck P 2010 *Phys. Rev.* C **81** 064310
[12] Wiringa R B S, Pieper C, Carlson J and Pandharipande V R 2000 *Phys. Rev.* C **62** 014001
[13] Dalfovo L, Giorgini S, Pitaevskii L D and Stringari S 1999 *Rev. Mod. Phys.* **71** 463
[14] Hoyle F, Dunbar D N F, Wenzel W A and Whaling W 1953 *Phys. Rev.* **92** 1095
[15] Cook C W, Fowler W A, Lauritsen C C and Lauritsen T B 1957 *Phys. Rev.* **107** 508
[16] Navrátil P, Vary J P and Barrett B R 2000 *Phys. Rev. Lett.* **84** 5728; Navrátil P, Vary J P and Barrett B R 2000 *Phys. Rev.* **C 62** 054311; Barrett B R, Mihaila B, Pieper S C and Wiringa R B 2003 *Nucl. Phys. News* **13** 17
[17] Fukushima Y and Kamimura M 1978 *Proc. Int. Conf. on Nuclear Structure* Ed. Marumori T *Suppl J. Phys. Soc. Japan* **44** 225; Kamimura M 1981 *Nucl. Phys.* A **351** 456
[18] Uegaki E, Okabe S, Abe Y and Tanaka H 1977 *Prog. Theor. Phys.* **57** 1262; Uegaki E, Abe Y, Okabe S and Tanaka H 1978 *Prog. Theor. Phys.* **59** 1031; 1979 **62**, 1621
[19] Horiuchi H 1974 *Prog. Theor. Phys.* **51** 1266; 1975 **53** 447
[20] Saito S 1968 *Prog. Theor. Phys.* **40**; 1969 **41** 705 1977 *Prog. Theor. Phys. Suppl.* **62** 11
[21] Morinaga H 1956 *Phys. Rev.* **101** 254 1966 *Phys. Lett.* **21** 78
[22] Tohsaki A, Horiuchi H, Schuck P and Röpke G 2001 *Phys. Rev. Lett.* **87** 192501
[23] Funaki Y, Tohsaki A, Horiuchi H, Schuck P and Röpke G 2003 *Phys. Rev.* C **67** 051306(R)
[24] A very early attempt to consider for ^{12}C a three α-wavefunction of somewhat similar kind as in Ref. [22], but without the condensate aspect and no RGM solution, has been undertaken in: Hutzelmeyer H Hackenbroich H H 1970 *Z. Phys.* **232** 356; see also discussion (p. 80) in *Clustering Phenomena in Nuclei* Wildermuth K and Kramer P editors 1977 Vol. **1**; Wildermuth K, Tang Y C 1977 *A Unified Theory of the Nucleus* Vieweg
[25] Yamada T and Schuck P 2005 *Eur. Phys. J.* A **26** 185
[26] Tohsaki A 1994 *Phys. Rev.* C **49** 1814
[27] Tohsaki A, Horiuchi H, Schuck P and Röpke G 2004 (*Nucl. Phys.* A **738** 259
[28] Matsumura H and Suzuki Y 2004 *Nucl. Phys.* A **739** 238
[29] The fact that this condensate state is actually an excited state of ^{12}C does not invalidate this picture, since, on nuclear scales, this state is long lived. Actually, also atomic Bose condensates in traps find themselves in meta-stable states, see e.g. [13].
[30] Funaki Y, Horiuchi H, Tohsaki A, Schuck P and Röpke G 2005 *Eur. Phys. J.* A **24** 321
[31] Itoh M *et al.* 2011 *Phys. Rev.* C **84** 054308
[32] Zimmerman W R *et al.* 2011 *Phys. Rev.* C **84** 027304
[33] Gai M for the UConn-Yale-Duke-Weizmann-PTB-UCL Collaboration 2011 J. Phys. G **267** 012046;for the UConn-Yale-Duke-Weizmann-PTB-UCL Collaboration 2011 Acta Phys. Pol. B **42** 775
[34] Kurakowa C and Kato K 2005*Phys. Rev.* C **71** 021301
[35] Funaki Y, Yamada T, Horiuchi H, Röpke G, Schuck and P Tohsaki A 2008 *Phys. Rev. Lett.* **101** 082502
[36] Ali S Bodmer A R 1966 *Nucl. Phys.* **80** 99
[37] Pitaevskii L P 1961 *Zh. Eksp. Teor. Fiz.* **40** 646 [1961 *Sov. Phys. JETP* **13** 451]; Gross E P 1961 *Nuovo Cimento* **20** 454; 1963 *J. Math. Phys.* **4** 195
[38] Yamada T and Schuck P 2004 *Phys. Rev.* C **69** 024309
[39] Kokalova Tz, Itagaki N, von Oertzen W and Wheldon C 2006 *Phys. Rev. Lett.* **96** 192502
[40] von Oertzen W et al. 2006 *Eur. Phys. J.* A **29** 133
[41] Ogloblin A A et al. 2005 *Proc. Int. Nuclear Physics Conference, Peterhof, Russia, June 28-July 2*
[42] Brenner M W et al. 2000 *Proceedings of the International Conference "Clustering Phenomena in Nuclear Physics"* St. Petersburg, published in Physics of Atomic Nuclei (Yadernaya Fizika)
[43] Harakeh M N 2012 private communication
[44] Freer M et al. 2004 *Phys. Rev.* C **70** 064311
[45] Raduta Ad R *et al.* 2011 *Phys. Lett.* B **705** 65
[46] Zarubin P I, Artemenkov D A, Orlova G I, Cechak T *et al.* (eds.) 2006 *Nuclear Science and Safety in Europe* 189 (Springer-Verlag, Berlin)
[47] Nozières P and Saint James D 1982 *J. Physique* **43** 1133
[48] Feldmeier H and Neff T 2005 *Physik Journal* **4**, Nr. **1**, 29 www.physik-journal.de
[49] Epelbaum E, Krebs H and Lee D Meissner U G 2011 *Phys. Rev. Lett.* **106** 192501
[50] Wiringa B R 2012 Argonne Nat. Lab. report
[51] Yamada T, Funaki Y, Horiuchi H, Röpke G, Schuck P and Tohsaki A 2011 arXiv:1103.3940, "Clusters in Nuclei" *Lecture Notes in Physics* Vol.2 Ed. by C. Beck (Springer Verlag, Berlin)

Fine structure of alpha decay of even-even trans-lead nuclei – an intriguing nuclear structure paradigm

D. Bucurescu, N.V. Zamfir

Horia Hulubei National Institute of Physics and Nuclear Engineering, P.O. Box MG-6, 077125 Bucharest - Măgurele, Romania

E-mail: bucurescu@tandem.nipne.ro, zamfir@tandem.nipne.ro

Abstract.

The systematics of the experimental data of alpha-decay fine structure in the even-even nuclei above lead are examined. The representation of relative branching ratios (b.r.) and hindrance factors (HF) for the $2_1^+, 4_1^+$, and 6_1^+ states within the valence correlation schemes (as a function of N_pN_n or $P = N_pN_n/(N_p + N_n)$) proves useful, as it provides smooth trajectories that can be discussed in parallel with the development of collectivity. It is shown that practically all recent theoretical calculations are not able to account for some of the most conspicuous features presented by the evolution of the branching ratios and hindrance factors. These are: a practically exponential increase of the b.r.'s and HF's for the 2_1^+ state of collective nuclei (above $P \approx 4.0$); a pronounced maximum around $P \approx 7.5$ of the same quantities for the 4_1^+ state, a region where these nuclei have usually been considered as well deformed rotors; and the decrease of these quantities for the 6_1^+ state in the interval of P from 4.0 to about 8.0. The observed evolutions of these fine structure quantities do not appear to be correlated with different structure indicators deduced from the low-lying level schemes, with the exception of the HF of the 2^+ state which is reasonably well correlated with a parameter that describes the nuclear rigidity, indicating the necessity of going beyond the rigid rotor description of the nuclei considered in this study. It is also necessary that a simultaneous description of both electromagnetic transitions and alpha decay fine structure data is achieved by the theoretical calculations.

1. Introduction

Although it is one of the oldest observed nuclear structure phenomena, alpha decay remains a very important experimental tool for the investigation of unstable nuclei, especially the superheavy ones. The fine structure of the alpha decay was experimentally observed by Salomon Rosenblum in 1929 [1], and soon after that explained by Gamow [2] as due to the population of the ground and excited states in the daughter nucleus. The population of the excited states in the residual nucleus is usually much weaker than that of the ground state, and a detailed theoretical understanding of the experimentally observed alpha decay branching ratios has not been achieved yet.

In this work we review the experimental data on the fine structure of the alpha decay of even-even trans-lead nuclei. This subject was partially covered in a previous paper where, however, only hindrance factors were considered [3]. Here, we consider the systematics of both relative branching ratios and of hindrance factors. Conspicuous features of these systematics are discussed, that are not fully understood theoretically at present. The usefulness of the

Journal of Physics: Conference Series **413** (2013) 012010 doi:10.1088/1742-6596/413/1/012010

presentation of these quantities within the valence correlation schemes, as proposed in [3], is also reinforced.

2. Systematics of branching ratios and hindrance factors

The considered alpha-decay data are those of the even-even daughter nuclei with $Z \geq 82$ for which at least the branch for the 2_1^+ state was measured besides that of the ground state. In addition to the 2_1^+ state, other states considered in our analysis were the 4_1^+ and 6_1^+ states that were measured in most of the rotational nuclei. For other excited states the measured data are less numerous and are not included in this study. The measured branching ratios and the corresponding hindrance factors for these states of the ground state band were collected and examined. Systematics of these quantities were published before by Ellis and Schmorak [4] (for $A \geq 229$), and by Akovali [5], but we have also checked these experimental data with the most recent ENSDF evaluations [6].

For self-consistency we review the experimental observables and different quantities based on these and used in the comparison with theoretical calculations. The quantities experimentally determined in alpha decay are the Q-values (which determine the excitation energies of the states populated in the daughter nucleus), the branching ratios $B_{r,i}$ (for the state i), and the half-life $T_{1/2}$ of the alpha-decaying state in the parent nucleus (in our case, the 0^+ ground state). For each state i one then defines a partial half-life $T_{1/2,i}$ and a partial width Γ_i:

$$T_{1/2,i} = \frac{T_{1/2}}{B_{r,i}}; \quad \Gamma_i = \frac{\hbar \ln 2}{T_{1/2,i}} \tag{1}$$

For comparison with theory one usually factorizes the widths as

$$\Gamma = \delta^2 P \tag{2}$$

where δ^2 is called reduced width, which mainly contains the nuclear structure information, and P is the penetrability of the alpha particle through the Coulomb barrier. The hindrance factors are defined as:

$$HF_i = \frac{{\delta_{gs}}^2}{{\delta_i}^2} = \frac{B_{r,gs}}{B_{r,i}}\frac{P_i}{P_{gs}} = \frac{\Gamma_{gs}}{\Gamma_i}\frac{P_i}{P_{gs}} \tag{3}$$

One sees that for the even-even nuclei, the hindrance factor defined for the ground state to ground state transition is, by definition, $HF = 1$. The hindrance factors are *model dependent* quantities, because they contain the penetrabilities that can be calculated in different ways. On the other hand, the reduced widths are essentially determined by an alpha formation amplitude, therefore they depend on the structure of the two states implied in the decay process. Definition (3) also shows that in calculating the hindrance factors one corrects for the barrier penetrabilities, leaving, in principle, mainly the dependence on the structure of the initial and final states. This makes the α-decay a strong spectroscopic tool for investigating the properties of the nuclear states. Actually, the structure information obtained from the alpha decay (the reduced widths) can be compared to the similar spectroscopic information obtained from the direct α-pickup (d,^{6}Li) reaction. However, for the set of nuclei that we study, there are only a small number of such studies [7, 8].

In ENSDF, the listed HF-values are based on the recipe of Preston (spherical case, rectangular potential well) [9]. Another definition often encountered is that of Rasmussen [10] (also spherical case, but using a more realistic alpha-nucleus potential). Deformation dependent penetrabilities, best suited to most of the nuclei in this set, were also recently considered (see, for example, Refs. [11, 12]).

In total, we have found experimental branching ratios and corresponding HF values for a number of 62 2_1^+ states, 29 4_1^+ states, and 17 6_1^+ states, respectively, in daughter nuclei with

Journal of Physics: Conference Series **413** (2013) 012010 doi:10.1088/1742-6596/413/1/012010

$Z \geq 82$. As already stated, preliminary results based on the examination of the hindrance factors were published in [3].

The evolution of the experimental branching ratios and hindrance factors can be investigated by representing them as functions of different variables, like mass number A, neutron number N, etc. Thus, in Ref. [4] their evolution with A was presented. We prefer, as in Ref. [3], to use the so-called valence correlation scheme representation, because although the A-representation shows a relatively smooth evolution above mass 220, where nuclei become collective, it does not distinguish between isobars. The VCS are based on the use of the N_pN_n [13, 14], or $P = N_pN_n/(N_p + N_n)$ [15] quantities, where N_p (N_n) represent the number of active protons (neutrons) counted with respect to the nearest magic number. It is known [13, 14, 15] that these type of representations give similar, rather compact trajectories for the evolution of different structure indicators (such as, e.g., $R(4/2) = E(4_1^+)/E(2_1^+)$, or $B(E2; 2_1^+ \rightarrow 0_1^+)$); also, they have the advantage that the observed evolutions can be correlated with the evolution of the collectivity (for example, it was shown that in all nuclear regions the transitional nuclei pass into deformed ones in the region $P \approx 4.0 - 5.0$ [15]). In our case, we have counted the active numbers of nucleons (particles or holes) in the daughter nucleus, with respect to the shell gap numbers 82 and 126 for protons, and 126 and 184 for the neutrons, respectively.

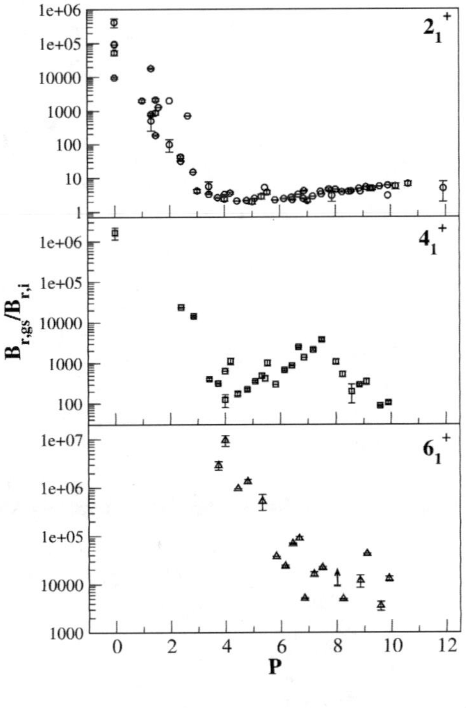

Figure 1. Experimental relative branching ratios for the first three excited states of the ground state (quasi)band, as a function of the P factor, where $P = N_pN_n/(N_p + N_n)$.

Fig. 1 shows the evolution of the relative branching ratios for the three considered excited states, as a function of the P-factor. Like in the case of the hindrance factors [3], rather compact trajectories are obtained, especially for the nuclei that start developing collective features (P larger than about 4.0). Conspicuous features (also remarked for the HF values) are: the exponential increase for the 2^+ state above $P \approx 6.0$, the pronounced maximum at $P \approx 7.5$ for the 4^+ state, and the decrease for the 6^+ state, almost out of phase with the 4^+ state variation.

Figures 2 and 3 show a comparison of these experimental data with the results of some recent theoretical calculations. In Fig. 2 are shown the results of calculations performed with a unified model for α-decay and α−capture (UMADAC) [16], and those of a Coulomb and proximity potential model for deformed nuclei (CPPMDN) [17]. Very similar results are provided by other calculations, like those based on the generalized liquid drop model [18], or those based on the Gamow theory with a square well potential barrier penetration [19]. One can see that a reasonable description is obtained with CPPMDN for the vibrational and transitional nuclei (P below ~4.0, while for the rest of the nuclei, especially the well deformed ones, none of these models is able to explain the main experimental features emphasized above.

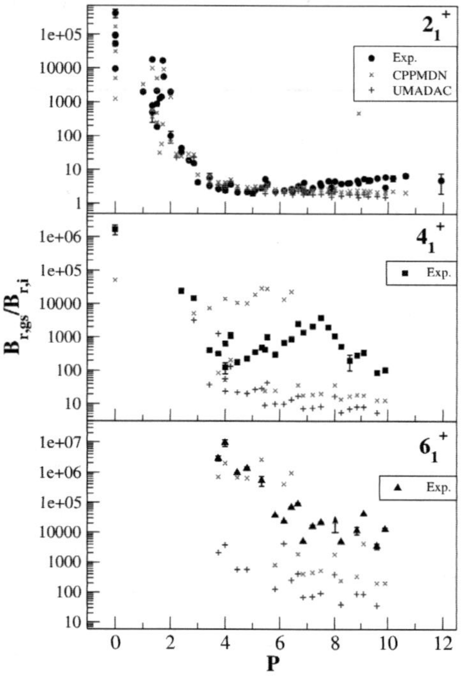

Figure 2. Comparison of experimental relative branching ratios with theoretical model calculations. CPPMDN – Coulomb and proximity potential model for deformed nuclei [17]; UMADAC – Unified model for alpha decay and alpha capture [16].

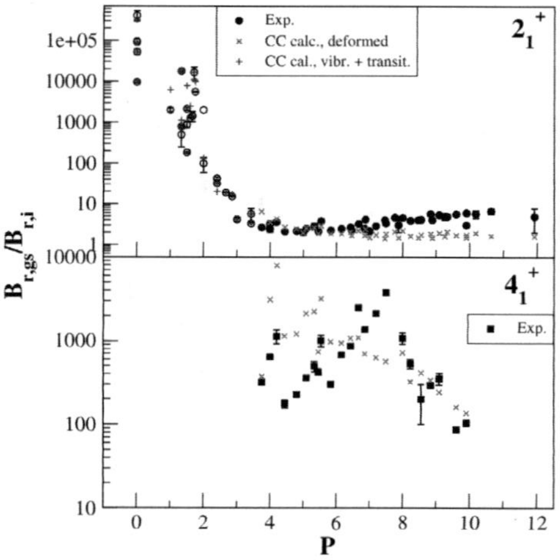

Figure 3. Same as Fig. 2, but for calculations performed with stationary coupled channel calculations for vibrational and transitional [20], and deformed [21] nuclei, respectively.

Figure 3 shows the same type of comparison with results of stationary coupled channels calculations [20, 21]. Again, while for the vibrational and transitional nuclei [20] a very good agreement is achieved, for the deformed nuclei [21] the exponential increase for the 2^+ state and the maximum observed for the 4^+ state are not accounted for. One should emphasize that a rigid rotor model was employed for the deformed nuclei.

Hindrance factors as adopted by ENSDF [6] were presented and discussed in [3]. We mention that hindrance factors calculated by using different calculations for the penetrabilities, such as those from Refs. [6, 10, 21] generally differ from each other just in absolute value, but present similar evolutions.

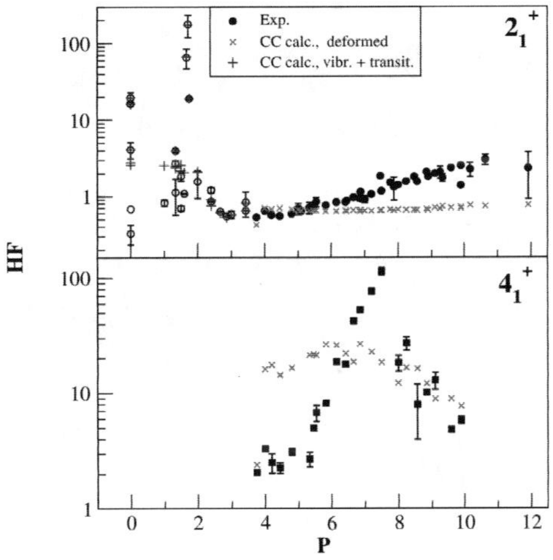

Figure 4. Comparison of experimental and calculated hindrance factors, data from Ref. [21].

As an example, Fig. 4 shows the experimental and theoretical hindrance factors calculated with the Coupled Channel method [20, 21]. As in the case of the branching ratios (Fig. 3), again a good agreement is obtained for the vibrational and transitional nuclei, while for the deformed ones there is a systematic deviation with respect to the experimental data.

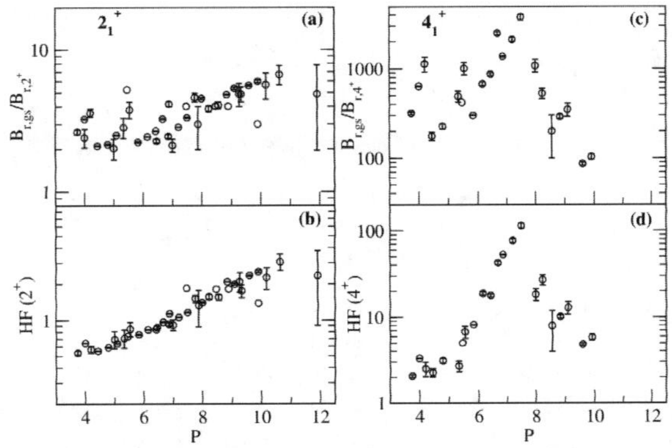

Figure 5. Relative branching ratios and hindrance factors (as calculated in Ref. [21]) for the 2^+ and 4^+ states, represented as a function of the VCS P-factor.

Fig. 5 shows, however, the clear advantage of the P-representation for the hindrance factors. The branching ratios show a larger scattering of the data points (for the 2^+ state one can even distinguish the isotopic chains). By correcting for the barrier penetrability and thus obtaining the HF mainly containing structure information, the scattering of the data is much reduced. For both 2^+ and 4^+ states one obtains raher compact trajectories for the collective nuclei (P above 4.0), therefore this kind of plot lends itself to the prediction of values for nuclei that are not measured yet.

3. Discussion

In Fig. 6 we present different structure indicators, extracted from the data of the low-lying excited states, and try to correlate their evolution with those observed for the α-decay fine structure quantities discussed above. These quantities comprise the following. (i) The moment

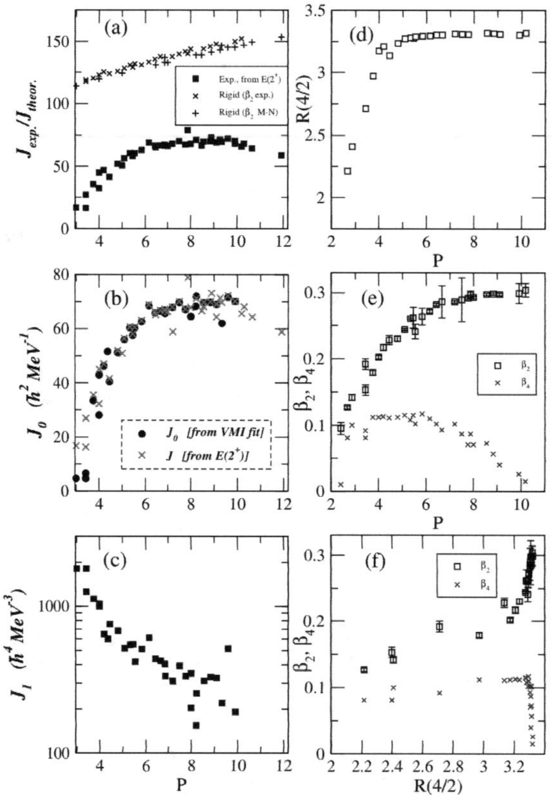

Figure 6. Different nuclear structure indicators of our set of nuclei: moment of inertia (a), as deduced from the 2^+ state energy with the rigid rotor formula; the J_0 and J_1 parameters of the fit with the Harris formula (4) (b) and (c); the $R(4/2)$ energy ratio (d); the experimental β_2 – quadrupole, and calculated [22] β_4 – hexadecapole, deformation parameters (e) and (f).

of inertia, as evaluated from the energy of the 2_1^+ state with the rigid rotor formula; in graph (a) it is also shown the rigid body of inertia calculated for an elipsoid of quadrupole deformation β_2 (either experimental, or calculated [22]). (ii) the J_0 and J_1 Harris parameters of the fit with the the variable moment of inertia (VMI) [23] formula

$$E = \frac{1}{2}\omega^2(J_0 + \frac{3}{2}\omega^2 J_1). \tag{4}$$

(iii) the ratio $R(4/2) = E(4_1^+)/E(2_1^+)$; (iv) the experimental quadrupole deformation parameter β_2, and the hexadecapole deformation parameter as predicted by Ref. [22].

An examination of these parameters shows that our set of nuclei undergoes a rather "normal" transition from vibrational to rotational nuclei. Thus, at $P \approx 5.0$, $R(4/2)$ reaches values that are close to the rotational limit 3.33, as shown in Ref. [14]. In the region of large P, the experimental moment of inertia stabilizes at about half the rigid body value, similar to the rare earths deformed region. The quadrupole deformation β_2 smoothly increases with P, with values above 0.25 for $P \geq 6.0$. The predicted [22] hexadecapole deformation β_4 has a sudden change of slope in the region of well deformed nuclei (at about $R(4/2) = 3.25$). An interesting feature is revealed by the examination of the energies of the yrast line with the VMI formula (4). The J_0 parameter is roughly identical with the moment of inertia extracted from the 2_1^+ state energy. The J_1 parameter decreases with the increase of P. Within the VMI [23], the inverse of J_1 is twice the value of a stiffness parameter C that measures the rigidity of the nucleus. Thus, relatively small J_1 values associated with large J_0 indicate an increased rigidity – a behavior that is closer to that of a rigid rotor (not in the sense of an increase of the moment of inertia, but in that the intrinsic structure of the nucleus changes little with increasing excitation energy, or rotation). We see that the rigidity of the nuclei considered here increases with increasing P.

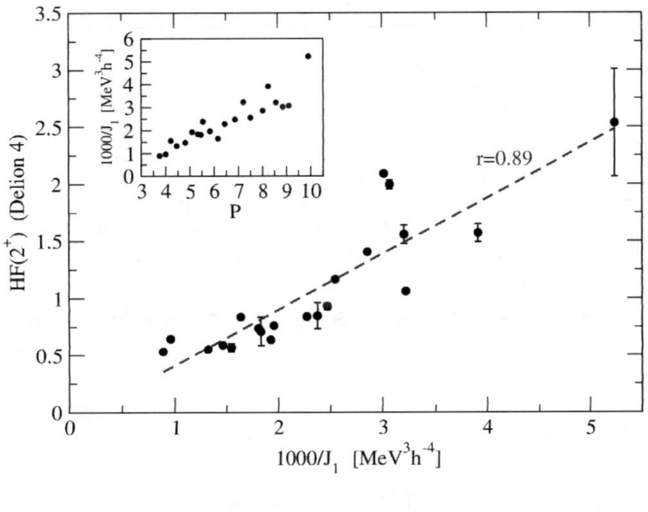

Figure 7. Correlation between the hindrance factors of the 2_1^+ state (as calculated in Ref. [21]) and the inverse of the J_1 Harris parameter of eq. (4), which is proportional (see text) with the rigidity paramater. The linear fit to the data gives a correlation coefficient $r = 0.89$.

We have tried to find correlations between the structure indicators discussed above and the α-decay fine structure quantities. The only useful correlation observed has been that between the HF of the 2^+ state and the stiffness parameter (the rigidity of the nucleus) – Fig. 7. This fact may indicate why by using the rigid rotor model the calculations (Figs. 3, 4) cannot

Journal of Physics: Conference Series **413** (2013) 012010 doi:10.1088/1742-6596/413/1/012010

reproduce the exponential increase experimentally observed for both branchings and HF's. On the other hand, the origin of the conspicuous maximum observed for the 4^+ state still remains very intriguing, especially as it is in a region of large P (about 7.5), where these nuclei are usually regarded as good rotors and no sudden changes are expected. Also, the decrease observed for the 6^+ state, which takes place in the same time with the strong increase for the 4^+ state (Figs. 1,2), has no simple explanation.

Recent coupled channel calculations performed by Ni and Ren [24] report an improvement in the description of the fine structure data. They consider coupling between five channels (up the the 8^+ state), and an improved consideration of the dynamics of the core. The results, essentially contained in their Fig. 1, show the following. There is still a certain underestimation of the branchings for the 2^+ state, that increases with the mass number (or P, in our case). There is also an improvement in the description of the 4^+ state values, however the data show a large scattering and do not seem to present the neat maximum around mass 240 (corresponding to the Pu isotopes with mass 236 to 244, with P around 7.5). The description of the 6^+ states is similar to that of the 4^+ state. While the general description of the fine structure data is improved, and clearly indicates that much of the observed effects may come from the couplings between channels, we lack a "simple" explanation of conspicuous effects such as the well defined maximum in the 4^+ state data. As observed in the present study, the consideration of a variable nuclear rigidity of the deformed nuclei in this region may be very important. It is also important that calculations that give a good description of the alpha decay fine structure data give also a good description of the electromagnetic transition probabilities of the excited states [25].

Acknowledgments

We thank Drs. D. Delion, Zhongzhou Ren, and P.-A. Söderström for stimulating discussion. This work was partially supported by UEFISCDI-Romania under program PN-II-IDEI Contract No. 127/2011.

4. References

[1] Rosenblum S 1929 *C. R. Acad. Sci. Paris* **188** 1401
[2] Gamow G 1930 *Nature* **126** 387
[3] Bucurescu D and Zamfir N V 2012 *Rom. J. Phys.* **57** 69
[4] Ellis Y A and Schmorak M R 1972 Nuclear Data Sheets **B8** 345
[5] Akovali Y A 1998 Nuclear Data Sheets **84** 1
[6] Evaluated Nuclear Structure Data File (ENSDF and XUNDL), maintained by the National Nuclear Data Center, Brookhaven National Laboratory: http://www.nndc.bnl.gov/ensdf
[7] Jänecke J, Becchetti F D, Overway D, Cossairt J D, and Spross R L 1981 *Phys. Rev.* **C23** 101
[8] Van den Berg A M, Blasi N, Siemssen R H, Sterrenburg W A, and Sujkowski Z 1984 *Nucl. Phys.* **A422** 45
[9] Preston M A 1947 *Phys. Rev.* **71** 865
[10] Rasmussen J O 1959 *Phys. Rev.* **113** 1593
[11] Delion D S, Peltonen S, and Suhonen J 2006 *Phys. Rev* **C73** 014315
[12] Karlgren D, Liotta R J, Wyss R, Huyse M, van de Vel K, and van Duppen P 2006 *Phys. Rev.* **C73** 044304
[13] Casten R F 1985 *Nucl. Phys.* **A443** 1
[14] Casten R F and Zamfir N V 1993 *Phys. Rev. Lett.* **70** 402
[15] Casten R F, Brenner D S, and Haustein P E 1987 *Phys. Rev. Lett.* **58** 658
[16] Denisov V Yu and Khudenko A A 2009 *Phys. Rev.* **C80** 034603
[17] Santhosh K P, Sahadevan S, and Joseph J G 2011 *Nucl Phys.* **A850** 34
[18] Wang Y Z, Zhang H F, Dong J M, and Royer G 2009 *Phys. Rev.* **C79** 014316
[19] Xu Chang and Ren Zhongzhou 2006 *Nucl. Phys.* **A778** 1
[20] Peltonen S, Delion D S, and Suhonen J 2007 *Phys. Rev.* **C54** 054301
[21] Peltonen S, Delion D S, and Suhonen J 2008 *Phys. Rev.* **C78** 034608
[22] Möller P, Nix J R, Myers W D, and Swiatecki W J 1995 *At. Data Nucl. Data Tables* **59** 185
[23] Scharff-Goldhaber G, Dover C B, and Goodman A L 1976 *Annu. Rev. Nucl. Sci.* **26** 239
[24] Ni Dongdong and Ren Zhongzhou 2011 *Phys. Rev.* **C83** 067302; and Z. Ren these proceedings
[25] Delion D S 2012 private communication

Journal of Physics: Conference Series **413** (2013) 012011 doi:10.1088/1742-6596/413/1/012011

Clustering features in decay processes

Doru S. Delion

Horia Hulubei National Institute for Physics and Nuclear Engineering, P.O. Box MG-6, RO-077125, Bucharest, Romania

E-mail: `delion@theory.nipne.ro`

Abstract. Decay widths in emission processes are described within the stationary scattering theory. We use a pocket-like potential between emitted fragments, centered on the nuclear surface. This picture predicts a linear dependence between the logaritm of the reduced width and fragmentation potential. It turns out that this law is satisfied by proton emission, alpha-decay and heavy cluster emission processes. In particular, we show that, by using surface Gaussian-like components in single particle orbitals mocking four body correlations, it becomes possible to enhance both electromagnetic and α transition probabilities up to the experimental values in ^{212}Po.

1. Introduction

The simplest Gamow approach of the α-decay process assumes a preformed α-particle moving around the core and penetrating the Coulomb barrier [1]. Later on, the cluster was built microscopically from two proton and two neutron single particle configurations [2, 3]. Anyway, the decay widths were by several orders of magnitude smaller than the experimental values. In the late 1970's it was shown that the inclusion of many configurations increases the value of the absolute decay width by more than four orders of magnitude [4]. Yet, the calculated widths were two orders of magnitude too small. Only a theory where a combined shell- and cluster-model configurations were considered could reproduce the absolute decay width [5, 6]. This may justify the so-called extreme cluster models, where the α-particle moves around the core [7, 8, 9]. This picture can be extended to the emission of heavier clusters, like C, O, Ne, Mg and Si [10].

In this presentation we will explain all emission processes within a simple approach, namely a pocket-like potential surounded by the Coulomb barrier. This picture is able to provide a simple analytical relation, connecting the reduced width to the fragmentation potential. It is valid for any kind of decay process induced by the strong interaction, except fission, and it is the next natural step beyond the Gamow picture. We will also show that a clustering component should be added to the single particle shell model orbitals in order to simultaneously describe α decay witdth and electromagnetic transitions in ^{212}Po.

2. Cluster description of decay processes

For the process $P(arent) \rightarrow D(aughter) + C(luster)$ one has a linear dependence between the logarithm of the half life and Coulomb parameter

$$\chi = \frac{2Z_D Z_C e^2}{\hbar v} , \qquad (1)$$

defined in tems of fragment electric charges and their asymptotic relative velocity v. This relation is called Geiger-Nuttal law [11] and it is valid for the cluster emission including α-decay, as it is shown in Fig. 1.

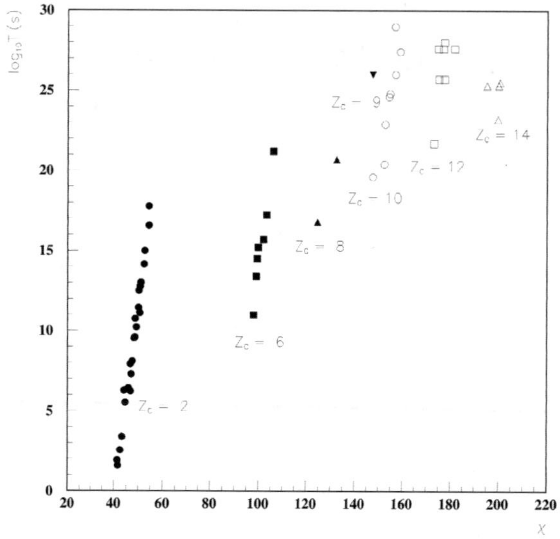

Figure 1. *Logarithm of the half life versus the Coulomb parameter (1) for various cluster emission processes.*

It can be explained within the stationary scattering theory [12]. For spherical emitters the decay width, which is inverse proportional with respect to the half life, can be written in terms of the so-called scattering amplitude N_0 as follows [12]

$$\Gamma \equiv \frac{\hbar ln2}{T} = \hbar v|N_0|^2 = \hbar v \left| \frac{Rf_0^{(int)}(R)}{G_0(\chi, R)} \right|^2 . \tag{2}$$

Notice that the above ratio does not depend on the cluster-daughter radius R, because both internal $f_0^{(int)}(R)$ and external Coulomb $G_0(\chi, R)/R$ wave functions should satisfy the same Schrödinger equation. Traditionally the decay width is rewritten

$$\Gamma = 2P_0\gamma^2 , \tag{3}$$

in terms of the penetrability and reduced width squared

$$P_0 = \frac{\kappa R}{|G_0(R)|^2} , \quad \gamma^2 = \frac{\hbar^2}{2\mu R}|f_0^{(int)}(R)|^2 , \tag{4}$$

where κ is the momentum and μ the reduced cluster-daughter mass. The monopole irregular Coulomb function $G_0(\chi, R)$, defining the penetrability, depends exponentially on the Coulomb parameter, explaining in this way the Geiger-Nuttall law.

Let us first analyze the α-decay process. The binding energy per nucleon for an α-particle is much larger with respect to their neighbours. Due to a smaller nuclear density at surface, an

Journal of Physics: Conference Series **413** (2013) 012011 doi:10.1088/1742-6596/413/1/012011

α-cluster structure is energetically more favourable in this region [13]. Therefore clustering is a surface effect and it can be simulated by assuming that the α-particle moves in a pocket-like potential [12]

$$\begin{aligned} V(r) &= \hbar\omega\frac{\beta(R-R_0)^2}{2} + v_0 \ , \quad R \le R_B \\ &= \frac{Z_D Z_C e^2}{R} \equiv V_C(R) \ , \quad R > R_B \end{aligned} \tag{5}$$

where R_0 is the surface pocket radius, as it is shown in Fig. 2.

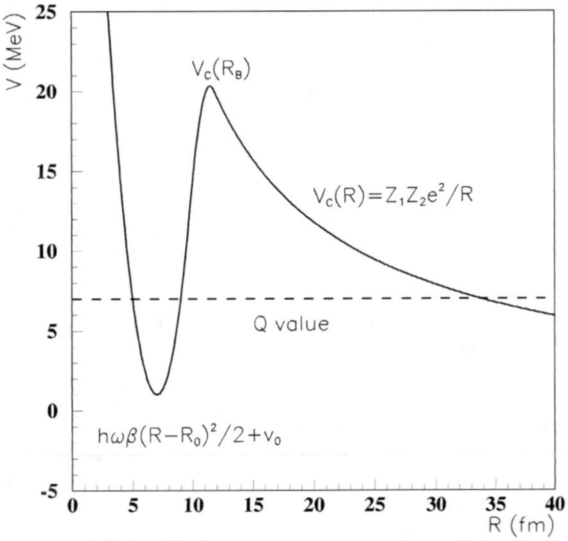

Figure 2. *The interaction potential between the cluster and daughter nucleus.*

By considering the Q-value as the energy of the first eigenstate in the shifted harmonic oscillator (ho) well $Q - v_0 = \frac{1}{2}\hbar\omega$, together with the continuity condition at the top of the barrier R_B, one obtains the following relation

$$\hbar\omega\frac{\beta(R_B-R_0)^2}{2} = V_{frag}(R_B) + \frac{1}{2}\hbar\omega \ , \tag{6}$$

where we introduced the so called fragmentation (or driving) potential, as the difference between the top of the Coulomb barrier and Q-value

$$V_{frag}(R_B) = V_C(R_B) - Q = Q\left[\frac{\chi}{\rho} - 1\right] \ , \tag{7}$$

in terms of the Coulomb parameter (1) and reduced radius $\rho = \kappa R$. Notice that the radial equation of the shifted ho potential is similar to the equation of the one-dimensional oscillator and therefore the wave function of the ground state is given by

$$f_0^{(int)}(R) = A_0 e^{-\beta(R-R_0)^2/2} \ . \tag{8}$$

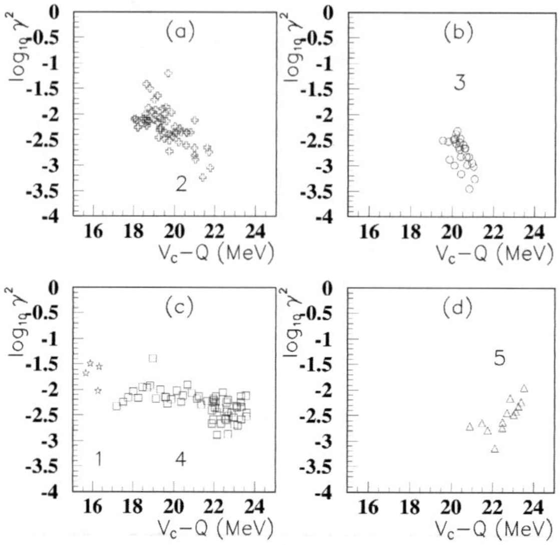

Figure 3. *Logarithm of the α-decay reduced width squared versus the fragmentation potential (7) for regions of the nuclear chart described by (10).*

By using Eq. (6) one obtains the following relation [14]

$$\log_{10}\gamma^2(R_B) = -\frac{log_{10}e^2}{\hbar\omega}V_{frag}(R_B) + \log_{10}\frac{\hbar^2 A_0^2}{2e\mu R_B} . \qquad (9)$$

The slope of this dependence has a negative value and it is connected with the shape of the interaction potential (ho energy $\hbar\omega$), while the free term gives information about the amplitude of the cluster wave function. Our calculation has shown that the linear relation (9), but with different coefficients, remains valid in the most general case of the double folding plus repulsive interaction between fragments, used in Refs. [15, 16].

We first analyzed Eq. (9) in α-decays connecting ground states of even-even nuclei. In Fig. 3 the data are divided into five regions of even-even α emitters as follows

$$\begin{array}{llll} 1) & Z < 82, & 50 < N < 82 & Fig.\ 3\ (c),\ stars\ ; \\ 2) & Z < 82, & 82 < N < 126 & Fig.\ 3\ (a),\ crosses\ ; \\ 3) & Z > 82, & 82 < N < 126 & Fig.\ 3\ (b),\ circles\ ; \\ 4) & Z > 82, & 126 < N < 152 & Fig.\ 3\ (c),\ squares\ ; \\ 5) & Z > 82, & N > 152 & Fig.\ 3\ (d),\ triangles\ . \end{array} \qquad (10)$$

In calculations it was used the value of the touching radius, i.e.

$$R_B = 1.2(A_D^{1/3} + A_C^{1/3}) . \qquad (11)$$

Notice that the regions 1-4 contain rather long isotopic chains, while in the last region 5 one has not more than two isotopes/chain. This is the reason why, except for the last region 5, the reduced width decreases with respect to the fragmentation potential, according to the theoretical

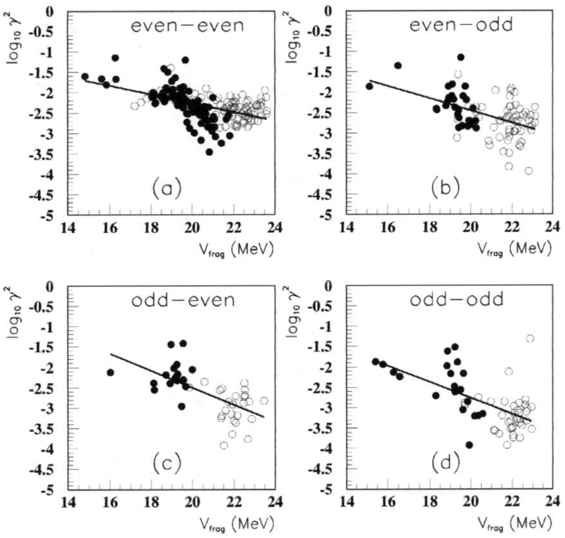

Figure 4. *Logarithm of the reduced width squared versus the fragmentation potential for even-even (a), even-odd (b), odd-even (c) and odd-odd (d) α-particle emitters.*

prediction given by (9). Eq. (9) is also fulfilled for α-decays from even-odd, odd-even and odd-odd emiister, as it is shown in Fig. 4.

In Fig. 5 we plotted the difference between the logarithm of the reduced width squared and fitting lines in Fig. 4. One sees that for most decays this law gives a reasonable description within a factor of three.

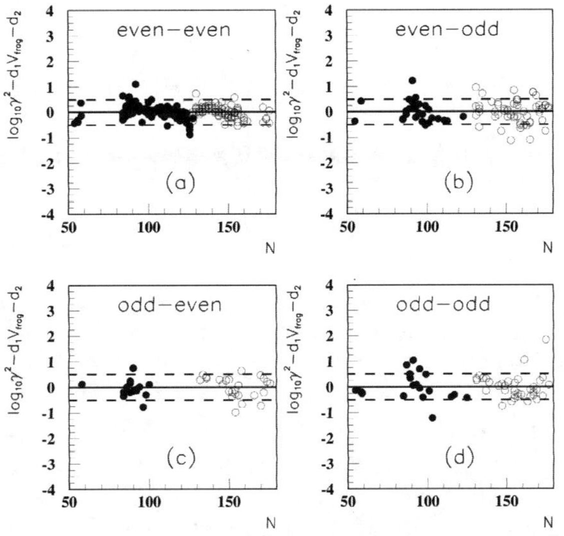

Figure 5. *The difference between the logarithm of the reduced width squared and the fitting lines in Fig. 4 for even-even (a), even-odd (b), odd-even (c) and odd-odd (d) α-particle emitters.*

The linear dependence between $log_{10}\gamma^2$ and the fragmentation potential (9) remains valid for any kind of cluster emission. This fact is nicely confirmed by heavy cluster emission processes in Fig. 6 (a). Here it is also plotted a similar dependence for α-decays corresponding to the same heavy cluster emitters. The straight line is the linear fit for cluster emission processes, except α-decays

$$log_{10}\gamma^2 = -0.586(V_C - Q) + 15.399 \ . \tag{12}$$

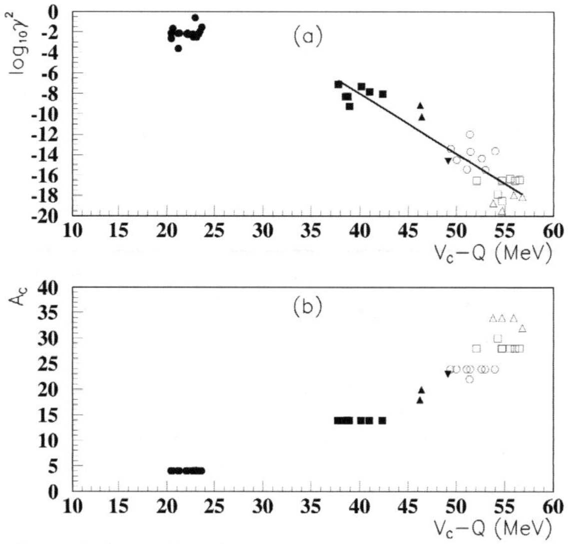

Figure 6. *(a) Logarithm of the reduced width squared versus the fragmentation potential (7). Different symbols correspond to various cluster decays. The straight line is the linear fit (12) for cluster emission processes, except α-decay. (b) Cluster mass number versus the fragmentation potential.*

The above value of the slope $-log_{10}e^2/\hbar\omega$ in Eq. (9) leads to $\hbar\omega \approx 1.5$ MeV, with the same order of magnitude as in the α-decay case.

Let us mention that a relation expressing the spectroscopic factor (proportional to the reduced width) for cluster emission processes was derived in Ref. [17]

$$S = S_\alpha^{(A_C-1)/3} \ , \tag{13}$$

where A_C is the mass of the emitted cluster and $S_\alpha \sim 10^{-2}$. As can be seen from figure 6 (b), between A_C and V_{frag} there exists a linear dependence and therefore the above scaling law can be easily understood in terms of the fragmentation potential.

Concerning the reduced widths of proton emitters, in Refs. [18, 19] it was pointed out the correlation between the reduced width and the quadrupole deformation. This fact can be seen in Fig. 7 (a), where the region with $Z < 68$ corresponds to $\beta > 0.1$ (open circles), while the other one with $Z > 68$ to $\beta < 0.1$ (dark circles). The two linear fits have obviously different slopes. Notice that the two dark circles with the smallest reduced widths correspond to the heaviest emitters with $Z > 80$.

At the same time one sees from Fig. 7 (b) that the same data are clustered into two regions, which can be directly related to the fragmentation potential (7). Here, the two linear fits in

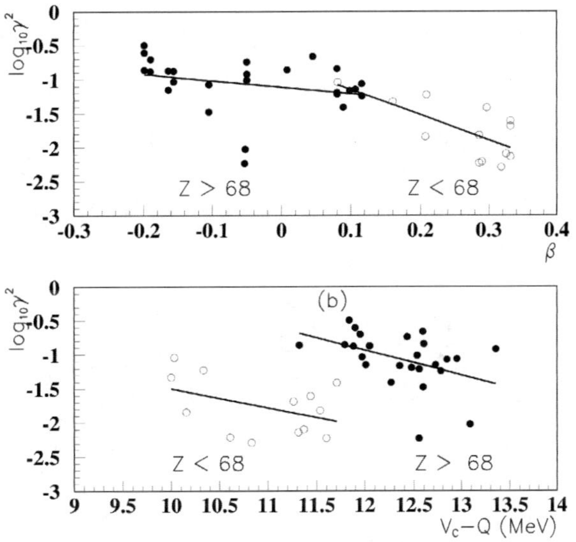

Figure 7. *(a) Logarithm of the reduced width squared for proton emission versus the quadrupole deformation. By open circles are given proton emitters with $Z < 68$, while by dark circles those with $Z > 68$. The two regression lines fit the corresponding data. (b) Logarithm of the reduced width squared versus the fragmentation potential (7). The symbols are the same as in (a).*

terms of the fragmentation potential, corresponding to the two regions of charge numbers, have roughly the same slopes, but different values in origin. Thus, the two diferrent lines seen in proton emission systematics [18] can be directly connected with similar lines in Fig. 7 (b). They correspond to different orders of magnitude of the fragmentation potential, giving different orders to wave functions and therefore to reduced widths.

3. Clustering features in ^{212}Po

The internal function $f_0^{(int)}(R)$ in Eq. (2) is called α-particle formation amplitude. For distances beyond the geometrical touching configuration, where the Pauli principle becomes less important, this quantity can be estimated by the following overlap integral [12]

$$f_0^{(int)}(R) = \langle \Psi_P | \Psi_D \Psi_\alpha \rangle \ , \tag{14}$$

in terms of parent, daughter and α-particle wave functions, respectivelly. In order to properly describe clustering properties, one has to consider at the single particle level two components of the wave function [20]

$$\psi_l(r) = \mathcal{N}_l^{(SM)} \psi_l^{(SM)}(r) + \mathcal{N}_l^{(clus)} \psi_l^{(clus)}(r) \ . \tag{15}$$

The shell model part has a standard spherical ho representation

$$\psi_l^{(SM)}(r) = \sum_n b_n \mathcal{R}_{nl}^{(\beta)}(r) \ , \tag{16}$$

where $\beta = M_N\omega/\hbar$ is the ho parameter and the principal quantum number has the values $N = 2n + l \leq 6$.

For the cluster part one uses the wave function of the shifted oscillator (8) expanded in terms of ho components

$$\psi_l^{(clus)}(r) = e^{-\beta_c(r-r_0)^2/2} = \sum_n c_n \mathcal{R}_{nl}^{(\beta)}(r) \ , \tag{17}$$

where N is spread around a larger value $N \sim 8-10$, as can be seen in Fig. 8.

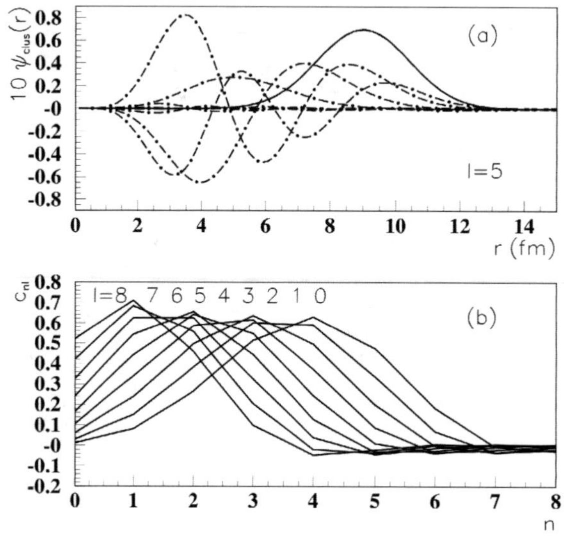

Figure 8. *(a) Gaussian distribution centered on the nuclear surface (solid line) and various expansion ho terms (dot-dashed lines). (b) Expanssion coefficients of a Gaussian centered on the nuclear surface given by Eq. (17).*

Notice that in Ref. [6] the ho parameter of the cluster part has a smaller value, in order to reduce the dimension of the single particle basis (15). Here, all coefficients are found by a diagonalization procedure of the Woods-Saxon mean field.

The typical example of a heavy nucleus where the α-clustering effects are very important is ^{212}Po. Here, the structure of low-lying states can be explained in terms of two proton and two neutron orbitals above the double magic inert core ^{208}Pb [21]. Anyway, by using the standard components of the two protons and two neutrons the decay width is underestimated by two orders of magnitude. It is necessary to consider the second cluster component in Eq. (15), with $\mathcal{N}_l^{(clus)} = 0.3$, in order to describe the absolute value of the α-decay width, as can be seen in Fig. 9 (b) (solid line). In spite of the fact that in figure 9 (a) the difference between the formation probability within the shell model approach (dashed line) and mixed approach (solid line) is rather small, the enhancement of the decay width is about two orders of magnitude. Notice that the same value was obtained in Ref. [5], by using a diagonalisation procedure of the residual interaction.

This picture is also supported by analyzing electromagnetic transitions. The electric transition probability of the multipolarity λ is proportional to the reduced matrix element of

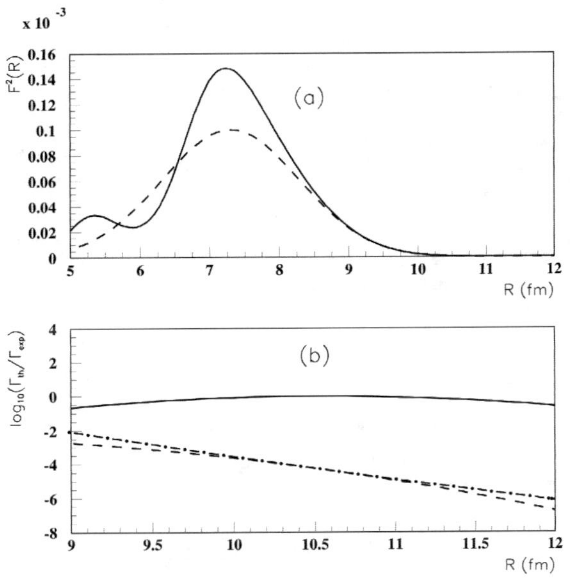

Figure 9. *(a) α-particle formation probability for* $^{212}Po \rightarrow ^{208} Pb + \alpha$*. Dashed line: shell model, Solid line: shell model + α-cluster.*
(b) Logarithm of the ratio between the decay width and the experimental value as a function of the daughter-α-particle radius

the transition operator squared, i.e.

$$B(E\lambda : J \rightarrow J') = \frac{1}{2J+1} \left| \langle J' || \hat{T}_\lambda || J \rangle \right|^2 , \tag{18}$$

where $|J\rangle$ and $|J'\rangle$ denote initial and final states and the transition operator is proportional to spherical harmonics

$$\hat{T}_{\lambda\mu} = r^\lambda Y_{\lambda\mu} . \tag{19}$$

The radial part of the quadrupole transition matrix element is proportional to the principal quantum number

$$\langle n_1 l_1 | r^2 | n_2 l_2 \rangle \sim N = 2n + l . \tag{20}$$

It is known that B(E2) values are satisfactory described within the shell model for ^{210}Pb, but they are by one order of magnitude less than the experimental values in ^{212}Po. They are of the same order of magnitude as those in ^{210}Pb [22]. In order to explain this discrepancy we used the mixed representation (15) with $\mathcal{N}_l^{(clus)} = 0.3$. From Table I one can see a nice agreement with respect to experimental data. The enhancement of the B(E2) value in ^{212}Po is explained by larger principal quantum numbers N involved by the cluster component (15) in transition matrix elements (20). This is a very convincing evidence in favour of the α-clustering structure of this nucleus.

Table I
Experimental and theoretical B(E2) values in ^{210}Po (second and third columns), ^{210}Pb (fourth and fifth columns) and ^{212}Po (sixth and last columns) in Weiskopf units.

	^{210}Po		^{210}Pb		^{212}Po	
$J \to J'$	$B(E2)_{exp}$	$B(E2)_{th}$	$B(E2)_{exp}$	$B(E2)_{th}$	$B(E2)_{exp}$	$B(E2)_{th}$
$2 \to 0$	0.56(12)	6.7	1.4(4)	3.9	—	9.2
$4 \to 2$	4.6(2)	12.9	3.2(7)	3.5	—	20.8
$6 \to 4$	3.0(1)	8.9	2.2(3)	2.4	13.5(36)	14.4
$8 \to 6$	1.18(3)	3.9	0.62(5)	1.0	4.60(9)	5.8

4. Conclusions

Based on a simple model of the cluster-core dynamics, namely a shifted harmonic oscillator potential surounded by the Coulomb interaction, it was derived an universal analytical relation expressing the logarithm of the reduced width squared as a linear function in terms of the fragmentation potential, defined as the difference between the Coulomb barrier and Q-value. It is fulfilled with a reasonable accuracy by all experimental decay data, describing transitions between ground states. In heavy nuclei this cluster component cannot be build from usual single particle orbitals because protons and neutron lie in different major shells. Thus, an additional gaussian centered on the nuclear surface, predicted by the above mentioned universal law, is necessary to be included in the single particle basis. We described both α-decay width and B(E2) value in ^{212}Po by using a mixed single particle basis, containing shell-model and α-cluster components.

Acknowledgments

This work was supported by a grant of the Romanian National Authority for Scientific Research, CNCS UEFISCDI, project number PN-II-ID-PCE-2011-3-0092.

References

[1] Gamow G 1928 *Z. Phys.* **51** 204
[2] Mang H J 1960 *Phys. Rev.* **119** 1069
[3] Sandulescu A 1962 *Nucl. Phys.* **37** 332
[4] Tonozuka I and Arima A 1979 *Nucl. Phys. A* **323** 45
[5] Varga K, Lovas R G and Liotta R J 1992 *Phys. Rev. Lett.* **69** 37
[6] Delion D S, Insolia A and Liotta R J 1996 *Phys. Rev. C* **54** 292
[7] Sandulescu A, Cusson R Y and Greiner W 1983 *Lett. Nuovo Cim.* **36** 321
[8] Buck B, Merchant A C and Perez S M 1994 *Phys. Rev. Lett.* **72** 1326
[9] Ohkubo S 1995 *Phys. Rev. Lett.* **74** 2176
[10] Sandulescu A, Poenaru D N and Greiner W 1980 *Sov. J. Part. Nucl.* **11** 528
[11] Geiger H and Nuttall J M 1911 *Philos. Mag.* **22** 613.
[12] Delion D S 2010 *Theory of particle and cluster emission* (Springer-Verlag, Berlin)
[13] Schuck P, Sogo T and Roepke G 2100 *Int. J. Mod. Phys. E* **20** 889
[14] Delion D S 2009 *Phys. Rev. C* **80** 024310
[15] Peltonen S, Delion D S and Suhonen J 2007 *Phys. Rev. C* **75** 054301
[16] Peltonen S, Delion D S and Suhonen J 2008 *Phys. Rev. C* **78** 034608
[17] Blendowske R, Fliessbach T, Walliser H 1991 *Z. Phys. A* **339** 121
[18] Delion D S, Liotta R J and Wyss R 2006 *Phys. Rev. Lett.* **96** 072501
[19] Medeiros E L, Rodrigues M M N, Duarte S B and Tavares O A P 2007 *Eur. J. Phys. A* **34** 417
[20] Delion D S, Liotta R J, Schuck P, Astier A and Porquet M -G 2012 *Phys. Rev. C* **85** 064306
[21] Delion D S and Suhonen J 2000 *Phys. Rev. C* **61** 024304
[22] Astier A, Petkov P, Porquet M -G, Delion D S and Schuck P 2010 *Eur. Phys. J. A* **47** 165

Int. Summer School for Advanced Studies 'Dynamics of open nuclear systems' (PREDEAL12) IOP Publishing
Journal of Physics: Conference Series **413** (2013) 012012 doi:10.1088/1742-6596/413/1/012012

Developments in radioactive decay during the last Century

R. J. Liotta

Department of Physics, Royal Institute of Technology, SE-10691 Stockholm, Sweden

E-mail: liotta@kth.se

Abstract. In this talk a review of the developments in radioactive decay processes that have taken place during the last Century, and the great outgrowths of these developments in Modern Physics, is presented.

1. Historical background

At the beginning of last Century a revolutionary transition in Physics was taking place. Already at the end of the 1800's the forms of radiation that nowadays are called *alpha* and *beta* were discovered. In 1896 Becquerel found that charged particles were emitted by uranium, and soon afterwards Marie and Pierre Curie confirmed these observations. In 1899 Rutherford discovered that these radiations could be divided in two separated strings according to the penetration power they had. He called this two forms of radiation alpha and beta. Alpha radiation could penetrate only thin foils of aluminium while beta rays could go through many millimeters of the same material. Only in 1900 the third form of radiation was found, which had even more penetration power than beta rays, and these were called gamma rays. It was soon found that the alpha ray was actually an alpha particle, that the beta ray was an electron and that the gamma ray was a photon.

The discovery that alpha radiation was actually the emission of ^{4}He nuclei was the basis of the *Chemistry* Nobel prize that Rutherford got in 1908. But Rutherford continued working at a very high level even after receiving the Nobel prize. In fact, he produced his most revolutionary work after receiving the Nobel prize, and that was the discovery of the nucleus. Due to this he was also known as the father of nuclear physics. But Rutherford, who introduced the concept of "half lives" in relation to decaying systems, was not only a great experimentalist. He was also a great teacher. He had many brilliant students, some of whom became themselves Nobel laurates, like James Chadwick. John Cockcroft, Ernest Walton and Edward Appleton. Others were strongly influenced by him, like Niels Bohr.

Among these brilliant students a number were from Germany, who preferred to visit Rutherford at the University of Manchester rather to be unemployed in the saturated market for physicists in German Universities at that time. Among those students there were H. Geiger and E. Mardsen, who together created the Geiger counter in 1909. At this time Rutherford proposed Geiger and J. M. Nuttall, an English student, to analyze the relation between the half lives of alpha emitting nuclei and their kinetic energies. They thus succeeded in 1911 to find the famous Geiger-Nuttall law of alpha radioactive decay [1], i. e.,

$$logT_{1/2} = \frac{a}{\sqrt{Q_\alpha}} + b \tag{1}$$

where a and b are Z-dependent constants and Q_α is the kinetic energy of the outgoing α-particle.

There has been an intense investigation on different aspects of the simple-looking Geiger-Nuttall law lasting for many years after it was formulated. For instance, much work was dedicated to find a generalization of the law valid for all Z. Thus, 55 years after Geiger-Nuttall, Viola and Seaborg found that, with a great precision, the dependence of the contants upon Z can be written as [2] $a(Z) = 2.11329Z - 48.9879$, $b(Z) = -0.390040Z - 16.9543$.

But the most important outcome from the many attemps performed in order to explain the origin of the Geiger-Nuttall law was the one due to Gamow, which profoundly changed the formulation and interpretation of physical laws. This we will now analyze.

2. Gamow treatment

Gamow considered the alpha-particle as a compact object which existed in the nucleus. He did not consider that the alpha particle actually consists of other more "elementary" particles, like neutrons and protons. He could not do that simply because the neutron was not discovered at that time yet, i. e. at the end of the 1920's. Gamow, as well as Rutherford and all other relevant physicist at that time, thought that the alpha particle was a conglomerate of four protons and two electrons. Gamow assumed that these six particles were strongly bound together and that this bunch of tightly bound particles existed inside the nucleus as a whole before decaying. He wrote the Schrödinger equation corresponding to the α-daughter system, assuming spherical symmetry and relative angular momentum l=0. He thus evaluated the penetrability of the α-particle through the Coulomb barrier, ignoring any other interaction [3]. Assuming that the local wave number does not change strongly inside the barrier, the correponding penetrability is,

$$P = \exp(-2\gamma) \tag{2}$$

and

$$\gamma = \frac{1}{\hbar}\int_{r_i}^{r_e} \sqrt{2m(\frac{Z_1 Z_2 e^2}{4\pi\epsilon_0 r} - Q)}dr \tag{3}$$

where the notation is standard. In particular r_i and r_e are the internal and external turning points.

Gamow interpreted the penetrability as the probability that the alpha particle went through the Coulomb barrier. This can not happen classically, since under the barrier the kinetic energy is negative, and therefore the velocity is imaginary. This was a great revolution in physics and confirmed the probabilistic interpretation of Quantum Mechanics, a feature that a number of distinguished physicists found to be weird. This is debated even today [4].

In order to get the right dimensions, Gamow used the very simple argument that the alpha particle is inside the nucleus without interacting with the protons and electrons that he assumed to be the nuclear constituents. That is, he considered the alpha particle as a little ball moving inside the nucleus, hitting the walls induced by the Coulomb barrier when arriving to the nuclear surface, going forward and back again. Within this assumption, the probability decay per second (i. e. the inverse of the mean life) is the probability of getting through the barrier (P) times the number of attempts to hit the wall of the nucleus. This number is v/R, where v is the velocity of the alpha particle and R is the nuclear radius. Therefore the mean life T is,

$$1/T = \frac{v}{R}P \tag{4}$$

Journal of Physics: Conference Series **413** (2013) 012012 doi:10.1088/1742-6596/413/1/012012

This expression for the mean life has had a tremendous success in explaining decay processes. One reason for this success is that it is extremely simple and can readily be applied to all forms of radioactive decay. Yet, it is an effective formula, without a firm theoretical background. One important quantity which is not included in Eq. (4) is the probability that the alpha particle is indeed present on the nuclear surface of the decaying nucleus. This probability is called "preformation factor" and is usually denoted by P_0. The value of P_0 was included in later analysis of the decay and its effective value was obtained through fittings of experimental data. This task was performed through many years of investigation, and the result has been of great help to the experimental interpretation of the decay data. We will see below how the decay can be analyzed within a complete theoretical framework. However, in strong contrast to the effective approach, thorough theoretical treatments are usually very complicated and difficult to apply. Therefore most radioactive decay studies are performed within effective approaches. In these approaches the flux $1/T = \lambda$ is usually called "decay constant", and the frequency of escape attemps is denoted by $v/R = \nu_0$. One thus gets,

$$\lambda = \nu_0 P_0 P \tag{5}$$

A recent study of determinations of ν_0 and P_0, obtained through the analysis of many decay processes, can be found in [5].

Gamow also realized that one can describe the decay process within an stationary formalim by assuming that the decaying state carries a complex energy. Thus, if the energy is $E = E_r - i\Gamma/2$, the wave function becomes

$$\varphi(r,t) = \varphi(r,0)e^{-iEt/\hbar} = \varphi(r,0)e^{-iE_t t/\hbar}e^{-\frac{\Gamma t}{2\hbar}} \tag{6}$$

and the probability density that the alpha particle (or any cluster) decays is

$$P(r,t) = P(r,0)e^{-\frac{\Gamma t}{\hbar}} \tag{7}$$

Which provides for the mean life the relation $T\Gamma = \hbar$, which sometimes is called, wrongly, the *time-energy uncertainty relation.*

It is convenient to point out that Eq. (7) would be valid only if all the imaginary parts are small, since only then one can interpret $Re(\varphi^2(r)) \approx |\varphi^2(r)|$ as a probability. A detailed discussion on this, and on the study of resonances by an extension of the formalism to the complex energy plane, can be found in [6].

So far we have considered effective theories that follow Gamow original formalism. There has also been a fully microscopical development of the theory. This started from the derivation of the width Γ as the residues of the R-matrix. In a rather difficult paper Thomas performed the corresponding calculation for the case of alpha decay [7]. But one can also get the width in a simple fashion by following scattering theory arguments [8]. To show this we consider a decaying cluster. At large distances only the centrifugal and Coulomb interactions are felt. Therefore the corresponding wave function is,

$$r\varphi_{lj}^{out}(r) = N_{lj}[G_{lj}(r) + iF_{lj}(r)] \tag{8}$$

where the notation is standard. The probability rate per second that the particle goes through a surface element $dS = r^2 \sin\theta d\theta d\phi$ is,

$$P_{lj} = |\varphi_{lj}^{out}(\mathbf{r})|^2 v dS \tag{9}$$

where v is the velocity of the cluster at large distances, i. e. $v = \hbar k/m$. Since at large distances $|G_{lj}(r) + iF_{lj}(r)|^2{=}1$, one obtains, by integrating over the angles, the decay probability per second as $1/T = |N_{lj}|^2 v$. Matching the out and the inner solution at R one gets $R\varphi_{lj}(R) = N_{lj}[G_{lj}(R) + iF_{lj}(R)]$ and

$$\Gamma_{lj} = \frac{\hbar}{T} = \frac{\hbar k}{m} \frac{R^2|\varphi_{lj}(R)|^2}{F_{lj}^2(R) + G_{lj}^2(R)} \tag{10}$$

This is the famous Thomas expression for the decay width, which he obtained as the residues of the R-matrix. $\varphi_{lj}(R)$ is the cluster formation amplitude and $kR/(F_{lj}^2(R) + G_{lj}^2(R))$ is the penetrability through the centrifugal and Coulomb barriers. It is important to notice that the width should *not* depend upon R if the calculation of the formation amplitude is properly performed. For the cluster (C) decay process $B \to A + C$ the formation amplitude is

$$\varphi(R) = \int d\xi_A d\xi_C \varphi_B^*(\xi_B)\varphi_A(\xi_A)\varphi_C(\xi_C) \tag{11}$$

This expression is the basis of all microscopical calculations of cluster decay. We will briefly review below the large amount of work performed through many years in this subject.

3. Shell Model, BCS and Rotational treatments

During the 1950's a revolution occurred in nuclear physics with the near simultaneous appeareance of the Shell Model, the Rotational Model and the BCS formalism. For alpha decay an equally important ingredient was the Thomas formulation leading to Eq. (11). This equation was applied to alpha decay from spherical nuclei by Rassmusen [9], Mang [10] and Sandulescu [19], and in deformed nuclei by Bohr and Mottelson, as described by Froman [20]. The main point of these calculations was the description of the mother nucleus B in terms of the daughter nuclei plus two neutrons and two protons, as described in detail in [21]. For the case in which the daughter nucleus A is a magic core the mother wave function can be written as,

$$\varphi_B(\xi_B) = \sum_{\pi\nu} X(\pi\nu; B)[\phi_{A+2}^{\pi}(\xi_\pi)\phi_{A+2}^{\nu}(\xi_\nu)]_B \phi_A(\xi_A) \tag{12}$$

For superconducting nuclei this equation is also valid but the amplitudes are,

$$\phi_{A+2}(\xi) = \sum_{\Omega_1 < \Omega_2} v_{\Omega_1}^{A+2} u_{\Omega_2}^{A} \mathcal{A}[\varphi_{\Omega_1}(r_1)\varphi_{\Omega_2}(r_2)] \tag{13}$$

For spherical nuclei this expression is relatively easy to compute, but for a deformed nucleus one has to take into consideration the different directions that the decay process may choose. The corresponding formalism is very involved, as can be seen in Ref. [20]. Moreover, the approximations involved in the derivations of the corresponding equations in this case implies that the results are not reliable for large deformations, as seen in Fig. 1.

With the computimg facilities of today it is more convenient to expand the wave function in spherical waves. The l-spherical component of the deformed wave function, g_l, satisfies the equation,

$$\left\{-\frac{\hbar^2}{2m_\alpha}\frac{d}{dr^2} + \frac{\hbar^2 l(l+1)}{2m_\alpha r^2}\right\} g_l(r) + \sum_{l'} V_{ll'}(r) g_{l'}(r) = E_\alpha g_l(r) \tag{14}$$

At large distances the l-component of the outgoing wave function has the form,

$$S_l^{(k)}(R) \to \delta_{kl}[G_l(R) + iF_l(R)] \tag{15}$$

and the deformed wave function at R is

$$g_l(R) = \sum_k S_l^{(k)}(R) C_k \to C_l[G_l(R) + iF_l(R)] \tag{16}$$

β	$\Gamma_{WKB}/\Gamma_{exp}$	Γ_{cc}/Γ_{exp}
0.0	0.65	0.65
0.1	0.73	0.69
0.2	1.02	0.75
0.3	1.86	0.98
0.4	4.01	1.52
0.5	9.32	2.46
0.6	27.87	3.48
0.7	101.50	4.97
0.8	450.80	6.90

Figure 1. Ratio between the theoretical and experimental decay widths corresponding to the transition $^{152}Dy \rightarrow^{148} Gd + \alpha$ as a function of the quadrupole deformation parameter. The width Γ_{WKB} corresponds to the expression given in Ref. [20], while Γ_{cc} is the one provided by the coupled channel equations (14). (Taken from reference [23]).

The fitting of the internal wave function $g_l(R)$ and the external one, $G_l(R) + iF_l(R)$, allows one to evaluate C_l and the corresponding decay width, as in the derivation of Eq. (10).

A detailed description of these equations, and some applications, can be found in Ref. [22].

The first microscopical calculations of alpha decay were wrong by some six orders of magnitude [9, 10]. This pioneers in the microscopic treatments of alpha decay realized that the reason of this hughe difference was that, due to computing limitations, only one configuration was included in the calculations. It was found much later that that was indeed the case, and that the inclusion of high lying configurations increased the value of the decay width by five orders of magnitude [11]. At about the same time it was realized that the reason of this increase was due to the clustering of the two neutrons and two protons, which eventually constitute the alpha particle, induced by the pairing force [12]. Yet the decay width was still too small. It was thought that this was due to the neutron-proton interaction, which was not taken into account in that work. However, it was shown that even including the neutron-proton interaction the decay width was too small by about one order of magnitude. This indicates that there is an intrinsic component of preformed alpha particle in the wave function, as indeed was shown in Ref. [13]. This feature was recently exploited to describe the decay pattern of ^{212}Po, where electromagnetic transitions compete with alpha decay in some cases [14].

There have been attempts to generalize the Geiger-Nuttall law including all types of radioactive decays, i. e. alpha, proton and cluster emissions as well as all possible angular momenta. This has not been achieved yet. However important steps in this direction have been taken. In particular, generalizations of proton decay laws [15] as well as of alpha and cluster decays [16, 17, 18] have been presented recently.

4. Cluster decay

In the 1970s a breakthrough in radioactive decay was achieved by the theoretical prediction of cluster decay. This was possible through a formalism developed by A. Sandulescu for radioactive decay as an extension of the one corresponding to nuclear fission [24]. This is not surprising, since decay and fission are similar processes which are determined by the Coulomb and centrifugal barriers. It was through the analysis of the penetrability shown in Fig. 2 that the decay of ^{224}Ra by emission of a ^{14}C cluster as well as the emission of ^{24}Ne clusters from ^{230}Th was predicted. This was because, as seen in the figure, the penetrability corresponding to the decay of these clusters is large.

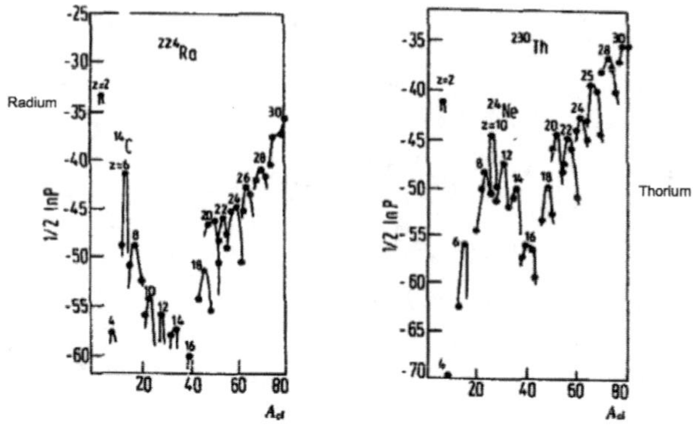

Figure 2. Logarithm of the penetrability as a function of the cluster mass corresponding to the systems $^{224}Ra +^{14} C$ and $^{230}Th +^{24} Ne$. (Taken from Ref. [25].

This prediction was confirmed experimentally several years afterward by Rose and Jones [26]. Naturally, it also induced many works performed to analyze this rare form of decay as well as the features in the nuclear structure that make the decay possible. It is beyond the scope of this talk to go through these developments in detail. For a review see Ref. [27]. However, it is worthwhile to mention that one of these investigations showed that the formation amplitudes of clusters (including protons) decreases exponentially with the mass of the cluster. This amplitude is of the order unity for protons, 10^{-2} for alpha particles, 10^{-4} for ^{8}Be, 10^{-5} for ^{14}C and 10^{-19} for ^{46}Ar [28].

One remarkable property of the penetrability shown in Fig. 2 is that it is extremely large, at least as large as for alpha decay, for clusters with mass number A=80. It is therefore not surprising the recent prediction that heavy clusters should be emittred with higher probability than alpha particles from superheavy nuclei [29]. This is an important property, since it may help to identify superheavy nuclei just by detecting the emitted heavy cluster.

Another important development in cluster radioactivity was the formulation of a model where the cluster (including alpha-particle) was assumed to move around the daughter nucleus (the core) under the influence of a central field. This potential was assumed to be very deep and therefore a harmonic oscillator approximation was adopted. Thus the cluster wave function contains a number of nodes given by the equivalent of the harmonic oscillator quantum number n. The principal quantum number is called G (global) and its value is G=2n+l. For decays into the lead region it is assumed G $\approx$ 20. There have been many publications related to this very successful model, which has been used even to study molecular structures in nuclei, as can be seen, e. g., in Refs. [30, 31] and references therein.

5. Summary

The study of radioactive decay processes have been fundamental in physics. To name but a few of its achievements one can mention,

It established the fundamental principle of the probabilistic interpretation of quantum mechanics.

It extended the study of physical processes to complex energies, This includes the introduction of propagators and S-matrix theories through which the treatment of resonances was possible.

For the case of many-body resonances one can mention the relatively recent shell model in the complex energy plane.

It has been a powerful tool to investigate the structure of nuclei and the correlations leading to nuclear clustering.

It has allowed one to determine the quantum numbers of nuclear levels and it seems that, in the near future, will also be a powerful tool to identify exotic, very heavy nuclei.

References

[1] Geiger H and Nuttall J M 1911 *Philos. Mag.* **22** 613
[2] Viola V E and T. Seaborg G T 1966 *J. Inorg. Nucl. Chem.* **28** 741
[3] Gamow G 1928 *Z. Phys.* **51** 204
[4] Hooft G, 2002 in *Conf. Proceedings "Quo Vadis Quantum Mechanics", Philadelphia arXiv : quant − ph*/0212095
[5] Zhang H F, Roger G and Li J Q 2011 **Phys. Rev.** C **84** 027303
[6] Betan R Id, Liotta R J, Sandulescu N, Vertse T and Wyss R 2005 *Phys. Rev.* C **72** 054322
[7] Thomas R G 1954 *Prog. Theor. Phys.* **12** 253
[8] Maglione E, Ferreira L S and Liotta R J 1998 *Phys. Rev.mLett.* **81** 538
[9] Rassmusen J O 1965 *Alpha-, Beta- and Gamma-Ray Spectroscopy* vol. 1 (North-Holland, Amsterdam)
[10] Mang H J 1960 *Phys. Rev.* **119** 1069
[11] Tonozuka I and Arima A 1979 *Nucl. Phys. A* **323** 45
[12] Janouch F and Liotta R J 1983 *Phys. Rev.* **27** 896
[13] Varga K, Lovas R G and Liotta R J 1992 *Phys. Rev. Lett.* **69** 37
[14] Delion D S, Liotta R J, Schuck P, Astier A and Porquet M -G 2012 *Phys. Rev.* **85C** 064306
[15] Delion D, Liotta R J and Wyss R 2006 *Phys. Rev. Lett.* **96** 072501
[16] Qi C, Xu F R, Liotta R J and Wyss R 2009 *Phys. Rev. Lett.* **103** 073501
[17] Poenaru D N, Gherghescu R A Greiner W 2012 *Phys. Rev.* **85C** 034615
[18] Ren Y Ren Z 2012 *Phys. Rev.* C **85** 044608
[19] Sandulescu A 1962 *Nucl. Phys.* **37** 332
[20] Froman A P O 1957 *Mat. Fys. Skr. Dan. Vid. Selsk.* **1** No. 3
[21] Lovas R G, Liotta R J, Insolia A, Varga K and Delion D S 1998 *Phys. Rep.* **294** 265
[22] D.S. Delion 2010 *Theory of particle and cluster emission* (Springer-Verlag, Berlin)
[23] Delion D S and Liotta R J 1998 *Phys. Rev.* C **58** 2073
[24] Sandulescu A 1977 *Heavy Ion Physics, Proc. Predeal International School* ed. V. Ceausescu and I. A. Dorobantu (Bucharest Central Institute of Physics) p 441
[25] Sandulescu A and Greiner W 1977 *J. Phys.* G **3** L189
[26] Rose H J and ones G A 1984 Nature **307** 245
[27] *Clusters in Nuclei I, Lecture Notes in Physics 818* 2011 (Springer, Berlin Ed. C. Beck)
[28] Iriondo M, Jerrestam M R. J. Liotta R J 1986 *Nucl. Phys.* **A454** 252
[29] Poenaru D N, Gherghescu R A and Greiner W 2011 *Phys. Rev. Lett.* **107** 062503
[30] Buck B, Merchant A C and Perez S M 1996 *Phys. Rev. Lett.* **76** 380
[31] Xu C *et al.* 2010 *Phys. Rev.* **81C** 054319

Int. Summer School for Advanced Studies 'Dynamics of open nuclear systems' (PREDEAL12) IOP Publishing
Journal of Physics: Conference Series **413** (2013) 012013 doi:10.1088/1742-6596/413/1/012013

Two center shell model description of superheavy element synthesis, fission and cluster decay

M. Mirea[1], **A. Sandulescu**[2,3] **and D.S. Delion**[1,3,4]

[1]Horia Hulubei National Institute for Physics and Nuclear Engineering, P.O. Box MG-6, Bucharest, Romania
[2]Institute for Advanced Studies in Physics, Romanian Academy, Bucharest, Romania
[3]Academy of Romanian Scientists, 54 Splaiul Independentei, Bucharest, Romania
[4]Bioterra University, 81 Garlei Str., Bucharest, Romania

E-mail: mirea@ifin.nipne.ro

Abstract. A two center shell model provides the possibility to characterize precisely scission configurations. Therefore, low energy disintegration processes could be described within such models. A Woods-Saxon two center shell model, recently elaborated, was used to investigate the cluster decay, the low energy fission and the synthesis of superheavy elements.

1. Introduction

The basic idea of the macroscopic-microscopic approach is that a macroscopic model, as the liquid drop one, describes quantitatively the smooth trends of the potential energy with respect to the particle number and deformation whereas a microscopic approach such as the shell model describes local fluctuations. The combined macroscopic-microscopic method should reproduce both smooth trends and local fluctuations. In this paper, the reliability of the latter approach is tested for cluster decay [1, 2, 3, 4, 5, 6], cold fission [7, 8] and superheavy element synthesis [9, 10, 11, 12] treated as cold rearrangement processes.

2. Two center shell model

In the macroscopic-microscopic method, the whole system is characterized by some collective coordinates that approximately determine the behavior of many other intrinsic variables. The basic ingredient in such an analysis is the shape parametrization that depends on several macroscopic degrees of freedom. The generalized coordinates associated with these degrees of freedom vary in time leading to a split of the nuclear system into two separate fragments. The macroscopic deformation energy is calculated within the liquid drop model. A microscopic potential must be constructed in order to be consistent with this nuclear shape parametrization. A microscopic correction is then evaluated using the Strutinsky procedure. The basic ingredient is the parametrization of the shape of the nuclei. In the following, an axial symmetric nuclear shape surface during the deformation process from one initial nucleus to the separated fragments is obtained by smoothly joining two spheroids of semi-axis a_i and b_i (i = 1,2) with a neck surface generated by the rotation of a circle around the axis of symmetry. By imposing the condition of volume conservation we are left with five independent generalized coordinates q_j (j = 1,5) that can be associated with five degrees of freedom: the elongation R given by the distance between the centers of the spheroids; the necking parameter $C = S/R_3$ related to the curvature of the neck, the eccentricities ϵ_i associated with the deformations of the nascent fragments and the mass asymmetry parameter $\eta = a_1/a_2$. A shape consistent two-center shell model with a Woods-Saxon potential was developed recently [13]. The Blomqvist-Walhlborn parametrization is adopted because it provides the same radius constant r_0 for the mean field and the pairing field. That ensures a consistency of the shapes of the two fields at hyperdeformations, i.e., two tangent ellipsoids. The Hamiltonian is obtained by adding the spin-orbit and the Coulomb terms to the Woods-Saxon potential. The eigenvalues are obtained by diagonalization of the Hamiltonian in the semi-symmetric harmonic two center basis [14, 15]. This model was extensively used in estimating the dissipated during disintegration processes [16, 17, 18, 19].

3. Cluster decay

The theoretical predictions [1, 2, 3, 4, 5, 6] for spontaneous emission of fragments with intermediate masses between fission products and alpha decay were confirmed experimentally in 1984 [20, 21, 22, 23]. Since that time, the exotic decay was investigated intensively. As evidenced in Ref. [24], unified approaches of binary disintegrations in a wide range of mass asymmetries as many body theories were used to reproduce the half-lives or to predict new decay modes [25]. All these cluster decay descriptions were made on a phenomenological basis in terms of experimental Q-values. The Q value gives a reference to the external barrier. From a theoretical point of view, the very good agreement obtained within the phenomenological models provides us three hints: the internal barrier obtained in the overlap region must be small, the effective mass of the reaction is well approximated by the reduced mass and the external barrier is well approximated within the Coulomb interaction.

Recently, the macroscopic-microscopic approach was developed to treat in a unitary manner the cluster decay and the cold fission process. The fine structure was explained within such models [26, 27, 28]. A magic valley in the ^{232}U and ^{238}Pu potential energy surfaces were evidenced [29, 30, 31, 32]. This valley belongs to a mass asymmetry consistent with the formation of the ^{208}Pb daughter and has its origin in the shell effects fluctuations. At the same time, a similar valley was obtained within the Hartree-Fock-Bogoliubov approximation [33, 34]. It is worth to mention that two valleys in the macroscopic-microscopic potential landscape were identified for the first time in Ref. [35]. One of these valleys corresponds to fission, while the second one, called fusion valley in Ref. [35] is related to a mass asymmetry compatible with the existence of a Pb daughter in a composite system. Our valley ressort from a calculation of the least action trajectory, by minimizing numerically the action integral from the ground state of the parent up to the exit point of the barrier. The method was initiated in Ref. [36] and was extensively used to describe the fission process [37, 38, 39, 40]. The results obtained for the ^{32}Si emission from ^{238}Pu are reproduced in the following, but the same qualitative behavior was obtained also for ^{24}Ne emission from ^{232}U. As a result of the minimization, the family of nuclear shapes for the cluster decay along the least action trajectory is plotted in Fig. 1. This figure reveals the fact that at a distance of 7 fm between the nascent fragments, the emitted cluster is clearly formed. A piece of matter of a volume amounting that of the emitted fragment begins to be separated in the compound system by a neck.

The neutron and proton single-particle diagrams are calculated along the minimal action trajectory, from the ground state of the parent nucleus and beyond the formation of two separated fragments. The neutron level scheme is plotted in Fig. 3. In Fig. 3, at $R \approx 0$, the parent nucleus is in a spherical configuration. For small deformations, the system evolves in a way similar to a Nilsson diagram for prolate deformations. In the left side of Fig. 3, the orbitals of the parent spherical nucleus are labeled by their spectroscopic notations. The levels of the emitted fragment can be identified and are plotted by thick lines. Both fragments are spherical after scission and their levels are bunched in shells. The levels of the light fragment are

Journal of Physics: Conference Series **413** (2013) 012013 doi:10.1088/1742-6596/413/1/012013

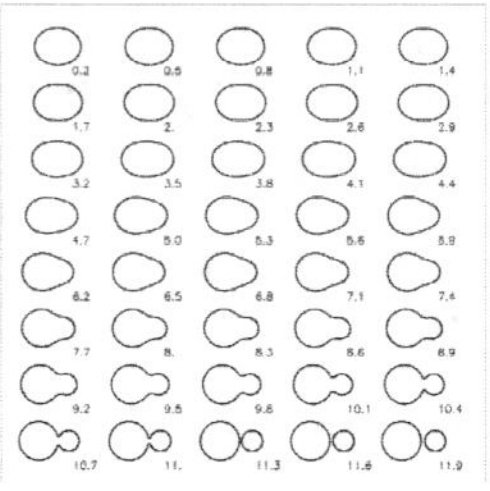

Figure 1. Family of shapes along the minimal action trajectory for cluster decay. The distance between the centers of the two fragments R in fm is marked for each shape.

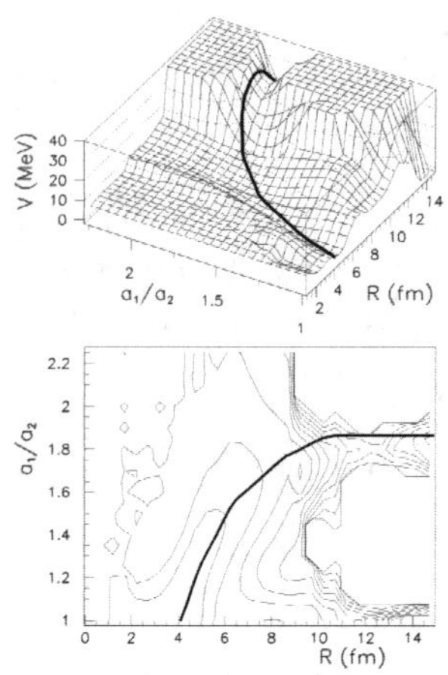

Figure 2. Upper part: potential energy surface V as function of the elongation R and the mass asymmetry $\eta = a_1/a_2$. Lower part: contour plot of the potential energy surface. The step between two equipotential curves is 4 MeV. The variations of the coordinates ϵ_1, ϵ_2, and C follow the least action path as function of R. The least action trajectory is plotted with a thick curve.

labeled by their spectroscopic notations. The shell effects reflect the large scale non-uniformity

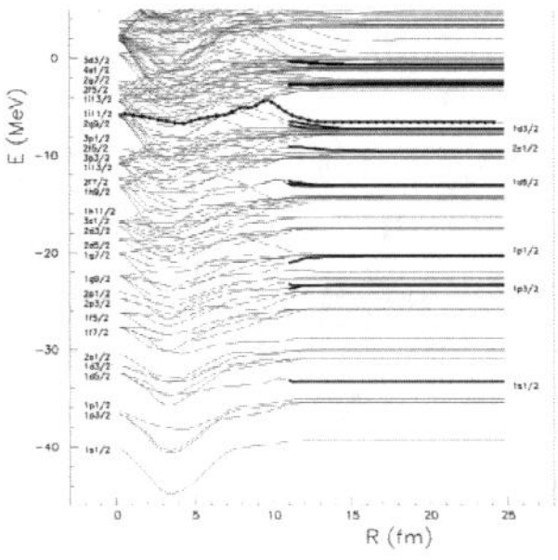

Figure 3. Neutron level diagram for ^{32}Si cluster decay from ^{238}Pu with respect to the elongation R. A dotted line is used for energy level that reaches asymptotically the N=126 single particle state of the daughter.

in the energy distribution of the individual particle states. For the nuclear binding the level distribution around the Fermi energy is of special importance. The nucleus is expected to be more bound if the level density around the Fermi energy is smaller. In Fig. 3 it is clear that a shell closure begins to appear at $R \approx$6-7 fm given by the gap that is formed around the N=126 daughter level. A similar phenomenon appears also in the case of the proton level scheme. This rearrangement of levels is translated in fluctuations of the shell effect energy values. The sum of shell and pairing corrections δU are displayed in Fig. 4. The ground state is stabilized at $R \approx$ 4 fm by these corrections. Notice that beyond $R \approx$ 7 fm, the total shell effect monotonically decreases up to the asymptotic final value. According to Fig. 1, at this elongation (7 fm) the emitted nucleus starts its preformation, that is the shape becomes necked in the median region of the nuclear shape.

The macroscopic-microscopic potential is plotted in the upper panel of Fig. 5 by a full line.

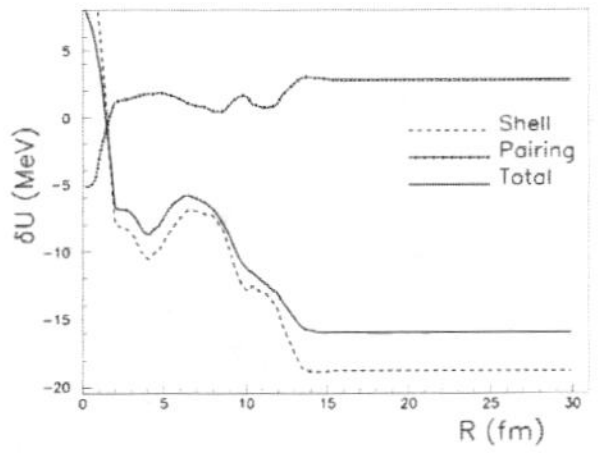

Figure 4. Total shell effects with pairing corrections δU as function of the internuclear distance R are plotted with a thick line. The partial contribution of shell effects alone is plotted with a dashed line while the pairing correction is represented with a dot dashed curve.

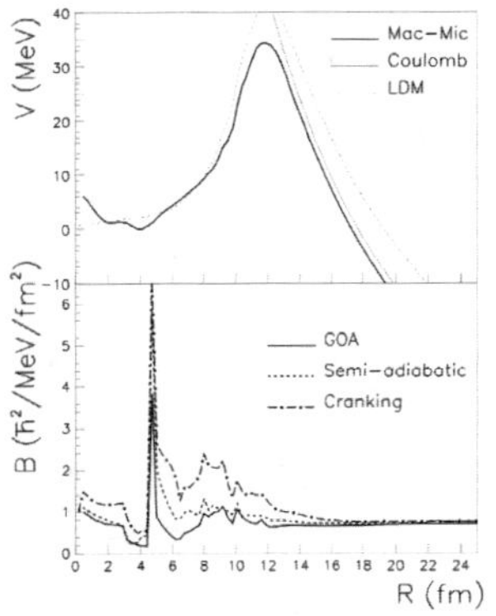

Figure 5. Upper panel: the macroscopic-microscopic potential barrier V for the ^{32}Si emission as a function of the elongation R is plotted by a thick curve. The thin curve represent the Coulomb energy $Z_1Z_2e^2/R - Q$ relative to the experimental Q-value of the process. The dashed line represents the barrier given by the liquid drop model, without shell effects. Lower panel: the effective mass along the minimal action trajectory calculated within the GOA (full curve), the semi-adiabatic (dashed curve) and the cranking model (dot-dashed curve).

The liquid drop energy is plotted by a dashed line. The Coulomb interaction corrected to the experimental Q-value, i.e. the phenomenological quantity $V_{COU} = Z_1Z_2e^2/R - Q$ is also plotted in order to test the validity of the model. In cluster decay, the major contribution of the action integral comes from the external region. The external barrier is very sensitive to the theoretical Q-value. In this way, it is possible to compare the external part of the barrier for the two cases and to estimate the deviation from the Q-value. The macroscopic-microscopic theoretical and the phenomenological values agree well enough in the external part of the barrier, the difference being less than 2 MeV. In the lower part of Fig. 5 the effective mass along the minimal action trajectory is plotted for three models [41, 42, 43]: cranking, Gaussian overlap approach (GOA) and semi-adiabatic formula. The three inertia exhibit a similar shell structure. The cranking model gives the larger values, the GOA the smaller ones, while the semi-adiabatic model has always intermediate values. The peak at $R \approx 4$ fm is especially due to the sudden variation of the asymmetry parameter a_1/a_2 as function of R. After the scission, the inertia reaches the values of the reduced mass.

Using the calculated values of the deformation energy and the inertia, the half-life of ^{32}Si emission was estimated by using a semi-empirical formula $T_{1/2} = 0.72 \times 10^{-21} P^{-1}$ $[s]$ where P is the WKB penetrability. We used the three above mentioned models for the inertia. The values of $\log(T_{1/2}[s])$ are 43.40, 26.73 and 31.89. They were obtained for the cranking model, the GOA and the semi-adiabatic formula, respectively. The experimental value is 25.30. Thus, a reasonable agreement is obtained when the inertia is calculated within the GOA.

We used this model to predict the best candidates for the cluster emission processes from

Table 1. Half-lives predictions for cluster decay.

Reaction	log(T[s]) Experiment [44]	log(T[s]) Theory	log(T[s]) Phenomenological [25]
^{24}Ne from ^{232}U	21.06	22.41	20.40
^{28}Mg from ^{232}U		23.13	24.50
^{22}Ne from ^{232}U		25.84	26.70
^{32}Si from ^{238}Pu	25.3	26.76	23.70
^{28}Mg from ^{238}Pu		27.68	24.80
^{30}Mg from ^{238}Pu		28.56	24.40

^{232}U and ^{238}Pu. The calculated half-lives are presented in table 1 and compared with phenomenological values. The theoretical estimations agree within two orders of magnitude with experimental data. These results represent a contribution towards an unitary treatment of fission and heavy ion emission.

4. Fission

Almost two decades ago, systematic measurements were performed to determine cold fission yields of ^{252}Cf [45, 46]. Our aim is to analyze these data and to show that the cold fission of ^{252}Cf is strongly connected with the cold valley of the double magic isotope ^{132}Sn, although the experimental cold yields have a maximum corresponding to a different charge number.

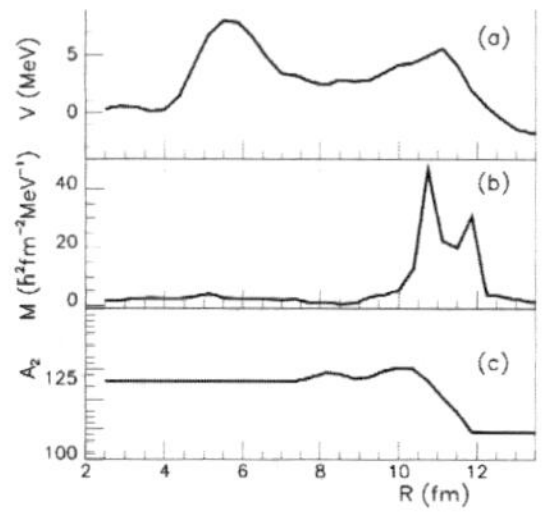

Figure 6. (a) Adiabatic fission barrier as function of the elongation R along the minimal action path. (b) Cranking inertia as function of R. (c) The estimated A_2 during the deformation as function of R.

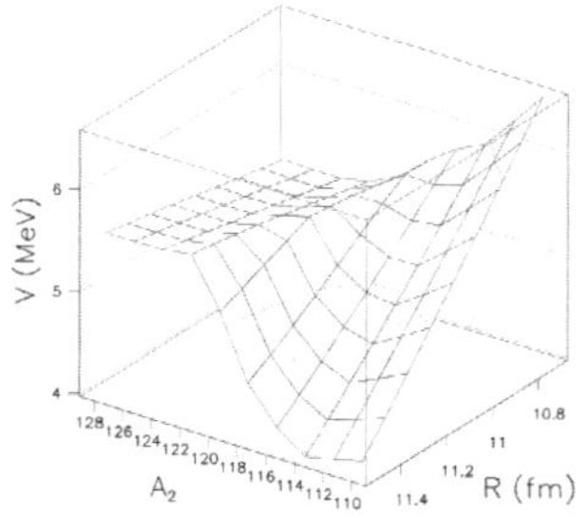

Figure 7. Deformation energy minimized statically with respect the eccentricity of the second fragment ε_2 and the mass asymmetry parameter η as function of R and A_2 in the second saddle region.

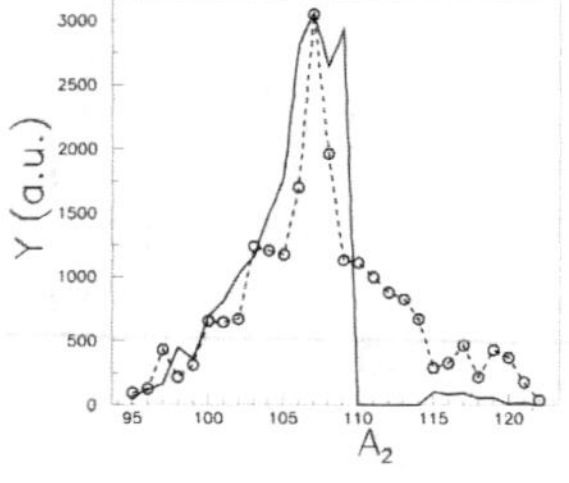

Figure 8. Experimental yields in arbitrary units (dashed line), compared with renormalized theoretical penetrabilities calculated within the microscopic-macroscopic model (solid line) with respect to A_2.

In the past the cold fission of ^{252}Cf was investigated within the double folding potential method [7, 8] emphasizing the role of higher deformations. In Ref [47], based on a macroscopic model by determining the tip distances for the exit point from the barrier for ground state deformed fragments, it was predicted that the major contribution in the yield distribution corresponds to the light fission fragment $A_2 \approx 100$, contradicting experimental data showing a peak at $A_2 = 107$. The root of this discrepancy can be understood in the following.

We extend the analysis performed in Ref. [48] to a more reliable microscopic approach to estimate the fission barrier, based on a new version of the Super Asymmetric Two Center Shell Model [13]. This approach was already used to describe the fusion/fission of some superheavy elements [49, 50] and of the dynamical effects in fission [13, 17].

For comparison with experimental data, the maximal values of the independent yields for a maximum excitation energy of 7 MeV were selected from Ref. [45]. The selected channels address binary partitions characterized by the following light fragments: ^{95}Rb, ^{96}Rb, ^{97}Sr, ^{98}Sr, ^{99}Y, ^{100}Y, ^{101}Zr, ^{102}Nb, ^{103}Zr, ^{104}Nb, ^{105}Mo, ^{106}Nb, ^{107}Mo, ^{108}Tc, ^{109}Mo, ^{110}Tc, ^{111}Ru, ^{112}Rh, ^{113}Ru, ^{114}Rh, ^{115}Rh, ^{116}Rh, ^{117}Pd, ^{118}Rh, ^{119}Pd, ^{120}Ag, ^{121}Cd and ^{122}Ag.

The adiabatic barrier in the multidimensional configuration space is determined by using the least action principle. In Fig. 6, the fission barrier V, the effective mass M and the estimated mass number A_2 are plotted along the fission path. The inertia is computed within the cranking approximation [42]. The second barrier top is located at R=11 fm and corresponds to a mass $A_2 \approx 120$. For a constant charge density, this ratio of the mass asymmetry addresses a heavy fragment with $A_1 - Z_1$=81 and Z_1=51, these values being close to magic numbers. Thus, the second saddle point corresponds to a partition that includes a double magic fragment. However, microscopic approaches to fission [51, 52] established that the second saddle point is asymmetrical with a value compatible with the observed mass ratio of the fragment distribution. Therefore, the dynamical saddle configuration obtained within our model is checked by minimizing statically the deformation energy around R=11 fm. The eccentricity ε_1 and the neck parameter C are kept constant. The detailed region is displayed in Fig. 7 confirming that the saddle point is located at $A_2 \approx 120$ and $R \approx 11$ fm.

Now we are interested in determining the fission barriers that address all the analyzed partitions. The asymptotic deformations of the two fragments are taken from the literature [53]. Thus, the shapes of the initial nucleus up to the second barrier and those of the final fragments are known. In order to avoid a complicated determination of the minimal action path for each partition, a linear variation from initial values of the generalized coordinates ε_i (i=1,2) and a_1/a_2 is postulated starting from the saddle of the second barrier configuration up to the final ones, characterizing the fragments at the end of the fission process.

With these ingredients, the calculated penetrabilities at zero excitation energies through the barrier are compared to experimental yields in Fig. 8. We implicitly assume that the penetration of the inner barrier is the same for all partitions, so that differences in the barrier transmission between channels are induced only by the external barrier.

We obtain a very good agreement between the theoretical penetrability distribution and the experimental yields for $A_2 < 110$. The maximum theoretical value is at A_2=107, while the maximum experimental yield is at the same value. A sudden drop of theoretical penetrabilities is theoretically obtained for channels with $A_2 > 110$. As previously noticed, this behavior is connected with the ground state shapes of fragments that become oblate for these channels. Perhaps these oblate shapes are not the best configurations during the penetration of the barrier, and the final ground state oblate configurations are obtained only after the exit from the fission barrier.

Concluding, we computed the cold fission path in the potential energy surface of ^{252}Cf by using the two center shell model, based on the idea of the cold rearrangements of nucleons during the cold fission process. We obtained a satisfactory agreement with experimental yields, by considering variable mass and charge asymmetry beyond the first barrier of the potential surface. We can see that the mass asymmetry changed from a symmetric to the asymmetric configuration in the vicinity of the second barrier, due to the influence of the magic numbers Z=50 and N=82, i.e. ${}^{120}_{47}Ag+{}^{132}_{51}Sb$ partition. It was shown that a good agreement with experimental data can be obtained only if the fission path proceeds through this saddle configuration. Thus, the cold fission process of ^{252}Cf can be called Sn-like radioactivity, similar with the Pb-like radioactivity, corresponding to various heavy cluster emission processes. We call all these processes shortly κ (cluster) decays. The peak in the final distribution corresponds to ^{107}Mo, due to the mass asymmetry degree of freedom, allowing a lower barrier from Sn to this nucleus.

5. Superheavy element synthesis and alpha decay

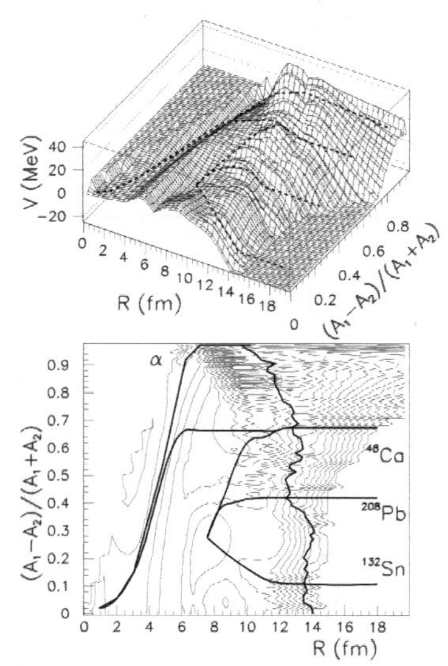

Figure 9. Fragmentation potential for the synthesis of the element $^{296}116$.

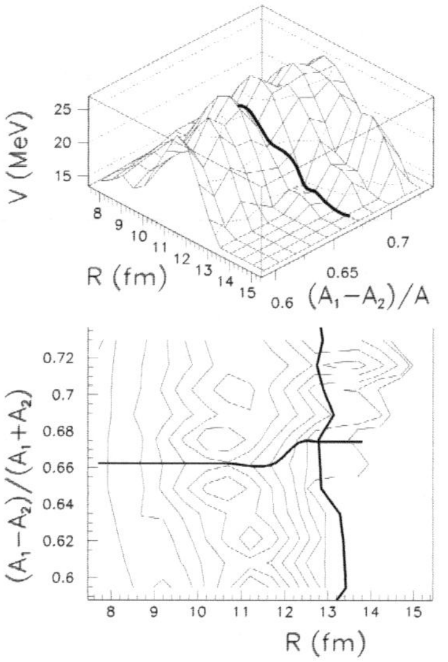

Figure 10. Potential energy for the entrance channel $^{248}Cm+^{48}Ca$ for the superheavy element synthesis.

The synthesis of superheavy elements beyond $Z = 104$, suggested by Flerov [54], was predicted within the so-called fragmentation theory in Ref. [9] by using the cold valleys in the potential energy surface between different combinations, giving the same compound nucleus. Soon it was shown in Refs. [12, 10, 11] that the most favorable combinations with $Z \geq 104$ are connected with the so-called Pb potential valley, i.e. the same valley of the heavy cluster emission [3].

Due to the double magicity of ^{48}Ca, similar with ^{208}Pb, in Ref. [10] it was proposed ^{48}Ca as a projectile on various transuranium targets. Indeed, the production of many superheavy elements with $Z \leq 118$ (corresponding to the last stable element Cf) during last three decades was mainly based on this idea [55, 56, 57, 58, 59, 60, 61, 62]. Another possibility was envisaged by using a Ni beam [63]. It was found that the fast fission channel is emphasized. Therefore, we decided to investigate the reaction $^{248}Cm+^{48}Ca$. The first step is to estimate the fragmentation potential with respect the internuclear distance for the projectile target optimum combinations. We computed the minimal value of the energy of all possible binary systems at the touching configuration that give rise to the $^{296}116$ superheavy element. The two interacting nuclei were considered at their ground state deformations. Afterwards, we computed the fragmentation

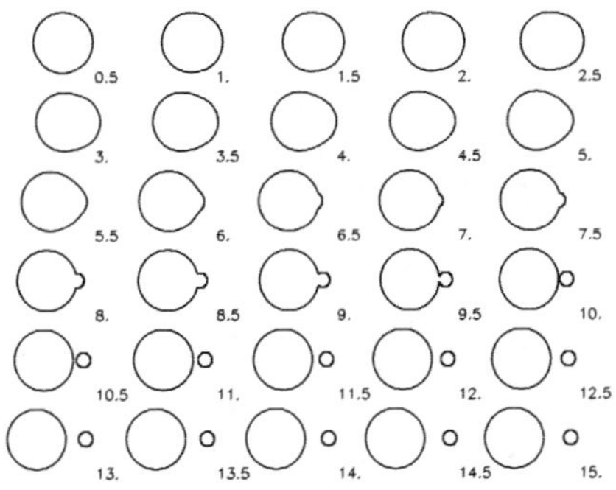

Figure 11. Family of shapes for the alpha-decay.

potential for the compound nucleus. The results are plotted in Fig. 9 as function of the distance between the centers of the fragments and the mass asymmetry. In the lower panel, a straight curve located between R=10-14 fm gives the location of the touching configuration for all possible partitions. It is possible to discern several minimums: the ground state minimum of the nearly spherical compound nucleus ($[A_1 - A_2]/[A_1 + A_2] \approx 0$ and $R \approx 0$ fm), an isomeric minimum for fission ($[A_1 - A_2]/[A_1 + A_2] \approx 0.3$ and $R \approx 7$ fm) and a molecular minimum for alpha decay ($[A_1 - A_2]/[A_1 + A_2] \approx 1$ and $R \approx 6$ fm). Several possible trajectories related to the formation of the superheavy element, the quasifission, the fission and the α-decay are also represented with full lines. These trajectories are plotted with dashed lines in the upper panel. It is interesting to note that the ^{48}Ca proceeds through a path that follows the minimal values of the potential energies up to the touching. When the two nuclei start to overlap, the trajectory shifts to smaller mass asymmetries to arrive in a pronounced valley. This valley starts from the ground state and reaches the molecular minimum for alpha decay. In Fig. 10, a detailed plot of the fragmentation potential in the region of the entrance channel ^{248}Cm+^{48}Ca is displayed. It can be observed that, due to shell effects, the maximal values of the fragmentation potential exhibits a structure that resemble to a serie of crenels. To form a superheavy element, the trajectory must proceed through a minimal value in this region, and the mass asymmetry changes towards lower values.

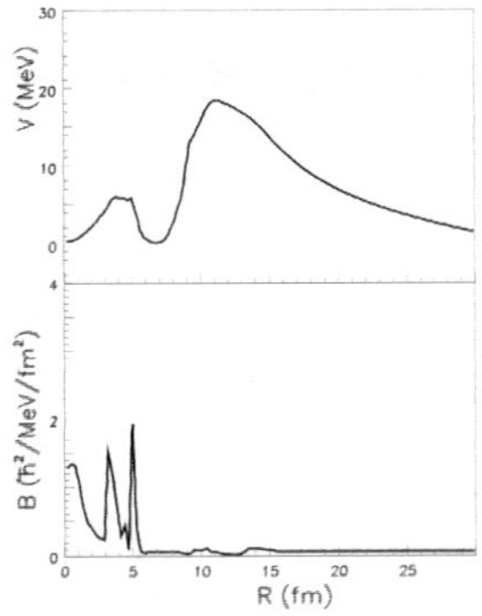

Figure 12. (a) Potential energy for the alpha emission along the alpha valley as function of the elongation R. (b) Effective mass.

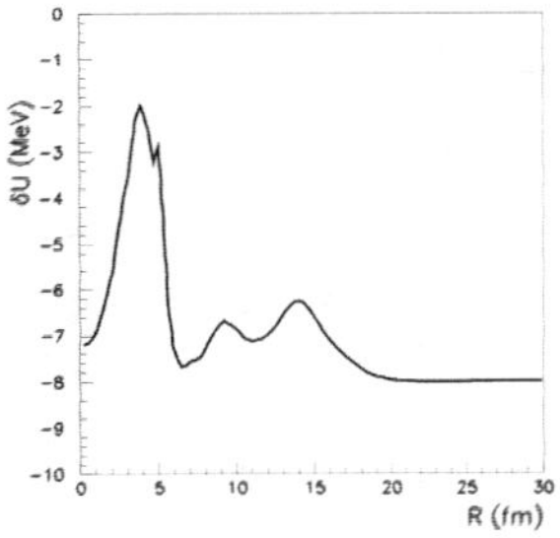

Figure 13. Shell effects as function of R.

Journal of Physics: Conference Series **413** (2013) 012013 doi:10.1088/1742-6596/413/1/012013

Table 2. Half-lives and Q-values of superheavy alpha decay.

	$^{290}116$	$^{291}116$	$^{292}116$	$^{293}116$	$^{296}116$
Q_α	11 MeV	10.89 MeV	10.80 MeV	10.67 MeV	9.95 MeV
T_α	15 ms	6.3 ms	18 ms	53 ms	206 s

The synthesis of superheavy elements and the fission was treated extensively in the literature within Hartree-Fock methods [64] and macroscopic-microscopic ones [65, 66]. It can be deduced from the potential landscape that the fission from the ground state of the compound nucleus proceeds through the deeper valley that belongs to Sn. The fission with Pb is obtained in peripheral reactions associated to quasi-fission. However, only few attempts exist for alpha-decay as superasymmetric fission [67, 68]. Most approaches compute the preformation probabilities and the penetration of the Coulomb barrier [69, 70, 71, 72, 73]. Therefore, we will focus on the alpha-decay valley. The shapes along the path denoted with the symbol α in Fig. 9 are plotted in Fig. 11. In the mentioned molecular minimum at $R \approx 6$ fm, an alpha particle begins to be formed on the surface of the daughter spherical nucleus. The alpha-decay barrier and the effective mass along the trajectory represented in Fig. 9 are displayed in Fig. 12. The origin of the molecular minimum can be understand if the shell effects are investigated. In Fig. 13 the Strutinsky corrections are plotted. A minimal value is obtained for the spherical parent nucleus. The values of the shell effects increase with the deformation. Once the alpha particle is formed on the surface, the nuclear system feels again a strong shell effect that is due to the spherical daughter, leading to a new energy minimum. Therefore, a molecular configuration is obtained. For larger R internuclear distances, the macroscopic part of the total energy increases and a barrier is created. The Q_α-value obtained with our model in 9.95 MeV. We computed phenomenologically the partial half-lifes and compared them with known experimental values obtained for other superheavy isotopes in Table 2. The predicted values for $^{296}116$ are well integrated in the systematics.

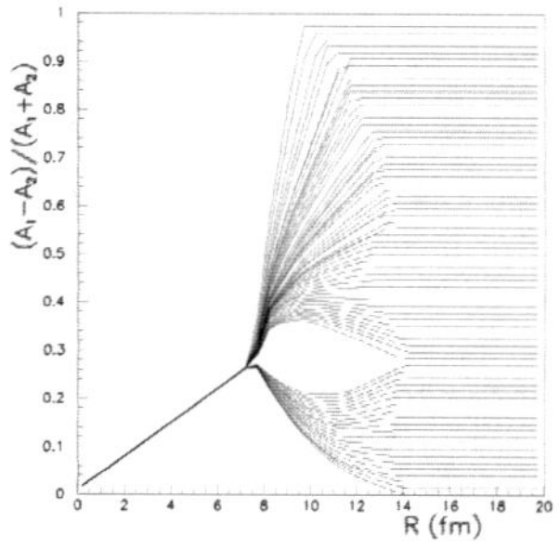

Figure 14. Selected trajectories between the isomeric minimum and the scission as function of the elongation and the mass asymmetry.

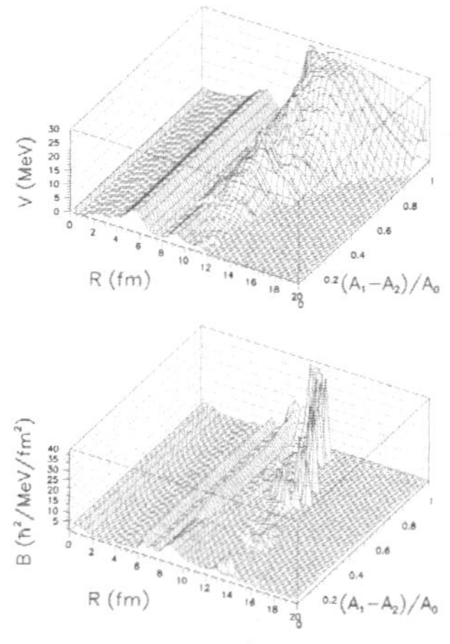

Figure 15. Deformation energy V and effective mass B in the cranking approximation for the trajectories plotted in Fig. 14.

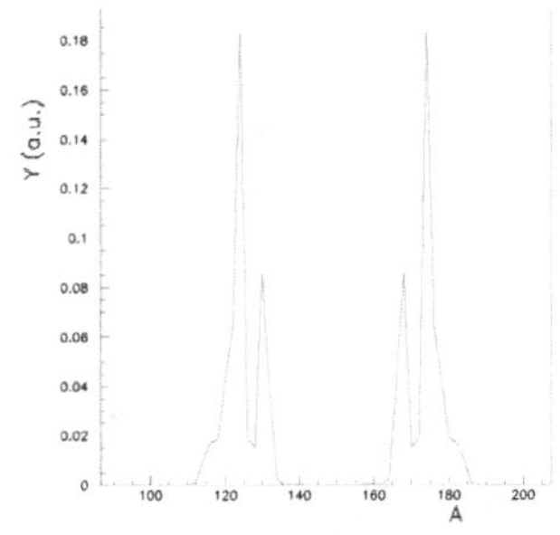

Figure 16. Normalized yields as function of the mass of the fragment.

In Ref. [74] the experimental mass distribution obtained in the reaction ^{48}Ca+^{248}Cm was measured at an excitation energy of 33 MeV. So, we check if the model can bring some information in this context. We constructed the possible paths that start from the isomeric minimum and connect the touching configurations plotted in Fig. 9. The paths are displayed in Fig. 14. We calculated the potential energy and the effective mass using the cranking approximation for these trajectories. Calculating the barrier, we take into account the fact that the shell effects vanish following a exponential dependence [75] $\delta E \exp(-T^2/T_0^2)$. It is known that T_0=1.5 MeV and $T \approx$1 MeV for and excitation energy of 33 MeV. The driving potentials and the effective masses are plotted in Fig. 15. The normalized yields for all these partitions were determined using the WKB method. The result is plotted in Fig. 16. So, the model evidence that a large yield in the mass distribution can be obtained around $A_2 \approx$125 if the system reaches the isomeric configuration. A such peak was found also in the experimental data [74].

Acknowledgments
Work supported by CNCS-UEFISCDI, project number PN-II-ID-PCE-2011-3-0068.

References

[1] Sandulescu A and Greiner W 1977 *J. Phys.* G **3** L189
[2] Sandulescu A, Lustig H J, Hahn J and Greiner W 1978 *J. Phys.* G **4** L279
[3] Sandulescu A, Poenaru D N and Greiner W 1980 *Sov. J. Part. Nucl.* **11** 528
[4] Sandulescu A 1989 *J. Phys.* G **15** 529
[5] Sandulescu A and Greiner W 1992 *Rep. Prog. Phys.* **55** 1423.
[6] Sandulescu A, Poenaru D N, Greiner W and Hamilton J H 1985 *Phys. Rev. Lett.* **54** 490
[7] Sandulescu A, Florescu A, Carstoiu F, Greiner W, Hamilton J H, Ramayya A V and Babu B R S (1996) *Phys. Rev.* C **54** 258
[8] Sandulescu A, Misicu S, Carstoiu F, Florescu A, and Greiner W 1998 *Phys. Rev.* C **57** 2321
[9] Sandulescu A, Gupta R K, Scheid W and Greiner W 1976 *Phys. Lett.* B **60** 225
[10] Gupta R K, Sandulescu A and Greiner W 1977 *Phys. Lett.* B **67** 257
[11] Gupta R K, Sandulescu A and Greiner W 1977 *Z. Naturforsch* **32a** 704
[12] Gupta R K, Parvulescu C, Sandulescu A and Greiner W 1977 *Z. Phys.* A **283** 217
[13] Mirea M 2008 *Phys. Rev.* C **78** 044618
[14] Mirea M 1996 *Phys. Rev.* C **54** 302
[15] Mirea M 2006 *Nucl. Phys.* A **780** 13
[16] Mirea M 2011 *Phys. Rev.* C **83** 054608
[17] Mirea M 2009 *Phys. Lett.* B **680** 316
[18] Mirea M 2012 *Phys. Lett.* B **717** 252
[19] Mirea M 2003 *Mod. Phys. Lett* A **18** 1809
[20] Rose H J and Jones G A 1984 *Nature* **307** 245
[21] Sandulescu A, Zamyatnin Yu S, Lebedev J A, Myasoedov B F, Tretyakova S P and Hasegan D 1984 *JINR Rapid Commun.* **5** 5
[22] Gales S, Hourani E, Hussonois M, Shapira H P and Vergnes M 1984 *Phys. Rev. Lett.* **53** 759

[23] Barwick S W, Price P B and Stevenson J D 1984 *Phys. Rev.* C **31** 1984
[24] D.S. Delion, *Theory of particle and cluster emission* (Springer-Verlag, Berlin, 2010)
[25] Poenaru D N, Greiner W, Depta K, Ivascu M and Sandulescu A 1986 *At. Data. Nucl. Data. Tables* **34** 423
[26] Mirea M 1999 *Eur. Phys. J* A **4** 335
[27] Mirea M 1998 *Phys. Rev* C **57** 2484
[28] Mirea M and Clapier F 1997 Europhys. Lett. **40** 509
[29] Mirea M, Sandulescu A and Delion D S 2011 *Nucl. Phys.* A **870-871** 23
[30] Mirea M, Sandulescu A and Delion D S 2011 *Proc. Rom. Acad. Series A* **12** 203
[31] Mirea M 2012 *Rom. J. Phys.* **57** 372
[32] Mirea M, Sandulescu A and Delion D S 2012 *Eur. Phys. J.* A **48** 86
[33] Warda M and Robledo L M 2011 *Phys. Rev.* C **84** 044608
[34] Warda M 2011 *Acta Phys. Pol.* B **42** 477
[35] Ichikawa T, Iwamoto A, Moller P and Sierk A J 2005 *Phys. Rev.* C. **71** 044608
[36] Poenaru D N, Mirea M, Greiner W, Cata I and Mazilu D 1990 *Mod. Phys. Lett.* A **5** 2101
[37] Companis I, Mirea M and Isbasescu A 2011 *Rom. J. Phys.* **56** 63
[38] Mirea M and Tassan-Got L 2011 *Centr. Eur. J. Phys* **9** 116
[39] Mirea M, Sandulescu A and Delion D S 2010 *Phys. Rev.* C **81** 044367
[40] Mirea M, Tassan-Got L, Stephan C, Bacri C O and Bobulescu R C 2007 *Phys. Rev.* C **76** 064608
[41] Mirea M 2012 *Int. J. Mod. Phys.* E **21** 1250035
[42] Mirea M 2011 *Rom. Rep. Phys.* **63** 676
[43] Mirea M and Bobulescu R C 2010 *J. Phys.* G **37** 055106
[44] Price P B 1989 *Nucl. Phys.* A **502** 41c
[45] Hambsch F-J, Knitter H-H, and Budtz-Jorgensen C 1993 *Nucl. Phys.* A **554** 209
[46] Knitter H-H, Hambsch F-J, and Budtz-Jorgensen C 1992 *Nucl. Phys.* A **536** 221
[47] Gonnenwein F and Borsig B (1991) *Nucl. Phys.* A **530** 27
[48] Delion D S and Sandulescu A 2007 *Rom. J. Phys.* **52** 43
[49] Mirea M, Sandulescu A and Delion D S 2009 *EPL* **85** 12001
[50] Stoica P 2011 *Rom. Rep. Phys* **63** 76
[51] Mirea M, Tassan-Got L, Stephan C, Bacri C O, Stoica P and Bobulescu R C 2005 *J. Phys.* G **31** 1165
[52] Mirea M, Tassan-Got L, Stephan C, Bacri C O and Bobulescu R C 2006 *Europhys. Lett.* **73** 705
[53] Moller P, Nix J R, Myers W D and Swyatecki W J 1995 *At. Data Nucl. Data Tables* **59** 185
[54] Flerov G N 1969 *Atom. Ener.* **26** 138
[55] Oganessian Y T 2001 *Nucl. Phys.* A **685** 17c
[56] Oganessian Y T *et. al.* 2004 *Nucl. Phys.* A **734** 109
[57] Oganessian Y T et al 2012 *Phys. Rev. Lett.* **108** 022502
[58] Oganessian Y T et al 2004 *Phys. Rev.* C **69** 021601(R)
[59] Oganessian Y 2007 *J. Phys.* G **34** R165
[60] Hofmann S and Munzemberg G 2000 *Rev. Mod. Phys.* **72** 733
[61] Hofmann S, Munzemberg, and Schadel M 2004 *Nucl. Phys. News* **14** 5
[62] Hofmann S 2012 *Rom. J. Phys.* **57** 214
[63] Kozulin E M et al 2010 *Phys. Lett.* B **686** 227
[64] Warda M and Egido J L 2012 *Phys. Rev.* C **86** 014322
[65] Misicu S and Greiner W 2002 *Phys. Rev.* C **66** 044606
[66] Zagrebaev V I 2001 *Phys. Rev.* C **64** 034606
[67] Mirea M 2001 *Phys. Rev.* C **63** 034603
[68] Mirea M 2010 *Buletinul Stiintific al Universitatii Politehnica din Timisoara* **55** 86
[69] Santhosh K P, Joseph J G, Priyanka B and Sahadevan S 2011 *J. Phys* G **38** 075101
[70] Silisteanu I and Budaca A I 2012 *Rom. J. Phys* **57** 493
[71] Budaca A I and Silisteanu I 2011 *Rom. Rep. Phys* **63** 1147
[72] Silisteanu I, Rizea A, Ciobanu B I and Neacsu A 2008 *Rom. J. Phys.* **53** 1191
[73] Mirzaei V and Miri-Hakimabad H 2012 *Rom. Rep. Phys.* **64** 50
[74] Itkis M G and al 2007 *Int. J. Mod. Phys. E* **16** 957
[75] Gupta R K and Greiner W, *Heavy Elements and Related New Phenomena*, p. 551 (World Scientific Singapore 1999)

$2\nu\beta\beta$ decay within a higher pnQRPA approach with the gauge symmetry preserved

A. A. Raduta[1,2)] **and C. M. Raduta**[1]

Institute for Physics and Nuclear Engineering, Atomistilor 407, RO-077125, POB-MG6, Bucharest, Romania

Academy of Romanian Scientists, 54 Splaiul Independentei, Bucharest 050094, Romania

E-mail: raduta@nipne.ro

Abstract. A many body Hamiltonian involving the mean field for a projected spherical single particle basis, the pairing interactions for alike nucleons, a repulsive dipole-dipole proton-neutron interaction in the particle-hole (ph) channel and an attractive dipole-pairing interaction is treated by a gauge restored and fully renormalized proton-neutron quasiparticle random phase approximation formalism. Application to the $2\nu\beta\beta$ decay rate show a good agreement with the corresponding data. The Ikeda sum rule is obeyed.

1. Introduction

The $2\nu\beta\beta$ process is interesting by its own but is also very attractive because it constitutes a test for the nuclear matrix elements (m.e.) which are used for the process of $0\nu\beta\beta$ decay. The discovery of this process may provide an answer to the fundamental question, whether neutrino is a Mayorana or a Dirac particle. The subject development is described by several review papers [1, 2]. The present talk refers to the $2\nu\beta\beta$ process, which is conceived as consisting of two consecutive virtual single β^- decays. The formalism yielding closest results to the experimental data is the proton-neutron random phase approximation ($pnQRPA$) which includes the particle-hole (ph) and particle-particle (pp) as independent two body interactions. The second leg of the $2\nu\beta\beta$ process is very sensitive to changing the relative strength of the later interaction, denoted hereafter by g_{pp}. It is worth mentioning that the ph interaction is repulsive while the pp one is attractive. Consequently, there is a critical value of g_{pp} for which the first root of the $pnQRPA$ equation vanishes. Actually, this is the signal that the $pnQRPA$ approach is no longer valid. Moreover, the g_{pp} value which corresponds to a transition amplitude which agrees with the corresponding experimental data is close to the mentioned critical value. That means that the result is not stable to adding corrections to the RPA picture. An improvement for the $pnQRPA$ was achieved by one of us (AAR), in collaboration, in Refs.[3], by using a boson expansion (BE) procedure. Another procedure, proposed in Ref.[4], renormalizes the dipole two quasiparticle operators by replacing the scalar components of their commutators with their average values. Such a renormalization is, however, inconsistently achieved since the scattering operators do not participate at the renormalization process. This lack of consistency was removed in Refs.[5] where a fully renormalized $pnQRPA$ ($FRpnQRPA$) is proposed. Unfortunately, all higher pnQRPA procedures mentioned above have the common drawback of violating the Ikeda sum rule (ISR) by an amount of about 20-30% [6]. It is believed that such a violation is caused by the gauge symmetry breaking. Consequently, a method of restoring this symmetry was formulated in Ref. [7].

Recently [8, 9], the results of Ref.[7] were improved in two respects: a) aiming at providing a unitary description of the process for the situations when the involved nuclei are spherical or deformed, here we use a projected spherical single particle basis; b) the space of proton-neutron dipole configurations is split in three subspaces, one being associated to the single β^- decay, one to the single β^+ process, and one spanned by the unphysical states. A set of *GRFRpnQRPA* (Gauge Restored and FRpnQRPA) equations is written down in the first two subspaces mentioned above, by linearizing the equations of motion of the basic transition operators corresponding to the two coupled processes.

Results are described according to the following plan. The approach is described in Section 2. Numerical applications and discussions are given in Section 3, while the final conclusions are drawn in Section 4.

2. Approximations and the model main assumptions

We suppose that the Gamow-Teller transitions dominate the Fermi ones which seems to be a reasonable hypothesis in medium and heavy nuclei. In the exact expression for the transition probability, the leptons energy is replaced by the average value: $\Delta E = mc^2 + \frac{1}{2}Q_{\beta\beta}$, where m denotes the rest mass of the emitted electron while $Q_{\beta\beta}$ the reaction heat of the process. Consequently, the half life is factorized:

$$\left[T^{2\nu}_{1/2}(0^+ \to 0^+)\right]^{-1} = F|M_{GT}|^2, \tag{1}$$

$$M_{GT} = \sqrt{3}\sum_m \frac{{}_i\langle 0||\beta^+||m\rangle_{ii}\langle m|m'\rangle_{if}\langle m'||\beta^+||0\rangle_f}{E_m + \Delta E_1},$$

where $\Delta E_1 = \Delta E + E_{1^+}$ and E_m are the *pnQRPA* energies. E_{1^+} denotes the experimental energy of the first 1^+ state. The GT transition operators for the single beta transitions are denoted by $\beta^{\mp}$. The model proposed by our group has two main ingredients:

a) The single particle basis is obtained from a deformed basis by projecting out the good angular momentum:

$$\begin{aligned}\Phi^{IM}_{nlj}(d) &= \mathcal{N}^I_{nlj}P^I_{MI}[|nljI\rangle\Psi_g] \equiv \mathcal{N}^I_{nlj}\Psi^{IM}_{nlj}(d), \\ \Psi_g &= exp[d(b^+_{20} - b_{20})]|0\rangle_b.\end{aligned} \tag{2}$$

The single particle energies are described by the average values of a particle-core Hamiltonian on the projected basis.

b)To describe the states involved in Eq.2 we used the following many body Hamiltonian:

$$\begin{aligned}H = &\sum_{\tau,\alpha,I,M}\frac{2}{2I+1}(\epsilon_{\tau\alpha I} - \lambda_{\tau\alpha})c^\dagger_{\tau\alpha IM}c_{\tau\alpha IM} \\ - &\sum_{\tau,\alpha,I,I'}\frac{G_\tau}{4}P^\dagger_{\tau\alpha I}P_{\tau\alpha I'} + 2\chi\sum_{pn;p'n';\mu}\beta^-_\mu(pn)\beta^+_{-\mu}(p'n')(-)^\mu \\ - &2X_{dp}\sum_{pn;p'n';\mu}\left(\beta^-_\mu(pn)\beta^-_{-\mu}(p'n') + \beta^+_\mu(pn)\beta^+_{-\mu}(p'n')\right)(-)^\mu.\end{aligned} \tag{3}$$

In the qp representation the Hamiltonian is expressed in terms of the dipole 2qp and dipole density operators:

$$\begin{aligned}A^\dagger_{1\mu}(pn) &= \sum C^{I_p\ I_n\ 1}_{m_p\ m_n\ \mu}a^\dagger_{pI_pm_p}a^\dagger_{nI_nm_n},\ B^\dagger_{1\mu}(pn) = \sum C^{I_p\ I_n\ 1}_{m_p\ -m_n\ \mu}a^\dagger_{pj_pm_p}a_{nI_nm_n}(-)^{I_n-m_n}, \\ A_{1\mu}(pn) &= \left(A^\dagger_{1\mu}(pn)\right)^\dagger,\ B_{1\mu}(pn) = \left(B^\dagger_{1\mu}(pn)\right)^\dagger.\end{aligned} \tag{4}$$

Linearized equations of motion of the above operators determine the dipole excitations of the many body system. Such equations are obtained by the mutual commutators:

$$\begin{aligned}\left[A_{1\mu}(k), A^\dagger_{1\mu'}(k')\right] &\approx \delta_{k,k'}\delta_{\mu,\mu'}\left[1-\frac{\hat{N}_n}{\hat{I}_n^2}-\frac{\hat{N}_p}{\hat{I}_p^2}\right],\\ \left[B^\dagger_{1\mu}(k), A^\dagger_{1\mu'}(k')\right] &\approx \left[B^\dagger_{1\mu}(k), A_{1\mu'}(k')\right]\approx 0,\\ \left[B_{1\mu}(k), B^\dagger_{1\mu'}(k')\right] &\approx \delta_{k,k'}\delta_{\mu,\mu'}\left[\frac{\hat{N}_n}{\hat{I}_n^2}-\frac{\hat{N}_p}{\hat{I}_p^2}\right],\ k=(I_p,I_n),\end{aligned}$$

with $\hat{N}_\tau$ denoting the quasiparticle number operator of type τ(=p,n). There are three distinct approximations for these equations: 1) *pnQRPA*; 2) Standard renormalized pnQRPA [4]; 3) Fully renormalized pnQRPA [5].

Denoting by $C^{(1)}_{I_p,I_n}$ and $C^{(2)}_{I_p,I_n}$ the averages of the right hand sides, with the renormalized pnQRPA vacuum state, the renormalized operators defined as

$$\bar{A}_{1\mu}(k)=\frac{1}{\sqrt{C_k^{(1)}}}A_{1\mu},\ \bar{B}_{1\mu}(k)=\frac{1}{\sqrt{|C_k^{(2)}|}}B_{1\mu}, \tag{5}$$

obey boson-like commutation relations:

$$\begin{aligned}\left[\bar{A}_{1\mu}(k), \bar{A}^\dagger_{1\mu'}(k')\right] &= \delta_{k,k'}\delta_{\mu,\mu'},\\ \left[\bar{B}_{1\mu}(k), \bar{B}^\dagger_{1\mu'}(k')\right] &= \delta_{k,k'}\delta_{\mu,\mu'}f_k,\ \ f_k=sign(C_k^{(2)}).\end{aligned} \tag{6}$$

Further, these operators are used to define the phonon operator:

$$C^\dagger_{1\mu}=\sum_k\left[X(k)\bar{A}^\dagger_{1\mu}(k)+Z(k)\bar{D}^\dagger_{1\mu}(k)\ -Y(k)\bar{A}_{1-\mu}(k)(-)^{1-\mu}-W(k)\bar{D}_{1-\mu}(k)(-)^{1-\mu}\right], \tag{7}$$

where $\bar{D}^\dagger_{1\mu}(k)$ is equal to $\bar{B}^\dagger_{1\mu'}(k')$ or $\bar{B}_{1\mu}(k)$ depending on whether f_k is + or -. The phonon amplitudes are determined by the equations:

$$\left[H, C^\dagger_{1\mu}\right]=\omega C^\dagger_{1\mu},\ \ \left[C_{1\mu}, C^\dagger_{1\mu'}\right]=\delta_{\mu\mu'}. \tag{8}$$

Unfortunately both the renormalized and fully renormalized pnQRPA violate the ISR by an amount of about 20-30%. The boson expansion procedure overestimate ISR, while the standard renormalized pnQRPA underestimate it. In Ref.[6] we have used a boson expansion formalism on the top of a renormalized pnQRPA. The result was that the departure of our predictions from ISR was diminished up to about 10%.

We believe that such a deviation from the ISR is caused by the fact that the renormalized ground state is not eigenstate of the nucleon total number operator.

Indeed, the state $C^\dagger_{1\mu}|0\rangle$, where $|0\rangle$ is the vacuum state for the phonon operator defined by the *FRpnQRPA* approach, with both the *ph* and *pp* interactions included, is a superposition of components describing the neighboring nuclei $(N-1,Z+1)$, $(N+1,Z-1)$, $(N+1,Z+1)$, $(N-1,Z-1)$. The first two components conserve the total number of nucleons (N+Z) but violate the third component of isospin, T_3. By contrast, the last two components violate the total number of nucleons but preserve T_3. Actually, the last two components are those which contribute to the ISR violation. However, one can construct linear combinations of the basic operators $A^\dagger, A, B^\dagger, B$ which excite the nucleus (N, Z) to the nuclei $(N-1, Z+1)$, $(N+1, Z-1)$, $(N+1, Z+1)$, $(N-1, Z-1)$, respectively. These operators are:

$$\begin{aligned}
\mathcal{A}^\dagger_{1\mu}(pn) &= -\left[c^\dagger_p c_{\tilde{n}}\right]_{1\mu}, \; \mathcal{A}_{1\mu}(pn) = -\left[c^\dagger_p c_{\tilde{n}}\right]^\dagger_{1\mu}, \\
\mathbf{A}^\dagger_{1\mu}(pn) &= \left[c^\dagger_p c^\dagger_n\right]_{1\mu}, \; \mathbf{A}_{1\mu}(pn) = \left[c^\dagger_p c^\dagger_n\right]^\dagger_{1\mu}.
\end{aligned} \tag{9}$$

In terms of the new operators, the many body model Hamiltonian is:

$$\begin{aligned}
H &= \sum_{\tau jm} E_{\tau j} a^\dagger_{\tau jm} a_{\tau jm} + 2\chi \sum_{pn,p'n';\,\mu} \sigma_{pn;p'n'} \mathcal{A}^\dagger_{1\mu}(pn)\mathcal{A}_{1\mu}(p'n') \\
&- X_{dp} \sum_{pn;p'n';\,\mu} \sigma_{pn;p'n'} (-)^{1-\mu} \left(\mathcal{A}^\dagger_{1\mu}(pn)\mathcal{A}^\dagger_{1,-\mu}(p'n') + \mathcal{A}_{1,-\mu}(p'n')\mathcal{A}_{1\mu}(pn)\right),
\end{aligned} \tag{10}$$

$$\sigma_{pn;p'n'} = \frac{2}{3\hat{I}_n \hat{I}_{n'}} \langle I_p||\sigma||I_n\rangle\langle I_{p'}||\sigma||I_{n'}\rangle. \tag{11}$$

The equations of motion of the operators involved in the phonon operator are determined by the commutation relations:

$$\left[\mathcal{A}_{1\mu}(pn), \mathcal{A}^\dagger_{1\mu'}(p'n')\right] \approx \delta_{\mu,\mu'}\delta_{j_p,j_{p'}}\delta_{j_n,j_{n'}} \left[U_p^2 - U_n^2 + \frac{U_n^2 - V_n^2}{\hat{I}_n^2}\hat{N}_n - \frac{U_p^2 - V_p^2}{\hat{I}_p^2}\hat{N}_p\right]. \tag{12}$$

The quasi-boson approximation replaces the r.h. side of the above equation by its average with the $GRFRpnQRPA$ vacuum state denoted by:

$$D_1(pn) = U_p^2 - U_n^2 + \frac{1}{2I_n+1}(U_n^2 - V_n^2)\langle\hat{N}_n\rangle - \frac{1}{2I_p+1}(U_p^2 - V_p^2)\langle\hat{N}_p\rangle. \tag{13}$$

Equations of motion show that the two qp energies are also renormalized:

$$E^{ren}(pn) = E_p(U_p^2 - V_p^2) + E_n(V_n^2 - U_n^2). \tag{14}$$

The space of the pn dipole states, $\mathcal{S}$, is written as a sum of three subspaces defined as:

$$\begin{aligned}
\mathcal{S}_+ &= \{(p,n)|D_1(pn) > 0, \; E^{ren}(pn) > 0,\}, \; \mathcal{S}_- = \{(p,n)|D_1(pn) < 0, \; E^{ren}(pn) < 0,\}, \\
\mathcal{S}_{sp} &= \mathcal{S} - (\mathcal{S}_+ + \mathcal{S}_-), \\
\mathcal{N}_\pm &= \dim(\mathcal{S}_\pm), \; \mathcal{N}_{sp} = \dim(\mathcal{S}_{sp}), \mathcal{N} = \mathcal{N}_+ + \mathcal{N}_- + \mathcal{N}_{sp}.
\end{aligned} \tag{15}$$

In $\mathcal{S}_+$ one defines the renormalized operators:

$$\bar{\mathcal{A}}^\dagger_{1\mu}(pn) = \frac{1}{\sqrt{D_1(pn)}}\mathcal{A}^\dagger_{1\mu}(pn), \bar{\mathcal{A}}_{1\mu}(pn) = \frac{1}{\sqrt{D_1(pn)}}\mathcal{A}_{1\mu}(pn), \tag{16}$$

while in $\mathcal{S}_-$ the renormalized operators are:

$$\bar{\mathcal{F}}^\dagger_{1\mu}(pn) = \frac{1}{\sqrt{|D_1(pn)|}}\mathcal{A}_{1\mu}(pn), \bar{\mathcal{F}}_{1\mu}(pn) = \frac{1}{\sqrt{|D_1(pn)|}}\mathcal{A}^\dagger_{1\mu}(pn). \tag{17}$$

An $pnQRPA$ treatment within $\mathcal{S}_{sp}$ would yield either vanishing or negative energies. The corresponding states are therefore spurious.

$FRpnQRPA$ with the gauge symmetry projected defines the phonon operator as:

$$\Gamma^{\dagger}_{1\mu} = \sum_k \left[X(k)\bar{\mathcal{A}}^{\dagger}_{1\mu}(k) + Z(k)\bar{\mathcal{F}}^{\dagger}_{1\mu}(k) \;-\; Y(k)\bar{\mathcal{A}}_{1-\mu}(k)(-)^{1-\mu} - W(k)\bar{\mathcal{F}}_{1-\mu}(k)(-)^{1-\mu}\right], \quad (18)$$

with the amplitudes determined by the $GRFRpnQRPA$ equations:

$$[H, \Gamma^{\dagger}_{1\mu}] = \omega\Gamma^{\dagger}_{1\mu}, \; [\Gamma_{1\mu}, \Gamma^{\dagger}_{1\mu'}] = \delta_{\mu,\mu'}. \quad (19)$$

In order to solve the $GRFRpnQRPA$ equations we need to know $D_1(pn)$ and, therefore, the averages of the qp's number operators, $\hat{N}_p$ and $\hat{N}_n$. These are written first in particle representation and then the particle number conserving term is expressed as a linear combination of $\mathcal{A}^{\dagger}\mathcal{A}$ and $\mathcal{F}^{\dagger}\mathcal{F}$ chosen such that their commutators with $\mathcal{A}^{\dagger}, \mathcal{A}$ and $\mathcal{F}^{\dagger}, \mathcal{F}$ are preserved. The final result is:

$$\langle\hat{N}_p\rangle = V_p^2(2I_p+1) + 3(U_p^2 - V_p^2)\Big(\sum_{\substack{n',k \\ (p,n')\in\mathcal{S}_+}} D_1(p,n')(Y_k(p,n'))^2 - \sum_{\substack{n',k \\ (p,n')\in\mathcal{S}_-}} D_1(p,n')(W_k^2)\Big),$$

$$\langle\hat{N}_n\rangle = V_n^2(2I_n+1) + 3(U_n^2 - V_n^2)\Big(\sum_{\substack{p',k \\ (p',n)\in\mathcal{S}_+}} D_1(p',n)(Y_k(p',n))^2 - \sum_{\substack{p',k \\ (p',n)\in\mathcal{S}_-}} D_1(p',n)(W_k^2)\Big).$$

$GRFRpnQRPA$ equations, the average qp numbers and the normalization factor equations are to be simultaneously considered and solved iteratively.

3. Results of the numerical analysis and discussions

The approach presented in the previous sections was applied for the transitions of fourteen double beta emitters. The parameters defining the single particle energies are those of the spherical shell model, the deformation parameter d and the parameter k are fixed as described in Ref.[10]. The proton and neutron pairing strengths are slightly different from those from the quoted reference since the dimension of the single particle basis used in the present paper is different from that from Ref.[10]. The strength χ was taken to be:

$$\chi = \frac{5.2}{A^{0.7}} MeV. \quad (5.1)$$

This expression was obtained by fitting the positions of the GT resonances in ^{40}Ca, ^{90}Zr and ^{208}Pb [13]. The strength for the attractive pn two-body interaction was chosen such that the result for the log ft value associated to one of the single beta decay of the intermediate odd-odd nucleus, be close to the corresponding experimental data. If the experimental data are missing, the restriction refers to the existent data in the neighboring region. Since for ^{100}Mo and ^{116}Cd, experimental data for the log ft values associated to the $\beta^{\pm}$ decays of the intermediate odd-odd nuclei ^{100}Tc and ^{116}In respectively, are available, the parameters χ and χ_1 were fixed such that the mentioned data are reproduced. For these cases, the results are compared with the data from [23] in Table 1. Let us just enumerate the results obtained with the formalism described above:

- The ISR is satisfied.
- We calculated the single $\beta^{\pm}$ strength distributions. For some of them experimental data are available. For example, β^- strength for the transitions ^{76}Ge→ ^{76}Se and ^{82}Se→ ^{82}Kr was extracted from the reactions ^{76}Ge(p,n)^{76}As, and ^{82}Se(p,n) ^{82}Br, respectively. The agreement of the calculated strength distribution and the corresponding experimental data is quite good. For illustration, four cases are presented in Fig. 1.

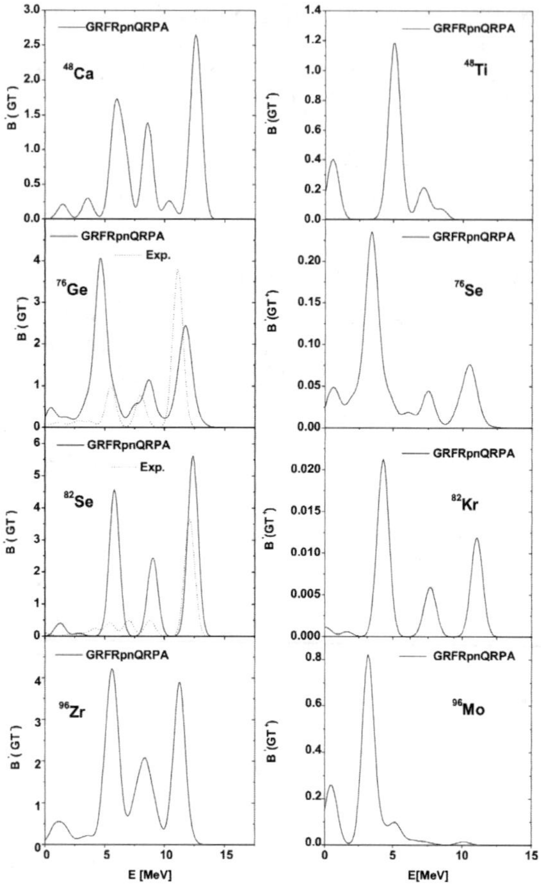

Figure 1. One third of the single β^- (left column) and one third of the β^+ (right column) strengths, denoted by $B'(GT^-)$ and $B'(GT^+)$, for the mother , ^{48}Ca, ^{76}Ge, ^{82}Se and ^{96}Zr, and daughter, ^{48}Ti, ^{76}Se, ^{82}Kr and ^{96}Mo, nuclei respectively, folded by a Gaussian function with a width of 1 MeV, are plotted as functions of the corresponding energies yielded by the present formalism. The experimental data for the β^- strengths of ^{76}Ge and ^{82}Se are also presented [14].

- Also, the summed single β^- and β^+ strengths, denoted conventionally by $\sum B(GT^-)$ and $\sum B(GT^+)$ respectively, were calculated and compared with the available experimental data. These single β decay total strengths quenched with a factor of 0.6 [12], accounting for the polarization effects on the single-β transition operator, ignored in the present paper, are listed in Table 1. Actually, the quenched values are to be compared with the experimental data, since the measured B(GT) strength represents about 60%-70% of the strength corresponding to the ISR.
The experimental value for the summed $B(GT^-)$ of ^{48}Ca is taken from Ref.[15], where from the total strength, which amounts about 15.3±2.2, the contribution of isovector spin monopole states was extracted. The result was obtained with the reaction ^{48}Ca(p,n)^{48}Sc, and corresponds to a large energy excitation interval, from 0 to 30 MeV.
In Ref.[14] the total GT strength, for ^{76}Ge and ^{82}Se, consists of the sum of the strength observed in the peaks plus the estimated contribution from the background. The experimental results correspond to 65 and 59% of the 3(N-Z) sum rule. According to Ref.[16], by adding to the GT cross section in discrete states the contribution from the background and that of continuum, the total strength magnitude is much improved to a better obey of the sum rule. We note a good agreement between the results of our calculations for the summed β^- strength and the corresponding experimental data.
The experimental data for the summed $B(GT^+)$ transition of ^{48}Ti, was taken from Ref.[15]. This result was obtained after extracting the contribution of the isovector spin monopole states from the total strength of 2.8±0.3. The reaction ^{48}Ti(n,p)^{48}Sc was used to study the

Table 1. The calculated summed strengths for the β^- strength associated to the mother nuclei and the summed β^+ strengths for the daughter nuclei, quenched by a factor 0.6, are are compared with the corresponding available data. Experimental data for total B(GT$^-$) are taken from Refs. [15] ($^{a)}$), [14] ($^{b)}$),[18] ($^{c)}$), [21] ($^{d)}$).

Nucleus	$0.6\sum B(GT)^-$	$\sum [B(GT)^-]_{exp}$
^{48}Ca	14.54	14.4±2.5 $^{a)}$
^{76}Ge	23.037	23.3 $^{b)}$
^{82}Se	25.372	24.6 $^{b)}$
^{96}Zr	29.163	-
^{104} Ru	32.921	-
^{110}Pd	32.932	-
^{128}Te	43.485	40.08 $^{b)}$
^{130}Te	47.432	45.90 $^{b)}$
^{148}Nd	51.74	-
^{150}Nd	54.11	-
^{154}Sm	54.68	-
^{160}Gd	57.93	-
Nucleus	$0.6\sum B(GT)^+$	$\sum [B(GT)^+]_{exp}$
^{48}Ti	3.666	1.9±0.5 $^{a)}$
^{76}Se	1.125	1.45±0.07 $^{c)}$
^{82}Kr	0.079	-
^{96}Mo	2.537	0.29±0.08 $^{d)}$
^{104}Pd	3.990	-
^{110}Cd	7.239	-
^{128}Xe	2.917	-
^{130}Xe	13.040	-
^{148}Sm	1.29	-
^{150}Sm	0.02	-
^{154}Gd	0.54	-
^{160}Dy	0.21	-

$B(GT^+)$ strength for excitation energies up to 30 MeV. This value for the total strength is larger than that reported by Alford *et al.*, in Ref. [17]

$$\sum B(GT^+) = 1.42 \pm 0.2. \tag{5.2}$$

where only contribution of states with excitation energies up to 15 MeV are taken into account. This comparison shows that, indeed, the B(GT) strength is sensitive to the magnitude of the considered energy interval. In this context we mention the results obtained through the charge exchange reactions (^{3}He,t) and (d,^{2}He) on ^{48}Ca and ^{48}Ni respectively [19], for $B(GT^-)$ and $B(GT^+)$ with an excitation energy interval $E_x \leq 5$ MeV: 1.43(38), 0.45.

The GT strength from the ^{76}Se(n,p)^{76}As reaction [18] is 1.45 ± 0.07 and corresponds to and excitation energy $E_x \leq 10MeV$. The authors used the multipole decomposition method. In Ref.[20] the $B(GT^+)$ strength was measured in a different reaction, ^{76}Se(d,^{2}He)^{76}As, and different excitation energy interval, $E_x \leq 4MeV$. The result reported is $\sum_{0-4MeV} B(GT^+) = 0.54 \pm 0.1$, which is smaller than that from Ref.[18]. The length of the energy intervals justifies the mentioned differences. We remark that the results

Table 2. The strengths B(GT) of the single β^- transitions from the mother nuclei to the intermediate odd-odd nuclei excited in the states of the two components, GTR1 and GTR2, of the GT giant resonance are listed. The experimental [23] (Exp.) and theoretical (Th.) values for the centroid energies are also specified.

Exc.	^{100}Tc			
st.	Ex [MeV]		B(GT)	
	Exp.	Th.	Exp.	Th.
G1	13.3	11.16	23.1± 3.8	15.63
G2	8.0	8.05	2.9±0.5	5.87
	^{116}In			
	Ex [MeV]		B(GT)	
	Exp.	Th.	Exp.	Th.
G1	14.5	12.37	25.8± 4.1	18.9
G2	8.9	7.87	6.6±1.1	7.2

for the summed β^+ strength in ^{48}Ti and ^{76}Se are in reasonable good agreement with the corresponding experimental data.
The last strength mentioned in Table 2 refers to the daughter nucleus ^{96}Mo. Through the reaction ^{96}Mo(d,^{2}He)^{96}Nb the strength taken mainly by a single state, placed at 0.69 MeV, was measured. However, from Fig.1 we note that, indeed, there is a state at 0.69 Mev which catch a certain β^+ strength, but that strength is smaller than that distributed among the states lying in the energy interval of 1.8 to 7.5 MeV. More complete measurement through a (p, n) reaction on ^{96}Mo and an energy range of 0-10 MeV is necessary in order to make a fair comparison with the results presented here.
The quenched values of the total β^- strength of ^{128,130}Te are compared with the experimental data since the measured $B(GT^-)$ strength, as we already mentioned before, represents about 56% and 59% respectively, of the strength corresponding to the ISR. There are some claims [16] saying that adding the strength carried by the states from the continuum, the total B(GT) strength are corrected up to 90% of the simple sum rule. We remark the good agreement between the calculated and experimental total strength. Note that if we replace the quenching factor by 0.56 for ^{128}Te and by 0.59 for ^{130}Te the results for the total strength would be 40.586 and 46.56 respectively which are closer to the experimental data. Unfortunately for the last four mother and for the last four daughter nuclei, there are no data available for the single β^- and single β^+ strengths, respectively.

- Experimental value [23] of the transition $0_i^+ \rightarrow 1^+$ m.e. describing the β^- strength of ^{100}Mo and ^{116}Cd was derived from the reactions output ^{100}Mo(^{3}He,t)^{100}Tc, and ^{116}Cd(^{3}He,t)^{116}In at $\theta_t \approx 0^0$, while the m.e. $1^+ \rightarrow 0_f^+$ from the corresponding experimental log ft value. These quantities are compared with the results of our calculations in Table 2.
- Transition amplitudes and half lives were calculated for 14 double beta emitters and the results are shown in Table 3.
- The $\log ft$ values associated with the single beta transitions of the intermediate odd-odd nucleus to the daughter and mother nuclei respectively, were calculated. Results are given in Table 4.

Table 3. The Gamow-Teller transition amplitude for the $2\nu\beta\beta$ process, in units of MeV^{-1}, and the corresponding half life ($T_{1/2}$), in units of yr, are listed. The references list for experimental data is given in Ref.[25, 22].

	M_{GT}	$T_{1/2}$[yr] present	Exp.	Klapdor *et al*
^{48}Ca	0.045	4.72×10^{19}	$4.2\pm1.2) \times 10^{19}$	3.2×10^{19}
^{76}Ge	0.177	0.938×10^{21}	$1.5\pm0.1 \times 10^{21}$	2.61×10^{20}
^{82}Se	0.083	1.293×10^{20}	$1.1^{+0.8}_{-0.3} \times 10^{20}$	0.85×10^{20}
^{96}Zr	0.115	1.59×10^{19}	$(1.4^{+3.5}_{-0.5}) \times 10^{19}$	5.2×10^{17}
^{100}Mo	0.221	8.79×10^{18}	$(8.0\pm0.16) \times 10^{18}$	2.9×10^{18}
^{104}Ru	0.453	2.26×10^{21}	-	1.8×10^{21}
^{110}Pd	0.188	3.11×10^{20}	-	1.2×10^{21}
^{116}Cd	0.160	2.02×10^{19}	$(3.2\pm0.3) \times 10^{19}$	5.1×10^{19}
^{128}Te	0.056	1.43×10^{24}	$(7.2\pm0.3) \times 10^{24}$	1.2×10^{23}
^{130}Te	0.023	1.56×10^{21}	$(1.5\text{-}2.8)\times10^{21}$	1.9×10^{19}
^{148}Nd	0.422	2.00×10^{19}	-	1.19×10^{21}
^{150}Nd	0.042	2.50×10^{19}	$\geq 1.8 \times 10^{19}$	1.66×10^{19}
^{154}Sm	0.303	2.02×10^{21}	-	1.49×10^{22}
^{150}Gd	0.111	1.02×10^{21}	-	2.81×10^{21}

Table 4. The $\log ft$ values characterizing the β^+/EC and β^- processes associated to the intermediate odd-odd nuclei are listed.

Mother			odd-odd		Daughter
^{48}Ca	Th.	8.44	^{48}Sc	4.63	^{48}Ti
^{76}Ge	Th.	4.57	^{76}As	6.13	^{76}Se
^{82}Se	Th.	8.11	^{82}Br	7.18	^{82}Kr
^{96}Zr	Th.	5.67	^{96}Nb	7.00	^{96}Mo
^{100}Mo	Exp.	$4.45^{+0.18}_{-0.30}$	^{100}Tc	4.66	^{100}Ru
	Th.	4.65		4.1	
^{104}Ru	Exp.	4.32	^{104}Rh	4.55	^{104}Pd
	Th.	4.71		6.47	
^{110}Pd	Exp.	4.08	^{110}Ag	4.66	^{110}Cd
	Th.	4.14		6.32	
^{116}Cd	Exp.	$4.45^{+0.18}_{-0.30}$	^{116}In	4.66	^{116}Sn
	Th.	4.65		4.1	
^{128}Te	Exp.	5.049	^{128}	6.061	^{128}Xe
	Th.	5.87		6.06	
^{130}Te	Th.	6.08	^{130}I	5.80	^{130}Xe
^{148}Nd	Th.	6.8	^{148}Pm	7.33	^{148}Sm
^{150}Nd	Th.	5.55	^{150}Pm	8.46	^{150}Sm
^{154}Sm	Th.	5.52	^{154}Eu	5.13	^{154}Gd
^{160}Gd	Th.	5.25	^{160}Tb	4.20	^{160}Dy

4. Summary and conclusions

Summarizing the results of this paper, one may say that *restoring the gauge symmetry from the fully renormalized pnQRPA provides a consistent and realistic description of the transition rate and, moreover, the ISR is obeyed.*

As shown in the quoted original paper, it seems that there is no need to include the *pp* interaction in the many body treatment of the process. Indeed, in the framework of a *pnQRPA* approach this interaction violates the total number of particles and consequently the gauge projection process makes it passive. The proton-neutron correlations in the ground state are however determined by an attractive dipole pairing interaction. The results of our calculations are compared with those obtained by different methods as well as with the available experimental data. The strength of the *ph* interaction was taken as given by Eq.(5.1), while the one for the dipole-pairing interaction was approximately fixed such that one decay branch of the intermediate odd-odd nucleus has the $\log ft$ value close to those known for the given nuclei or for the nuclei belonging to the neighboring region. Small deviations of the predicted and experimental *GT* resonance centroids suggest that the parameter χ should be fixed by fitting the centroids within the *GRFRpnQRPA*. By contrast to the standard *pnQRPA* models where the strength of the *pp* interaction is not affecting the position of the GT resonance centroids, here the attractive interaction contributes to the distribution of the β^- strength. Therefore, the two strengths should be fixed at a time by fitting two data, either the GT resonance centroid and the log ft value of one decay of the intermediate odd-odd nuclei or by fixing the log ft values corresponding to the single beta decays of the odd-odd intermediate nucleus.

Acknowledgments

This work was supported by the Romanian Ministry for Education Research Youth and Sport through the CNCSIS project ID-2/5.10.2011.

References

[1] Suhonen J *et al.*, 1998, *Phys. Rep.* **300**, 123.
[2] Raduta A A, 2002, *Prog. Part. Nucl. Phys.* **48**, 233.
[3] Raduta A A *et al*, 1991, *Nucl. Phys.* **A534**, 149.
[4] Toivanen J *et al*, 1995, *Phys. Rev. Lett.* **75**, 410.
[5] Raduta A A *et al*, 1998, *Nucl. Phys.* **A 634**, 497.
[6] Raduta A A *et al*, 2000, *Jour. Phys.* **G 26**, 793.
[7] Raduta CM *et al*, 2005, *Nucl. Phys.* **A 756**, 153.
[8] Raduta C M *et al*, 2010, *Phys. Rev.* **C 82**, 068501.
[9] Raduta C M and Raduta A A, 2011, *Journal of Physics G; Nucl. Part. Physics,* **38**, 055102.
[10] Raduta A A *et al*, 2005, *Phys. Rev.* **C 71**, 164.
[11] Rodin V *et al*, 2002, *Phys. Rev.* **C 66**, 051303(R).
[12] Zamick L *et al*, 1982, *Phys. Rev. C* 26, 2185.
[13] Homma H *et al*, *Phys. Rev.* **C 54**, 2972.
[14] Madey R *et al*, Phys. Rev. C **40**, 540 (1989).
[15] Yako K *et al*, 2009, *Phys. Rev. Lett.* **103**, 012503.m.
[16] Anderson B D *et al*,1985, *Phys. Rev.* **C 31**, 1161.
[17] Alford W P *et al*, 1990, *Nucl. Phys.* **A 514**, 49.
[18] Helmer R *et al*, 1997, *Phys. Rev.* **C 55**, 2802.
[19] Grewe E -W *et al*, 2007, *Phys. Rev.* **C 76**, 054307.
[20] Grewe E -W *et al*, 2008, *Phys. Rev.* **C 78**, 044301.
[21] Dohmann H *et al*, 2008, *Phys. Rev.* **C 78**, 041602(R).
[22] Barabash A S, 2010, *Phys. Rev.* **C 81**, 035501.
[23] Akimune H *et al*, 1997, *Phys. Lett.* **B 394**, 23.
[24] Klapdor H Vit et al, Phys. Lett. **142**, 323; Hirsch M *et al*, 1994, *Phys. Rep.* **242** (1994).
[25] Raduta C M *et al*, 2011, *Phys. Rev.* **C82**, 064322.

Systematics of fine structure in the α decay of deformed even-even nuclei

Dongdong Ni[1] **and Zhongzhou Ren**[1,2,3]
[1]Department of Physics, Nanjing University, Nanjing 210093, China
[2]Kavli Institute for Theoretical Physics China, Beijing 100190, China
[3]Center of Theoretical Nuclear Physics, National Laboratory of Heavy-Ion Accelerator, Lanzhou 730000, China

E-mail: dongdongnick@gmail.com and zren@nju.edu.cn

Abstract.
We present detailed systematics of the results for α transitions to excited 2^+, 4^+, and 6^+ states in the α decay of deformed even-even nuclei. Both the semiclassical and coupled-channel results are analyzed in the valence nucleon scheme. This provides a good test of the reliability of the various theoretical models for the α-decay fine structure. It is found that the coupled-channel results agree well with the experimental data in both systematic behavior and magnitude, while the semiclassical results have relatively large deviations from the experimental data. This confirms the good reliability of the coupled-channel studies.

1. Introduction

Alpha decay has been investigated for a long time in heavy and superheavy nuclei. The decay energies pose a tough test for nuclear mass models. The half-lives provide information on the stability of nuclides especially for superheavy nuclei. Since the pioneering work of Gamow [1], the experimental α-decay half-lives have been interpreted with improved accuracy for both spherical and deformed α-emitters [2, 3, 4, 5, 6, 7, 8, 9, 10, 11, 12]. Here we take deformed emitters for example. The semiclassical methods, based on the one-dimensional Wentzel-Kramers-Brillouin (WKB) approximation, evaluate the direction-dependent penetration probability of α particles and then average the probability along all directions [5, 6, 8]. The coupled-channel methods, based on the three-dimensional Schrödinger equation, use various channel wave functions with outgoing wave boundary conditions to determine partial widths and then total α-decay half-lives [11, 12]. These two groups, in spite of different consideration, give similarly precise descriptions of total α-decay half-lives. Nevertheless, many deformed emitters show a line spectrum of α-groups corresponding to α transitions to various daughter states, which is confirmed by the fact that the energy differences between various α-groups fit with γ-rays [13]. It was discovered by Rosenblum in 1929 [14], namely, α-decay fine structure. Sine the fine structure is a rich source of nuclear structure information, extensive theoretical studies have been devoted to pursuing a quantitative description of such an interesting phenomenon. Recently the fine structure in the α decay of deformed even-even nuclei has been widely investigated from both semiclassical and coupled-channel sides [15, 16, 17, 18, 19, 20, 21, 22, 23]. In addition, systematics of hindrance factors in the α decay of even-even trans-lead nuclei has been achieved as a function of the collectivity indicators [24].

Within the coupled-channel framework, it has been clearly shown that the α transitions to excited states are closely related to the structure properties of daughter nuclei such as nuclear deformation and excitation spectrum [20, 22, 23]. In other words, the fine structure observed in α decay is an important and sensitive probe of the structure properties of daughter nuclei. From the viewpoint of nuclear structure, it has been known that the evolution of nuclear structure is closely related to the integrated strength of the residual valence p-n interaction, and the simple valence nucleon product N_pN_n is employed to gauge this interaction [25]. Various structure quantities have been displayed to follow certain simple trends in the N_pN_n scheme [26, 27, 28, 29], such as nuclear deformation, $B(E2)$ values, ground band energies of even-even nuclei, core cluster decompositions in the rare earth region, yrast energies of even-even nuclei, and so on. With this in mind, it is expected that the α-decay fine structure, associated with nuclear structure, also follows one certain trend in the N_pN_n scheme. The objective of this article is to perform systematics of various theoretical results for the α-decay fine structure within the valence nucleon scheme. By comparison with systematics of the experimental data, one can discern the reliability of the theoretical models developed for the α-decay fine structure.

2. Various theoretical models for fine structure observed in α decay

Within the semiclassical framework, the partial decay width can be easily evaluated as [5, 6, 7, 8, 9]

$$\Gamma(Q_I,\ell) = P_\alpha F \exp\Big(-\frac{2}{\hbar}\int_a^b \sqrt{2\mu[V(r)+\ell(\ell+1)\hbar^2/(2\mu r^2)-Q_I]dr}\Big), \tag{1}$$

where P_α represents the α-preformation probability, F is the frequency of the α cluster inside the barrier, $V(r)$ is the α-nucleus potential, the decay energy Q_I is determined from the decay energy for the transition to ground states and the excitation energy of the final state, $Q_I = Q_0 - E_I$, and ℓ is the angular momentum carried by the α particle. If one takes into account nuclear deformation, the expression (1) would depend on deformation and orientation due to the deformed potential $V(r)$, and the partial width is determined by averaging along all orientations [5, 6, 8]. In terms of different α-nucleus potentials used in the calculations, there are various models such as the simple WKB barrier penetration approach [15], generalized liquid drop model (GLDM) [3, 16], unified model for α decay and α capture (UMADAC) [6, 17], Coulomb and proximity potential model for deformed nuclei (CPPMDN) [18].

The starting point of the coupled-channel study is the coupled equations for radial parts [19, 20, 21, 22, 23]

$$\Big[-\frac{\hbar^2}{2\mu}\Big(\frac{d^2}{dr^2}-\frac{\ell(\ell+1)}{r^2}\Big)-Q_I\Big]u_\alpha(r)+\sum_{\alpha'}V_{\alpha,\alpha'}(r)u_{\alpha'}(r)=0. \tag{2}$$

In this equation, u^J_α [$\alpha \equiv (n\ell I)$ labels the channel quantum numbers] is the cluster radial function describing the relative radial motion of the α particle with respect to the daughter nucleus, and $V_{\alpha,\alpha'}(r)$ is the matrix element of the interaction $V(r)$ taken between channels α and α'. The matrix elements could be evaluated using two different techniques, the multipole expansion of interaction potentials [19, 20, 23, 30, 31] and diagonalization of the operator matrix [21, 22, 32]. After solving the equations (2), one can express the partial width as

$$\Gamma_{Q_I,\ell}(R) = \frac{\hbar^2 k_I}{\mu}\frac{|u_{n\ell I}(R)|^2}{G_\ell(k_I R)^2+F_\ell(k_I R)^2}, \tag{3}$$

where R denotes large distances beyond the range of the nuclear potential and beyond the distance where the Coulomb potential can be regarded as spherically symmetric. It should be particularly noted that the results of $\Gamma_{Q_I,\ell}(R)$ show rather weak sensitivity to the choice of R.

No matter it concerns the semiclassical or coupled-channel picture, the branching ratio (BR) for the transition from ground states to excited states I is determined as

$$\mathrm{BR}_{I\ell} = \Gamma(Q_I, \ell) \Big/ \sum_{I\ell} \Gamma(Q_I, \ell) \times 100\%. \tag{4}$$

In some cases, the hypothesis of the Boltzmann distribution (BD) for daughter states, $\rho(E_I) = \exp(-cE_I)$ is proposed [22, 23, 33, 34, 35], and hence the BR is calculated as

$$\mathrm{BR}_{I\ell} = \rho(E_I)\Gamma(Q_I, \ell) \Big/ \sum_{I\ell} \rho(E_I)\Gamma(Q_I, \ell) \times 100\%. \tag{5}$$

To gain better insight into the α-decay fine structure, one also define the quantity [12, 19, 20]

$$\chi_{I\ell} = \log_{10}(\mathrm{BR}_{00}/\mathrm{BR}_{I\ell}), \tag{6}$$

which represents the relative intensity of different channels with respect to favored channels.

3. Numerical results and discussion

In the following, we will focus on the α transitions from ground 0^+ states of an even-even nucleus to various members of the ground-state rotational band in the daughter nucleus (i.e., 0^+, 2^+, 4^+, 6^+...). The experimental data for the BRs are taken from [36]. Actually, there are other α transitions to highly excited states rather than the ground-state rotational band, but they usually have quite small contributions to α decay [36].

First, we examine the theoretical results which are obtained using the semiclassical WKB calculations such as the UMADAC [17] and CPPMDN [18]. Figure 1 displays the calculated relative intensities $\chi_{I\ell}$ for excited 2^+, 4^+, and 6^+ states in the $N_pN_n/(N_p+N_n)$ scheme, compared with the systematics of the experimental data. As one can see, the results for 2^+ states are generally smaller than the experimental values especially in the larger $N_pN_n/(N_p + N_n)$ region. Within the CPPMDN, there is also an abnormal point corresponding to the α decay of ^{248}Fm where the calculated value has a sudden increasing by about two orders of magnitude with respect to the neighboring emitters [18]. This suggests that the CPPMDN calculations for ^{248}Fm should be checked. For the case of 4^+ states, the experimental data show a striking peak around $N_pN_n/(N_p + N_n) \approx 7.5$ and the peak value is larger than the minimum by roughly two orders of magnitude [24]. However, the WKB results do not exhibit this behavior. For the case of 6^+ states, the experimental data show a decreasing trend in the interval of $N_pN_n/(N_p+N_n) = 4 \sim 7$, by more than one order of magnitude [24]. By contrast, the CPPMDN results decrease by more than three orders of magnitude in the same interval of $N_pN_n/(N_p + N_n)$, while the UMADAC results decrease by about one order of magnitude. Besides, one can also notice that there is a clear difference between the CPPMDN and UMADAC results for 4^+ and 6^+ states in the interval of $N_pN_n/(N_p + N_n) = 4.5 \sim 6.5$. In a word, the differences between the experimental data and the WKB results are considerable. This is not so surprising because some important effects associated with nuclear structure are ignored in the WKB calculations, for example, the coupling effect between decay channels resulting from nuclear deformations.

Next, we transfer our attention from the semiclassical results to the coupled-channel studies. Delion and coworkers expanded the double-folding potentials into spherical multipoles and introduced the repulsive core for the monopole component [19, 20]. They performed three-channels calculations for deformed even-even nuclei, showing that the α-decay fine structure is closely related to the deformation parameters of daughter nuclei. We used the diagonalization technique to evaluate the interaction matrix elements and performed four-channels calculations for deformed even-even α-emitters [21, 22]. Recently we also used the multipole expansion to deal

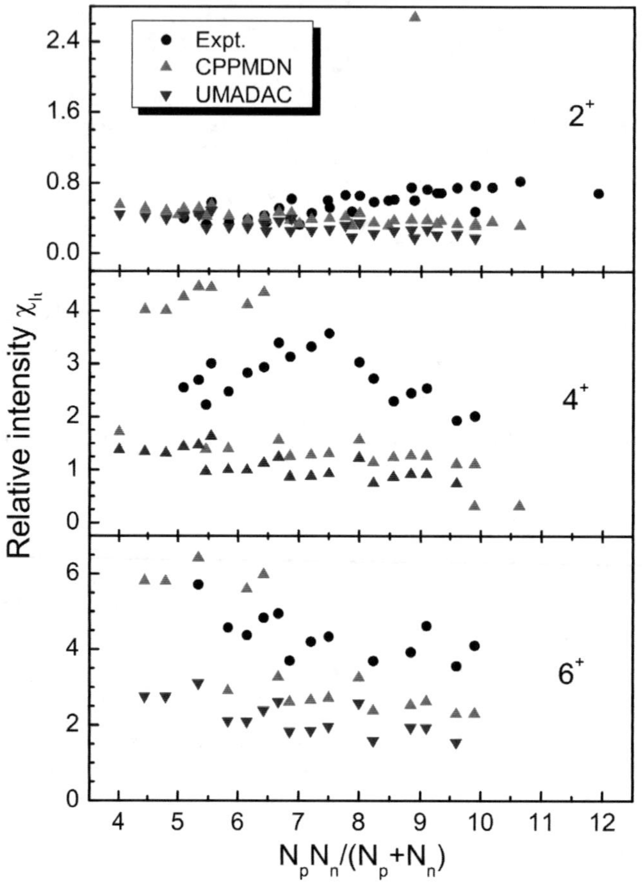

Figure 1. (Color online) Relative intensities $\chi_{I\ell}$ for excited 2^+, 4^+, and 6^+ states as a function of the quantity $N_pN_n/(N_p + N_n)$. The theoretical results are obtained using the semiclassical methods and they are taken from [17, 18].

with the interaction matrix, where the dynamic effects of core nuclei are included [23]. The five-channels analysis showed that the excitation spectrum of daughter nuclei also plays an important role in the α-decay fine structure in addition to the deformation of daughter nuclei. In a similar fashion, the detailed numerical results of these works are showed in figure 2. One can see that the theoretical results for 2^+ states show good agreement with the experimental data. In details, the four-channels and five-channels calculations show similarly good results and the three-channels analysis is slightly worse in the larger $N_pN_n/(N_p + N_n)$ region. For the case of 4^+ states, the results of the three cases have a tendency to form a peak around $N_pN_n/(N_p + N_n) \approx 7.5$ like the experimental results, but this is not evident especially in the smaller $N_pN_n/(N_p + N_n)$ region (corresponding to the outset of U, Pu, and Cm isotopic chains). This can be understood since in these α decays members of negative-parity and excited-state rotational bands, such as 1^-, 3^-, 0_2^+, and 2_2^+, emerge with low excitation energies in the daughter nuclei [36], leading to significant effects on the α-decay fine structure. The consideration of these states requires

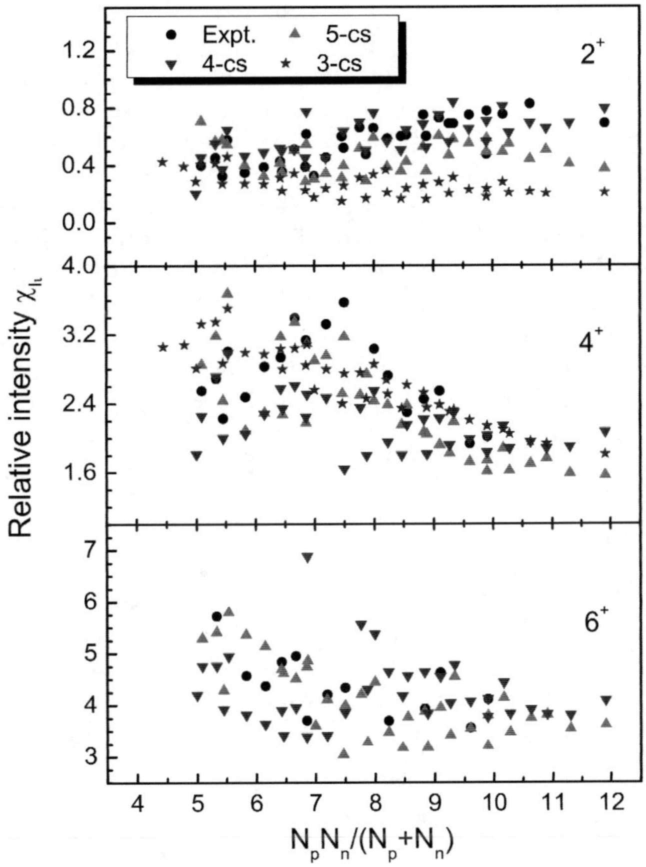

Figure 2. (Color online) Same as in figure 1, but the theoretical results are obtained using the coupled-channel methods, containing the three-channels (3-cs) [20], four-channels (4-cs) [22], and five-channels (5-cs) [23] calculations.

suitable theoretical schemes to describe the coupling between different rotational bands. Efforts towards this goal are being made. For the case of 6^+ states, the results of the four-channels and five-channels calculations also follow the experimental points well in both the systematic behavior and the magnitude: as mentioned above, the relative intensities decrease in the interval of $N_pN_n/(N_p + N_n) = 4 \sim 7$ and show a smooth increasing in the region $N_pN_n/(N_p + N_n) > 7$. An obvious derivation is the four-channels result for the emitter ^{244}Pu. This is worth us further investigation. On the whole, there is good agreement between experiment and theory for the coupled-channel studies of the α-decay fine structure, as shown in figure 2. This in itself not only gives a strong support to the applicability and reliability of the coupled-channel approach, but also gives us some guaranty for reliable predictions of the α-decay properties of superheavy nuclei. In addition, we would like to note that both the semiclassical and coupled-channel methods can give the results for 8^+ states. But there are quite few experimental data for the transitions to excited 8^+ states; that is, about six data are available in experiments [36]. So in the present analysis we do not perform systematics of the results for 8^+ states.

4. Summary

In summary, the theoretical results for the α-decay fine structure in deformed even-even nuclei, obtained from both the semiclassical and coupled-channel methods, are investigated as a function of the collectivity indicator $N_pN_n/(N_p + N_n)$. And they are compared with the systematics of the experimental data. It is found that the results obtained from the semiclassical models deviate from the experimental data evidently in both the systematic behavior and the magnitude, while the results obtained from the various coupled-channel calculations agree well with the experimental data. This confirms the good reliability of the coupled-channel studies so that coupled-channel predictions are allowed for the α-decay fine structure in the heavier mass region. Although the systematic behaviors of the results for excited 4^+ and 6^+ states are clearly seen, theoretical explanations are still not available, calling for more detailed structure investigations. In future, it will be of particular interest to use reliable structure models to achieve more information on the α-decay fine structure, which could be useful for future structure researches on superheavy nuclei.

Acknowledgments

We are thankful to Prof. Bucurescu for helpful discussion and valuable suggestions. This work is supported by the National Natural Science Foundation of China (Grants No. 10735010, No. 10975072, No. 11035001, and No. 11120101005), by the 973 National Major State Basic Research and Development of China (Grants No. 2010CB327803 and No. 2013CB834400), by CAS Knowledge Innovation Project No. KJCX2-SW-N02, by Research Fund of Doctoral Point (RFDP), Grants No. 20100091110028, and by a Project Funded by the Priority Academic Program Development of Jiangsu Higher Education Institutions (PAPD).

References

[1] Gamow G 1928 *Z. Phys.* **51** 204
[2] Buck B, Merchant A C, and Perez S M 1993 *At. Data Nucl. Data Tables* **54** 53
[3] Royer G 2000 *J. Phys. G* **26** 1149
[4] Poenaru D N, Plonski I H, and Greiner W 2006 *Phys. Rev.* C **74** 014312
[5] Xu C and Ren Z 2006 *Phys. Rev.* C **73** 041301(R) Xu C and Ren Z 2006 *Phys. Rev.* C **74** 014304
[6] Denisov V Yu and Ikezoe H 2005 *Phys. Rev.* C **72** 064613
[7] Mohr P 2006 *Phys. Rev.* C **73** 031301(R)
[8] Pei J C, Xu F R, Lin Z J and Zhao E G 2007 *Phys. Rev.* C **76** 044326
[9] Kelkar N G and Castañeda H M 2007 *Phys. Rev.* C **76** 064605
[10] Chowdhury P R, Samanta C, and Basu D N 2008 *Phys. Rev.* C **77** 044603
[11] Delion D S, Sandulescu A, and Greiner W 2004 *Phys. Rev.* C **69** 044318
[12] Ni D and Ren Z 2009 *Phys. Rev.* C **80** 051303(R) Ni D and Ren Z 2010 *Phys. Rev.* C **81** 024315
[13] Heyde K 1994 *Basic ideas and concepts in nuclear physics* (Bristol and Philadelphia: Insitute of Phsics Publishing)
[14] Rosenblum R 1929 *C. R. Acad. Sci. Pairs* **188** 1401
[15] Xu C and Ren Z 2006 *Nucl. Phys.* A **778** 1
[16] Wang Y Z, Zhang H F, Dong J M, and Royer G 2009 *Phys. Rev.* C **79** 014316
[17] Denisov V Yu and Khudenko A A 2009 *Phys. Rev.* C **80** 034603
[18] Santhosh K P, Sahadevan S, and Joseph J G 2011 *Nucl. Phys.* A **850** 34
[19] Delion D S, Peltonen S, and Suhonen J 2006 *Phys. Rev.* C **73** 014315
[20] Peltonen S, Delion D S, and Suhonen J 2008 *Phys. Rev.* C **78** 034608
[21] Ni D and Ren Z 2010 *Phys. Rev.* C **81** 064318
[22] Ni D and Ren Z 2011 *Sci. China: Phys. Mech. Astron.* **54** s24
[23] Ni D and Ren Z 2011 *Phys. Rev.* C **83** 067302
[24] Bucurescu D and Zamfir N V 2012 *Rom. Journ. Phys.* **57** 69
[25] Casten R F 1985 *Phys. Rev. Lett.* **54** 1991
[26] Casten R F and Zamfir N V 1993 *Phys. Rev. Lett.* **70** 402

[27] Saha M and Sen S 1994 *Phys. Rev. C* **49** 2460
[28] Buck B, A. C. Merchant A C, and Perez S M 2005 *Phys. Rev. Lett.* **94** 202501
[29] Ha E and Hong S W 2009 *Phys. Rev. C* **79** 037301
[30] Kruppa A T, Barmore B, Nazarewicz W, and Vertse T 2000 *Phys. Rev. Lett.* **84** 4549 Barmore B, Kruppa A T, Nazarewicz W, and Vertse T 2000 *Phys. Rev.* C **62** 054315
[31] Suzuki Y and Ohkubo S 2010 *Phys. Rev.* C **82** 041303(R)
[32] Hagino K, Rowley N, and Kruppa A T 1999 *Comput. Phys. Comm.* **123** 143
[33] Einstein A 1917 *Phys. Z* **18** 121
[34] Stewart T L, Kermode M W, Beachey D J *et al* 1996 *Phys. Rev. Lett.* **77** 36
[35] Delion D S 2009 *Phys. Rev.* C **80** 024310
[36] NNDC of the Brookhaven National Laboratory, http://www.nndc.bnl.gov.

Double beta decay: an interface between nuclear, particle and atomic physics

Jouni Suhonen[1]

[1] Department of Physics, University of Jyväskylä, PO Box 35, FIN-40014, Jyväskylä, Finland

E-mail: `jouni.suhonen@phys.jyu.fi`

Abstract. General properties of the nuclear matrix elements (NMEs) related to the various modes of neutrinoless double β decays are examined and analyzed. The decays include the electron-emitting double beta-minus decays $\beta^-\beta^-$ and the various positron-emitting/electron capture decays. Special interest is devoted to the neutrinoless double electron capture decay with a resonance condition.

PACS number(s): 21.60.Ev; 21.60.Jz; 23.40.Bw; 23.40.Hc; 27.50.+e; 27.60.+j
Keywords: Double beta decay; Nuclear matrix elements; Resonant neutrinoless double electron capture

1. Introduction

Numerous neutrino-oscillation experiments have provided a lot of information about the neutrino masses and mixings [1]. In particular, the mixing picture has been completed by the latest results on the so-called θ_{13} mixing angle. Only the squared neutrino mass differences can be accessed in these experiments, not the absolute mass scale. Complementary experiments are needed to learn about the absolute mass scale and mass hierarchy of the neutrinos. The neutrinoless double beta ($0\nu\beta\beta$) decay of atomic nuclei is a process that can provide us with this information, supplemented with the possibility to identify the neutrino as a Majorana particle, provided that the $0\nu\beta\beta$ really occurs in Nature. Possible detection of this decay mode would have several fundamental implications for particle physics.

Double beta decay experiments have thus far mainly concentrated on measuring the electron-emitting mode of $0\nu\beta\beta$ decays, hereafter denoted as $0\nu\beta^-\beta^-$ decays. These decays have favorable decay energies (Q values) and are thus of great experimental interest at the present. Of secundary interest are the positron emitting/electron capture modes of double beta decays. There are three different types of such modes, namely the double positron emission ($0\nu\beta^+\beta^+$ decay), positron emission together with electron capture ($0\nu\beta^+$EC decay) and the double electron capture (0νECEC decay). The last mode deserves particular attention since it cannot be an ordinary phase-space decay but needs either a resonance condition or higher-order mechanisms in order to spend the decay energy released in the transition.

All the above described modes of neutrinoless double beta decay can run through the Majorana mass of the neutrino. There are several other possibilities through which the decay can occur [1] but here we concentrate on the mass mode as inspired by the detection of the neutrino mass by the oscillation experiments. Decay rate of the mass mode is proportional to the square of the so-called effective neutrino mass, $\langle m_\nu \rangle$, being a linear combination of the mass eigenstates weighted by the two independent Majorana phases and the elements of the electron row of the neutrino mixing matrix. Determination of the magnitude of $\langle m_\nu \rangle$ is the key issue in experiments on double beta decays. The effective mass can be extracted from the measured double-beta half-lives via the so-called nuclear matrix elements (NMEs) that intimately intertwine with the lepton aspects of the various modes of double beta decays [2, 3]. The NMEs contain the information on the nuclear wave functions of the initial and final states of the transition, in addition to the form of the effective operator(s) mediating the transition. To compute the NME we need to study details of nuclear structure by using suitable nuclear models, either microscopic or semi-microscopic [2]. These models are addressed in section 4 of this article.

2. Basic features of the neutrinoless double beta decays

Here we give the basic ingredients of the formalism of the neutrinoless double beta decays in order to enable a later discussion of the differences in the various formalisms used to describe these decays.

The $0\nu 2\beta$-decay half-life can be written as [2, 4, 5]

$$\left[t_{1/2}^{(0\nu)}(0_i^+ \rightarrow 0_f^+)_\alpha\right]^{-1} = G_\alpha^{(0\nu)} \left(M^{(0\nu)\prime}\right)^2 |\langle m_\nu \rangle|^2 , \tag{1}$$

where $\langle m_\nu \rangle$ is the effective neutrino mass [2]. The symbol $\alpha = \beta^-\beta^-, \beta^+\beta^+, \beta^+\mathrm{EC}$ lists the various possible modes of neutrinoless double beta decays: double electron emission ($\beta^-\beta^-$), double positron emission ($\beta^+\beta^+$) and positron emission combined with electron capture ($\beta^+\mathrm{EC}$) [2, 6]. The nuclear matrix element of (1) can be written as a linear combination of the Gamow–Teller, Fermi and tensor terms [7, 8], i.e.

$$M^{(0\nu)\prime} = \left(\frac{g_\mathrm{A}}{g_\mathrm{A}^\mathrm{b}}\right)^2 \left[M_\mathrm{GT}^{(0\nu)} - \left(\frac{g_\mathrm{V}}{g_\mathrm{A}}\right)^2 M_\mathrm{F}^{(0\nu)} + M_\mathrm{T}^{(0\nu)}\right] , \tag{2}$$

where $g_\mathrm{A}^\mathrm{b} = 1.25$ is the bare-nucleon value of the weak-interaction axial-vector coupling constant. Values for the phase-space factors $G_\alpha^{(0\nu)}$ are given in [2, 4, 9, 10] for the value $g_\mathrm{A} = 1.25$, as is required by the definition of the NME $M_\alpha^{(0\nu)\prime}$ in the above equations. Details of the associated NMEs are given e.g. in [2, 11].

The neutrinoless double electron capture 0νECEC cannot occur as such due to missing leptonic degrees of freedom in the final state. Instead, the resonant neutrinoless double electron capture (R0νECEC) can occur and it was studied in the works [12, 13] from the lepton aspects points of view. There it was suggested that the fulfillment of a resonance condition in this decay could enhance the decay rates up to a factor of a million. The R0ν2EC decay proceeds between two atomic states as

$$e^- + e^- + (A, Z) \rightarrow (A, Z-2)^* \rightarrow (A, Z-2) + \gamma + 2X, \tag{3}$$

where the capture of two atomic electrons leaves the final nucleus in an excited state that decays by one or more gamma-rays (in case of an excited state) and the atomic vacancies are filled by outer electrons with emission of X-rays. The corresponding half-life for the 0^+ final states can be written as

$$\left[T_{1/2}^{\mathrm{R}0\nu\mathrm{ECEC}}\right]^{-1} = G_{0\nu}^{\mathrm{ECEC}} \left|\frac{1}{R_A} M^{(0\nu)\prime}\right|^2 \frac{|\langle m_\nu \rangle|^2 \Gamma}{(Q-E)^2 + \Gamma^2/4}, \tag{4}$$

where $R_A = 1.2A^{1/3}$ fm is the nuclear radius and the NME is given in (2). The difference $Q - E$ is the degeneracy of the initial and final states, Q being the difference between the masses of the initial and final atoms (decay Q value) and E is the total energy of the excited state in the final atom (consisting of the possible nuclear excitation energy, the excitation energy of the two holes in the electronic shells and the Coulomb repulsion between the holes). The quantity Γ is the decay width of the two holes in the atomic shells [12]. The atomic factor $G_{0\nu}^{\mathrm{ECEC}}$ is given in [14, 15].

3. Ingredients of the pnQRPA calculations

The pnQRPA formalism is reviewed nicely in [2, 16, 17]. The pnQRPA was successfully used to explain the long two-neutrino double beta ($2\nu\beta\beta$) decay half-lives in [18] by suppression of the associated NME relative to its simple single-particle estimate. The explanation called for the use of the strength parameter g_{pp} of the particle-particle part of the proton-neutron interaction in the 1^+ channel. This parameter governs the relative magnitudes of the particle-particle and particle-hole terms in the nuclear Hamiltonian. While the couplings of the particle-hole channel can be fixed by fitting the observed energy of the 1^+ GTGR, there is no direct way of fixing g_{pp}. Since the original work a lot of studies of the effects of the g_{pp} have been done, e.g., in [17, 19, 20, 21, 22, 23].

The idea of fixing the value of the strength parameter g_{pp} by the data on half-lives of $2\nu\beta\beta$ decays (see the recent compilation [24]) was advocated in [25] and adopted in many subsequent works dealing with $0\nu\beta^-\beta^-$ decays. A shortcoming of the method is that it can be used only for those nuclei for which the $2\nu\beta\beta$ decay half-life is known experimentally. An alternative way to fix the value of g_{pp} would be to use the measured $\log ft$ value of the β^- decay of the lowest 1^+ state in the intermediate nucleus of the double beta decay. This method is the only alternative for those nuclei where the $2\nu\beta\beta$ decay half-life is not known [8, 26, 27, 28].

The short-range correlations between the two decaying nucleons in the $0\nu\beta^-\beta^-$ processes cause an effective repulsion between the two nucleons so as to prevent their overlap in the 1 fm region of their relative distance. This distance corresponds to the large average momentum exchange ($\sim$ 200 MeV) occurring in the propagation of the virtual Majorana neutrino between the two decay vertices. The use of the traditional Jastrow correlator [29] causes an unrealistically large suppression of the $0\nu\beta^-\beta^-$ NMEs as noticed in [30]. In this article it was proposed that the Jastrow short-range correlations should be replaced by the correlations effected by the unitary correlation operator method (UCOM) [31]. As pointed out in [30], the UCOM is a softer way to account for the short-range correlation effects in the double beta decays than the Jastrow method. Further refinements in the computation of the $0\nu\beta\beta$ decays were introduced in [7] in the form of finite-size dipole form factors of nucleons and higher-order nucleon currents, including the interference between the vector, axial-vector and induced pseudo-scalar contributions. All the latest calculations include these corrections in the $0\nu\beta\beta$ NMEs.

4. Comparison of the $0\nu\beta^-\beta^-$ NMEs calculated by different formalisms

Here we analyze the similarities and differences of the ground-state-to-ground-state $0\nu\beta^-\beta^-$ NMEs calculated in different theoretical formalisms. In Figs. 1 and 2 the NMEs calculated by the pnQRPA are compared with the corresponding NMEs of the interacting shell model (ISM) [32], the (proton-neutron) interacting boson model (IBA-2) [33, 34], the Gogny-based energy-density functional approach (EDF) [35] and the projected Hartree-Fock-Bogoliubov mean-field scheme (PHFB) [36]. The pnQRPA NMEs have been taken from [8, 37, 38, 39]. Here the unquenched value $g_{\rm A} = 1.25$ for the axial-vector coupling constant has been adopted. In Fig. 1 the Jastrow short-range correlations are used and in Fig. 2 the UCOM short-range correlations are adopted. It should be noted that in the case of the decay of ^{48}Ca an old formalism of Refs. [40, 41, 42] has been used and thus an update would be welcome. This is why a direct comparison of the pnQRPA results with the results of the other models is not straightforward in the case of ^{48}Ca.

¿From Figs. 1 and 2 it is easy to deduce that the overall magnitudes of the NMEs for the pnQRPA and the IBA-2 are quite similar (magnitudes and spread) whereas the NMEs of the ISM are much smaller with very small spread. The NMEs of the EDF have similar magnitudes as those of the pnQRPA but a smaller spread. The NMEs (Jastrow and UCOM correlated) of the PHFB are by far the largest ones with clearly the largest spread. It seems that in bulk the NMEs calculated by using the pnQRPA, IBA-2 and EDF frameworks have quite similar magnitudes whereas the magnitudes of the ISM-calculated and the PHFB-calculated NMEs deviate notably from them, the ISM NMEs being quite small and the PHFB NMEs being quite large. The IBA-2 model is an advanced form of the generalized-seniority model and thus very closely related to the ISM in spirit. Also the valence spaces used in both models are restricted to one closed major shell. Hence, it is a big mystery why there is such a large difference between the magnitudes of the $0\nu\beta^-\beta^-$ NMEs produced by the two models.

All the different theoretical approaches (IBA-2, ISM, EDF and PHFB), except the pnQRPA, use the closure approximation in the calculations. This may introduce

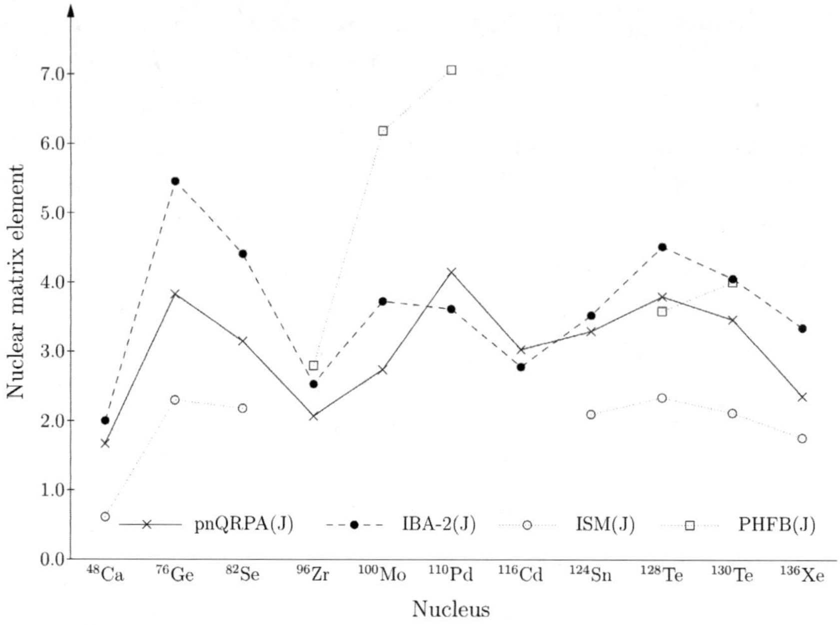

Figure 1. Ground-state-to-ground-state NMEs of the discussed $0\nu\beta^-\beta^-$ decays for the four indicated calculations. All NMEs are calculated using the Jastrow short-range correlations. The mother nucleus is indicated on the x axis. The value $g_A = 1.25$ is adopted.

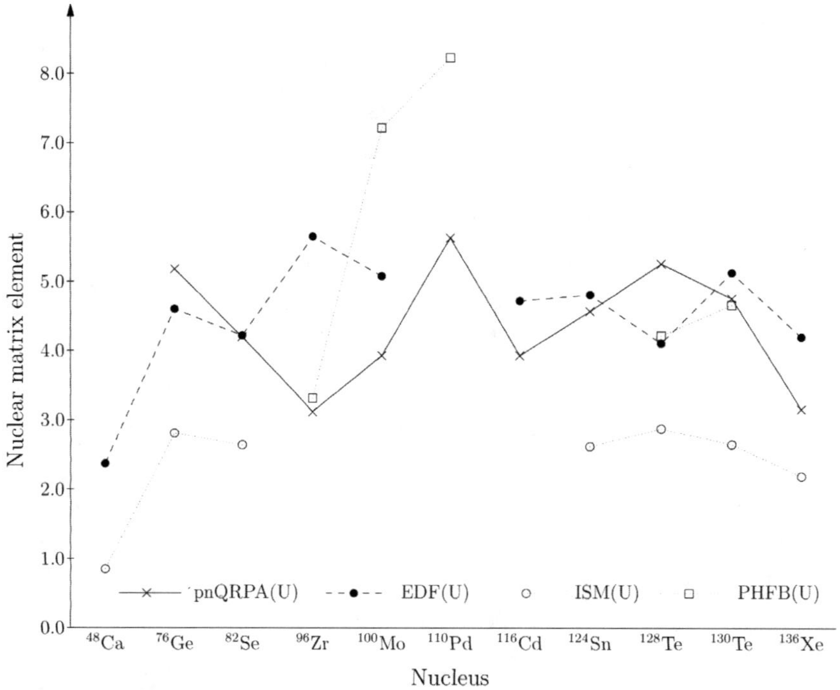

Figure 2. The same as Fig. 1 but for different calculations and using the UCOM short-range correlations.

an error of some 10 percent in the calculations [43]. As noted above, the pnQRPA calculations avoid this complication by including all intermediate states in the $0\nu\beta^-\beta^-$ NMEs. The inclusion of all the spin-orbit partners in the single-particle basis is essential for the pnQRPA results to satisfy the Ikeda $3(N-Z)$ sum rule [16]. In [39] the different aspects of including single-particle states beyond the minimal one nuclear major shell were discussed and illustrated for the $0\nu\beta^-\beta^-$ ground-state-to-ground-state decays of ^{76}Ge, ^{82}Se, ^{128,130}Te and ^{136}Xe. A similar study was performed in [44, 45] for the $0\nu\beta^-\beta^-$ decays of ^{76}Ge, ^{82}Se and ^{136}Xe to the first excited 0^+ states in ^{76}Se, ^{82}Kr and ^{136}Ba. In all these studies stressed was the importance of including all spin-orbit partners beyond the simple one valence major shell used in the shell-model calculations [46] of double-beta processes. Omission of these spin-orbit partners and the other single-particle states around the valence shell could cause a serious underestimation of the magnitudes of the $0\nu\beta^-\beta^-$ NMEs. This drawback not only affects the ISM calculations but also the IBA-2 calculations that are done using only one complete closed major shell. In relation to this defect it would be crucial for the ISM and IBA-2 calculations to analyze their computed rates for beta decays in medium-heavy and heavy nuclei. It should be further noted that in [39, 44, 45, 47] was performed a thorough study of the effects on the $0\nu\beta\beta$ NMEs by the use of different occupancies of the valence single-particle orbitals. Large effects could be found only for ^{76}Ge.

Let us study next the importance of shell effects on the $0\nu\beta^-\beta^-$ NMEs. From Fig. 1 one immediately perceives the stunning similarity between the trends of the pnQRPA, IBA-2 and the ISM NMEs. Indeed the three models are one-to-one for the mass ranges $A \leq 82$ and $A \geq 124$. The striking similarity of the pnQRPA and the IBA-2 continues even beyond that, to the masses $96 \leq A \leq 116$. The only qualitative deviation is at ^{110}Pd where a much weaker peaking of the NME is obtained in the IBA-2 model than in the pnQRPA. From Fig. 2 it is seen that the trends of the pnQRPA and the two mean-field based models (EDF and PHFB) are quite different. To study more closely these matters, we can take as a paradigm the generalized seniority (GS) scheme which is a reasonable approximation to a more complete shell-model calculation. The generalized seniority prescription advocated in [33] describes the $0\nu\beta^-\beta^-$ NMEs in the $N = 50-82$ major shell as smooth concave arches for different proton numbers Z. In this way one obtains the ratio

$$R = M^{(0\nu)'}/M^{(0\nu)'}_{\rm GS} , \qquad (5)$$

which gives a measure how close the calculated NMEs of a nuclear model are to those of the simple generalized seniority scheme.

The ratios (5) are summarized in Table 1 for all the discussed nuclear models in the range $124 \leq A \leq 136$, i.e. between ^{124}Sn and ^{136}Xe. As can be seen the IBA-2, ISM and pnQRPA conform excellently to the GS scheme. It is even more striking to see that, in fact, the pnQRPA obeys by far the best the seniority scheme. On the other hand the EDF and PHFB are drastically off from the generalized-seniority trends which may point to missing shell effects in these two mean-field based schemes. In particular,

Table 1. Ratios (5) for the indicated nuclei in the five different model frameworks (first column).

Model	$A = 124$	$A = 128$	$A = 130$	$A = 136$
IBA-2	0.912	1.00	1.00	1.13
ISM	1.05	1.00	1.01	1.15
pnQRPA	1.01	1.00	1.02	0.95
EDF	1.36	1.00	1.39	1.56
PHFB	-	1.00	1.24	-

the GS scheme predicts for the ratio

$$R(\text{Te}) = M^{(0\nu)\prime}(^{128}\text{Te})/M^{(0\nu)\prime}(^{130}\text{Te}) \tag{6}$$

the value $R_{\text{GS}}(\text{Te}) = 1.11$. For the respective nuclear models this ratio is 1.11 (IBA-2), 1.10 (ISM), 1.10 (pnQRPA), 0.80 (EDF) and 0.90 (PHFB), thus indicating that the trend is not correct for the EDF and PHFB models.

5. Double positron-emission/electron-capture decays

The double positron-emission/electron-capture decays have generally smaller decay Q values than the double electron-emission decays [2]. On the positron-emission/electron-capture side the decay Q value depends on the mode.

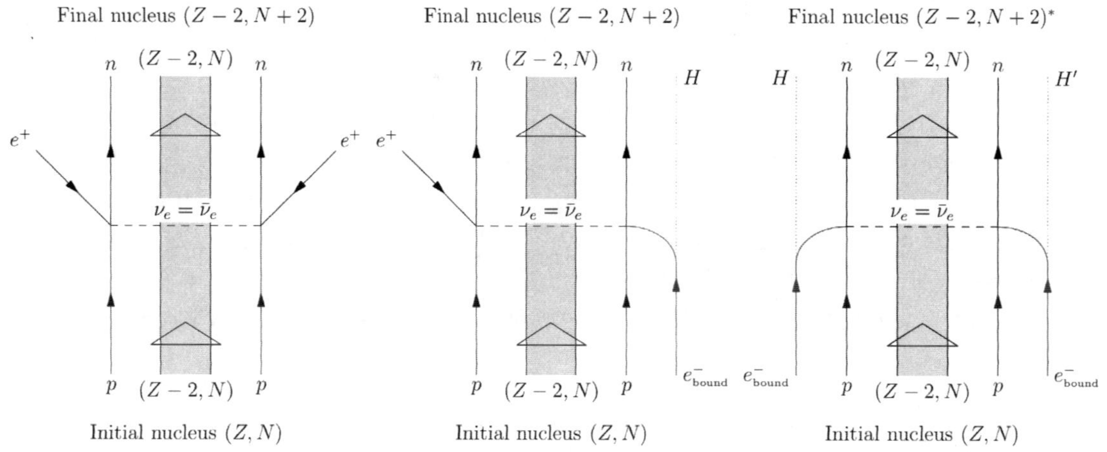

Figure 3. Schematic representation of the $0\nu\beta^+\beta^+$ decay (left panel), the $0\nu\beta^+$EC decay (middle panel) and the R0νECEC decay (right panel). In the last decay an atomic resonance state matches the decay energy. The horizontal dashed line represents the exchanged Majorana neutrino.

In Fig. 3 the two-positron emission ($0\nu\beta^+\beta^+$), the positron-emission plus electron capture ($0\nu\beta^+$EC) and the resonant neutrinoless double electron capture (R0νECEC) modes of decay are displayed schematically. In these decays a Majorana neutrino is exchanged as displayed by the horizontal dashed line in the figures. In all modes two protons are changed to two neutrons and one has to consider the short-range correlations between the decaying nucleons in the same way as done for the $0\nu\beta^-\beta^-$ decays [30]. The Q value is larger for the $0\nu\beta^+$EC mode than for the $0\nu\beta^+\beta^+$ mode since there is one less positron in the final state and the rest-mass energy of the captured electron (minus the electron binding in the atomic orbital) can be exploited as decay energy. The neutrinoless ECEC cannot proceed as such since there are no leptons in the final state and hence the decay energy cannot be absorbed by them. Instead, the decay energy can be wasted by hitting a resonance in the final atom, as discussed in the section 2. This leads to the R0νECEC decay.

The orders of magnitude of the half-lives associated with the various positron-emitting transitions in nuclei have been assessed recently in [14] for the decay of ^{106}Cd and in [48] for the decay of ^{96}Ru. Furthermore, in Table 2 is presented an overview of those resonant neutrinoless ECEC decays for which the associated decay Q values have been measured by exploiting the Penning-trap techniques.

Table 2. R0νECEC decay transitions with the degeneracy parameters $|Q-E|$ derived from Q-value measurements cited in the last column. Also the involved atomic orbitals are given. The second last column lists the currently available theoretical half-life estimates.

Transition	J^π_f	$Q-E$ [keV]	Atomic orbitals	$C^{\rm ECEC}$	Ref.
^{74}Se $\to$ ^{74}Ge	2^+	2.23	L_2L_3	$(0.2-100)\times10^{43}$	[49]
^{96}Ru $\to$ ^{96}Mo	2^+	8.92(13)	L_1L_3		[50]
	0^+?	−3.90(13)	L_1L_1	$(4.4-19)\times10^{31}$	This work
^{102}Pd $\to$ ^{102}Ru	2^+	75.26(36)	KL_3		[51]
^{106}Cd $\to$ ^{106}Pd	0^+?	8.39	KK	$(2.1-5.7)\times10^{30}$	[14]
	$(2,3)^-$	−0.33(41)	KL_3		[51]
^{112}Sn $\to$ ^{112}Cd	0^+	−4.5	KK	$>5.9\times10^{29}$	[52]
^{136}Ce $\to$ ^{136}Ba	0^+	−11.67	KK	$(3-23)\times10^{32}$	[53]
^{144}Sm $\to$ ^{144}Nd	2^+	171.89(87)	KL_3		[51]
^{152}Gd $\to$ ^{152}Sm	$0^+_{\rm gs}$	0.91(18)	KL_1	$\sim1\times10^{26}$	[54]
^{156}Dy $\to$ ^{156}Gd	1^-	0.75(10)	KL_1		[55]
	0^+	0.54(24)	L_1L_1		[55]
	2^+	0.04(10)	M_1N_3		[55]
^{162}Er $\to$ ^{162}Dy	2^+	2.69(30) keV	KL_3		[50]
^{168}Yb $\to$ ^{168}Er	(2^-)	1.52(25) keV	M_1M_3		[50]

In the table are also listed the theoretically estimated half-lives for the cases for which such exist. In the table a quantity $C^{\rm ECEC}$ is given and it ties to the R0νECEC half-life through

$$T_{1/2}^{\rm R0\nu ECEC} = \frac{C^{\rm ECEC}}{(\langle m_\nu\rangle[{\rm eV}])^2}\ {\rm years}\ , \qquad (7)$$

where the effective neutrino mass should be given in units of eV.

6. Conclusions

In this paper I have presented the ranges of values of the nuclear matrix elements corresponding to the double β^- decays of nuclei of masses $48 \leq A \leq 136$. All these nuclear matrix elements are analyzed with respect to the use of the closure approximation, size of the single-particle model space, short-range correlations and shell effects. One can conclude that thus far only the pnQRPA manages to avoid the use of the closure approximation in the NME calculations. Some of the recent models are mean-field based and seem to have trouble in reproducing shell effects. The IBA-2 and the ISM use only one closed major shell as their model space whereas the pnQRPA and the mean-field based models allow larger valence spaces. Although the IBA-2 maps to the seniority scheme and is thus closely related to the ISM, the magnitudes of the NMEs of the two models deviate notably from each other. These models also suffer of the lack of explicit inclusion of the spin-orbit partner orbitals in the calculations. Summarizing the above features one could argue that the pnQRPA is still the scheme that is most complete and best suited for reliable calculation of the values of the double-beta-decay nuclear matrix elements.

For the positron-emission/electron-capture decays there remains a whole lot of theoretical and experimental work to be done, especially in the case of the resonant neutrinoless ECEC decays.

Acknowledgments

This work has been partially supported by the Academy of Finland under the Finnish Centre of Excellence Programme 2012-2017 (Nuclear and Accelerator Based Programme at JYFL).

References

[1] Avignone III F T, Elliott S R and Engel J 2008 *Rev. Mod. Phys.* **80** 481–516
[2] Suhonen J and Civitarese O 1998 *Phys. Rep.* **300** 123–214
[3] Faessler A and Šimkovic F 1998 *J. Phys. G* **24** 2139–2178
[4] Doi M and Kotani T 1993 *Prog. Theor. Phys.* **89** 139–159
[5] Doi M, Kotani T and Takasugi E 1985 *Prog. Theor. Phys. Suppl.* **83** 1–175
[6] Doi M and Kotani T 1992 *Prog. Theor. Phys.* **87** 1207–1231
[7] Šimkovic F, Pantis G, Vergados J D and Faessler A 1999 *Phys. Rev. C* **60** 055502
[8] Suhonen J 2011 *Nucl. Phys. A* **864** 63–90
[9] Aunola M and Suhonen J 1998 *Nucl. Phys. A* **643** 207–221
[10] Suhonen J and Aunola M 2003 *Nucl. Phys. A* **723** 271–288
[11] Kortelainen M and Suhonen J 2007 *Phys. Rev. C* **76** 024315
[12] Bernabeu J, De Rujula A and Jarlskog C 1983 *Nucl. Phys. B* **223** 15–28
[13] Sujkowski Z and Wycech S 2004 *Phys. Rev. C* **70** 052501
[14] Suhonen J 2011 *Phys. Lett. B* **701** 490–495
[15] Suhonen J 2012 *Eur. Phys. J. A* **48** 51
[16] Suhonen J 2007 *From Nucleons to Nucleus: Concepts of Microscopic Nuclear Theory* (Berlin: Springer)

[17] Suhonen J, Taigel T and Faessler A 1988 *Nucl. Phys A* **486** 91–117
[18] Vogel P and Zirnbauer M R 1986 *Phys. Rev. Lett.* **57** 3148–3151
[19] Civitarese O, Faessler A and Tomoda T 1987 *Phys. Lett. B* **194** 11–14
[20] Civitarese O, Faessler A, Suhonen J and Wu X R 1990 *Phys. Lett. B* **251** 333–337
[21] Civitarese O, Faessler A, Suhonen J and Wu X R 1991 *Nucl. Phys. A* **524** 404–424
[22] Toivanen J and Suhonen J 1995 *Phys. Rev. Lett.* **75** 410–413
[23] Toivanen J and Suhonen J 1997 *Phys. Rev. C* **55** 2314–2323
[24] Barabash A S 2010 *Phys. Rev. C* **81** 035501
[25] Rodin V A, Faessler A, Šimkovic F and Vogel P 2006 *Nucl. Phys. A* **766** 107–131
[26] Suhonen J 2005 *Phys. Lett. B* **607** 87–95
[27] Civitarese O and Suhonen J 1998 *Phys. Rev. C* **58** 1535–1538
[28] Civitarese O and Suhonen J 1999 *Nucl. Phys. A* **653** 321–337
[29] Miller G A and Spencer J E 1976 *Ann. Phys. (NY)* **100** 562–606
[30] Kortelainen M, Civitarese O, Suhonen J and Toivanen J 2007 *Phys. Lett. B* **647** 128–132
[31] Feldmeier H, Neff T, Roth R and Schnack J 1998 *Nucl. Phys. A* **632** 61–95
[32] Menéndez J, Poves A, Caurier E and Nowacki F 2009 *Nucl. Phys. A* **818** 139–151
[33] Barea J and Iachello F 2009 *Phys. Rev. C* **79** 044301
[34] Iachello F, Barea J and Kotila J 2011 *AIP Conf. Proc.* **1417** 62–68
[35] Rodríguez T R and Martínez-Pinedo G 2010 *Phys. Rev. Lett.* **105** 252503
[36] Rath P K, Chandra R, Chaturvedi K, Raina P K and Hirsch J G 2010 *Phys. Rev. C* **82** 064310
[37] Suhonen J 1993 *J. Phys. G* **19** 139–160
[38] Suhonen J and Kortelainen M 2008 *Int. J. Mod. Phys. E* **17** 1–11
[39] Suhonen J and Civitarese O 2010 *Nucl. Phys. A* **847** 207–232
[40] Suhonen J, Khadkikar S B and Faessler A 1990 *Phys. Lett. B* **237** 8–13
[41] Suhonen J, Khadkikar S B and Faessler A 1991 *Nucl. Phys. A* **529** 727–740
[42] Suhonen J, Khadkikar S B and Faessler A 1991 *Nucl. Phys. A* **535** 509–547
[43] Tomoda T 1991 *Rep. Prog. Phys.* **54** 53
[44] Suhonen J 2011 *Int. J. Mod. Phys. E* **20** 451–458
[45] Suhonen J 2011 *Nucl. Phys. A* **853** 36–60
[46] Menéndez J, Poves A, Caurier E and Nowacki F 2009 *Phys. Rev. C* **80** 048501
[47] Suhonen J and Civitarese O 2008 *Phys. Lett. B* **668** 277–281
[48] Suhonen J 2012 *Phys. Rev. C* **86** 024301
[49] Kolhinen V, Elomaa V, Eronen T, Hakala J, Jokinen A, Kortelainen M, Suhonen J and Äystö J 2010 *Phys. Lett. B* **684** 17–21
[50] Eliseev *et al* S 2011 *Phys. Rev. C* **83** 038501
[51] Goncharov *et al* M 2011 *Phys. Rev. C* **84** 028501
[52] Rahaman S, Elomaa V, Eronen T, Hakala J, Jokinen A, Kankainen A, Rissanen J, Suhonen J, Weber C and Äystö J 2009 *Phys. Rev. Lett.* **103** 042501
[53] Kolhinen V S, Eronen T, Gorelov D, Hakala J, Jokinen A, Kankainen A, Rissanen J, Suhonen J and Äystö J 2011 *Phys. Lett. B* **697** 116–120
[54] Eliseev *et al* S 2011 *Phys. Rev. Lett.* **106** 052504
[55] Eliseev *et al* S 2011 *Phys. Rev. C* **84** 012501(R)

Journal of Physics: Conference Series **413** (2013) 012017 doi:10.1088/1742-6596/413/1/012017

Cluster decay calculations based on mean-field potentials

F R Xu, J C Pei and S M Wang

State Key Laboratory of Nuclear Physics and Technology, School of Physics, Peking University, Beijing 100871, China

E-mail: frxu@pku.edu.cn

Abstract. Based on mean-field model, an effective cluster potential has been suggested and applied successfully to the calculations of various cluster decays including diproton, α, ^{6}He and heavier cluster decays. One of the advantages of the new potential is no new parameter introduced. All the parameters keep the same values as in mean-field model, which leads to self-consistent descriptions for both single-particle and cluster motions.

1. Introduction

Cluster decays have been observed widely from light to heavy nuclei. In light nuclei, cluster structures are particularly favored when nuclei are excited to the vicinities of the cluster decay thresholds [1]. In heavy nuclei, cluster decays can happen in their ground states. Cluster decays have long played an important role in the studies of nuclear structures. For example, fine α-decay measurements in neutron-deficient Pb-Hg region can provide unique shell structure information [2, 3] and evidence for shape coexistence [4]. Especially in the syntheses of superheavy nuclei, the α-decay measurement is the unique method to identify new elements at present.

More exotic is molecular-type cluster structures in which there are two or more clusters with similar masses. Early as 1960s, the resonant states observed in the ^{12}C+^{12}C scattering were explained for molecular structures of the compound nucleus ^{24}Mg [5, 6]. Recent experiment [7] and theory [8] gave interesting molecular structure in ^{12}Be with two α-particles and four covalent neutrons. The famous ^{12}C Hoyle state which is at an excitation energy of 7.65 MeV is considered to have three α molecular structure. However, it is still unclear that the Hoyle state has three α-particles in chain or in triangle.

In theory, cluster decay is considered a process of quantum tunneling of a cluster particle through a potential barrier, which is called the Gamow model [9]. A right barrier is crucial for the calculation of decay width or lifetime. Experimentally, α-cluster potentials can be extracted from α scatterings but are generally energy- and method-dependent. Theoretically, Phenomenological α-cluster potentials have been suggested for the calculations of α-decay half-lives and spectroscopic properties [10, 11, 12]. Microscopic cluster potentials obtained from the double folding method have also been successfully applied to the investigations of α-decays and α-scatterings [13, 14, 15, 16, 17].

2. Model

In our previous works [18, 19, 20, 21], we suggested cluster potentials based on mean-field models. One of the advantages of such mean-field cluster potentials is that all the potential parameters can keep the same values as in the mean-field potentials [18, 19, 20, 21], which gives self-consistent descriptions for both single-particle and cluster motions and also makes predictions more reliable. Our mean-field cluster potentials have been successfully applied to the calculations of various cluster decays including diproton, α, heavier cluster decays [18, 19] and molecular structures as well [19, 21].

In the spherical case, the cluster potential can be simply written as (e.g., [11])

$$V(r) = V_N(r) + V_C(r) + \frac{\hbar}{2\mu r^2}(L + \frac{1}{2})^2, \tag{1}$$

that contains the nuclear potential $V_N(r)$, the Coulomb potential $V_C(r)$ and the Langer modified centrifugal potential with L and μ for the angular momentum carried by the cluster and the reduced mass of the cluster-core system, respectively [18]. We construct the nuclear potential between the cluster and the remaining core as follows [18, 19],

$$V_N(r) = \lambda[N_c v_n(r) + Z_c v_p(r)], \tag{2}$$

where λ is a folding factor which scales the nuclear potential to give right decay energy [18, 19]. N_c and Z_c are the neutron and proton numbers of the cluster, respectively. $v_n(r)$ and $v_p(r)$ are the single-neutron and -proton potentials (excluding the Coulomb potential) respectively, generated by the core. The single-particle potentials can be obtained by mean-field models, such as the Skyrme-Hartee-Fock (SHF) [18, 19, 20] or the simpler Woods-Saxon potential [21]. The Coulomb potential $V_C(r)$ is well defined in physics and should not be folded. We approximated the Coulomb potential by $V_C(r) = Z_c v_c(r)$ in the SHF potential calculation (where $v_c(r)$ is the single-proton Coulomb potential obtained with the density of the protons of the core) [18, 19, 20] . Such a cluster potential actually assumes that the cluster is structureless.

The folding factor λ can be determined using the Bohr-Sommerfeld quantization condition,

$$\int_{r_1}^{r_2} \sqrt{\frac{2\mu}{\hbar^2}|Q_L^* - V(r)|} = (2n+1)\frac{\pi}{2} = (G - L + 1)\frac{\pi}{2}, \tag{3}$$

where r_1, r_2 (and r_3) are classical turning points obtained by $V(r) = Q_L^*$ (the decay energy). The global quantum number $G = 2n+L$ (n is the node number in the radial wave function of the cluster tunneling motion) is determined by the Wildermuth rule [22], giving $G = \sum_{i=1}^{A_C} g_i$ where A_c is the nucleon number of the cluster anf g_i is the oscillator quantum number of a cluster nucleon orbiting the core. The g_i numbers can be easily obtained from the Nilsson diagram. For the lowest-energy cluster decay, for example, cluster nucleons should occupy single-particle orbits immediately above the Fermi levels of the daughter nucleus. For a cluster rotational spectrum with a given intrinsic structure, the G number is fixed.

The cluster decay energy from an excited state is obtained by

$$Q_L^* = Q_0 + E_J^*, \tag{4}$$

where Q_0 is the cluster Q value for the ground state, and E_J^* is the excitation energy of the given state with the spin J. The decay process can occur only if the state has a positive Q_L^* value (i.e., above the decay threshold). The spin L carried by the cluster is determined by the coupling rule of the spins of mother and daughter nuclei. One can also assume a same folding factor for a given cluster structure in the excited and ground states of the mother nucleus, i.e., the folding factor λ is determined in the case of the ground state and used for the calculations of cluster decays from excited states. Since the decay calculation is very sensitive to the Q_L^* value, the experimental Q_0 and E_J^* have been used in numerical calculations.

The partial cluster decay width is calculated by [11, 19]

$$\Gamma = P\frac{\hbar^2}{4\mu}\frac{\exp[-2\int_{r_2}^{r_3} k(r)dr]}{\int_{r_1}^{r_2} dr/2k(r)}, \tag{5}$$

where $k(r) = \sqrt{\frac{2\mu}{\hbar^2}|Q_L^* - V(r)|}$ is the wave number, and P is the preformation factor of the cluster being formed in the mother. For even-even nuclei, it has been testified that the $P = 1$ assumption under the use of the Bohr-Sommerfeld condition can well reproduce the experimental half-lives of various cluster decays [11, 18]. The decay half-life is obtained by $T_{1/2} = \hbar ln2/\Gamma$.

3. Calculations and discussions

Before the calculations of cluster decays, we test the validity of the proposed cluster potential by investigating the volume integral J_R per interacting nucleon pair for α-cluster decays or α-cluster scatterings [13]

$$J_R = \frac{4\pi}{A_\alpha A_d}\int_0^\infty V_N(r)r^2dr, \tag{6}$$

where A_α and A_d are the mass numbers of the α and daughter, respectively. In the potential calculations, we take the SkI4 force which has been developed with a good isospin property and can well reproduce the bulk properties of nuclei. For the α decay from the ground state of ^{212}Po, for example, we obtained a folding factor $\lambda = 0.595$ by using the Bohr-Sommerfeld condition, which leads to a volume integral, $J_R^{cal.} = 325$ MeV fm^3 [18]. Actually, obtained volume integrals are quite stable for a given mass region: $J_R^{cal.} = 330$ MeV fm^3 for trans-lead nuclei and $J_R^{cal.} = 290$ MeV fm^3 for trans-tin nuclei, agreeing well with the experimental values of $J_R^{expt.} \approx 300 - 350$ MeV fm^3 for a wide range of nuclei, obtained by α scattering experiments (see e.g., Ref. [13] for experimental data).

The α decays from the ground states of even-even trans-tin and trans-lead nuclei have been well studied experimentally. We have performed systematical calculations for the α decays of these two mass regions. The calculated results are quite encouraging [18]. We obtained quite stable folding factor which is $\lambda \sim 0.5$ for trans-tin mass region and $\lambda \sim 0.6$ for trans-lead region [18]. The calculated half-lives agree with experimental data within a factor of 2 [18].

We have also investigated heavier cluster decays from the ground states of trans-lead nuclei. Experiments have well established the decays, such as ^{14}C, ^{20}O, ^{24}Ne, ^{28}Mg and ^{32}Si. The present parameter-free calculations for the partial half-lives of the heavy cluster decays agree with experimental data within one order of magnitude [18]. As an exotic example, we have investigated the diproton radioactivity which has been observed experimentally. In our model, we assume the diproton as a cluster with a folding potential of $V_{2p} = \lambda \times 2v_p(r) + 2v_c(r)$. The typical examples of diproton decays from the ground states of ^{16}Ne and ^{38}Ti have been calculated. The folding factors determined with using experimental decay energies are $\lambda = 0.66$ and 0.75 for ^{16}Ne and ^{38}Ti, respectively [18]. The calculations give quite reasonable half-lives of the diproton emissions of the two nuclei, within one order of magnitude compared with data [18].

In Ref. [20], we gave the detailed and systematical calculations of α decays for nuclei from $Z = 78$ to $Z = 118$, compared with available data. Our calculations reproduce systematically the observed decay properties, giving evidence to testify the validity of the mean-field-based cluster potential. In the paper [20], we also studied the deformation effect on the cluster decay by calculating an averaged decay width, $\Gamma = \int_0^{\pi/2}\Gamma(\theta)sin\theta d\theta$ (θ is the angle between

Journal of Physics: Conference Series **413** (2013) 012017 doi:10.1088/1742-6596/413/1/012017

the cluster-emission direction and the deformation symmetry axis of the nucleus [18]. The α-decay calculations also reflect well the shell closure around $N = 126$.

The suggested mean-field-based cluster potential has also been successfully applied to the calculations of cluster decays from excited states, studying so-called molecular structures [19]. We calculated the α-decay widths of excited states which belong to the sequence 0^+, 2^+, $4^+ \cdots$ in ^{8}Be, well reproducing the measured decay widths [19]. ^{20}Ne is an interesting nucleus in which excited states at excitation energies higher than ≈ 4.7 MeV (i.e., the α-decay threshold) are considered to have the molecular structure of α+^{16}O. There have already been good experimental measurements on the resonant widths of the molecular states with positive or negative parities. Our calculations can well reproduce the experimental widths within a factor of about two for 7 observed resonant states at excitation energies ranging from 5.8 MeV to 22.9 MeV [19].

Compared with beryllium and neon nuclei discussed above, carbon isotopes have more complicated geometric structures with three α-particles which can form chain or triangular cluster structure. Our calculations show these two structures lead to different decay widths or lifetimes [19]. In the numerical calculations of the α-decay widths of excited states in carbon isotopes, we assumed that the remaining cores (i.e., two α-particles + possible covalent neutrons) are well deformed with an approximate axis ratio of 2:1. The core produces a cluster potential which can be obtained also from a mean-field model. We have made detailed calculations for ^{12,13,14}C, giving reasonable results compared with data.

In the carbon isotopes, ^{12}C is an especially interesting nucleus. The large abundance of ^{12}C in the universe is attributed to the second 0^+ state which accelerate the fusion process from helium to carbon. The 7.6 MeV 0_2^+ excited state (just above the threshold of the α + ^{8}Be fusion) is called the Hoyle state predicted by Hoyle in 1950s. Though 60 years passed, its structure which is chain or triangular is still unclear. For the decay to α + ^{8}Be, our calculation with a 3α chain gave a decay width of 5.9 eV [19] which is close to the experimental data of 8.5 ± 1.0 eV. The calculation with assuming a 3α triangular configuration decaying to α + ^{8}Be shows a smaller width of 1.6 eV. From the comparison between calculated and experimental decay widths, the chain structure seems to be more likely. However, we cannot exclude the triangular structure. It is also possible that the 3α-chain and -triangular configurations might coexist. Indeed, our preliminary calculations based on a 3-body Hamiltonian with a Gausian force shows that these two configurations coexist in the Hoyle state, while the 2_2^+ state built on the Hoyle state is predicted to be more triangular [23]. The 2_2^+ excited Hoyle state was observed recently with an excitation energy of $E_x \sim 10$ MeV [24].

4. Summary

We have suggested a cluster potential which is based on single-particle mean field. The advantage of the potential is no new free parameter introduced, which makes predictions more reliable. This also leads to self-consistent descriptions for both single-particle and cluster motions. The potential includes automatically the isospin dependence. In most of our numerical calculations, the microscopic Skyrme force has been adopted. With such a mean-field-based cluster potential, we have successfully investigated various cluster decays including diproton, α and heavier cluster decays, with calculated decay widths or lifetimes agreeing well with available experimental data. Some useful structure information has been obtained. For example, the calculations of the ^{12}C Hoyle state gave a large component of the 3α-chain structure. As a verification of the validity of the cluster potential, we have calculated the α-decay volume integral of interaction, giving right values agreeing well with experimental data obtained by α scatterings.

Acknowledgments

This work has been supported by the National Key Basic Research Program of China under Grant 2013CB834400, and the National Natural Science Foundation of China under Grant

Nos. 11235001 and 10975006.

References

[1] Ikeda K, Takigawa N and Horiuchi H 1968 *Prog. Theor. Phys. (Japan)* **(Suppl. Extra Number)** 464
[2] Toth K *et al.* 1984 *Phys. Rev. Lett.* **53** 1623
[3] Wauters J *et al.* 1994 *Phys. Rev. Lett.* **72** 1329
[4] Andreyev A N *et al.* 2000 *Nature (London)* **405** 430
[5] Bromley D A, Kuehner J A and Almqvist E 1960 *Phys. Rev. Lett.* **4** 365
[6] Vogt E and McManus H 1960 *Phys. Rev. Lett.* **4** 518
[7] Freer M *et al.* 1999 *Phys. Rev. Lett.* **82** 1383
[8] Ito M, Itagaki N, Sakurai H and Ikeda K 2008 *Phys. Rev. Lett.* **100** 182502
[9] Gamow G 1928 *Z. Phys.* **51** 204
[10] Buck B, Merchant A C and Perez S M 1992 *Phys. Rev.* C **45** 2247
[11] Buck B, Merchant A C and Perez S M 1993 *At. Data Nucl. Data Tables* **54** 53
[12] Xu C and Ren Z 2004 *Phys. Rev.* C **69** 024614
[13] Mohr P 2000 *Phys. Rev.* C **61** 045802
[14] Atzrott U, Mohr P, Abele H, Hillenmayer C and Staudt G 1996 *Phys. Rev.* C **53** 1336
[15] Xu C and Ren Z Z 2005 *Nucl. Phys.* A **760** 303
[16] Ren Z Z, Xu C and Wang Z 2004 *Phys. Rev.* C **70** 034304
[17] Chowdhury P R, Samanta C and Basu D N 2006 *Phys. Rev.* C **73** 014612
[18] Xu F R and Pei J C 2006 *Phys. Lett.* B **642** 322
[19] Pei J C and Xu F R 2007 *Phys. Lett.* B **650** 224
[20] Pei J C, Xu F R, Lin Z J and Zhao E G 2007 *Phys. Rev.* C **76** 044326.
[21] Xu C, Qi C, Liotta R J, Wyss R, Wang S M, Xu F R, and Jiang D X 2010 *Phys. Rev.* C **81** 054319
[22] Wildermuth K and Tang Y C 1977 *A Unified Theory of the Nucleus* (New York: Academic Press)
[23] Wang S M *et al.* to be published.
[24] Itoh M *et al.* 2011 *Phys. Rev.* C **84** 054308.

Exploring the nuclear matter phase diagram with heavy ion reactions

M Colonna[1], **V Baran**[2], **M Di Toro**[1,3], **C Rizzo**[1]
[1]Laboratori Nazionali del Sud, INFN, via Santa Sofia 62, I-95123, Catania, Italy
[2]Physics Faculty, University of Bucharest, Romania
[3]Physics-Astronomy Dept., University of Catania, Italy
E-mail: colonna@lns.infn.it

Abstract. We review recent results concerning heavy ion reactions between charge asymmetric systems from low up to intermediate energies.

Solving numerically self-consistent transport equations for neutrons and protons, we focus on isospin sensitivite observables, aiming at extracting information on the density dependence of the nuclear symmetry energy. For reactions close to the Coulomb barrier, we explore the structure of pre-equilibrium dipole oscillations, rather sensitive to the low-density behavior of the isovector component of the nuclear effective interaction. In the Fermi energy regime, we investigate the interplay between dissipation mechanisms, fragmentation and isospin effects. At intermediate energies, where regions with higher density and momentum are reached, we discuss collective flows and their sensitive to neutron and proton effective masses.

1. Introduction

The Equation of State (EOS) of nuclear matter is of crucial importance for the understanding of a large variety of phenomena in nuclear physics and astrophysics. Transient states of nuclear matter far from normal conditions can be created in terrestrial laboratories and many experimental and theoretical efforts have been devoted to the study of nuclear reactions, from low to intermediate energies, as a possible tool to learn about the behavior of nuclear matter and its EOS. In particular, the availability of exotic beams has opened the way to explore, in laboratory conditions, new aspects of nuclear structure and dynamics up to extreme ratios of neutron (N) to proton (Z) numbers. Over the past years, measurements of isoscalar collective vibrations, collective flows and meson production have contributed to constrain the EOS for symmetric matter for densities up to five time the saturation value [1]. However, the EOS of asymmetric matter has comparatively few experimental constraints: The isovector part of the nuclear effective interaction (Asy-EOS) and the corresponding symmetry energy are largely unknown far from normal density.

This information is essential also in the astrophysical context, for the understanding of the properties of compact objects such as neutron stars, whose crust behaves as low-density asymmetric nuclear matter [2] and whose core may touch extreme values of density and asymmetry. Moreover, the low-density behavior of the symmetry energy also affects the structure of exotic nuclei and the appearance of new features involving the neutron skin [3].

Over the past years, several observables which are sensitive to the Asy-EOS and testable experimentally, have been suggested [4, 5, 6, 7].

In this article we will discuss dissipative collisions in a wide range of beam energies, from just above the Coulomb barrier up to the $AGeV$ range. Low to Fermi energies will bring information on the symmetry term around (below) normal density, while intermediate energies will probe high density regions.

2. Transport theories and symmetry energy

Nuclear reactions are modeled by solving transport equations based on mean field theories, with correlations included via hard nucleon-nucleon elastic collisions and via stochastic forces, selfconsistently evaluated from the mean phase-space trajectory, see [8, 5]. Stochasticity is essential in order to get distributions as well as to allow for the growth of dynamical instabilities.

In the energy range up to a few hundred $AMeV$, the appropriate tool is the so-called Boltzmann-Langevin equation (BLE) [8]:

$$\frac{df}{dt} = \frac{\partial f}{\partial t} + \{f, H\} = I_{coll}[f] + \delta I[f], \tag{1}$$

where $f(\mathbf{r}, \mathbf{p}, t)$ is the one-body distribution function, or Wigner transform of the one-body density, $H(\mathbf{r}, \mathbf{p}, t)$ the mean field Hamiltonian, I_{coll} the two-body collision term incorporating the Fermi statistics of the particles, and $\delta I[f]$ the fluctuating part of the collision integral. Here we follow the approximate treatment of the BLE introduced in Ref.[9], the Stochastic Mean Field (SMF) model. Effective interactions and the nuclear matter EOS can be considered as ingredients of the transport codes and from the comparison with experimental data one can finally get some hints on nuclear matter properties.

We recall that the symmetry energy E_{sym} appears in the energy density $\epsilon(\rho, \rho_3) \equiv \epsilon(\rho) + \rho E_{sym}/A(\rho_3/\rho)^2 + O(\rho_3/\rho)^4 + ..$, expressed in terms of total ($\rho = \rho_p + \rho_n$) and isospin ($\rho_3 = \rho_p - \rho_n$) densities. E_{sym} gets a kinetic contribution directly from basic Pauli correlations and a potential part, $C(\rho)$, from the highly controversial isospin dependence of the effective interactions:

$$\frac{E_{sym}}{A} = \frac{E_{sym}}{A}(kin) + \frac{E_{sym}}{A}(pot) \equiv \frac{\epsilon_F}{3} + \frac{C(\rho)}{2\rho_0}\rho. \tag{2}$$

The sensitivity of the simulation results is tested against different choices of the density and momentum dependence of the isovector part of the Equation of State. We will consider for the potential part of the symmetry energy, $C(\rho)$, two opposite density parametrizations (Asy-EOS) of the mean field [5, 10]: i) $\frac{C(\rho)}{\rho_0} = 482 - 1638\rho$ $(MeV fm^3)$ for Asysoft EOS: $E_{sym}/A(pot)$ has a weak density dependence close to the saturation, with an almost flat behavior below the saturation density ρ_0 and even decreasing at suprasaturation; ii) a constant coefficient, $C = 32\ MeV$, for the Asystiff EOS choice, see figure 1. The isoscalar section of the EOS is the same in both cases, fixed requiring that the saturation properties of symmetric nuclear matter with a compressibility around $220 MeV$ are reproduced.

The role of the isospin degree of freedom on the dynamics of a nuclear reaction can be discussed in a compact way by means of the chemical potentials for protons and neutrons as a function of density ρ and asymmetry $I = (N - Z)/A$ [11]. The p/n currents can be expressed as

$$\mathbf{j}_{p/n} = D^{\rho}_{p/n} \nabla \rho - D^{I}_{p/n} \nabla I, \tag{3}$$

with $D^{\rho}_{p/n}$ the drift, and $D^{I}_{p/n}$ the diffusion coefficients for transport, which are given explicitely in Ref. [11]. Of interest for the study of isospin effects are the differences of currents between protons and neutrons which have a simple relation to the density dependence of the symmetry energy

$$\begin{aligned} D^{\rho}_{n} - D^{\rho}_{p} &\propto 4I \frac{\partial E_{sym}}{\partial \rho}, \\ D^{I}_{n} - D^{I}_{p} &\propto 4\rho E_{sym}. \end{aligned} \tag{4}$$

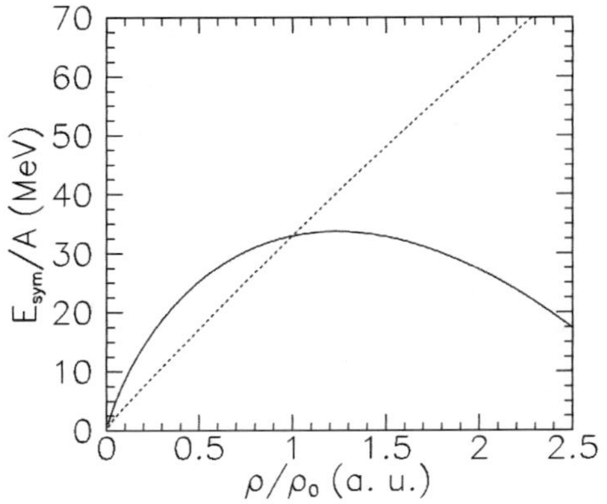

Figure 1. Two representative effective parameterizations of the symmetry energy : Asystiff (dotted line) and Asysoft (full line).

Thus the isospin transport due to density gradients (isospin migration) depends on the slope of the symmetry energy, or the symmetry pressure, while the transport due to isospin concentration gradients (isospin diffusion) depends on the absolute value of the symmetry energy. Hence transport phenomena in nuclear reactions are directly linked to the EOS properties.

3. The prompt dipole γ-ray emission in dissipative collisions

The low-density behavior of the symmetry energy can be explored looking at dipole excitations in dissipative charge asymmetric reactions around 10 $AMeV$. The possibility of an entrance channel bremsstrahlung dipole radiation due to an initial different N/Z distribution was suggested at the beginning of the nineties [12]. After several experimental evidences, in fusion as well as in deep-inelastic reactions, [13, 14, 15] and references therein, the process is now well understood and stimulating new perspectives are coming from the use of radioactive beams.

During the charge equilibration process taking place in the first stage of dissipative reactions between colliding ions with different N/Z ratios, a large amplitude dipole collective motion develops in the composite dinuclear system, the so-called Dynamical Dipole mode. This collective dipole gives rise to a prompt γ-ray emission which depends on the absolute value of the initial amplitude, $D(t=0)$, on the fusion/deep-inelastic dynamics and on the symmetry term, below saturation, that is acting as a restoring force. Indeed this oscillation develops in the low density interface between the two colliding ions (neck region).

A detailed description is obtained in mean field transport approaches [16]. One can follow the time evolution of the dipole moment in the r-space, $D(t) = \frac{NZ}{A}(R_Z - R_N)$ and in $p-$space, $DK(t) = (\frac{P_p}{Z} - \frac{P_n}{N})$, being R_p, P_p (R_n, P_n) the centers of mass in coordinate and momentum space for protons (neutrons). A nice "spiral-correlation" clearly denotes the collective nature of the mode, see figure 2. The "prompt" photon emission probability, with energy $E_\gamma = \hbar\omega$, can be estimated applying a bremsstrahlung approach to the dipole evolution given from the BLE approach [16]:

$$\frac{dP}{dE_\gamma} = \frac{2e^2}{3\pi\hbar c^3 E_\gamma}|D''(\omega)|^2, \tag{5}$$

where $D''(\omega)$ is the Fourier transform of the dipole acceleration $D''(t)$. We remark that in this way it is possible to evaluate, in *absolute* values, the corresponding pre-equilibrium photon emission.

We must add a couple of comments of interest for the experimental selection of the Dynamical Dipole: i) The centroid is always shifted to lower energies (large deformation of the dinucleus);

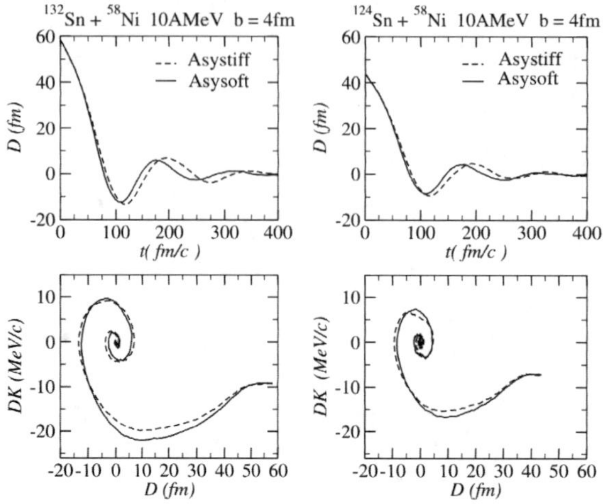

Figure 2. Dipole Dynamics at 10 $AMeV$, $b = 4fm$ centrality. Left panels: exotic "132" system. Upper panel: time evolution of dipole moment D(t) in real space; Lower panel: dipole phase-space correlation (see text). Right panels: same as before for the stable "124" system. Solid lines correspond to Asysoft EOS, the dashed to Asystiff EOS.

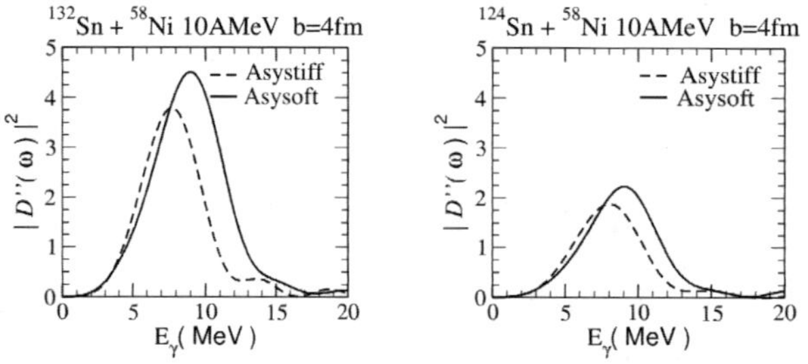

Figure 3. Left panel: exotic "132" system. Power spectra of the dipole acceleration at $b = 4$fm (in c^2 units). Right panel: corresponding results for the stable "124" system. Solid lines correspond to Asysoft EOS, the dashed to Asystiff EOS.

ii) A clear angular anisotropy should be present since the prompt mode has a definite axis of oscillation (on the reaction plane) at variance with the statistical giant dipole resonance (GDR). These features have been observed in recent experiments [14].

The use of unstable neutron rich projectiles would largely increase the effect, due to the possibility of larger entrance channel asymmetries.

One can notice in figure 2 the large amplitude of the first oscillation for the "132" system, but also the delayed dynamics for the Asystiff EOS related to a weaker isovector restoring force. In figure 3 (left panel) we report the power spectrum, $\mid D''(\omega) \mid^2$ in semicentral "132" reactions, for different Asy-EOS choices. The gamma multiplicity is simply related to it, see Eq.(5). The corresponding results for the stable "124" system are drawn in the right panel. As expected from the larger initial charge asymmetry, we clearly see an increase of the Prompt Dipole Emission for the exotic n-rich beam. Such entrance channel effect allows also for a better observation of the Asy-EOS dependence. We remind that in the Asystiff case we have a weaker restoring force for the dynamical dipole in the dilute "neck" region, where the symmetry energy is smaller [17]. This is reflected in the lower value of the centroid, ω_0, as well as in the reduced total yield, as shown in figure 3. The sensitivity of ω_0 to the stiffness of the symmetry energy is more easily identified increasing $D(t_0)$, i.e. using exotic, more asymmetric beams. Moreover, a good sensitivity to the symmetry energy is also observed for dipole oscillations in more peripheral collisions [18], where low-density surface contributions become more important.

The prompt dipole radiation angular distribution is the result of the interplay between the collective oscillation life-time and the dinuclear rotation. In this sense one expects also a sensitivity to the Asy-EOS of the anisotropy, in particular for high spin event selections [17].

4. Isospin equilibration and fragmentation mechanisms at Fermi energies

In this energy range (30-60 $AMeV$), reactions between charge asymmetric systems are charecterized by a direct isospin transport in binary events. This process also involves the low density neck region and is sensitive to the low density behavior of E_{sym}, see Refs.[19, 20] and references therein. Moreover, it is now quite well established that the largest part of the reaction cross section for dissipative collisions at Fermi energies goes through the *Neck Fragmentation* channel, with intermediate mass fragments (IMF) directly produced in the interacting zone in semiperipheral collisions on short time scales [21]. It is possible to predict interesting isospin transport effects also for this fragmentation mechanism since clusters are formed still in a dilute asymmetric matter but always in contact with the regions of the projectile-like and target-like remnants almost at normal densities.

Results on these mechanisms, obtained with the SMF model, are discussed below.

4.1. The isospin transport ratio

In peripheral and semi-peripheral reactions, it is interesting to look at the asymmetries of the various parts of the interacting system in the exit channel: emitted particles, projectile-like (PLF) and target-like fragments (TLF), and in the case of ternary (or higher multiplicity) events, IMF's. In particular, one can study the so-called isospin transport ratio, which is defined as

$$R^x_{P,T} = \frac{2(x^M - x^{eq})}{(x^H - x^L)} \ , \tag{6}$$

with $x^{eq} = \frac{1}{2}(x^H + x^L)$. Here, x is an isospin sensitive quantity that has to be investigated with respect to equilibration. We consider primarily the asymmetry $\beta = I = (N - Z)/A$, but also other quantities, such as isoscaling coefficients, ratios of production of light fragments, etc, can be of interest [6]. The indices H and L refer to the symmetric reaction between the heavy (n-rich) and the light (n-poor) systems, while M refers to the mixed reaction. P, T denote the rapidity region, in which this quantity is measured, in particular the PLF and TLF rapidity regions. Clearly, this ratio is ± 1 in the projectile and target regions, respectively, for complete transparency, and oppositely for complete rebound, while it is zero for complete equilibration.

In a simple model one can show that the isospin transport ratio mainly depends on two quantities: the strength of the symmetry energy and the interaction time between the two reaction partners. Let us take, for instance, the asymmetry β of the PLF (or TLF) as the quantity x. At a first order approximation, in the mixed reaction this quantity relaxes towards its complete equilibration value, $\beta_{eq} = (\beta_H + \beta_L)/2$, as

$$\beta^M_{P,T} = \beta^{eq} + (\beta^{H,L} - \beta^{eq})\, e^{-t/\tau}, \tag{7}$$

where t is the time elapsed while the reaction partners stay in contact (interaction time) and the damping τ is mainly connected to the strength of the symmetry energy [20]. Inserting this expression into Eq.(6), one obtains $R^{\beta}_{P,T} = \pm e^{-t/\tau}$ for the PLF and TLF regions, respectively. Hence the isospin transport ratio can be considered as a good observable to trace back the strength of the symmetry energy from the reaction dynamics provided a suitable selection of

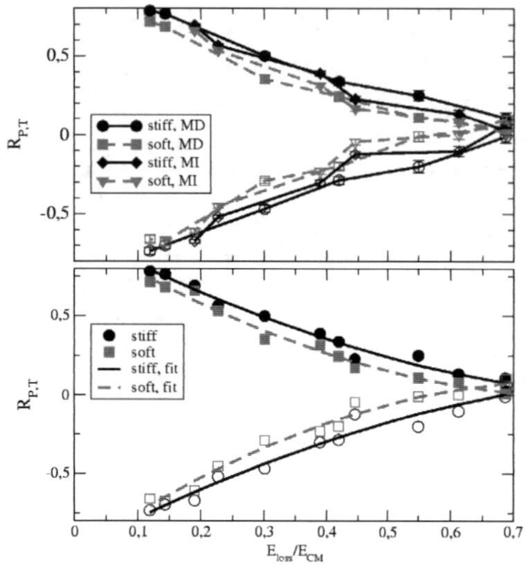

Figure 4. Isospin transport ratios as a function of relative energy loss. Upper panel: separately for stiff (solid) and soft (dashed) Asy-EOS, and for two parameterizations of the isoscalar part of the interaction: MD (circles and squares) and MI (diamonds and triangles), in the projectile region (full symbols) and the target region (open symbols). Lower panel: quadratic fit to all points for the stiff (solid), resp. soft (dashed) Asy-EOS.

the interaction time is performed. The centrality dependence of the isospin ratio, for Sn + Sn collisions at 35 and 50 $AMeV$, has been investigated in experiments as well as in theory [19, 20, 22], and information about the stiffness of the symmetry energy has been extracted from the analysis presented in [22], based on other transport models.

Here we investigate more in detail the relation between charge equilibration, interaction times and thermal equilibrium. Longer interaction times should be correlated to a larger dissipation. It is then natural to look at the correlation between the isospin transport ratio and the total kinetic energy loss. In this way one can also better disentangle dynamical effects of the isoscalar and isovector part of the EOS, see [20].

It is seen in figure 4 (top) that the curves for the asysoft EOS (dashed) are generally lower in the projectile region (and oppositely for the target region), i.e. show more equilibration, than those for the asystiff EOS, due to the higher value of the symmetry energy at low density. To emphasize this trend, all the values for the stiff (circles) and the soft (squares) Asy-EOS, corresponding to different impact parameters, beam energies and also to two possible parametrizations of the isoscalar part of the nuclear interaction (with and without momentum dependence, MD and MI), are collected together in the bottom part of the figure. One can see that all the points essentially follow a given line, depending only on the symmetry energy parameterization adopted. It is seen, that there is a systematic effect of the symmetry energy of the order of about 20 percent, which should be measurable. Moreover, we notice that the quantity R is a rapidly decreasing function of the degree of dissipation, E_{loss}, reached in the collision. This can be explained in terms of dissipation mechanisms maily due to mean-field effects, as predicted by the SMF model. Indeed, according to a mean-field picture, a significant degree of thermal equilibrium (i.e. a considerable E_{loss}) would imply a rather long contact time between the two reaction partners, thus certainly leading to isospin equilibration, that needs a shorter time scale to be reached. The correlation suggested in figure 4 should represent a general feature of isospin diffusion, as expected on the basis of dominant mean-field mechanisms, and it would be of great interest to verify it experimentally.

4.2. Isospin dynamics in neck fragmentation at Fermi energies

In presence of density gradients, as the ones occurring when a low-density neck region is formed between the two reaction partners in semi-peripheral collisions, the isospin transport is mainly ruled by the density derivative of the symmetry energy and so we expect a larger neutron flow to the neck clusters for a stiffer symmetry energy around saturation [5]. This mechanism leads to the neutron enrichment of the neck region (isospin migration). This is shown in figure 5 (left), where the asymmetry of the neck and PLF-TLF regions, obtained in neutron-rich reactions at 50 $AMeV$, are plotted for two Asy-EOS choices.

From the experimental point of view, a new analysis has been recently published on Sn+Ni data at 35 $AMeV$ by the Chimera Collab.[23]. A strong correlation between neutron enrichement and fragment alignement (when the short emission time selection is enforced) is seen, that points to a stiff behavior of the symmetry energy ($L \approx 70\ MeV$), for which a large neutron enrichment of neck fragments is seen (left).

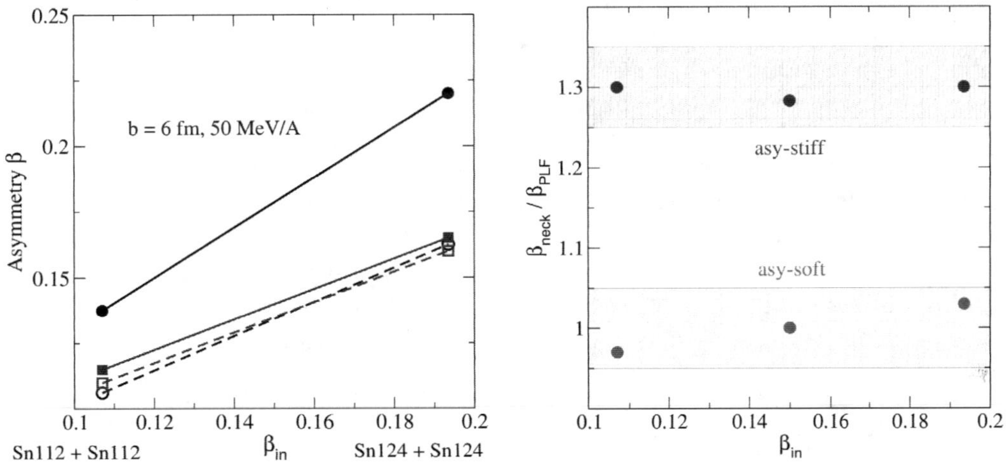

Figure 5. Left panel: asymmetry of IMF's (circles) and PLF-TLF (squares), as a function of the system initial asymmetry, for two Asy-EOS choices: asystiff (full lines) and asysoft (dashed lines). Right panel: Ratio between the neck IMF and the PLF asymmetries, as a function of the system initial asymmetry. The bands indicate the uncertainty in the calculations.

In order to build observables less affected by secondary decay effects, in figure 5 (right) we consider the ratio of the asymmetries of the IMF's to those of the residues (β_{res}) for stiff and soft Asy-EOS. This quantity can be estimated on the basis of simple energy balance considerations. By imposing to get a maximum (negative) variation of E_{sym} when transfering the neutron richness from PLF and TLF towards the neck region, one obtains:

$$\frac{\beta_{IMF}}{\beta_{res}} = \frac{E_{sym}(\rho_R)}{E_{sym}(\rho_I)} \tag{8}$$

From this simple argument the ratio between the IMF and residue asymmetries should depend only on symmetry energy properties and, in particular, on the difference of the symmetry energy corresponding to the residue and neck densities (ρ_R and ρ_I), as appropriate for isospin migration. It should also be larger than one, more so for the asystiff than for the asysoft EOS. It is seen indeed in figure 5 (right part), that this ratio is nicely dependent on the Asy-EOS only (being larger in the asystiff case) and not on the system considered. If final asymmetries were affected in the same way by secondary evaporation in the case of neck and PLF fragments, then one could directly compare the results of figure 5 (right) to data. However, due to the different size and temperature of the neck region with respect to PLF or TLF sources, de-excitation effects should be carefully checked with the help of suitable decay codes.

5. Asy-EOS at supra-saturation density: collective flows

The problem of Momentum Dependence in the Isovector channel ($Iso - MD$) of the nuclear interaction is still very controversial and it would be extremely important to get more definite experimental information, see the recent Refs. [24, 25]. Momentum dependent effective interactions essentially lead to the concept of effective masses. If also the isovector component of the interaction is momentum dependent, one observes different effective masses (i.e. effective mass splitting) for neutrons and protons. Exotic Beams at intermediate energies (100-500 $AMeV$) are of interest in order to have high momentum particles and to test regions of high baryon (isoscalar) and isospin (isovector) density during the reaction dynamics. Transport codes are usually implemented with different (n, p) momentum dependences, see [24, 25]. This allows one to follow the dynamical effect of opposite n/p effective mass (m^*) splitting while keeping the same density dependence of the symmetry energy [20].

For central collisions in the interacting zone baryon densities about $1.7 - 1.8\rho_0$ can be reached in a transient time of the order of 15-20 fm/c. The system is quickly expanding and the Freeze-Out time is around 50 fm/c. Here it is very interesting to study again the collective response of the system. Collective flows are very good candidates since they are expected to be rather sensitive to the momentum dependence of the mean field, see [26, 5]. The transverse flow, $V_1(y, p_t) = \langle \frac{p_x}{p_t} \rangle$, where $p_t = \sqrt{p_x^2 + p_y^2}$ is the transverse momentum and y the rapidity along the beam direction, provides information on the anisotropy of the nucleon emission on the reaction plane. Very important for the reaction dynamics is the elliptic flow, $V_2(y, p_t) = \langle \frac{p_x^2 - p_y^2}{p_t^2} \rangle$. The sign of V_2 indicates the azimuthal anisotropy of emission: on the reaction plane ($V_2 > 0$) or out-of-plane ($squeeze - out$, $V_2 < 0$) [26].

Let us consider semicentral ($b/b_{max} = 0.5$) collisions of ^{197}Au+^{197}Au at 400 $AMeV$. As

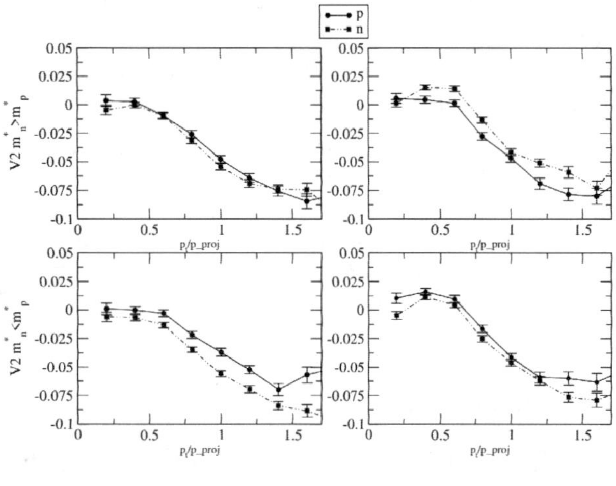

Figure 6. Proton (thick) and neutron (thin) V_2 flows in a semi-central reaction Au+Au at 400 $AMeV$. Transverse momentum dependence at midrapidity, $| y_0 |< 0.3$. Upper curves for $m_n^* > m_p^*$, lower curves for the opposite splitting $m_n^* < m_p^*$. Left: Asystiff. Right: Asysoft.

seen from figure 6, the neutron squeeze-out is larger (i.e. more negative) in the $m_n^* < m_p^*$ case, independently of the stiffness of the symmetry term. In particular, in the Asysoft case we see an inversion of the neutron/proton squeeze-out at mid-rapidity for the two effective mass-splittings. Good data seem to be suitable to disentangle $Iso - MD$ potentials [27, 28]. The results are also slightly depending on the Asy-stiffness, with large neutron *squeeze* – *out* effects in the Asystiff case. Due to the difficulties in measuring neutrons, one could investigate the difference between light isobar flows, like ^{3}H vs. ^{3}He. We still expect to see effective mass splitting effects [29].

6. Conclusions

We have reviewed some aspects of the rich phenomenology associated with nuclear reactions, from which interesting hints are emerging to constrain the nuclear EOS and, in particular, the largely debated density behavior of the symmetry energy.

Information on the low density region can be accessed in reactions from low to Fermi energies, where collective excitations and fragmentation mechanisms are dominant. Results on isospin sensitive observables have been presented. In particular, we have concentrated our analysis on the charge equilibration mechanism (and its relation to energy dissipation) and on the neutron-enrichment of the neck region in semi-peripheral reactions. From the study of the latter mechanism, for which new experimental evidences have recently appeared [23], hints are emerging towards a stiff behavior of the symmetry energy around normal density. This is compatible with recent results from structure data, see for instance the review article [30].

The greatest theoretical uncertainties concerns the high density domain, that has the largest impact on the understanding of the properties of neutron stars. This regime can be explored in terrestrial laboratories by using intermediate energy and relativistic heavy ion collisions of charge asymmetric nuclei. Collective flows and meson production are promising observables.

A considerable amount of work has already been done in the symmetry energy domain. In the near future, thanks to the availability of both stable and rare isotope beams, more selective analyses, also based on new exclusive observables, are expected to provide further stringent constraints.

Acknowledgements

This work for V. Baran was supported by a grant of the Romanian National Authority for Scientific Research, CNCS - UEFISCDI, project number PN-II-ID-PCE-2011-3-0972.

References

[1] Danielewicz P, Lacey R and Lynch WG 2002 *Science* **298** 1592
[2] Lattimer JM and Prakash M 2007 *Phys.Rep* **442** 109
[3] Carbone A *et al* 2010 *Phys. Rev.* C **81** 041301
[4] *Isospin Physics in Heavy-ion Collisions at Intermediate Energies*, Eds. Li BA and Schröder WU, Nova Science Publishers (2001, New York)
[5] Baran V, Colonna M, Greco V and Di Toro M 2005 *Phys. Rep.* **410** 335
[6] Colonna M and Tsang MB 2006 *Eur. Phys. J* A **30** 165, and references therein
[7] Li BA, Chen LW and Ko CM 2008 *Phys. Rep.* **465** 113
[8] Chomaz P, Colonna M and Randrup J 2004 *Phys. Rep.* **389** 263
[9] M. Colonna et al, Nucl. Phys. A **642**, 449 (1998)
[10] Baran V *et al* 2002 *Nucl. Phys.* A **703** 603
[11] Baran V *et al* 2005 *Phys.Rev.* C **72** 064620
[12] Chomaz P, Di Toro M and Smerzi A 1993 *Nucl. Phys.* A **563** 509
[13] Pierroutsakou D *et al* 2005 *Phys. Rev.* C **71** 054605
[14] Martin B, Pierroutsakou D *et al* (Medea Collab.) 2008 *Phys. Lett.* B **664** 47
[15] Sandulescu A, Petrovici M, Pop A *et al* 1981 *Jou. Phys.* G **7** L55-L61
[16] Baran V, Brink DM, Colonna M and Di Toro M 2001 *Phys. Rev. Lett.* **87** 182501
[17] Baran V, Rizzo C, Colonna M, Di Toro M and Pierroutsakou D 2009 *Phys. Rev.* C **79** 021603

[18] Rizzo C *et al* 2011 *Phys Rev* C **83** 014604
[19] Tsang MB *et al* 2004 *Phys. Rev. Lett.* **92** 062701
[20] Rizzo J *et al* 2008 *Nucl.Phys.* A **806** 79
[21] Di Toro M, Olmi A and Roy R 2006 *Eur. Phys. Jour.* A **30** 65
[22] Tsang MB *et al* 2009 *Phys. Rev. Lett.* **102** 122701; Sun ZY *et al* 2010 *Phys. Rev.* C **82** 051603
[23] De Filippo E *et al* (Chimera Collab.) 2012 *Phys. Rev.* C **86** 014610
[24] Li BA, Das Champak B, Das Gupta S and Gale C 2004 *Nucl. Phys.* A **735** 563
[25] Rizzo J, Colonna M and Di Toro M 2005 *Phys. Rev.* C **72** 064609
[26] Danielewicz P 2000 *Nucl. Phys.* A **673** 375
[27] Russotto P *et al* 2011 *Phys. Lett.* B **697** 471
[28] Cozma MD 2011 *Phys. Lett.* B **700** 139
[29] Giordano V *et al* 2010 *Phys. Rev.* C **81** 044611
[30] Tsang MB *et al* 2012 *Phys. Rev.* C **86** 015803

Some aspects of the phase diagram of nuclear matter relevant to compact stars

F Gulminelli[1], Ad R Raduta[2], J Margueron[3], P Papakonstantinou[3] and M Oertel[4]

[1] CNRS, ENSICAEN, UMR6534, LPC ,F-14050 Caen cédex, France
[2] NIPNE, Bucharest-Magurele, POB-MG6, Romania
[3] IPNO, CNRS/IN2P3, Université Paris - Sud 11, F-91406 Orsay cedex, France
[4] LUTH, CNRS, Observatoire de Paris, Université Paris Diderot, 5 place Jules Janssen, 92195 Meudon, France

E-mail: gulminelli@lpccaen.in2p3.fr

Abstract. Dense matter as it can be found in core-collapse supernovae and neutron stars is expected to exhibit different phase transitions which impact the matter composition and the equation of state, with important consequences on the dynamics of core-collapse supernova explosion and on the structure of neutron stars. In this paper we will address the specific phenomenology of two of such transitions, namely the crust-core solid-liquid transition at sub-saturation density, and the possible strange transition at super-saturation density in the presence of hyperonic degrees of freedom. Concerning the neutron star crust-core phase transition at zero and finite temperature, it will be shown that, as a consequence of the presence of long-range Coulomb interactions, a clusterized phase is expected which is not accessible in the grand-canonical ensemble. A specific quasi-particle model will be introduced and some quantitative results relevant for the supernova dynamics will be shown. The opening of hyperonic degrees of freedom at higher densities corresponding to the neutron stars core also modifies the equation of state. The general characteristics and order of phase transitions in this regime will be analyzed in the framework of a self-consistent mean-field approach.

1. Introduction

Triggered by the first data taking of X and gamma satellites, as well as by the improved capabilities of radio telescopes and high and ultra-high gamma ray detectors, there has been in the last ten years an impressive accumulation of observational data on supernova and neutron stars in different environments and at different evolution stages. Among the most important recent results one should mention the precise measurement of a very massive two solar-mass neutron star [1] which challenges a number of theoretical equations of state of dense matter, and the measurement of the cooling curve of the young pulsar Cassiopeia A [2], which gives strong constraints on the super-fluid properties of neutron star cores.

Understanding these data requires a good modelization of the supernova dynamics and of the formation process of neutron stars [3, 4], but also a precise and detailed control of the relevant microphysics. Sophisticated two and three-dimensional core-collapse supernova simulations show that matter in the supernova core explores a very large interval of baryonic densities, ranging from about $\rho > 10^{10}$ g $\cdot$ cm$^{-3} \approx 10^{-4}\rho_0$ to several times the normal nuclear density ρ_0 ,

Journal of Physics: Conference Series **413** (2013) 012019 doi:10.1088/1742-6596/413/1/012019

temperatures between some hundreds of keV and around 80 MeV, and proton fractions between ≈ 0.5 and ≈ 0.1.

Both the equation of state and the neutrino cross sections are strongly influenced by the possible presence of phase transitions in these wide thermodynamic conditions. In particular, first-order phase transitions lead to a softening of the baryonic pressure, while second order phase transitions lead to a dramatic increase of the electroweak interaction probability [5]. This is why a control of the phase diagram of stellar matter is of great importance in our understanding of these complex astrophysical process.

In this contribution, we will limit ourselves to the thermodynamic conditions where matter can be described by nucleonic or hadronic degrees of freedom and will examine two phase transitions which are expected to play a role in supernova physics, namely the crust-core phase transition at sub-saturation density and the possible phase transition towards strange matter at super-saturation density.

2. Sub-saturation densities and the quenching of the first-order phase transition

Since the early days of the equation of state modelization it was clearly recognized that, because of the strong similarities between the effective nucleon-nucleon potential and molecular interactions, the phase diagram of diluted nuclear matter should present a first-order phase transition of the liquid-gas type, terminating at high temperature and density in a critical point [6]. This conjecture has been confirmed in all phenomenological [7, 8] or microscopic [9] modelizations of symmetric and asymmetric nuclear matter with realistic effective interactions compatible with recent nuclear data.

In these calculations, nuclear matter is defined as an idealized bulk medium of neutrons and protons where the electromagnetic interaction is artificially switched off in order to achieve a thermodynamic limit. In the physical situation of stellar matter, however, charge neutrality is achieved by the screening effect of electrons on the proton charge, , which induces Coulomb effects that drastically modify the liquid-gas phase transition associated to uncharged nuclear matter. Because of the very different mass between electrons and protons, the compressibility of electron and proton matter is very different. Therefore, in all density and temperature conditions relevant for neutron star physics, the electron charge can be safely considered as uniformly distributed [10]. This means that any baryonic density fluctuation (which is correlated to a proton density fluctuation because of the repulsive symmetry energy) induces a fluctuation in the electric charge. A well known consequence of the resulting Coulomb correlations is that the low density phase at zero temperature is not a gas, but a Wigner crystal of nuclei immersed in the homogeneous electron background [11]. At finite temperature, the presence of the electron background leads to a quenching of the first order liqui-gas transition, which is replaced by a continuous transition through a clusterised medium [12]. This Coulomb frustration effect has been confirmed by different microscopic models [13] . A continuous evolution from the Coulomb lattice to a homogeneous nuclear fluid, passing through the formation of clusters of different sizes and strongly deformed dis-homogeneous structures close to the saturation density has been reported, both at zero and at finite temperature.

These calculations are numerically very heavy and a complete thermodynamic characterization of stellar matter under the frustrated phase transition in a microscopic model has not been done yet. Such a task can however be performed in the phenomenological so-called nuclear statistical equilibrium (NSE) approaches, which treat the bound states of nucleons as new species of quasi-particles [14, 15, 16, 17].

2.1. A phenomenological model of dilute stellar matter

Within NSE, the baryonic component of the stellar matter is regarded as a statistical equilibrium of neutrons and protons, the electric charge of the latter being screened by a homogeneous electron background. As a first approximation, one can consider that the system of interacting nucleons is equivalent to a system of non-interacting clusters, with the nuclear interaction being completely exhausted by clusterization. This simple model can describe only diluted matter at $\rho \ll \rho_0$ as it can be found in the outer crust of neutron stars, while nuclear interaction among nucleons and clusters has to be included for applications at higher density,when the average inter-particle distance becomes comparable to the range of the force. A simple possibility [18, 19] is to take into account interactions among composite clusters in the simplified form of a hard-sphere excluded volume, and interactions among nucleons in the self-consistent Hartree-Fock approximation with a phenomenological realistic energy functional fitted on a large number of measured mass and radii. In this version of the model, the thermodynamics can be calculated in the ensemble (β, μ_I, ρ) defined by inverse temperature $\beta = T^{-1}$, isovector chemical potential $\mu_I = \mu_n - \mu_p$ and baryonic density $\rho = \rho_n + \rho_p$. The total entropy density is given by

$$\sigma^{can}_{\beta,\mu_I}(\rho) = \ln \mathcal{Z}^{a=1}_{\beta,\mu_I}(\rho_f) + \lim_{V\to\infty} \frac{1}{V} \ln \mathcal{Z}^{a>1}_{\beta,\mu_I}(V\rho_{cl}) \quad (1)$$

where the density repartition between the clustered ρ_{cl} and unbound ρ_f component $\rho = \rho_{cl} + \rho_f$ is uniquely defined by the condition of having a single chemical potential μ for both components. The unbound nucleons partition sum is calculated in the mean-field approximation:

$$\ln \mathcal{Z}^{a=1}_{\beta,\mu_I}(\rho) = \ln \mathcal{Z}^0_{\beta,\mu,\mu_I} - \beta\left(\frac{\partial}{\partial\beta}\ln \mathcal{Z}^0_{\beta,\mu,\mu_I} + V\epsilon^{HF}\right) - \beta\mu\rho V, \quad (2)$$

In this equation ϵ^{HF} is the Hartree-Fock energy and the non-interacting part of the partition sum can be expressed as a functional of the kinetic energy density τ_q for neutrons (q) or protons $(q = p)$:

$$\ln \mathcal{Z}^0_{\beta,\mu,\mu_I} = \frac{2\beta V}{3} \sum_{q,p} \tau_q \frac{\partial\epsilon^{HF}}{\partial\tau_q}, \quad (3)$$

where τ_q are self-consistently derived as a function of the associate chemical potential μ_q and effective mass m^*_q.

The partition sum of clusters for a total particle number $A = V\rho_{cl}$ is calculated via a recursive relation:

$$\mathcal{Z}_{\beta,\mu_I}(A) = \frac{1}{A}\sum_{a=2}^{A} a\omega_a \mathcal{Z}_{\beta,\mu_I}(A-a). \quad (4)$$

where the weight of the different cluster size is given by:

$$\omega_a = V_F \sum_{i=-a}^{a} \left(\frac{2\pi a m_0}{\beta h^2}\right)^{3/2} \exp\left(-\beta f^{\beta}_{a,i}\right) \exp(\beta i \mu_I). \quad (5)$$

Here, $f^{\beta}_{a,i}$ is a phenomenological free energy functional including the screening effect of electrons in the Wigner-Seitz approximation, and $V_F = V(1 - \rho_{cl}/\rho_0)$ is the volume fraction available to the clusters [18, 19] .

The thermodynamics of the extended NSE model was studied in ref. [19]. As it is well-known when studying phase transitions with long range interactions [20], the different statistical ensembles are not equivalent and it is very important to use the modelization adapted to the physical situation under study. In the case of compact stars, the density profile is imposed by the gravitational field and the correct ensemble is the canonical one.

The phase transition quenching and the effect of the statistical ensemble is illustrated in the left part of Figure 1, which shows the cluster distributions at a given temperature. As a function

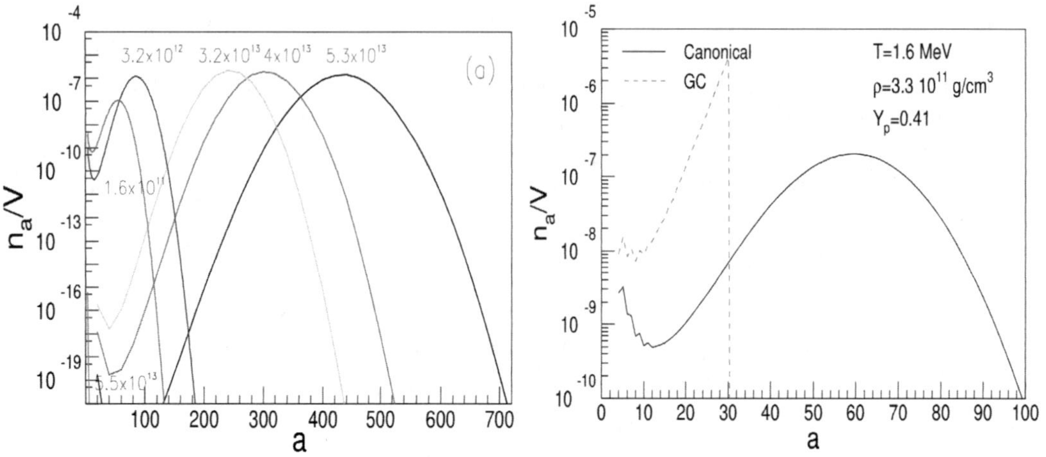

Figure 1. Left side: Cluster number densities as a function of the cluster size a at different densities (expressed in g/cm^3) of the extended NSE model at T = 1.6 MeV and asymmetry chemical potential μ_I = 1.68 MeV. Right side: Comparison between canonical (full line) and grand-canonical (dashed line) predictions for the cluster distribution in a specific thermodynamic condition relevant for supernova dynamics. Figure taken from ref. [19].

of the density, the average cluster size is continuously changing and can never be assimilated to a portion of the fluid phase. The right part of the same figure compares the grandcanonical and canonical cluster distribution in a thermodynamic situation relevant for supernova physics. We can see that accounting for the correct thermodynamics has important consequences on the matter composition. Since the cluster abundances determine the electron capture rate, which in turn is a capital ingredient for the size of the homologous core and the cooling process, it is clear that a control of the phase diagram is important for supernova physics. It is also interesting to remark that the most probable abundances concern not so neutron-rich nuclei which are potentially accessible in laboratory experiments: a detailed knowledge of the mass, level density and electron capture cross section for these nuclei is thus needed to have a reliable description of the equation of state.

2.2. In-medium effects

In the extended NSE model presented in the previous section clusters are treated as an ideal gas characterized by an empirical energy functional which is modified with respect to nuclei in the vacuum only through the excluded volume mechanism. This is an improvement compared to standard NSE [14, 15, 16, 17] in which the Coulomb screening by the electron background is the only interaction accounted for, but still represents a crude approximation of the effect of the medium [21]. Microscopic modifications of the cluster binding energies due to Pauli blocking shifts have been computed [9] and shown to correctly describe the Mott dissolution of deuterons in dense matter, but this heavy numerical approach cannot be extended to massive clusters. In the original Lattimer-Swesty model [22] the assumption is made that the effect of the medium consists in a modification of the cluster surface tension, the bulk energy being unaffected. However some interpretations of multifragmentation data suggest that bulk terms might also be affected [23].

To microscopically account for in-medium modifications of the cluster binding energies, we have realized a systematic set of Hartree-Fock calculations in the Wigner-Seitz cell for various isotopic chains with a total neutron number ranging from the stability line up to $N \approx 3000$. The

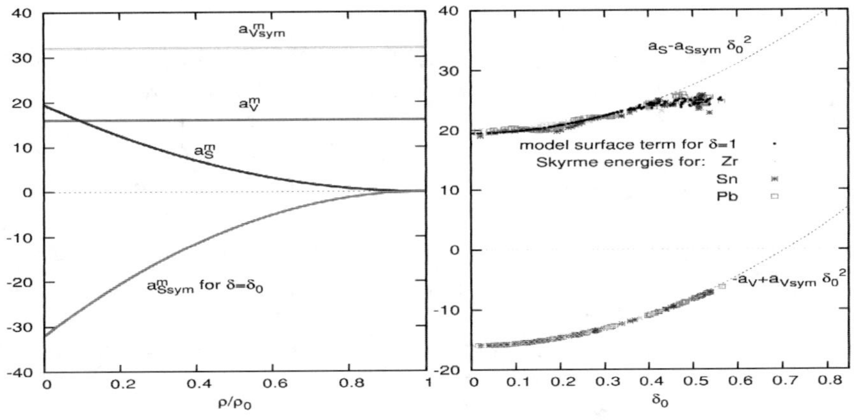

Figure 2. Left part: isoscalar and isovector bulk and isoscalar surface term of the in-medium modified mass formula extracted from Hartree-Fock calculations in the Wigner-Seitz cell. Right part: comparison between the microscopic Hartree-Fock volume energy per particle and surface energy per $A^{2/3}$ and the parameterization eq.(6) for three isotopic chains as a function of the bulk proton-neutron asymmetry. The model surface term is given by $\left[a^m_S(\rho) - a^m_{Ssym}(\rho,\delta){\delta_0}^2\right]$, see eq. (6).

resulting microscopic nuclear energy, consisting of a volume and a surface term, is parameterized with an in-medium modified mass formula

$$E_{HF} = \epsilon^{HF}(\rho_n,\rho_p)(V_{WS}-V_{cl}) - \left[a^m_V(\rho) - a^m_{Vsym}(\rho,\delta)\delta_0^2\right]A + \left[a^m_S(\rho) - a^m_{Ssym}(\rho,\delta){\delta_0}^2\right]A^{2/3} \quad (6)$$

where V_{WS} is the volume of the calculation cell and $\rho_n, \rho_p, \rho, \delta$ is the neutron, proton, total density and asymmetry $\delta = (\rho_n - \rho_p)/\rho$ of the free nucleons constituting the dripping gas. The cluster mass and charge A, Z and the gas densities are extracted through a Wood-Saxon fit of the local density. The bulk density ρ_0 and asymmetry $\delta_0 = (\rho_{0n} - \rho_{0p})/\rho_0$ in the model are approximated by analytical functions of A, Z, while $V_{cl} = A/\rho_0$. The resulting in-medium parameters obtained from a preliminary fit of the microscopic Hartree-Fock results through eq.(6) are shown in the left part of Figure 2, while the quality of the fit is displayed in the right part of the same figure.

To obtain the fit shown in Fig.2 we have employed the ansatz

$$a^m_S(\rho) = a_S\left(1 - \frac{\rho}{\rho_0}\right)^{\gamma_1} \quad ; \quad a^m_{Ssym}(\rho,\delta) = a_{Ssym}\left(1 - \frac{\rho\delta}{\rho_0\delta_0}\right)^{\gamma_2} \quad (7)$$

with $\gamma_1 = \gamma_2 = 2$ and with $a_{Ssym} = -32$ MeV. A dependence of a^m_{Ssym} on ρ_0,δ_0 is implicit. We can see that the in-medium modified liquid drop expression eq.(6) provides a reasonably accurate description of the energy modification due to the external gas. In agreement with the conjecture of ref. [22], the bulk terms are completely unaffected by the external medium ($a^m_V(\rho) = a_V$=cst., $a^m_{Vsym}(\rho,\delta) = a_{Vsym}$=cst.), while important effects on the surface properties of nuclei are seen. The implementation of such an improved energy functional in the extended NSE model is in progress.

3. Supra-saturation densities and the strangeness phase transition

If it is well admitted that hyperonic and deconfined quark matter could exist in the inner core of neutron stars, a complete understanding of the composition and equation of state of dense matter is far from being achieved. Concerning hyperons, simple energetic considerations suggest that they should be present at high density [24]. However, in the standard picture the opening of hyperon degrees of freedom leads to a considerable softening of the equation of state [24, 25, 26], which in turns leads to maximum neutron star masses smaller than the observed masses of many neutron stars, in particular the high values obtained in recent observations [1]. This puzzling situation implies that the hyperon-hyperon and hyperon-nucleon couplings must be much more repulsive at high density than presently assumed [27, 28, 29, 30, 31, 32], and/or that something is missing in the present modelization.

The generic presence of attractive and repulsive couplings suggests the existence, in a model-independent manner, of a phase transition involving strangeness. Let us consider a simplified situation where matter is uniquely composed of neutrons and Λ-hyperons. The energy density $\epsilon = \epsilon_{kin} + \epsilon_{pot}$ at zero temperature is obtained with the energy density functional proposed by Balberg and Gal [33]

$$\begin{aligned}\epsilon_{pot}(\rho_n, \rho_\Lambda) &= \frac{1}{2}\left[(a_{NN} + b_{NN})\rho_n^2 + c_{NN}\rho_n^{\delta+1}\right] + \frac{1}{2}\left[a_{\Lambda\Lambda}\rho_\Lambda^2 + c_{\Lambda\Lambda}\rho_\Lambda^{\gamma+1}\right] \\ &+ a_{\Lambda n}\rho_n\rho_\Lambda + c_{\Lambda n}\frac{\rho_n\rho_\Lambda}{\rho_n + \rho_\Lambda}\left[\rho_n^\gamma + \rho_\Lambda^\gamma\right], \end{aligned} \tag{8}$$

Due to the attractive part of the $n\Lambda$ coupling, a mixture of the two particle species admits a bound state at finite density. This rather general feature of the energy density is found within other parameterizations and other models, e.g. the parameterization of the energy density functional from G-matrix calculations by Vidana et al. [34]. The existence of a minimum in the energy functional for symmetric matter means that such a state represents the stable matter phase at zero temperature. On the other side, the vanishing density gas phase is always the stable phase in the limit of infinite temperature. This implies that a dilute-to-dense phase change in strange matter has to be expected on very general grounds. It is intertesting to remark that the minimum, for a certain interval of neutron densities, is also observed as a function of the Λ density. This is again due to the interplay between the attractive $\Lambda - \Lambda$ interaction required by hypernuclear data, and the repulsion at high density required by the two-solar mass neutron star observation. This information means that the expected phase change is expected to have strangeness contributing to the order parameter.

The presence of a minimum is however not sufficient to discriminate between a smooth cross-over and a phase transition. In the next section we will therefore demonstrate the existence of a phase transition by explicitly calculating the $n\Lambda$ phase diagram.

3.1. Neutron-Λ phase diagram

First-order transitions are signaled by an instability or concavity anomaly in the mean-field thermodynamic potential, which has to be cured by means of the Gibbs phase equilibrium construction at the thermodynamic limit. At zero temperature, one thermodynamic potential is given by the total energy

$$\epsilon_{tot}(\rho_n, \rho_\Lambda) = \epsilon_{pot}(\rho_n, \rho_\Lambda) + \epsilon_{kin}(\rho_n, \rho_\Lambda) + (\rho_n m_n + \rho_\Lambda m_\Lambda)c^2 \tag{9}$$

The spinodal region is then recognized as the locus of negative curvature of the energy surface, $C_{min} < 0$. The corresponding eigenvector defines a direction in the density space given by

$$\frac{\rho_n}{\rho_\Lambda} = \frac{C_{n\Lambda}}{C_{min} - C_{nn}} = \frac{C_{min} - C_{\Lambda\Lambda}}{C_{\Lambda n}} \tag{10}$$

with $C_{ij} = \partial^2 e_{tot}/\partial\rho_i\partial\rho_j$. This instability direction physically represents the chemical composition of density fluctuations which are spontaneously and exponentially amplified in

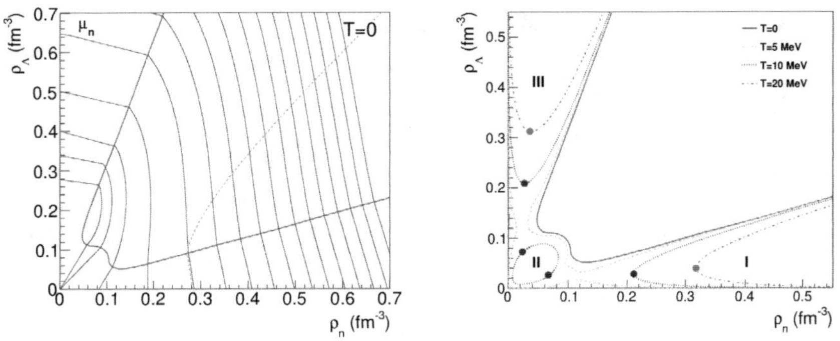

Figure 3. Left part: (n, Λ) mixture at $T = 0$: Borders of the phase coexistence region (red thick lines), constant-μ_n paths after Maxwell construction (blue thin lines) and $\mu_n = \mu_\Lambda$ trajectory after Gibbs construction (dark green dotted line). Right part: borders of the phase coexistence region of the $n\Lambda$ mixture for different temperatures, T=0, 5, 10, 20 MeV. The full circles mark the critical points. Figures taken from ref. [35].

the unstable region in order to achieve phase separation, and gives the order parameter of the associated phase transition.

The zero temperature instability region of the $n\Lambda$ mixture is shown in the left part of Fig. 3. We can see that a large portion of the phase diagram is concerned by the instability. We can qualitatively distinguish three regions characterized by different order parameters. Below nuclear saturation density, we observe an isoscalar $\rho_n \approx \rho_\Lambda$ instability, very close to ordinary nuclear liquid-gas, with Λ's playing the role of protons. Increasing neutron density the phase separation progressively changes towards the strangeness $\rho_S = -\rho_\Lambda$ direction: the two stable phases connected by the instability have close baryon densities but a very different fraction of Λ's. For high ρ_Λ we observe the same behavior with the roles of neutrons and Λ-hyperons exchanged. The part of the phase diagram at high neutron density comprises the region physically explored by supernova and neutron star matter, which is characterized by chemical equilibrium for reactions implying strangeness, $\mu_S = \mu_n - \mu_\Lambda = 0$. The strangeness equilibrium trajectory is represented by a dashed line in Fig.3. This physically corresponds to the sudden opening of strangeness, observed in many modelizations of neutron star matter (see e.g. [33, 26]).

Let us stress that the existence of a strangeness phase transition at high density is not a general model-independent feature, although many models show it. There are others, for instance the G-matrix models of ref. [34], which do not show an instability in this region. The reason is essentially due to the absence of a $\Lambda\Lambda$ interaction in these microscopic calculations. Owing to the lack of reliable information on the hyperonic interactions, which would discriminate between different models, we cannot affirm the existence of the strangeness phase transition, related to this instability, but, turning the argument around, the presence of this phase transition in a

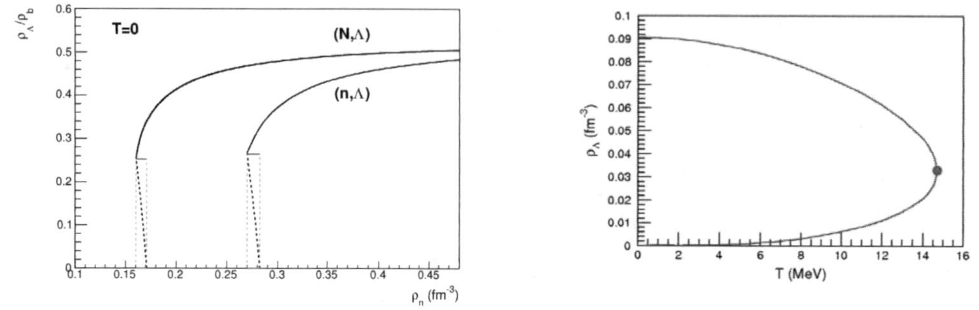

Figure 4. Left: Λ fraction as a function of the baryon density within the condition $\mu_n = \mu_\Lambda$ for a $n\Lambda$ and, respectively, a symmetric $\rho_n = \rho_p$ $np\Lambda$ mixture at T=0. The full lines correspond to the stable mean-field results; the dashed lines illustrate the Gibbs constructions; the horizontal lines indicate the relative amount of Λ-hyperons at the frontiers of the phase-coexistence domain. Right: Λ density for the two coexisting phases as a function of temperature for $\mu_S = 0$. Figures taken from ref. [35].

physical system would allow to learn much about the shape of the interaction.

Up to now we have only presented results at zero temperature. The extension to finite temperature, as it is needed for supernova matter, is relatively simple in the mean-field approximation. The appropriate thermodynamic potential at non-zero temperature is given by the Helmholtz free energy

$$f_T(\rho_n, \rho_\Lambda) = \epsilon_{tot}(\rho_n, \rho_\Lambda) - Ts(\rho_n, \rho_\Lambda) \tag{11}$$

where s is the mean-field entropy density. Chemical potentials can be obtained by differentiating this expression. Then the Gibbs construction is performed from the combined analysis of $\mu_n(\rho_n)$ at constant μ_Λ, and $\mu_\Lambda(\rho_\Lambda)$ at constant μ_n.

The phase diagram as a function of temperature is presented in the right panel of Fig. 3. This phase diagram exhibits different interesting features. We can see that the three regions that we have tentatively defined at zero temperature appear as distinct phase transitions at finite temperature. The first phase transition (zone II in Fig.3 - right panel) separates a low density gas phase from a high density more symmetric liquid phase, very similar to ordinary liquid-gas. The second one (zone III in Fig.3 - right panel) reflects the instability of dense strange matter towards the appearance of neutrons and has an almost symmetric counterpart (zone I in Fig.3 - right panel) in the instability of dense neutron matter towards the formation of Λ-hyperons. Up to a certain temperature, this latter phase transition is explored by the $\mu_n = \mu_\Lambda$ trajectory, meaning that it is expected to occur in neutron stars and supernova matter. At variance with other known phase transitions in nuclear matter, this transition exists at any temperature and is not limited in density; it is always associated with a critical point, which moves towards high density as the temperature increases. This means that criticality should be observed in hot supernova matter, at a temperature which is estimated as $T_c = 14.8$ MeV in the present schematic model.

The consequences of these findings for the composition of neutron star matter are drawn in Fig. 4. The left panel shows the Λ fraction $Y_\Lambda = \rho_\Lambda/(\rho_n + \rho_\Lambda)$ as a function of the baryon density under the condition $\mu_n = \mu_\Lambda$. The crossing of the mixed-phase region with increasing neutron density implies that, as soon as the lower density transition border is crossed, the system has to be viewed as a dis-homogeneous mixture of macroscopic regions composed essentially of neutrons, with other macroscopic regions with around 25% of hyperons. The extension of the Λ-rich zone increases with density until the system exits the coexistence zone, and becomes homogeneous again. The density values associated to the phase transition slightly change with the energy functional assumed. To give a simple example, we present on the same Fig.4 a calculation where Λ's are put in equilibrium with symmetric $\rho_n = \rho_p$ non-strange matter, that is eq.(8) is still used, but the symmetry energy in the nucleon sector b_{NN} is put to zero [33]. We can see that only small differences are observed in the strange phase transition considering pure neutron matter or symmetric matter.

On the right panel of Fig. 4 the Λ-densities for the two coexisting phases are displayed as a function of temperature, again under the condition of $\mu_n = \mu_\Lambda$, physically relevant for neutron star and supernova matter. The extension of the coexistence region decreases with increasing temperature. The latter finally disappears at $T_c = 14.8$ MeV. This well illustrates the fact that the critical point of the strangeness phase transition moves to higher density, crossing the physical line $\mu_n = \mu_\Lambda$ at a given temperature.

4. Conclusions

In this contribution we have analyzed the phase diagram of stellar matter as it is produced in core-collapse supernovae and proto-neutron stars in the thermodynamical conditions where such matter can be described by nucleonic and hadronic degrees of freedom.

At baryonic densities lower than nuclear saturation, the first-order phase transition of neutral nuclear matter is quenched in stellar matter due to the effect of Coulomb frustration. As a consequence, matter is expected to be clusterized even at finite temperature in the whole sub-saturation regime.

The composition and thermodynamics of such clusterized matter has been explicitly worked out in the framework of a quasi-particle model where inter-particle interactions are modeled through excluded volume. Coulomb correlations with the neutralizing electron gas are accounted for in the Wigner-Seitz approximation and residual nuclear correlations are included in an effective way through an in-medium modification of the cluster energy functional with parameters fitted from Hartree-Fock calculations with a realistic effective interaction. The same formalism is used to evaluate the free nucleon self-energies.

The expectation of the phase transition quenching is confirmed by this model, which shows that the single nucleus approximation widely used is a poor approximation at finite temperature and a large distribution of cluster species has to be included in supernova modeling.

In the supersaturation regime, we have used the same mean field approximation with effective phenomenological energy functionals in order to describe the characteristics of the phase diagram in the presence of strangeness.

We have shown that already a simple system composed of neutrons and Λ hyperons in thermal equilibrium presents a complex phase diagram with first and second order phase transitions. Some of these phase transitions are probably never explored in physical systems. However, a possible phase transition at super-saturation baryon densities, from non-strange to strange matter is expected to be observed both in the inner core of neutron star and in the dense regions of core-collapse supernova. For this latter phenomenology, a critical point is predicted and the associated critical opalescence could have an impact on supernova dynamics [5]. The immediate consequence of that is that the opening of hyperon channels at high density should not be viewed as a continuous (though abrupt) increase of strangeness in the matter, observed in many models of hyperonic matter, cf e.g. [26], but rather as the coexistence of hyperon-poor and hyperon-rich macroscopic domains.

An extension of this simple model to include all possible hyperons and resonances is in progress.

Journal of Physics: Conference Series **413** (2013) 012019 doi:10.1088/1742-6596/413/1/012019

References

[1] Demorest P et al 2010 *Nature* **467** 1081; Ozel F et al 2010 *Astrophys. J.* **724 L** 199.
[2] Page D et al 2011 *Phys. Rev. Lett.* 106 081101; P.S.Shternin et al 2011 *Monthly Notices of the RAS: Letters* **412** L108 .
[3] Nakazato K et al 2012 *Astrophys. J.***745** 197.
[4] Marek A and Janka H Th 2009 *Astrophys. Journ.* **694** 664.
[5] Margueron J, Navarro J, and Blottiau P 2004 *Phys. Rev.* C **70** 028801.
[6] Siemens P J 1983 *Nature* **305** 410.
[7] Ducoin C, Chomaz P, Gulminelli F 2006 *Nucl. Phys.* A **771** 68.
[8] Rios A 2010 *Nucl. Phys.* A **845** 58.
[9] Typel S et al 2010 *Phys. Rev.* C **81** 015803.
[10] Maruyama T et al 2005 *Phys. Rev.* C **72** 015802.
[11] Lattimer J M and Prakash M 2004 *Science* Vol. **304** no. 5670 536. %bibitemshen Shen H, Toki H, Oyamatsu K and Sumiyoshi K 1998*Prog. Theor. Phys.* **100** 1013.
[12] Napolitani P et al 2007 *Phys. Rev. Lett.* **98** 131102; Ducoin C et al 2007 *Phys. Rev.* C **75** 065805.
[13] Watanabe G 2010 *Prog. Theor. Phys. Suppl.* **186** 45.
[14] Phillips A C 1994 *The Physics of Stars* (Chichester: John Wiley & Sons).
[15] Botvina A S and Mishustin I N 2010 *Nucl. Phys.* A **843** 98.
[16] Blinnikov S I, Panov I V, Rudzsky M A and Sumiyoshi K 2011 *Astron. Astrophys.* **535** A37.
[17] Souza S R, Steiner A W, Lynch W G, Donangelo R and Famiano M A 2009 *Astrophys. J.* **707** 1495.
[18] Raduta A R and Gulminelli F 2010 *Phys. Rev.* C **82** 065801.
[19] Gulminelli F and Raduta A R 2012 *Phys. Rev.* C **85** 025803.
[20] Campa A et al 2009 *Phys.Rep.* **480** 57.
[21] Hempel M and Schaffner-Bielich J 2010 *Nucl. Phys.* A **837** 210.
[22] Lattimer J M and Swesty F D 1991 *Nucl. Phys.* A **535** 331.
[23] Shetty D V et al *Preprint* nucl-ex/0406008.
[24] Glendenning N 1982 *Phys.Lett.* B **114** 392.
[25] Vidana I, Polls A, Ramos A, Engvik L and Hjorth-Jensen M 2000 *Phys. Rev.* C **62** 035801.
[26] Massot E, Margueron J and Chanfray G 2012 *Europhys. Lett.* **97** 39002.
[27] Hofmann F, Keil C and Lenske H 2001 *Phys. Rev.* C **64** 025804.
[28] Bonanno L and Sedrakian A 2012 *Astron. Astrophys.* **539** A16.
[29] Weissenborn S, Chatterjee D and Schaffner-Bielich J 2012 *Nucl. Phys.* A **881** 62; 2012 *Phys. Rev.* C **85** 065802.
[30] Bednarek I, Haensel P, Zdunik J L, Bejger M and Manka R *Preprint* astro-ph/1111.6942.
[31] Lastowiecki R, Blaschke D, Grigorian H and Typel S 2012 *Acta Phys. Polon. Supp.* **5** 535.
[32] Oertel M, Fantina A and Novak J 2012 *Phys. Rev.* C **85** 055806.
[33] Balberg S and Gal A 1997 *Nucl. Phys.* A **625** 435.
[34] Vidana I, Logoteta D, Providencia C, Polls A and Bombaci I *Preprint* astro-ph/1004.3958.
[35] Gulminelli F, Raduta A R and Oertel M 2012 *Phys.Rev.* C **86** 025805.

Int. Summer School for Advanced Studies 'Dynamics of open nuclear systems' (PREDEAL12) IOP Publishing
Journal of Physics: Conference Series **413** (2013) 012020 doi:10.1088/1742-6596/413/1/012020

Nuclear Structure for Reactions and Decays

Mihai Horoi

Department of Physics, Central Michigan University, Mount Pleasant, MI 48859, USA

E-mail: mihai.horoi@cmich.edu

Abstract. Shell Model techniques are powerful tools to describe the nuclear structure effects observed in reactions and decays. This contribution describes several relevant shell model applications, including double beta decays, compound nuclear reactions, and direct nuclear reactions. In particular it discusses the role of spectroscopic factor extracted from transfer reactions in revealing proton-neutron correlations effects in nuclei. It also identifies the elements of the effective shell model Hamiltonian that are responsible for these strong correlations.

1. Introduction

Reactions and decays are generally viewed as processes with many particles in continuum. For structureless particles, the scattering theory is a very reliable tool. Many decays and reaction of structureless particles can be describe using the penetrability factor know since the early days of quantum mechanics and alpha decay theories. However, most of the nuclei have complicated structure, and the nuclear structure information is essential to realistically describe their reactions and decays. Shell Model (ShM) techniques are powerful tools to describe the nuclear structure effects observed in reactions and decays.

In some reactions, such as compound nuclear reactions that goes through excitation energy region involving a high density of states, mere knowledge of the density of states for given spin and parity would suffice. Indeed, those cases, a statistical theory due to Hauser and Feshbach [1] turned out to be very useful. In recent years, we proposed and advanced approach to the ShM spin- and parity-dependent nuclear level densities (NLD) [2, 3, 4, 5, 6, 7], and we started to use this approach to describe reaction cross-sections and reaction rates for nuclear astrophysics [8]. Due to the limited space, this approach will not be further developed here, but we direct the interested reader to the our papers listed in this paragraph.

Direct reactions, such as break-up and transfer reactions, are also of large interest in nuclear structure studies, nuclear astrophysics, nuclear medicine, and nuclear energy applications. The structureless cross section are usually calculated using a distorted wave Born approximation. The comparison with the experimental data requires a multiplicative spectroscopic factor that contains information about the nuclear structure of the transferred particle relative to the core. This information can be further used to tune the nuclear structure models, such as the ShM Hamiltonian. In this contribution we describe the spectroscopic factors of the N=28 isotopes, and we show that they contain essential information about the proton-neutron (pn) correlations in nuclei and about the pn part of the ShM Hamiltonian.

Neutrinoless double beta ($0\nu\beta\beta$) decay, which can only occur by violating the conservation of the total lepton number, if observed it will unravel physics beyond Standard Model (SM), and it will represent a major milestone in the study of the fundamental properties of neutrinos

[10]-[16]. Indeed, its discovery would decide if neutrinos are their own antiparticles [17], and would provide a hint on the scale of their absolute masses. That is why there are intensive investigations of this process, both theoretical and experimental. Recent results from neutrino oscillation experiments have demonstrated that neutrinos have mass and they can mix [18]-[20]. However, the neutrino oscillations experiments cannot be used to determine the lowest neutrino mass and the neutrino mass hierarchy. Neutrinoless double beta decay is viewed as one of the best routes to decide these unknowns. A key ingredient for extracting the absolute neutrino masses from $0\nu\beta\beta$ decay experiments is a precise knowledge of the nuclear matrix elements (NME) for this process.

There are potentially many mechanisms that could contribute to the neutrinoless double beta decay process, which will be briefly reviewed below. Several of these mechanisms do not provide contributions to the decay rate that explicitly depend on the neutrino masses, but their contributions will vanish if the neutrinos are not massive Majorana particles [17]. In this review we will concentrate to those mechanisms for which the decay rate explicitly depends on the neutron masses. In all cases the half-lives depend on the nuclear matrix elements that need to be accurately calculated using low-energy nuclear structure models. The two neutrino double beta ($2\nu\beta\beta$) decay is an associate process that is allowed by the Standard Model, and it was observed in about ten isotopes. Therefore, a good but not sufficient test of nuclear structure models would be a reliable description of the $2\nu\beta\beta$ half-lives.

Since most of the $\beta\beta$ decay emitters are open shell nuclei, many calculations of the NME have been performed within the pnQRPA approach and its extensions [21]-[32]. However, the pnQRPA calculations of the more common two-neutrino double beta decay half-lives, which were measured for about 10 cases [33], are very sensitive to the variation of the so called g_{pp} parameter (the strength of the particle-particle interactions in the 1^+ channel) [21]-[23], and this drawback still persists in spite of various improvements brought by its extensions [24]-[29], including higher-order QRPA approaches [30]-[32]. The outcome of these attempts was that the calculations became more stable against g_{pp} variation, but at present there are still large differences between the values of the NME calculated with different QRPA-based methods, which do not yet provide a reliable determination of the two-neutrino double beta decay half-life. Therefore, although the QRPA methods do not seem to be suited to predict the $2\nu\beta\beta$ decay half-lives, one can use the measured $2\nu\beta\beta$ decay half-lives to calibrate the g_{pp} parameters that are further used to calculate the $0\nu\beta\beta$ decay NME [34]. Another method that recently provided NME for most $0\nu\beta\beta$ decay cases of interest is the Interactive Boson Model (IBA-2) [35, 36].

On the other hand, the progress in computer power, numerical algorithms, and improved nucleon-nucleon effective interactions, made possible large scale shell model calculations of the $2\nu\beta\beta$ and $0\nu\beta\beta$ decay NME [37]-[39]. The main advantage of the large scale shell model (LSSM) calculations is that they seem to be less dependent on the effective interaction used, as far as these interactions are consistent with the general spectroscopy of the nuclei involved in the decay. Their main drawback is the limitation imposed by the exploding shell model dimensions on the size of the valence spaces that can be used. The most important success of the large scale shell model calculations was the correct prediction of the $2\nu\beta\beta$ decay half-life for ^{48}Ca [37, 40]. In addition, these calculations did not have to adjust any additional parameters, i.e. given the effective interaction and the Gamow-Teller (GT) quenching factor extracted from the overall spectroscopy in the mass-region (including beta decay probabilities and charge-exchange form factors), one can reliably predict the $2\nu\beta\beta$ decay half-life of ^{48}Ca.

Clearly, there is a need to further check and refine these calculations, and to provide more details on the analysis of the NME that could be validated by experiments. We have recently revisited [42] the $2\nu\beta\beta$ decay of ^{48}Ca using two recently proposed effective interactions for this mass region, GXPF1 and GXPF1A, and we explicitly analyzed the dependence of the double Gamow-Teller sum entering the NME on the excitation energy of the 1^+ states in

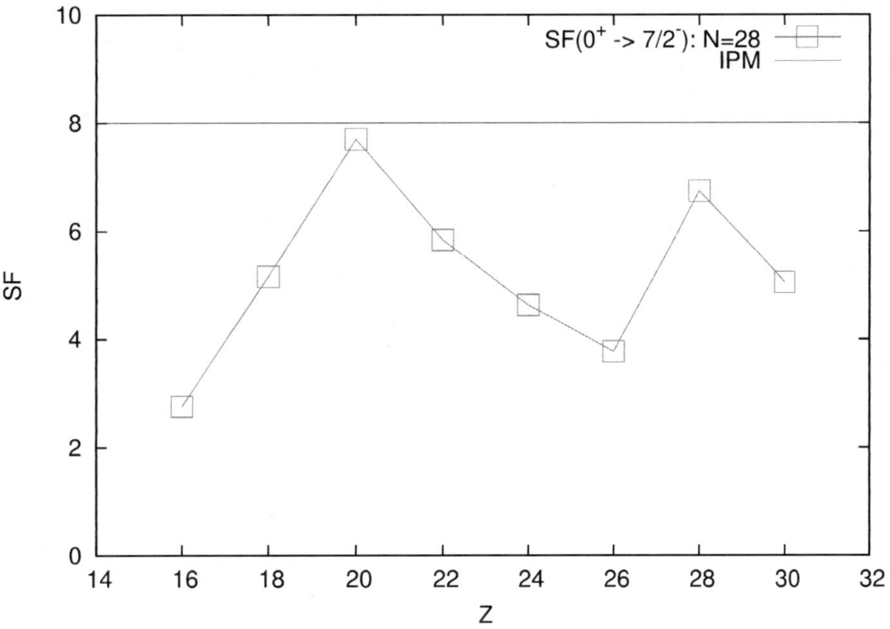

Figure 1. (Color online) The ShM prediction for the SF of the N=28 isotones.

the intermediate nucleus ^{48}Sc. This sum was recently investigated experimentally [43], and it was shown that the incoherent summation of the absolute values of the Gamow-Teller matrix elements would provide an incorrect NME, thus validating our prediction. We have also corrected by several orders of magnitude the probability of transition of the g.s. of ^{48}Ca to the first excited 2^+ state of ^{48}Ti. Future experiments on double beta decay of ^{48}Ca (CANDLES [44] and CARVEL [45]) may reach the required sensitivity of measuring such transitions, and our results could be useful for planning these experiments.

2. Signal of pn correlations in the spectroscopic factor of the N=28 isotones

Direct nuclear reactions, such as break-up and transfer reactions, are also largely dependent on the nuclear structure information, such as the spectroscopic factors. In reverse, information from direct nuclear reaction can be used to better understand the nuclear structure of exotic nuclei, such as the pn correlations and their effect on changes in the single particle energies and magic numbers. As an example we show here the case of the N=28 isotones.

The precise definition of the spectroscopic factor (SF) and its relation to DWBA cross-sections can be found in many references, e.g. [47, 48]. The SF is not a genuine observable, and it is useful to gauge it to some simple model, such as the independent particle model (IPM), which assumes that one removes/add nucleons to a spherical single particle shell, say $f_{7/2}$, and only paring interaction act to remove degeneracy. If an even number n of nucleons are filling the shell, the SF for removing one nucleon is n. For example, if one removes one neutron from any of the N=28 isotones, for which the $f_{7/2}$ shell is fully filled with 8 neutrons, then the IPM SF is 8. Indeed, experimental results and full ShM calculation in pf shell for Ca isotopes shows that the IMP predictions are observed [47]. However, if one includes or subtracts proton pairs, full ShM calculation in pf and $sd - pf$ model spaces indicate large deviations from the simple IPM predictions as shown in Fig. 1. In particular, one can see a significant drop in the SF for ^{54}Fe and ^{46}Ar, while the SF prediction for ^{56}Ni is almost back to the IPM value. Clearly, this behavior is related to correlations between the proton pairs filling the $d_{3/2}$ and $f_{7/2}$ shell and the N=28 neutron system. In fact, the ^{54}Fe SF was measured long time ago [51], and the ^{46}Ar and ^{56}Ni SFs were recently measured, e.g. Refs. [49] and [50], respectively. The experimental data clearly backs-up the theoretical predictions.

We further investigate the pn mechanisms responsible for the behavior of the SF in Fig. 1. We found out that the mechanism responsible for the 40pn correlations between the protons filling up the $f_{7/2}$ shell and the neutrons in the $f_{7/2}$ shell. Indeed, reducing to zero the $f_{7/2}$ pn matrix elements coupled to J=0, 1, and 7, one gets the SF of ^{54}Fe close to its IPM value. These changes clearly break the isospin symmetry. Therefore, one could also say that isospin symmetry is responsible for these strong pn correlations effects [51]. What is new here is that we identified the specific pn pieces of the effective Hamiltonian responsible for this dramatic effect: a 50% reduction of the cross section for the neutron break-up of ^{54}Fe.

The mechanism behind the pn effects in the SF of ^{46}Ar is different. Here, there are clear correlations between the proton pairs filling the $d_{3/2}$ and the neutrons in the $f_{7/2}$ shell. We found out that by reducing by 0.6 MeV the pn part of the T=0 monopole matrix elements between the $d_{3/2}$ and $f_{7/2}$ shells one pushes down the position of the $d_{3/2}$ level, wile keeping the pf single particle energies unchanged. Therefore, this change is reducing the interaction of the $d_{3/2}$ proton paris with the neutrons in the $f_{7/2}$ shell. As a consequence, the ^{46}Ar SF is going up towards the IPM value. Details will be presented elsewhere [52].

3. Two-neutrino double beta decay

LSSM calculations of $2\nu\beta\beta$ decay NME can now be carried out rather accurately for many nuclei [55]. In the case of ^{48}Ca, Ref. [37] reported for the first time a full pf-shell calculation of the NME for the $2\nu\beta\beta$ decay mode, for both transitions to the g.s. and to the 2_1^+ excited state of ^{48}Ti. As an effective interaction it was used the Kuo-Brown G-matrix [56] with minimal monopole modifications, KB3 [57]. In Ref. [42] we use the recently proposed GXPF1A two-body effective interaction, which has been successfully tested for the pf shell [58]-[60], to perform $2\nu\beta\beta$ decay calculations for ^{48}Ca. Our goal was to obtain the values of the NME for this decay mode, both for transitions to the g.s. and to the 2_1^+ state of ^{48}Ti, with increased degree of confidence, which will allow us in the next future to address similar calculations for the $0\nu\beta\beta$ decay mode of this nucleus [39]. The $2\nu\beta\beta$ transitions to excited states have longer half-lives, as compared with the transitions to the g.s., due to the reduced values of the corresponding phase space factors, but they are measurable in some cases, such as ^{100}Mo [61].

For the $2\nu\beta\beta$ decay mode the relevant NME are of Gamow-Teller type, which has the following expressions for the decay to states in the grand-daughter that has the angular momentum $J = 0, 2$ [10]-[15],

$$M_{GT}^{2\nu}(J^+) = \frac{1}{\sqrt{J+1}} \sum_k \frac{\langle J_f^+||\sigma\tau^-||1_k^+\rangle\langle 1_k^+||\sigma\tau^-||0_i^+\rangle}{(E_k+E_J)^{J+1}} \tag{1}$$

Here E_k is the excitation energy of the 1_k^+ state of intermediate odd-odd nucleus, and $E_J = \frac{1}{2}Q_{\beta\beta}(J^+)+\Delta M$. $Q_{\beta\beta}(J^+)$ is the Q-value corresponding to the $\beta\beta$ decay to the final J_f^+ state of the grand-daughter nucleus, and ΔM is the mass difference between the parent and the grand-daughter. The most common case is to look to decay to the 0^+ g.s. of the grand-daughter, but decays to the first excited 0^+ and 2^+ states are also investigated.

The $2\nu\beta\beta$ decay half-life expression is given by

$$\left[T_{1/2}^{2\nu,J}\right]^{-1} = G_J^{2\nu}|M_{GT}^{2\nu}(J)|^2 \tag{2}$$

where $G_J^{2\nu}$ are $2\nu\beta\beta$ phase space factors. Specific values of $G_J^{2\nu}$ for different $2\nu\beta\beta$ decay cases can be found in different reviews, such as Ref. [11]. For a recent analysis of $G_J^{2\nu}$ see Ref. [36]. In Ref. [42] we explicitly analyzed the dependence of the double-Gamow-Teller sum entering the NME Eq. (1) on the excitation energy of the 1^+ states in the intermediate nucleus ^{48}Sc. This sum was recently investigated experimentally [43], and it was shown that indeed, the incoherent sum (using only absolute values of the Gamow-Teller matrix elements) would provide an incorrect NME, validating our prediction. We have also corrected by several orders of magnitude the probability of transition of the g.s. of ^{48}Ca to the first excited 2^+ state of ^{48}Ti.

In Ref. [42] we fully diagonalized 250 1^+ states in the intermediate nucleus to calculate the $2\nu\beta\beta$ decay NME for ^{48}Ca. This procedure can be used for somewhat heavier nuclei using the J-scheme shell model code NuShellX [41], but for large dimension cases one needs an alternative method. The pioneering work on ^{48}Ca [37] used a strength-function approach that converges after a small number of Lanczos iterations, but it requires large scale shell model diagonalizations when one wants to check the convergence. Ref. [46] proposed an alternative method, which converges very quickly, but it did not provide full recipes for all its ingredients, and it was never used in practical calculations. Recently [62], we proposed a simple numerical scheme to calculate all coefficients of the expansion proposed in Ref. [46]. Following Ref. [46], we choose as a starting Lanczos vector, $L_1^{\pm}$, either the initial or final state in the decay (only 0^+ to 0^+ transitions are considered), on which we apply the Gamow-Teller operator. This approach is very efficient for large model spaces, as for example the $jj55$ space (consisting of the $0g_{7/2}$, $1d, 2s$, and $h_{11/2}$ orbits), which for ^{128}Te leads to m-scheme dimensions of the order of 10 billions. In the calculation of ^{48}Ca decay we use the standard quenching factor, 0.77, for the Gamow-Teller operator $\sigma\tau$. We checked the result reported in Ref. [42] using this alternative method and we found the same result. The novel result report here for the first time is for the transition to the first excited 0^+ state in ^{48}Ti at 2.997 MeV. The matrix element we found is 0.05, very close to that for the transition to the g.s. Using the phase space factor from Ref. [11] we found the the half-life for this transition is 1.6×10^{24} y. We recall here that our results reported in [42] for the half-lives of the transitions to g.s. and to the first 2^+ excited state are 3.3×10^{19} y and 8.5×10^{23} y, respectively.

We also calculated using the same techniques the $2\nu\beta\beta$ decay NME for ^{76}Ge and ^{82}Se in $jj44$ valence space ($0f_{5/2}$, $1p$, $0g_{9/2}$) using the JJ44PN effective interaction briefly described in Ref. [53], and with JUN45 interaction [63]. Our results are about 25-40% larger than the corresponding NMEs extracted from experimental data [33]. The experimental NMEs could be reproduced if a small quenching factor, 0.6, is used. Such a reduction of the GT quenching factor would require some deeper understanding. An alternative explanation could be related to the missing spin orbit partners of the $0f_{5/2}$ and $0g_{9/2}$ orbits, which lead to a significant violation ($\sim$50%) of the Ikeda sum-rule. We could show that by adding the $0f_{7/2}$ and $0g_{7/2}$ orbits to the valence space for the intermediate nucleus on can fully recover the sum-rule. In addition, preliminary investigations in this larger valence space indicate a decrease of the NME towards the experimental value. We also found a similar behavior for the $2\nu\beta\beta$ matrix elements of ^{136}Xe, ^{130}Te, and ^{128}Te using the MC interaction of Ref. [64] in the $jj55$ model space. An even more drastic decrease of the effective quenching factors for these nuclei are reported by Caurier et al. (see e.g. Table 2 of Ref. [65]). The half-life of ^{136}Xe was recently measured by different groups [66, 67, 68]. Using the the value of 2.23×10^{21} y [67] and the newer phase space factors [36] one gets 0.022 MeV^{-1} for $M_{GT}^{2\nu}(0_{gs}^+)$. Typical extra-quenching, as for the ^{76}Ge and ^{82}Se cases, is necessary to describe this small matrix element. The Ikeda sum-rule can be satisfied adding the $0g_{9/2}$ and $0h_{9/2}$ orbits to the $jj55$ model space for the intermediate nucleus ^{136}Cs. However, the issues related to the choice of the effective interaction and the removal of the center-of-mass spurious contributions have to be resolved.

4. Neutrinoless double beta decay

The $0\nu\beta\beta$ decay $(Z, A) \to (Z+2, A) + 2e^-$ requires the neutrino to be a massive Majorana fermion, i.e. it is identical to the antineutrino [17]. We already know from the neutrino oscillation experiments that in the weak interaction some of the neutrinos have mass and the mass eigenstates are mixed by the PNMS matrix U_{lk}, where l is the lepton flavor and k is the mass eigenstate number (see e.g. [69]). However, the neutrino oscillations experiments cannot decide the mass hierarchy, the mass of the lightest neutrino, and some of the CP non-conserving phases of the PNMS matrix (assuming that neutrinos are Majorana particles).

Considering only contributions from the exchange of light, left-handed(chirality), Majorana neutrinos [16], the $0\nu\beta\beta$ decay half-live is given by

$$\left(T^{0\nu}_{1/2}\right)^{-1} = G^{0\nu} \mid M^{0\nu} \mid^2 \left(\frac{\mid \langle m_{\beta\beta}\rangle \mid}{m_e}\right)^2 , \tag{3}$$

Here, $G^{0\nu}$ is the phase space factor, which depends on the $0\nu\beta\beta$ decay energy, $Q_{\beta\beta}$ and the nuclear radius [11, 70]. The effective neutrino mass, $\langle m_{\beta\beta}\rangle$, is related to the neutrino mass eigenstates, m_k, via the lepton mixing matrix, U_{ek},

$$\langle m_{\beta\beta}\rangle /m_e \equiv \eta_{\nu L} = \sum_k m_k U^2_{ek} \,/m_e. \tag{4}$$

m_e is the electron mass. The NME, $M^{0\nu}$, is given by

$$M^{0\nu} = M^{0\nu}_{GT} - \left(\frac{g_V}{g_A}\right)^2 M^{0\nu}_F - M^{0\nu}_T , \tag{5}$$

where $M^{0\nu}_{GT}$, $M^{0\nu}_F$ and $M^{0\nu}_T$ are the Gamow-Teller (GT), Fermi (F) and tensor (T) matrix elements, respectively. Using closure approximation these matrix elements are defined as follows:

$$\begin{aligned} M^{0\nu}_\alpha &= \left\langle 0^+_f \mid \sum_{m,n} \tau_{-m}\tau_{-n} O^\alpha_{mn} \mid 0^+_i \right\rangle \\ &= \sum_{j_p j_{p'} j_n j_{n'} J_\pi} TBTD\left(j_p j_{p'}, j_n j_{n'}; J^\pi\right) \left\langle j_p j_{p'}; J^\pi T \mid \tau_{-1}\tau_{-2} O^\alpha_{12} \mid j_n j_{n'}; J^\pi T\right\rangle_a , \end{aligned} \tag{6}$$

where O^α_{mn} are $0\nu\beta\beta$ transition operators, $\alpha = (GT,\ F,\ T)$, $\mid 0^+_i >$ is the g.s. of the parent nucleus, and $\mid 0^+_f >$ is the final 0^+ state of the grand daughter nucleus. The two-body transition densities (TBTD) can be obtained from ShM calculations [54]. Expressions for the anti-symmetrized two-body matrix elements (TBME) $\langle j_p j_{p'}; J^\pi T \mid \tau_{-1}\tau_{-2} O^\alpha_{12} \mid j_n j_{n'}; J^\pi T\rangle_a$ can be found elsewhere, e.g. Refs. [71, 54]. One should also notice that due to rotational invariance of the TBME, transitions to final 2^+ of grand daughter are forbidden in closure approximation and strongly suppressed in general. Assuming that one unambiguously measures a $0\nu\beta\beta$ half-life, and one can reliably calculate the NME for that nucleus, one could use Eqs. (3) and (4) to extract information about the lightest neutrino mass and the neutrino mass hierarchy [69]. In addition, one could consider the contribution from the right handed currents to the effective Hamiltonian, which can mix light and heavy neutrons of both chiralities (L/R)

$$\begin{aligned} \nu_{eL} &= \sum_{k=light} U_{ek}\nu_{kL} + \sum_{k=heavy} U_{ek}N_{kL} \\ \nu_{eR} &= \sum_{k=light} V_{ek}\nu_{kR} + \sum_{k=heavy} V_{ek}N_{kR} , \end{aligned} \tag{7}$$

where N_k are the heavy neutrinos that are predicted by several see-saw mechanisms for neutrino masses [69]. U_{lk} and V_{lk} are the left and right-handed components of the unitary matrix that diagonalizes the neutrino mass matrix. One should also mention that there are several other mechanisms that could contribute to the $0\nu\beta\beta$ decay, such as the exchange of supersymmetric (SUSY) particles (e.g. gluino and squark exchange [76]), etc, whose contributions are not directly related to the neutrino masses, but indirectly via the Schechter-Valle theorem [17]. Assuming that the masses of the light neutrinos are smaller than 1 MeV and the masses of the heavy neutrinos, M_k, are larger than 1 GeV, the particle physics and nuclear structure parts get separated, and the inverse half-life can be written as

$$\begin{aligned}\left(T^{0\nu}_{1/2}\right)^{-1} &= G^{0\nu} \mid \eta_{\nu L} M^{0\nu} + <\lambda> \tilde{X}_\lambda + <\eta> \tilde{X}_\eta + (\eta_{NL}+\eta_{NR})\, M^{0N} \\ &+ \eta_{\lambda'} M^{0\nu}_{\lambda'} + \eta_{\tilde{q}} M^{0\nu}_{\tilde{q}} + \eta_{KK} M^{0\nu}_{KK} \mid^2, \end{aligned} \tag{8}$$

where $\eta_{\nu L}$ was defined in Eq. (4), and

$$\begin{aligned}\eta_{NL} &= \sum_{k=heavy} U^2_{ek} \frac{m_p}{M_k}, \quad \eta_{NR} \approx \left(\frac{M_{W_L}}{M_{W_R}}\right)^4 \sum_{k=heavy} V^2_{ek} \frac{m_p}{M_k} \\ <\lambda> &= \epsilon \sum_{k=light} U_{ek} V_{ek}, \quad <\eta> = \left(\frac{M_{WL}}{M_{WR}}\right)^2 \sum_{k=light} U_{ek} V_{ek} \ . \end{aligned} \tag{9}$$

Here ϵ is the mixing parameter for the right heavy boson W_R and the standard left-handed heavy boson W_L, $W_R \approx \epsilon W_1 + W_2$, M_{WR} and M_{WL} are their respective masses, and m_p is the proton mass. The $\eta_{\lambda'}$ and $\eta_{\tilde{q}}$ are the R-parity violation contributions in supersymmetric (SUSY) Grand Unified Theories (GUT) related to the long range gluino exchange and squark-neutrino mechanism, respectively [69]. Finally, the η_{KK} term is due to possible Kaluza-Klein (KK) neutrino exchange in extra-dimensional model [72]. The set of nuclear matrix elements $M^{0\nu}$, $\tilde{X}_\lambda$, $\tilde{X}_\eta$, M^{0N}, $M^{0\nu}_{\lambda'}$, and $M^{0\nu}_{\tilde{q}}$ are discussed in many reviews, e.g. Ref. [69]. The $M^{0\nu}_{KK}$ analysis can be found in Ref. [72]. In particular, using the factorization ansatz [72] one gets

$$\eta_{KK} M^{0\nu}_{KK} = \frac{<m>_{SA}}{m_e} M^{0\nu} + m_p <m^{-1}> M^{0N} \tag{10}$$

where $<m>_{SA}$ and $<m^{-1}>$ KK masses depend on the brane shift and bulk radius parameters, and are given in Table II of [72]. One can see that the mass parameters $<m>_{SA}/m_e$ and $m_p<m^{-1}>$ has the effect of modifying $\eta_{\nu L}$ and η_{NR} respectively. $\mid m_p <m^{-1}> \mid < 10^{-8}$ and it could in principle compete with η_{NR}. $\mid <m>_{SA}/m_e \mid$ varies significantly with the model parameters and it could also compete with $\eta_{\nu L}$. One needs to go beyond the factorization ansatz, and use information from several nuclei [73] to discern any significant contribution from the KK mechanism.

Constraints from colliders experiments suggest that terms proportional with the mixing angles, ϵ, $U_{ek(heavy)}$, and $V_{ek(light)}$ are very small [77]. Information from colliders also puts some limits on $(M_{W_R},\ M_N) \sim (2.5GeV,\ 1.4GeV)$. Based on this information and the present limit on the $0\nu\beta\beta$ decay of ^{76}Ge one can estimate that $\mid \eta_{\nu L} \mid < 10^{-6}$, and $\mid \eta_{NR} \mid < 10^{-8}$. They also suggest that $\mid <\lambda> \mid < 10^{-8}$ and $\mid <\eta> \mid < 10^{-9}$. In addition, these contributions should be signaled by the different angular and energy distribution of the outgoing electron, signals which are under investigation at SuperNEMO [74]. Here we assume that these contributions are small and can be neglected. In addition, if $<\lambda>$ is small, Eq. (9) suggests that η_{NL} is small. Then, the half-life can be written as

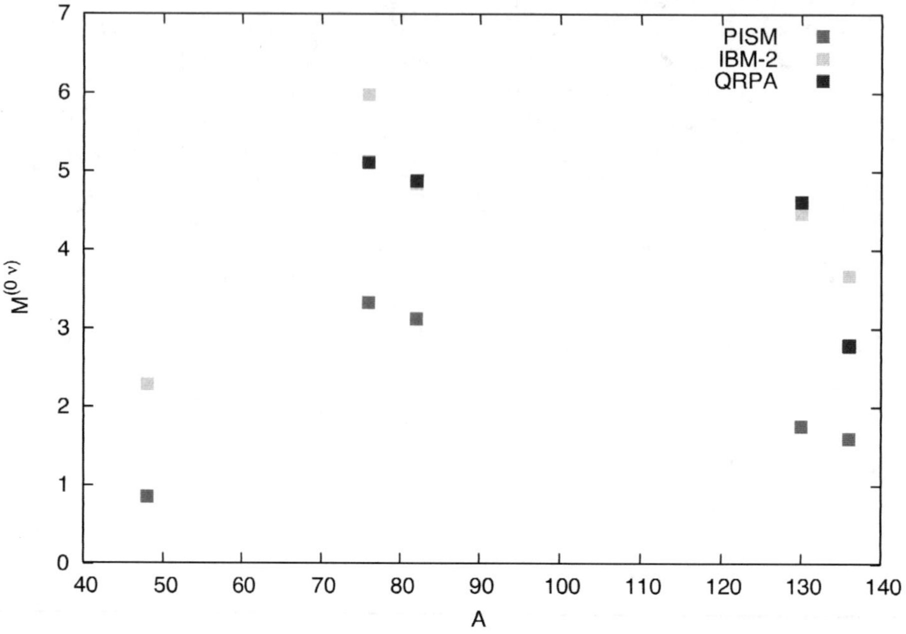

Figure 2. (Color online) The light neutrino-exchange NME for a set of relevant double-beta decay emiters. PISM are the results of present work. IBM-2 results are taken from [36], and QRPA results are taken from [76].

$$\left(T^{0\nu}_{1/2}\right)^{-1} = G^{0\nu}\left(\mid M^{0\nu}\mid^2\mid \eta_{\nu L}\mid^2 + \mid M^{0N}\mid^2\mid \eta_{NR}\mid^2\right) , \tag{11}$$

where we used the fact that the interference between the left-handed terms and the right-handed terms is negligible [69]. We also neglect the SUSY and KK contribution until a hint of their existence is provided by colliders experiments, or future results of $0\nu\beta\beta$ decay experiments show that these contributions are necessary [73].

The structure of the M^{0N} is as described in Eqs. (5)-(8), with slightly different form factors H_α (see e.g. page 68 of Ref. [69]). A similar treatment can be observed for the contributions from R-parity breaking SUSY mechanisms [76]. A detailed description of the matrix elements of O^{α}_{12} for the jj-coupling scheme consistent with the conventions used by modern shell model effective interactions is given in Ref. [54]. One should also mention that our method [54] of calculating the TBTD. Eq. (6), is different from that used in other shell model calculations [39]. We included in the calculations the recently proposed higher order terms of the nucleon currents, three old and recent parametrization of the short-range correlations (SRC) effects, finite size (FS) effects, intermediate states energy effects, and we treated careful few other parameters entering the into the calculations. We found very small variation of the NME with the average energy of the intermediate states, and FS cutoff parameters, and moderate variation vs the effective interaction and SRC parametrization. We could also show that if the ground state wave functions of the initial and final nucleus can be accurately described using only the valence space orbitals, the contribution from the core orbitals can be neglected. This situation is different from that of the nuclear parity-nonconservation matrix elements [78], for which the "mean-field" type contribution from the core orbitals could be significant [79].

Comparisons of light neutrino-exchange matrix elements, $M^{0\nu}$, can be found in several recent publications (see, e.g. [36]). Our results are similar to those of other calculations based on the shell model. We include here in Fig. 2 our results for ^{48}Ca, ^{76}Ge, ^{82}Se, ^{130}Te, and ^{136}Xe of the light neutrino-exchange matrix elements, M^{0N}, in comparison with the QRPA and IBA-2 results. Our corresponding ShM results for heavy neutrino-exchange, M^{0N}, were presented in a recent publication [75]. To our knowledge, no other results of shell model calculations for these matrix elements were reported so far. Based on these calculations one can extract "single-mechanism" limits for $\eta_{\nu L}$, η_{NR} from the experimental results or, assuming that both mechanisms contributing to the half-life in Eq. (11) compete, one can use the experimental data from two isotopes to assess the contribution of each mechanism [76]. This analysis will be reported elsewhere [80]. If the exchange of light neutrino would be determined as the dominant mechanism, then our results could possible decide the light neutrino mass hierarchy and the lowest neutrino mass [69].

5. Conclusions and outlook

In conclusion, we emphasized the role of nuclear structure for reactions and decays. For compound nucleus reactions the spin- and parity-dependent NLD are essential ingredients, besides the traditional transmission probabilities used in reaction theories for point-like particles. We showed that we developed advanced tools based on the statistical spectroscopy to efficiently and accurately calculate ShM NLD.

Direct nuclear reactions are also largely dependent on the nuclear structure information, such as the spectroscopic factors. In reverse, information from direct nuclear reaction can be used to better understand the nuclear structure of exotic nuclei, such as the pn correlations and their effect on changes in the single particle energies and magic numbers. In the case of N=28 isotones we found strong pn correlations between the number of proton pairs filling the $f_{7/2}$ and $d_{3/2}$ shells and the SF for breaking up a neutron, which are reflected in the unusually small SF for the cases of ^{54}Fe and ^{46}Ar. We found that in both cases pieces of the pn part of the effective ShM Hamiltonian are responsible for these effects. In principle one should not be worried, because these pieces are present in any realistic ShM Hamiltonian, but they might be missing in most, if not all, mean-field approaches.

We also analyzed the $2\nu\beta\beta$ and $0\nu\beta\beta$ decays of nuclei with masses $A = 48 - 136$ using shell model techniques. We described very efficient techniques to accurately calculate the $2\nu\beta\beta$ NME in cases that involve large shell model dimension, which were tested for the case of ^{48}Ca, and we used them to make some predictions for ^{76}Ge, ^{82}Se, ^{128}Te, ^{130}Te, and ^{136}Xe. We conclude the $A > 48$ analysis by emphasizing the potential role of including all spin-orbit partners and satisfying the Ikeda sum-rule.

We reviewed the main contributing mechanisms to the $0\nu\beta\beta$ decay, and we showed that based on the present constraints from colliders one could reduce the contribution to the $0\nu\beta\beta$ half-life to the relevant terms described in Eq. (11). A reliable analysis of the $0\nu\beta\beta$ decay experimental data requires an accurate calculations of the associated NME for light and heavy neutrino-exchange contributions. We extended our recent analysis [54] of the $0\nu\beta\beta$ NME for ^{48}Sc to the experimentally relevant cases of ^{76}Ge, ^{82}Se, ^{130}Te, and ^{136}Xe. We also presented for the first time shell model results for the heavy neutrino-exchange nuclear matrix elements necessary to the analysis. However, more effort has to be done to include all spin-orbit partners in the valence space, eliminate the center-of-mass spurious contribution, and better understand the changes in the effective $0\nu\beta\beta$ transition operators [81]. In addition, the closure approximation used in the shell model calculations and by other methods (e.g. IBA-2 [35], PHFB [82], and GSM [83]) needs to be further checked for accuracy, especially for the heavy neutrino exchange NME.

Acknowledgments

The author had useful conversations with M.B. Tsang, J. Lee, B.A. Brown, and S. Stoica (acknowledging project IDEI-PCE Nr. 58/28.10/2011). Support from the USA NSF grant

PHY-1068217 is acknowledged.

References

[1] Hauser W and Feshbach H 1952 *Phys. Rev.* **87** 366
[2] Horoi M, Kaiser J and Zelevinsky V 2003 *Phys. Rev. C* **67** 054309
[3] Horoi M, Ghita M and V. Zelevinsky V *Phys. Rev. C* **69** 041307(R)
[4] Horoi M, Ghita M and Zelevinsky V 2005 *Nucl. Phys.* **A785** 142c
[5] Horoi M and Zelevinsky V 2007 *Phys. Rev. Lett.* **98** 262503
[6] Scott M and Horoi M 2010 *EPL* **91** 52001
[7] Sen'kov R A and Horoi M 2010 *Phys. Rev. C* **82** 024304
[8] Horoi M and Sen'kov R A 2010 *PoS(NIC-X)222* http://pos.sissa.it/ .
[9] Sen'kov M, Horoi M and Zelevinsky V 2013 *Comp. Phys. Comm.* **184** 215 DOI: 10.1016/j.cpc.2012.09.006
[10] Haxton W C and Stephenson G J 1984 *Prog. Part. Nucl. Phys.* **12** 409
[11] Suhonen J and Civitarese O 1998 *Phys. Rep.* **300** 123
[12] Faessler A and Simkovic F 1998 *J. Phys. G: Nucl. Part. Phys.* **24** 2139
[13] Klapdor-Kleingrothaus H V 2001 *"60 Years of Double-Beta Decay - From Nuclear Physics to Beyond Standard Model Particle Physics"* World Scientific Singapore
[14] Elliot S R and Engel J 2004 *J. Phys. G* **30** R183
[15] Elliot S R and Vogel P 2002 *Annu. Rev. Nucl. Part. Sci.* **52** 115
[16] Avignone F T , Elliott S R and Engel J 2008 *Rev. Mod. Phys.* **80** 481
[17] Schechter J and Valle J W F 1982 *Phys. Rev. D* **25** 2951
[18] Aharmim B *et al* 2005 *Phys. Rev. C* **72** p 055502.
[19] Arsepella C *et al* 2008 *Phys. Lett. B* **658** 101; Araki T *et al* 2005 *Phys. Rev. Lett.* **94** 081801
[20] Schwetz T 2008 *Nucl. Phys. B (proc. Suppl)* **188** 158
[21] Vogel P and Zirnbauer M R 1986 *Phys. Rev. Lett.* **57** 3148
[22] Grotz K and Klapdor H V 1986 *Nucl. Phys.* **A 460** 395
[23] Suhonen J, Taigel T and Faessler A 1988 *Nucl. Phys.* **A 486** 91
[24] Staudt A, Muto K and Klapdor-Kleingrothaus H V 1990 *EPL* **13** 31
[25] Civitarese O, Faessler A, Suhonen J and Wu X R 1990 *Phys. Lett. B* **251** 333
[26] Stoica S and Kaminski W A 1993 *Phys. Rev. C* **47** 867; Stoica S 1994 *Phys. Rev. C* **49** 2240
[27] Raduta A A, Delion D S and Faessler A 1995 *Phys. Rev. C* **51** 3008
[28] Pantis G, Simkovic F, Vergados J D and Faessler A 1996 *Phys. Rev. C* **53** 695
[29] Bobyk A, Kaminski W A and Zareba P 1999 *Eur. Phys. J.* **A 5** 385
[30] Raduta A A, Faessler A, Stoica S and Kaminski W A 1991 **Phys. Lett. B 254** 7
[31] Toivanen J and Suhonen J 1995 **Phys. Rev. Lett. 75** 410; 1997 *Phys. Rev. C* **55** 2314
[32] Simkovic F, Schwieger J, Veselsky M, Pantis G and Faessler A 1997 *Phys. Lett. B* **393** 267
[33] Barabash A S 2010 *Phys. Rev. C* **81** 035501
[34] Rodin V A , Faessler A, Simkovic F and Vogel P 2006 *Nucl. Phys.* **A766** 107, erratum ibidem
[35] Barea J and Iachello F 2009 *Phys. Rev. C* **79** 044301
[36] Barea J, Kotila J and Iachello F 2012 *Phys. Rev. Lett.* **109** 042501
[37] Caurier E, Poves A and Zuker A P 1990 *Phys. Lett.* **B252** 13
[38] Caurier E, Nowacki F, Poves A and Retamosa J 1996 *Phys. Rev. Lett.* **77** 1954
[39] Retamosa J, Caurier E and Nowacki F 1995 *Phys. Rev. C* **51** 371
[40] Balysh A *et al* 1996 *Phys. Rev. Lett.* **77** 5186
[41] http://www.garsington.eclipse.co.uk
[42] Horoi M, Stoica S and Brown B A 2007 *Phys. Rev. C* **75** 034303
[43] Yako K *et al* 2009 *Phys. Rev. Lett.* **103** 012503
[44] Umehara S *et al* 2006 *J. Phys.: Conference Series* **39** 356
[45] Zdesenko Yu G *et al* 2005 *Astropart. Phys.* **23** 249
[46] Engel J, Haxton W and VogelP 1992 *Phys. Rev. C* **46** 2153R
[47] Tsang M B *et al* 2005 *Phys. Rev. Lett.* **95** 222501
[48] Tsang M B *et al* 2009 *Phys. Rev. Lett.* **102** 062501
[49] Lee J *et al* 2010 *Phys. Rev. Lett.* **104** 112701
[50] Tsang M B *et al* to be submitted
[51] Suehiro T, Finck J E and Nolen J A 1979 *Nucl. Phys. A* **313** 141
[52] Horoi M *et al* in preparation
[53] Lee J, Tsang M B, Lynch W G, Horoi M and Su S C 2009 *Phys. Rev. C* **79** 054611
[54] Horoi M and Stoica S 2010 *Phys. Rev. C* **81** 024321
[55] Poves A 2009 *invited talk at the ECT* workshop* Trento, Italy, July 27-32, 2009

[56] Kuo T T S and Brown G E 1968 *Nucl. Phys.* **A114** 235
[57] Poves A and Zuker A P 1981 *Phys. Rep.* **71** 141
[58] Honma M, Otsuka T, Brown B A and Mizusaki T 2004 *Phys. Rev. C* **69** 034335
[59] Honma M, Otsuka T, Brown B A and Mizusaki T 2005 *Eur. Phys. J. A* **25** Suppl. 1 499
[60] Horoi M, Brown B A, Otsuka T, Honma M and Mizusaki T 2006 *Phys. Rev. C* **73** 061305(R)
[61] Hornish M J, De Braeckeleer L, Barabash A S and Umatov V I 2005 *arXiv:nucl-ex/0512030*
[62] Horoi M 2010 *AIP Procs.* **1304** 106
[63] M. Honma M, T. Otsuka T, T. Mizusaki T and Hjorth-Jensen M 2009 *Phys. Rev. C* **80** 064323
[64] Qi C and Xu Z X 2012 *arXiv:1207.5056 [nucl-th]*
[65] Caurier E, Nowacki F and Poves A 2012 *arXiv:1112.5039v2 [nucl-th]*
[66] Gavriljuk Ju M *et al* 2011 *arXiv:1112.0859v2 [nucl-ex]*
[67] Auger M *et al* 2012 *Phys. Rev. Lett.* **109** 032505
[68] Gondo A *et al* 2012 *Phys. Rev. C* **85** 045504
[69] Vergados J D, Ejiri H and Simkovic F 2012 *arXiv:1205.0649v2 [hep-ph]*
[70] Kotila J and Iachello F 2012 *Phys. Rev. C* **85** 034316
[71] Neacsu A, Stoica S and Horoi M2012 *arXiv:1208.5728 [nucl-th]*
[72] Bhattacharyya G, Kalpdor-Kleingrothaus H V and Kovalenko S G 2003 *Phys. Rev. D* **67** 113001
[73] Deppisch F and Päs H 2007 *Phys. Rev. Lett.* **98**232501
[74] Arnold R *et al* 2010 *arXiv:1005.1241 [hep-ex]*
[75] Horoi M 2012 to be published in the Proceedings of the Carpathian Summer School of Physics 2012, Sinaia, Romania, June 24 - July 7, 2012, *AIP Proceedings*
[76] Faessler A, Meroni A, Petcov S T, Simkovic F and J. Vergados J D 2011 *Phys. Rev. D* **83** 113003
[77] Das S P, Deppisch F F, Kittel O and Valle J W F 2012 *Phys. Rev. D* **86** 055006
[78] Horoi M and Brown B A 1995 *Phys. Rev. Lett.* **74** 231
[79] Adelberger E G and Haxton W C 1985 *Annu. Rev. Nucl. Part. Sci.* **35** 501
[80] Horoi M 2012 in preparation
[81] Engel J and Hagen G 2009 *Phys. Rev. C* **79** 064317
[82] Rath P K, Chandra R, Chaturvedi K, Raina P K and Hirsch J G 2010 *Phys. Rev. C* **82** 064310
[83] Rodrigez T R and Martinez-Pinedo G 2010 *Phys. Rev. Lett.* **105** 252503

Journal of Physics: Conference Series **413** (2013) 012021 doi:10.1088/1742-6596/413/1/012021

Probing Nuclear Symmetry Energy and its Imprints on Properties of Nuclei, Nuclear Reactions, Neutron Stars and Gravitational Waves

Bao-An Li[1], Lie-Wen Chen[1,2], Farrukh J. Fattoyev[1], William G. Newton[1] and Chang Xu[1,3]

[1]Department of Physics and Astronomy, Texas A&M University-Commerce, Commerce, Texas 75429-3011, USA

[2]Department of Physics, Shanghai Jiao Tong University, Shanghai 200240, China

[3]Department of Physics, Nanjing University, Nanjing 210008, China

E-mail: Bao-An.Li@Tamuc.edu

Abstract. Significant progress has been made in recent years in constraining nuclear symmetry energy at and below the saturation density of nuclear matter using data from both terrestrial nuclear experiments and astrophysical observations. However, many interesting questions remain to be studied especially at supra-saturation densities. In this lecture note, after a brief summary of the currently available constraints on nuclear symmetry energy near the saturation density we first discuss the relationship between the symmetry energy and the isopin and momentum dependence of the single-nucleon potential in isospin-asymmetric nuclear medium. We then discuss several open issues regarding effects of the tensor force induced neutron-proton short-range correlation (SRC) on nuclear symmetry energy. Finally, as an example of the impacts of nuclear symmetry energy on properties of neutron stars and gravitational waves, we illustrate effects of the high-density symmetry energy on the tidal polarizability of neutron stars in coalescing binaries.

1. Introduction

The Equation of State (EOS) of neutron-rich nucleonic matter can be written within the parabolic approximation in terms of the binding energy per nucleon at density ρ as

$$E(\rho,\delta) = E(\rho,\delta=0) + E_{\rm sym}(\rho)\delta^2 + \mathcal{O}(\delta^4)\ , \tag{1}$$

where $\delta \equiv (\rho_n - \rho_p)/(\rho_p + \rho_n)$ is the neutron-proton asymmetry and $E_{\rm sym}(\rho)$ is the density-dependent nuclear symmetry energy. The latter is among the most uncertain properties of dense, neutron-rich nucleonic matter. It has important ramifications for many interesting questions in both astrophysics and nuclear physics. Thanks to the hard work of many people in both astrophysics and nuclear physics communities, significant progress has been made in recent years in constraining the density dependence of nuclear symmetry energy using data from both terrestrial nuclear experiments and astrophysical observations. A deeper understanding about the underlying physics governing the density dependence of nuclear symmetry energy was also obtained from various theoretical studies. However, many challenging questions remain to be answered. As an illustration of our current understanding about nuclear symmetry

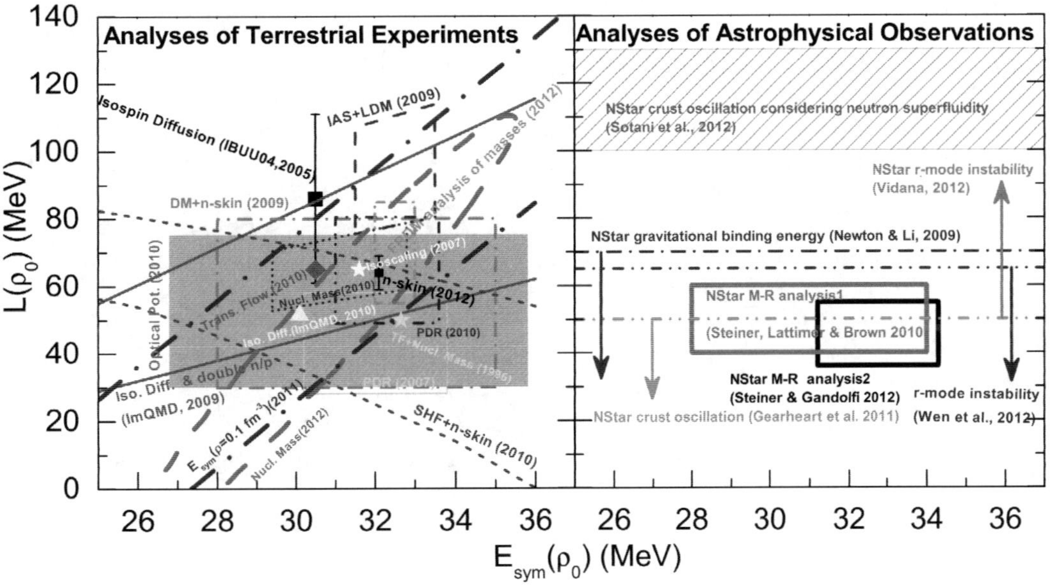

Figure 1. Density slope versus the magnitude of the symmetry energy at saturation density extracted from analyzing terrestrial experiments (left window) and astrophysical observations (right window). The analyses of terrestrial experiments include (1) analyses of isospin diffusion experiments with $^{124}Sn+^{112}Sn$ at 50 MeV/A within the Isospin-Dependent Boltzmann-Uehling-Ulenbeck (IBUU04-2005)[1, 2], (2) the isospin diffusion and neutron/proton ratio of pre-equilibrium nucleon emissions in $^{124}Sn+^{112}Sn$ reactions at 50 MeV/A within the Improved Molecular Dynamics (ImQMD-2009) model [3, 4, 5], (3) $^{124}Sn+^{112}Sn$ reactions at 35 MeV/A within the Improved Molecular Dynamics (ImQMD-2010) model [6], (4) isoscaling (isoscaling-2007) [7], (5) energy shift of isobaric analogue states within liquid drop model (IAS+LDM-2009)[8], (6) neutron-skins of several heavy nuclei using the droplet model (DM+n-skin (2009))[9, 10], or the Skyrme-Hartree-Fock (SHF+n-skin) approach [11], or the phenomenological approach (n-skin 2012) [12], (7) pygmy dipole resonances (PDR 2007 and 2010) in ^{209}Pb, ^{68}Ni and ^{132}Sn [13, 14], (8) the nucleon global optical potentials (Optical Pot. 2010) [15], (9) atomic masses analyzed by Myers and Swiatecki using the Thomas-Fermi model (TF+Nucl. Mass (1996) [16], (10) atomic masses analyzed by Möller et al. using the finite-range droplet model (FRDM) [17], (11) atomic masses analyzed by Liu et al. (Nucl. Mass (2010) [18], (12) atomic masses analyzed by Lattimer and Lim (Nucl. Mass (2012) [19], (13) Anti-symmetrized Molecular Dynamics (AMD) analyses of transverse flow of inter mediate mass fragments (Trans. Flow (2010)) [20], (14) empirical value of the symmetry energy at $\rho = 0.1\,\mathrm{fm}^{-3}$ ($E_{\mathrm{sym}}(\rho = 0.1\,\mathrm{fm}^{-3})$(2011) [21]. The analyses of astrophysical observations include (15) the mass-radius correlation of neutron stars (NStar analysis1 and analysis2) [22, 23], (16) gravitational binding energy of neutron stars (Newton & Li, 2009) [24], (17) torsional oscillations of neutron star crust analyzed by Gearheart et al. [25] and Sotani et al. [26], (18) the r-mode instability of neutron stars analyzed by Wen et al. [27] and Vidana [28]. Similar plots from selecting different sets of constraints available in the literature at the time can be found in Refs. [15, 29, 30, 31].

energy near saturation density ρ_0, shown in Fig. 1 are the available constraints on the slope $L(\rho_0) \equiv 3\rho \frac{\partial E_{\rm sym}(\rho)}{\partial \rho}|_{\rho_0}$ versus symmetry energy $E_{\rm sym}(\rho_0)$ at ρ_0 from analyses of both terrestrial nuclear experiments and astrophysical observations. Besides experimental error bars, there are some model dependences in most analyses and not all model assumptions are equally valid. Even assuming all published results are equally physical, given the still widely scattered constraints it is difficult to calculate a community-average of the $E_{\rm sym}(\rho_0)$ and $L(\rho_0)$ with physically meaningful error bars at this time. With all due respects to conclusions others may have drawn, in our obviously biased opinion, $E_{\rm sym}(\rho_0) = 31 \pm 2$ MeV and $L(\rho_0) = 50 \pm 20$ MeV are probably the best empirical values with the optimistic error bars we can currently conclude. In the following, using three examples we illustrate some recent progress in understanding the underlying physics governing the density dependence of nuclear symmetry energy, why the symmetry energy is still very uncertain, and how to probe the high density behavior of nuclear symmetry energy. The materials presented here are mostly taken from our recent work published originally in Refs. [15, 25, 32, 33, 34, 35, 36, 37, 38, 39, 40].

2. The relationship between nuclear symmetry energy and single-nucleon mean-field potential

What is the direct relationship between the symmetry energy and the isoscalar and isovector parts of the single-nucleon potential $U_{\rm n/p}(\rho, \delta, k)$? An answer to this question helps us better understand why the symmetry energy is still very uncertain. To our best knowledge, this question was first studied by Brueckner, Dabrowski and Haensel [41, 42, 43, 44] using K-matrices within the Brueckner theory in the 1960's. More recently, it was studied by Xu et al. [33] and Chen et al. [34] using the Hugenholtz-Van Hove (HVH) theorem [45]. This is an important question for several reasons. First of all, in both nuclear physics and astrophysics, one of the ultimate goals is to understand the isospin dependence of strong interaction. Both the symmetry energy and the single-particle potential are determined by the same underlying strong interaction. Their relationship can thus help us better understand why the symmetry energy is still very uncertain, and connect with the QCD theory of nuclear strong interaction. Theoretically, one usually derives both the single-nucleon potential and the symmetry energy from a model energy density functional constrained by empirical properties of nuclear matter and finite nuclei. However, the single-nucleon potential is often the one directly tested by comparing model calculations with experimental data. For example, the single-particle potential is the input for shell model calculations of nuclear structure and transport model simulations of nuclear reactions. Therefore, being able to know directly the corresponding symmetry energy from the single-particle potential without first going through the procedure of constructing the energy density functional is advantageous. For example, from nucleon-nucleus scattering and (p, n) charge exchange experiments one can directly extract from the data both the isoscalar and isovector nucleon optical potentials at normal density. One can then easily calculate the symmetry energy and its density slope at normal density directly from the optical potentials as demonstrated recently in Ref. [15].

According to the well-known Lane potential [46], the neutron/proton single-particle potential $U_{n/p}(\rho, k, \delta)$ can be well approximated by

$$U_{\rm n/p}(\rho, k, \delta) \approx U_0(\rho, k) \pm U_{\rm sym}(\rho, k)\delta \ , \quad (2)$$

where the $U_0(\rho, k)$ and $U_{\rm sym}(\rho, k)$ are, respectively, the isoscalar and isovector (symmetry) potentials for nucleons with momentum k in nuclear matter of isospin asymmetry δ at density ρ. It has been shown that the nuclear symmetry energy can be explicitly expressed as

[15, 33, 34, 41, 42, 43, 44]

$$E_{\rm sym}(\rho) = \frac{1}{6}\frac{\partial(t+U_0)}{\partial k}\Big|_{k_{\rm F}} \cdot k_{\rm F} + \frac{1}{2}U_{\rm sym}(\rho,k_{\rm F})\ , \tag{3}$$

where $t(k) = \hbar^2k^2/2m$ is the kinetic energy and $k_{\rm F} = (3\pi^2\rho/2)^{1/3}$ is the nucleon Fermi momentum in symmetric nuclear matter at density ρ. The slope of nuclear symmetry energy at an arbitrary density ρ can be written as [15, 33, 34]

$$\begin{aligned} L(\rho) &\equiv 3\rho\frac{\partial E_{\rm sym}(\rho)}{\partial\rho}\Big|_{\rho} = \\ &= \frac{1}{6}\frac{\partial(t+U_0)}{\partial k}\Big|_{k_{\rm F}} \cdot k_{\rm F} + \frac{1}{6}\frac{\partial^2(t+U_0)}{\partial k^2}\Big|_{k_{\rm F}} \cdot k_{\rm F}^2 + \frac{3}{2}U_{\rm sym}(\rho,k_{\rm F}) + \frac{\partial U_{\rm sym}}{\partial k}\Big|_{k_{\rm F}} \cdot k_{\rm F}\ . \end{aligned} \tag{4}$$

Moreover, the relative neutron-proton effective mass is

$$\begin{aligned} \frac{m_{\rm n}^* - m_{\rm p}^*}{m} &= -2\delta\frac{m}{\hbar^2 k_{\rm F}}\frac{dU_{\rm sym}}{dk}\Big|_{k_{\rm F}} \Big/ \left[1 + 2\frac{m}{\hbar^2 k_{\rm F}}\frac{dU_0}{dk}\Big|_{k_{\rm F}}\right] \\ &= -2\delta\frac{m}{\hbar^2 k_{\rm F}}\frac{dU_{\rm sym}}{dk}\Big|_{k_{\rm F}} \Big/ \left[1 + 2\left(\frac{m}{m_0^*} - 1\right)\right]\ . \end{aligned} \tag{5}$$

We emphasize that this relationship is valid only at the mean-field level. Taking into account the tensor force induced neutron-proton short range correlation, the kinetic part of the symmetry energy might be reduced significantly from the Fermi gas model prediction as we shall discuss in detail in the next section. The above expressions for $E_{\rm sym}(\rho)$ and $L(\rho)$ in terms of the isoscalar and isovector single-particle potentials are particularly useful for extracting the symmetry energy and its density slope from terrestrial nuclear laboratory experiments. While the density and momentum dependence of the isoscalar potential $U_0(\rho,k)$ has been relatively well determined up to about 4 to 5 times the normal nuclear matter density ρ_0 using nucleon global optical potentials from nucleon-nucleus scatterrings as well as kaon production and nuclear collective flow in relativistic heavy-ion collisions [47, 48], our current knowledge about the symmetry potential $U_{\rm sym}(\rho,k)$ is rather poor especially at high density and/or momenta [40, 49, 50, 51, 52, 53]. Experimentally, there is some constraints on the symmetry potential only at normal density for low energy nucleons up to about 100 MeV obtained from nucleon-nucleus and (p, n) charge exchange reactions [15]. Shown in the left window of Fig. 2 are all the energy dependent symmetry potentials in the literature [15]. Assuming that these various global energy dependent symmetry potentials are equally accurate and all have the same predicting power beyond the original energy ranges in which they were analyzed, an averaged symmetry potential of

$$U_{\rm sym}(\rho_0, E) = 22.75 - 0.21E \tag{6}$$

was obtained (thick solid line in the left window of Fig. 2). It represents the best fit to the global symmetry potentials constrained by the world data up to date. With this best estimate for the $U_{\rm sym}(\rho_0,E)$, Xu et al. found that $E_{\rm sym}(\rho_0) = 31.3 \pm 4.5$ MeV and $L(\rho_0) = 52.7 \pm 22.5$ MeV. Shown in the right window of Fig. 2 are the various contributions to the $E_{\rm sym}(\rho_0)$ and $L(\rho_0)$. The $E_{\rm sym}(1) = \frac{1}{3}\frac{\hbar^2k_{\rm F}^2}{2m_0^*}$ denotes the kinetic energy term with the effective mass m_0^* and the $E_{\rm sym}(2) = \frac{1}{2}U_{\rm sym}(\rho_0,k_{\rm F})$ is the symmetry potential contribution. It is seen that the two terms are comparable. Their respective uncertainties are marked by the red boxes. The $L(\rho_0)$ has three terms: $L(1) = \frac{2}{3}\frac{\hbar^2k_{\rm F}^2}{2m_0^*}$, $L(2) = \frac{3}{2}U_{\rm sym}(\rho_0,k_{\rm F})$ and $L(3) = \frac{\partial U_{\rm sym}(\rho,k)}{\partial k}\big|_{k_{\rm F}} k_{\rm F}$. While both the $L(1)$

Journal of Physics: Conference Series **413** (2013) 012021 doi:10.1088/1742-6596/413/1/012021

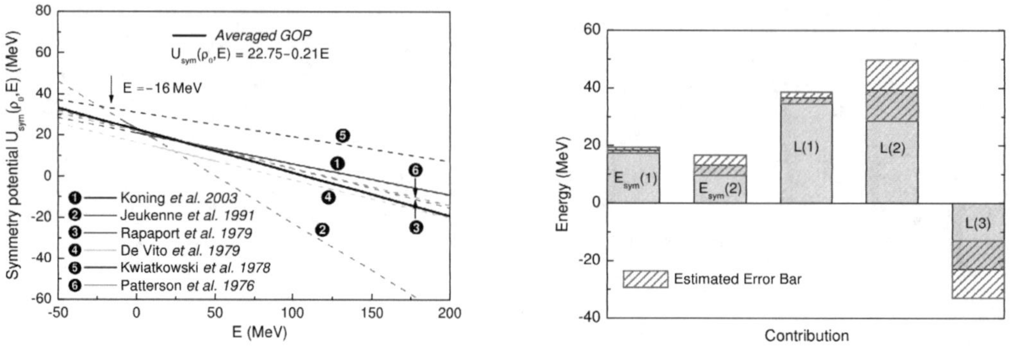

Figure 2. Left Window: Energy dependence of the nuclear symmetry potential $U_{\rm sym}(\rho_0, E)$ at normal density from different global optical model fits. The solid lines are in the energy ranges where the original analyses were made while the dashed parts are extrapolations. **Right Window**: The magnitude of each term in the nuclear symmetry energy $E_{\rm sym}(\rho_0)$ and its density slope $L(\rho_0)$ at normal nuclear density. Taken from Ref. [15].

and $L(2)$ are positive, the $L(3)$ is negative because of the decreasing symmetry potential with increasing energy. To our best knowledge, extracting the symmetry energy and its density slope directly from the optical potential is probably the most straight forward approach available in the literature. However, the major challenge of using this approach is our poor knowledge about the momentum dependence of the isovector potential. From the symmetry potential given in Eq. (6), Xu et al. extracted a neutron-proton effective mass splitting of $(m^*_{\rm n}-m^*_{\rm p})/m = (0.32\pm0.15)\delta$. Interestingly, from the dispersive optical model anylysis of some new data on neutron-nucleus scattering, R. Charity recently also found a value of $(m^*_{\rm n} - m^*_{\rm p})/m = 0.32\delta$ with an error bar to be determined [54]. The extracted constraint on $E_{\rm sym}(\rho_0)$ versus $L(\rho_0)$ from the global nucleon optical potentials are compared with constraints extracted from other approaches [31] in Fig. 1. It is consistent with the ones from most other approaches.

3. Effects of tensor force induced neutron-proton short-range correlation on nuclear symmetry energy

From Fig. 1 and the related references, it is clear that there are still appreciable uncertainties about the density dependence of nuclear symmetry energy even around the saturation density. Moreover, at supra-saturation densities, even the tendency of the symmetry energy remains controversial [40]. So, why is the nuclear symmetry energy, especially at supra-saturation densities, so uncertain? Of course, the answer itself is model dependent. Generally, besides our poor knowledge about the isospin dependence of strong interaction in dense neutron-rich medium, different approaches used in treating nuclear many-body problems in various models contribute to the divergence of the predicted symmetry energy especially at supra-saturation densities. Nonetheless, there are several key and commonly used physics ingredients that can affect the predicted $E_{\rm sym}(\rho)$ in all theories. For instance, the symmetry energy has a kinetic part. Often, it is assumed to be the one predicted by the free Fermi gas model, i.e., $E^{\rm kin}_{\rm sym}({\rm FG})(\rho) \equiv (2^{\frac{2}{3}} - 1)(\frac{3}{5}\frac{\hbar^2 k_{\rm F}^2}{2m}) \approx 12.5(\rho/\rho_0)^{2/3}$. Interestingly, it was first found recently within a phenomenological model [55] that the tensor force induced high momentum tail in the single-nucleon momentum distribution in symmetric nuclear matter (SNM) reduces significantly the $E^{\rm kin}_{\rm sym}(\rho)$ to values much smaller than the $E^{\rm kin}_{\rm sym}({\rm FG})(\rho)$. In fact, the $E^{\rm kin}_{\rm sym}(\rho)$ can become zero or even negative if more than about 15% nucleons populate the high-momentum tail above the Fermi surface as indicated by the recent experiments done at the Jefferson National Laboratory

(J-Lab) by the CLAS Collaboration [56]. This finding was recently confirmed qualitatively by three independent studies using the state-of-the-art microscopic many-body theories [57, 58, 59]. As discussed in detail by Xu et al. in Ref. [55], the high momentum tail in SNM increases the

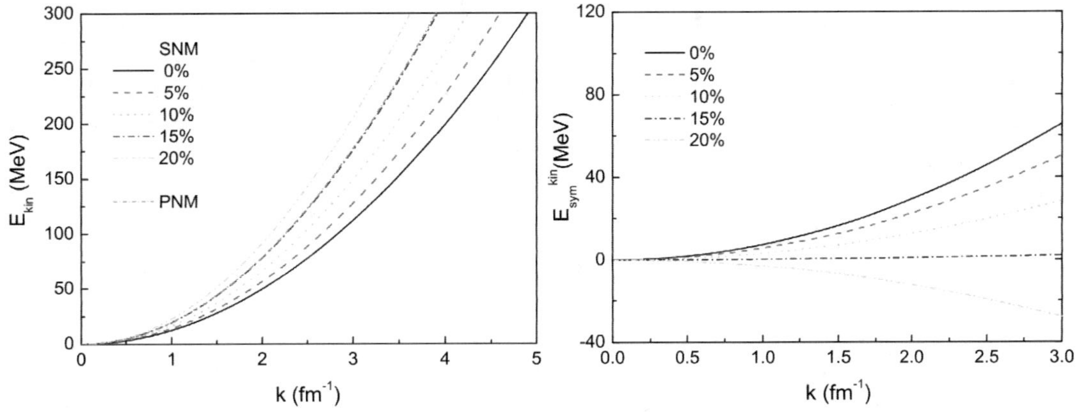

Figure 3. Left Window: The average kinetic energy per nucleon $E_{\rm kin}$ for pure neutron matter and symmetric nuclear matter with different percentages of correlated nucleons ($\theta_{k>k_{\rm F}}$) as a function of Fermi momentum. **Right Window**: The kinetic energy part of nuclear symmetry energy with different percentages of correlated nucleons ($\theta_{k>k_{\rm F}}$) as a function of Fermi momentum. Taken from Ref. [55].

average kinetic energy of nucleons to values above the free Fermi gas model prediction. While in PNM (pure neutron matter), the Fermi gas prediction is a good approximation. Shown in the left window of Fig. 3 is the the average kinetic energy as a function of Fermi momentum for PNM and SNM with the percentage of high momentum nucleons to be $\theta_{k>k_{\rm F}} = 0\%$, 5%, 10%, 15%, and 20%, respectively. As one expects, the SRC increases the $E_{\rm kin}$ significantly for SNM. More quantitatively, for SNM at the saturation density corresponding to $k_{\rm F} = 1.33$ fm^{-1}, the $E_{\rm kin}$ with $\theta_{k>k_{\rm F}} = 20\%$ ($E_{\rm kin}(k_{\rm F}) \simeq 40$ MeV) is about twice of that ($E_{\rm kin}(k_{\rm F}) \simeq 22$ MeV) for the free Fermi gas. However, the $E_{\rm kin}$ for PNM is the same as for the free Fermi gas. Consequently, the tensor force induced high momentum tail in SNM affects the kinetic part of the nuclear symmetry energy. In particular, if about 15% nucleons in SNM are in the high momentum tail, it is seen that the average kinetic energy is about the same in PNM and SNM. This leads to an approximately zero kinetic symmetry energy as shown in the right window of Fig. 3. In many studies in both nuclear physics and astrophysics, it is customary to write the total symmetry energy as $E_{\rm sym}(\rho) = 12.5(\rho/\rho_0)^{2/3} + E_{\rm sym}^{\rm pot}(\rho)$ where the first term is the Fermi gas prediction for the $E_{\rm sym}^{\rm kin}(\rho)$ and the $E_{\rm sym}^{\rm pot}(\rho)$ is the potential contribution. In doing so, however, one neglects completely effects of the tensor force on the $E_{\rm sym}^{\rm kin}(\rho)$. For example, in transport model analyses of heavy-ion reactions, the $E_{\rm sym}^{\rm pot}(\rho)$ is normally parameterized as a function of density. The corresponding single-particle potential based on some energy density functions is used as input to transport models. Thus, heavy-ion reactions test directly the single nucleon potential. The kinetic part of the symmetry energy based on the Fermi gas model prediction is normally added by hand to the $E_{\rm sym}^{\rm pot}(\rho)$ in fixing parameters in the EOS. The results shown in Fig. 3 raise serious questions about this practice. Moreover, many interesting questions regarding effects of the tensor-force induced isospin dependence of short-range nucleon-nucleon correlations and the uncertain short-range behavior of tensor force due to the ρ meson exchange on the density dependence of nuclear symmetry energy remain to be studied more systematically and self-consistently [32, 60, 61].

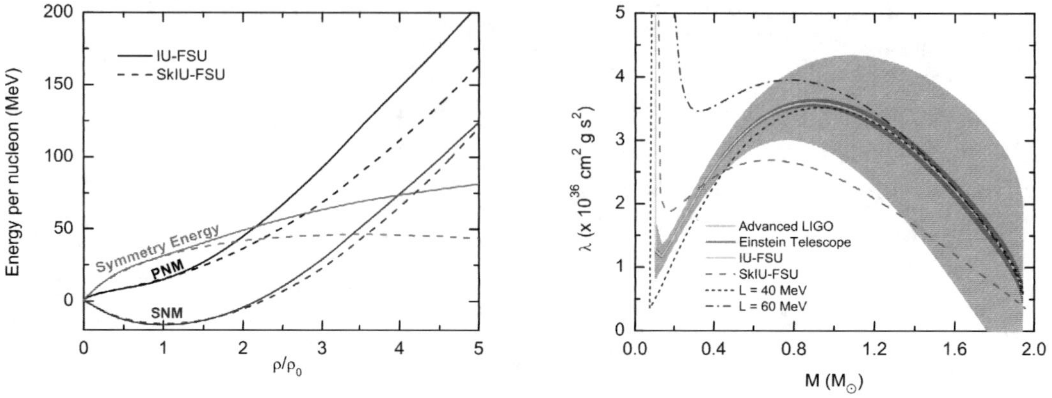

Figure 4. Left Window: The EOS of symmetric nuclear matter and pure neutron matter as well as the symmetry energy as a function of density obtained within the IF-FSU RMF model and the SHF approach using the SkIU-FSU parameter set. **Right Window**: Tidal polarizability λ of a single neutron star as a function of neutron-star mass for a range of EOS that allow various stiffness of symmetry energies. A crude estimate of uncertainties in measuring λ for equal mass binaries at a distance of $D = 100$ Mpc is shown for the Advanced LIGO (shaded light-grey area) and the Einstein Telescope (shaded dark-grey area) . Taken from Ref. [39].

4. Probing the high-density symmetry energy with the tidal polarizability of neutron stars

The high-density behavior of nuclear symmetry energy has long been regarded as the most uncertain property of dense neutron-rich nucleonic matter [40, 62, 63]. While several observables have been proposed [40] and some indications of the high-density symmetry energy have been reported based on terrestrial nuclear experiments [64, 65], unfortunately, the conclusions remain controversial. Interestingly, it was recently proposed that the late time neutrino signal from a core collapse supernova [66] and the tidal polarizability [39] of canonical neutron stars in coalescing binaries are very sensitive probes of the high-density behavior of nuclear symmetry energy. Coalescing binary neutron stars are among the most promising sources of gravitational waves (GW). One of the most important features of the binary mergers is the tidal deformations of neutron stars, which give us precious information about the neutron-star matter EOS [67, 68, 69, 70, 71, 72, 73, 74]. At the early stage of an inspiral tidal effects may be effectively described through *the tidal polarizability* parameter λ [67, 68, 69, 72] defined via $Q_{ij} = -\lambda\mathcal{E}_{ij}$, where Q_{ij} is the induced quadrupole moment of a star in binary, and $\mathcal{E}_{ij}$ is the static external tidal field of the companion star. As an example, shown in the left window of Fig. 4 are two models chosen to have the same EOSs for both SNM and PNM using the IU-FSU RMF model and the SHF using the SkIU-FSU parameter set [38], i.e., they have the same symmetry energy at and below the saturation density. At supra-saturation densities, however, the symmetry energy with the IU-FSU RMF is significantly more stiff above about $1.5\rho_0$. It is seen from the right window of Fig. 4 that the two models predict significantly different polarizability (λ) values in a broad mass range from 0.5 to 2 $M_\odot$. More quantitatively, for a canonical neutron star of 1.4 $M_\odot$, a 41.41% change from $\lambda = 2.828 \times 10^{36}$ (IU-FSU) to $\lambda = 1.657 \times 10^{36}$ (SkIU-FSU) is observed. As it was discussed in detail in Ref. [39], the observed symmetry energy effect on the tidal polarizability is as strong as its effect on the late time neutrino flux from the cooling of proto-neutron stars [66]. Moreover, it is interesting to note that the narrow uncertain range for the proposed Einstein Telescope will enable it to tightly constrain the symmetry energy especially at supra-saturation densities.

5. Summary

In summary, significant progress has been made in recent years in constraining the symmetry energy mostly around and below the saturation density using both terrestrial nuclear laboratory data and astrophysical observations. However, to fully understand the nature of neutron-rich nucleonic matter and its equation of state especially at supra-saturation densities many interesting questions remain to be studied. In particular, the high-density behavior of the symmetry energy is still among the most uncertain properties of neutron-rich matter. Besides continuing the search of sensitive observables in terrestrial experiments and astrophysical observations, it is important to understand the underlying physics leading to the divergent predictions for the high-density symmetry energy. In particular, effects of the tensor-force induced short-range neutron-proton correlation and the short-range behavior of the tensor force itself on the high-density behavior of the symmetry energy deserve some special attention. Given the strong ongoing efforts and close collaborations of many people in both the nuclear physics and astrophysics communities, it is expected that more stringent constraints on the symmetry energy over a broad density range will come soon.

6. Acknowledgments

This work is supported in part by the US National Aeronautics and Space Administration under grant NNX11AC41G issued through the Science Mission Directorate, the US National Science Foundation under Grants No. PHY-0757839, No. PHY-1062613 and No. PHY-1068022, the NNSF of China under Grant Nos. 10975097, 11135011, 11275125 and 11175085, the Shanghai Rising-Star Program under grant No. 11QH1401100, the "Shu Guang" project supported by Shanghai Municipal Education Commission and Shanghai Education Development Foundation, the Program for Professor of Special Appointment (Eastern Scholar) at Shanghai Institutions of Higher Learning, and the Science and Technology Commission of Shanghai Municipality (11DZ2260700).

References

[1] Chen L -W, Ko C M and Li B -A 2005 *Phys. Rev. Lett.* **94** 032701
[2] Li B -A and Chen L -W 2005 *Phys. Rev. C* **72** 064611
[3] Tsang M B *et al.* 2004 *Phys. Rev. Lett.* **92**, 062701
[4] Tsang M B *et al.* 2009 *Phys. Rev. Lett.* **102**, 122701
[5] Famiano M A *et al.* 2006 *Phys. Rev. Lett.* **97**, 052701
[6] Sun Z Y *et al.* 2010 *Phys. Rev. C* **82**, 051603 (R)
[7] Shetty D V, Yennello S J and Souliotis G A 2007 *Phys. Rev. C* **76**, 024606
[8] Danielewicz P and Lee J 2009 *Nucl. Phys. A* **818**, 36
[9] Centelles M, Roca-Maza X, Vinas X and Warda M 2009 *Phys. Rev. Lett.* **102**, 122502
[10] Warda M, Vinas X, Roca-Maza X and Centelles M 2009 *Phys. Rev. C* **80**, 024316
[11] Chen L -W, Ko C M, Li B -A and Xu J 2010 *Phys. Rev. C* **82**, 024321
[12] Agrawal B K, De J N and Samaddar S K 2012 *Phys. Rev. Lett.*, in press, *arXiv:1212.0292*
[13] Klimkiewicz A *et al.* 2007 *Phys. Rev. C* **76**, 051603(R)
[14] Carbone A *et al.* 2010 *Phys. Rev. C* **81**, 041301(R)
[15] Xu C, Li B -A and Chen L -W 2010 *Phys. Rev. C* **82**, 054607
[16] Myers W D and Swiatecki W J 1996 *Nucl. Phys. A* **601**, 141
[17] Moller P, Myers W D, Sagawa H and Yoshida S 2012 *Phys. Rev. Lett.* **108**, 052501
[18] Liu M, Wang N, Li Z and Zhang F 2010 *Phys. Rev. C* **82**, 064306
[19] Lattimer J M and Lim Y 2012 *arXiv:1203.4286*
[20] Kohley Z *et al.* 2010 *Phys. Rev. C* **82**, 064601
[21] Chen L -W 2011 *Phys. Rev. C* **83**, 044308
[22] Steiner A W, Lattimer J M and Brown E F 2010 *Astrophys. J.* **722**, 33
[23] Steiner A W and Gandolfi S 2012 *Phys. Rev. Lett.* **108**, 081102

[24] Newton W G and Li B -A 2009 *Phys. Rev. C* **80**, 065809
[25] Gearheart M, Newton W G, Hooker J and Li B -A 2011 *Mon. Not. Roy. Astron. Soc.*, **418**, 2343
[26] Sotani H *et al.* 2012 *Mon. Not. Roy. Astron. Soc. Lett.*, in press, *arXiv:1210.0955*
[27] Wen D -H, Newton W G and Li B -A 2012 *Phys. Rev. C* **85**, 025801
[28] Vidaa I 2012 *Phys. Rev. C* **85**, 045808
[29] Tsang M B *et al.* 2012 *Phys. Rev. C* **86**, 024605
[30] Lattimer J 2012 *Annu. Rev. Nucl. Part. Sci.* **62**, 485
[31] Chen L -W 2012 *in Proceedings of the 14th National Conference on Nuclear Structure in China*, World Scinetific, in press, *arXiv:1212.0284*
[32] Xu C, and Li B -A 2010 *Phys. Rev. C* **81**, 064612
[33] Xu C, Li B -A, Chen L -W and Ko C M 2011 *Nucl. Phys. A* **865**, 1
[34] Chen R, Cai B -J, Chen L -W, Li B -A, Li X -H and Xu C 2012 *Phys. Rev. C* **85**, 024305
[35] Xu C, Li A and Li B -A 2013 *J. Phys.: Conf. Series*, in press, *arXiv:1207.1639*
[36] Newton W G, Gearheart M and Li B -A 2012 *Astrophys. J.*, in press, *arXiv:1110.4043*
[37] Newton W G, Gearheart M, Hooker J and Li B -A 2012 *in the book "Neutron Star Crust"*, edited by C. A. Bertulani and J. Piekarewicz. Nova Science Publishers, Inc., *arXiv:1112.2018*
[38] Fattoyev F J, Newton W G, Xu J and Li B -A 2012 *Phys. Rev. C* **86**, 025804
[39] Fattoyev F J, Carvajal J, Newton W G, Li B -A 2012 *arXiv:1210.3402*
[40] Li B -A, Chen L -W, and Ko C M 2008 *Phys. Rep.* **464**, 113
[41] Brueckner K A and Dabrowski J 1964 *Phys. Rev.* **134**, B722
[42] Dabrowski J and Haensel P 1972 *Phys. Lett. B* **42**, 163
[43] Dabrowski J and Haensel P 1973 *Phys. Rev. C* **7**, 916
[44] Dabrowski J and Haensel P 1974 *Can. J. Phys.* **52**, 1768
[45] Hugenholtz N M and van Hove L 1958 *Physica* **24**, 363
[46] Lane A M 1962 *Nucl. Phys.* **35**, 676
[47] Danielewicz P, Lacey R and Lynch W G 2002 *Science* **298**, 1592
[48] Fuchs C 2006 *Prog. Part. Nucl. Phys.* **56**, 1
[49] van Dalen E N E, Fuchs C and Faessler A 2005 *Phys. Rev. C* **72**, 065803
[50] Zuo W, Cao L G, Li B -A, Lombardo U and Shen C W 2005 *Phys. Rev. C* **72**, 014005
[51] Chen L -W, Ko C M and Li B -A 2005 *Phys. Rev. C* **72**, 064606.
[52] Li Z -H, Chen L -W, Ko C M, Li B -A and Ma H -R 2006 *Phys. Rev. C* **74**, 044613
[53] Giordano V, Colonna M, Di Toro M, Greco V and Rizzo J 2010 *Phys. Rev. C* **81**, 044611
[54] Charity R 2011 *private communications*
[55] Xu C, Li A and Li B -A 2012 *J. Phys.: Conf. Series*, in press, *arXiv:1207.1639*
[56] Egiyan K S et al. (CLAS Collaboration) 2006 *Phys. Rev. Lett.* **96** 082501
[57] Vidana I, Polls A, Providencia C 2011 *Phys. Rev. C* **84** 062801
[58] Carbone A, Polls A, Rios A 2012 *EPL* **97** 22001
[59] Lovato A 2011 *private communications*
[60] Lee H K, Park B -Y and Rho M 2011 *Phys. Rev. C* **83**, 025206
[61] Li A and Li B -A 2011 *arXiv:1107.0496*
[62] Kutschera M 1994 *Phys. Lett. B* **340**, 1
[63] Kubis S and Kutschera M 1999 *Acta Phys. Pol. B* **30**, 2747; 2003 *Nucl. Phys. A* **720**, 189
[64] Xiao Z, Li B -A, Chen L -W, Yong G -C and Zhang M 2009 *Phys. Rev. Lett.* **102**, 062502
[65] Russotto P et al. 2011 *Phys. Lett. B* **697**, 471
[66] Roberts L F et al. 2012 *Physical Review Letters* **108**, 1103
[67] Flanagan É É and Hinderer T 2008 *Phys. Rev. D* **77**, 021502
[68] Damour T and Nagar A 2009 *Phys. Rev. D* **80**, 084035
[69] Damour T and Nagar A 2010 *Phys. Rev. D* **81**, 084016
[70] Damour T, Nagar A and Villain L 2012 *Phys. Rev. D* **85**, 123007
[71] Hinderer T 2008 *Astrophys. J.* **677**, 1216
[72] Hinderer T, Lackey B D, Lang R N and Read J S 2010 *Phys. Rev. D* **81**, 123016
[73] Pannarale F, Rezolla L, Ohme F and Read J S 2011 *Phys. Rev. D* **84**, 104017
[74] Postnikov S, Prakash M and Lattimer J M 2010 *Phys. Rev. D* **82**, 024016

Journal of Physics: Conference Series **413** (2013) 012022 doi:10.1088/1742-6596/413/1/012022

Centrality dependence of mid-rapidity charged particles density in relativistic heavy ion collisions - energy scaling

M. Petrovici, I. Berceanu, A. Pop, C. Andrei, A. Herghelegiu
Horia Hulubei National Institute for Physics and Nuclear Engineering, R-077125 Bucharest, Romania

E-mail: mpetro@nipne.ro

Abstract. As far as details on the "Relativistic Heavy Ion Collisions - Expectations, Experimental Facts and Future Plans" lecture could be found on the Summer School web site, the present contribution will summarize the results considered to be of general interest, i.e. the energy scaling of centrality dependence of mid-rapidity charged particles density in relativistic heavy ion collisions, presented in the second part of the lecture, in the section dedicated to the global observables. The comparison of charged particle multiplicity in heavy ion central collisions and e^+e^- as a function of $\sqrt{s}$, including the latest results obtained at LHC in Pb+Pb collisions at 2.76 is followed by the energy scaling of the pseudo-rapidity charged particle density in central rapidity as a function of centrality.

1. Introduction

The lecture "Relativistic Heavy Ion Collisions - Expectations, Experimental Facts and Future Plans" , accessible on the Summer School web site [1], has three main chapters. An overview of theoretical considerations on which are based the expectations on the properties of matter in extreme conditions of temperature and density which could be obtained in colliding two heavy ions at relativistic and ultra-relativistic energies is presented in the first part. The second part is dedicated to the latest results in the region of high temperature and vanishing chemical potential in Pb + Pb collisions at 2.76 TeV at CERN LHC compared with the previous Au + Au at 0.2 TeV RHIC. Results and expectations evidencing collective type phenomena in high multiplicity and nearly azimuthal isotropic events selected using event shape global observables, Directivity, Sphericity, Trust and Fox-Wolfram moments, in p+p and p+Pb at LHC energies are discussed at the end of this second chapter. In the last chapter is presented the physics program on detailed studies of the QCD phase diagram in the region of highest net baryon density of the CBM experiment at FAIR relative to the recent measurements at RHIC within the beam energy scan program. This chapter ends with a short overview on the main R&D activities in developing highly performant high counting rate Transition Radiation Detectors (TRD) and Multi Strip Multi Gap Resistive Plate Counters (MSMGRPC) for CBM Experiment. Due to the limited number of pages and considering to be a subject of general interest, we decided to concentrate in this written version on the energy scaling of centrality dependence of mid-rapidity charged particles density in relativistic heavy ion collisions, presented in the second part of the lecture, in the section dedicated to the global observables. Pseudo-rapidity charged particle density in central rapidity and its dependence on the collision geometry is an information sensitive to the underlying mechanism of particle production and properties and dynamics of strongly interacting matter in extreme conditions of temperature and density produced in relativistic heavy ion collisions. In the second chapter will be presented a comparison of charged particle multiplicity in heavy ion central collisions and e^+e^- as a function of $\sqrt{s}$, including the latest results obtained at LHC in Pb+Pb collisions at 2.76 TeV. An energy scaling of the pseudo-rapidity charged particle density in central rapidity as a function of centrality is discussed in the third chapter. Conclusions are presented in chapter 4.

2. Charged particle multiplicity in heavy ion central collisions

Normalized average total multiplicity of primary charged particles, $< N_{ch} >/< N_{part}/2 >$, as a function of centrality was measured by PHOBOS Collaboration at RHIC energies in Au+Au collsions at $\sqrt{s_{NN}} = 19.6$, 130 and 200 GeV[2]. Within the error bars the $< N_{ch} >/< N_{part}/2 >$ does not show any centrality dependence starting from N_{part} of about 100. In the same paper it was shown that above $\sqrt{s} \sim 20$ GeV, for central events, the total average multiplicity per participating nucleon pairs as a funtion of $\sqrt{s_{NN}}$ is the same as in e^+e^- data [3, 4]. Recently ALICE Collaboration measured the average total multiplicity as a function of centrality in Pb+Pb collsions at 2.76 TeV [5, 6, 7]. On the same centrality range, i.e. 125 - 375 N_{part}, the normalized average total charged particle multiplicity, $< N_{ch} >/< N_{part}/2 >$, shows a small increase, i.e about $\sim 20\%$, as a fuction of centrality [6].

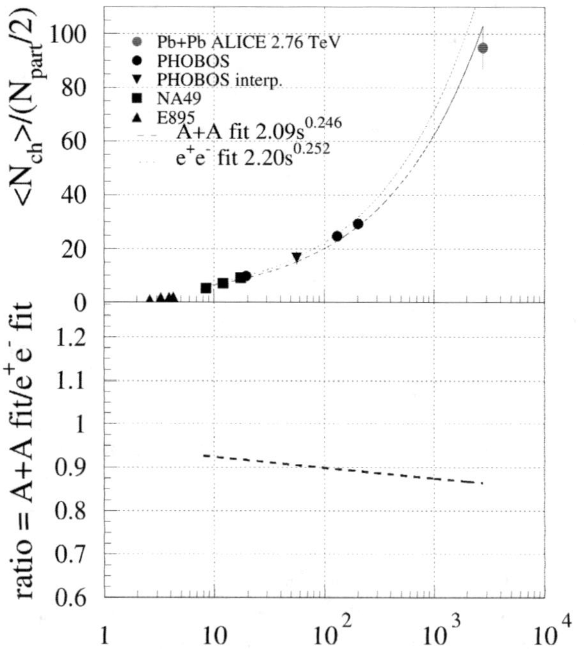

Figure 1. Upper plot - The average total multiplicity of primary charged particles normalized to the participating nucleon pairs, $< N_{ch} >/< N_{part}/2 >$. Dotted line - $< N_{ch} > = \mathrm{a} \cdot s^b$ with a and b parameters found by fitting the charged particle multiplicity in e^-e^+ annihilation. Dashed line -the results of the fit of normalized average total multiplicity of primary charged particles in A+A collisions. Bottom plot -The ratio between the results of the fits for A + A relative to e^+e^+ data

Journal of Physics: Conference Series **413** (2013) 012022 doi:10.1088/1742-6596/413/1/012022

In Fig.1 is represented the average total multiplicity of primary charged particles normalized to the participating nucleon pairs, $< N_{ch} >/< N_{part}/2 >$, for Au + Au at AGS ($\sqrt{s_{NN}}$ = 2.6-4.3) [8], Pb+Pb at SPS ($\sqrt{s_{NN}}$ = 8.6, 12.2, 17.3 GeV) [9], Au+Au at RHIC ($\sqrt{s_{NN}}$ = 19.6, 56, 130, 200 GeV) [2] and Pb+Pb at $\sqrt{s_{NN}}$ = 2.76 TeV [5, 6, 7] as a function of $\sqrt{s_{NN}}$. The dotted line corresponds to the parametrization $< N_{ch} > = a \cdot s^b$ with a=2.20±0.03 and b=0.252±0.002 found by fitting the charged particle multiplicity in e^-e^+ annihilation up to 91 GeV [4] . As could be seen, the fit follows rather well the normalized averaged total multiplicity in A+A collisions above 20 GeV, including RHIC data, as it was shown already in Ref. [2]. It has to be mentioned that the authors of Ref. [2] used a parametrization suggested by leading logarithm QCD [10]. In Ref [4] three parametrizations were used, i.e. $< N_{ch} > = a \cdot s^b$ inspired by the fireball and hydrodinamical models for hadron-hadron interactions [11], empirical expression used in the analysis of p + p data [12] $< N_{ch} > = a + b \cdot s + c \cdot lns^2$ and $< N_{ch} > = a + b \cdot exp[c\sqrt{ln(s/Q_0^2)}]$ [10] and it was shown that they give similar values for χ^2/d.o.f. if the fits do not include energies lower than 5 GeV. This was the main argument in deciding to use in the present analysis the first expression. Extrapolated to the highest energy, the fit of charged particle multiplicity in e^-e^+ annihilation is slightly above the ALICE data measured at 2.76 TeV.

The result of the fit of normalized average total multiplicity of primary charged particles, $< N_{ch} >/< N_{part}/2 >$, in A+A collisions above 20 GeV, including the recent result of ALICE, using the same parametrization, is represented in Fig.1 by the dashed line. The parameters of the above parametrization have the following values: a=2.096±0.161 and b=0.246±0.008. The ratio between the results of the fits using the same parametrization for A + A relative to e^+e^+ can be followed in the bottom of Fig.1. In the average a difference of about 10% could be observed with a slight decrease of the ratio from 0.925 at 20 GeV to 0.865 at 2.76 TeV.

3. Energy scaling of the pseudo-rapidity charged particle density in central rapidity as a function of centrality

One of the observables which are sensitive to the particle production mechanism and partonic structure of the colliding objects is the charged particle density and its dependence on the collision geometry. Detailed information on centrality dependence of charged particle density at midrapidity in A + A collisions exists starting from AGS [13, 14], at SPS [15, 16] and RHIC [11], [17]-[23] energies. The systematic compilation of these results [23] and the recent data from ALICE Collaboration at 2.76 TeV [24] are represented in Fig.2.

At SPS and AGS energies, within the systematic errors of the measurements, the normalized charged particle density $dN_{ch}/d\eta/(0.5N_{part})$ is rather constant as a function of centrality, i.e. N_{part}, estimated based on the Glauber model [25]-[27]. Although the shape of $dN_{ch}/d\eta/(0.5N_{part})$ dependence on N_{part} is quite the same at 130 and 200 GeV, a slight increase of charged particle density from peripheral towards central collision can be observed. This trend is rather enhanced going to the LHC energy of 2.76 TeV where the data published by ALICE [24], ATLAS [28] and CMS [29] are in a remarkable agreement. The data points for 2.76 TeV, represented by full dots in Fig.2, are taken from Ref. [24].

As it is expected and confirmed by the Monte Carlo Glauber approach, in very peripheral collisions where the two diffuseness zones of the colliding nuclei overlap, essentially binary collisions take place, therefore the normalized charged particle density should have the same value as the charged particle density in nucleon + nucleon inelastic collisions. Charged particle density in p + p inelastic collisions was mesured at ISR [30, 31], RHIC [32] and LHC energies [33, 34] and the results are presented in Fig.3 [34].

Based on these data one could obtain scaling factors $(dN_{ch}/d\eta)^{pp}_{2.76TeV}/(dN_{ch}/d\eta)^{pp}_{0.2TeV}$ and $(dN_{ch}/d\eta)^{pp}_{2.76TeV}/(dN_{ch}/d\eta)^{pp}_{0.0196TeV}$. The charged particle pseudorapidity density at 19.6 GeV (dark stars) and at 200 GeV (dark full squares) from Fig.2 scaled by the corresponding above

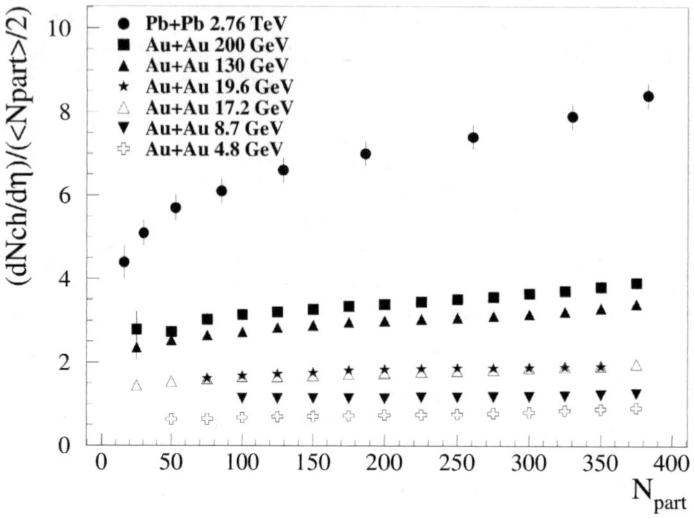

Figure 2. Charged particle density at midrapidity in A + A collisions at AGS [13, 14], SPS [15, 16] and RHIC [11], [17]-[23] energies

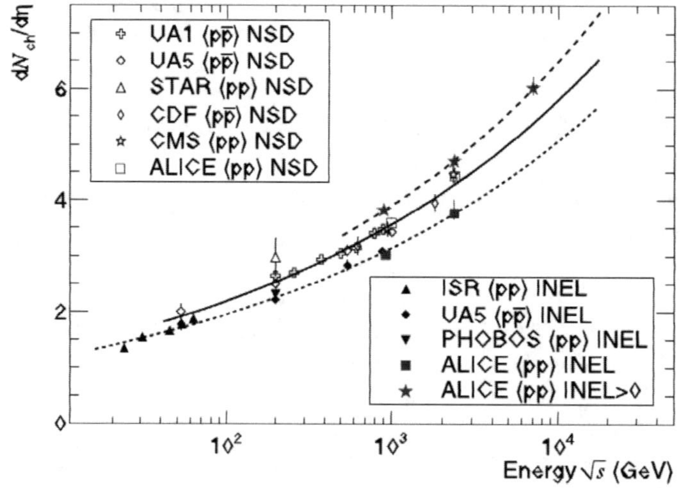

Figure 3. Charged-particle pseudorapidity density in the central pseudorapidity region —η—¡0.5 for inelastic (full black simbols and red squares) and non-single-diffractive collisions (open symbols). Red open stars represent the results for innelastic collisions in —η—¡1 defined as collisions in which at least one charged and particle in —η— ¡ 1 was produced. The results of power low fits are indicated by dotted, continuous and dashed lines respectivly [34]

ratios are represented in Fig.4 by blue stars and full blue squares, respectively. As it should be, based on the argument above, for very peripheral collisions, the charged particle pseudorapidity densities at 2.76 TeV and the ones at 19.6 GeV and 200 GeV multiplied by the corresponding scaling factors are the same within the experimental uncertainties. Towards higher centralities the difference between the 2.76 TeV data and the scaled 200 and 19.6 GeV results is increasing.

In Fig.5 are presented the average number of collisions suffered by a wounded nucleon $< N_{coll} >$ in Au+Au collisions at 19.6 and 200 GeV and in Pb+Pb at 2.76 TeV as a function

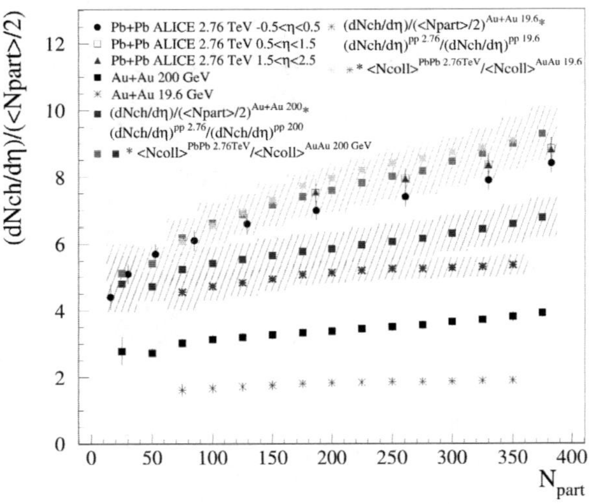

Figure 4. The charged particle pseudorapidity density at 19.6 GeV (dark stars), 200 GeV (dark full squares) and 2.76 TeV (black dots $|\eta|<0.5$ [24], open black squares $0.5|\eta|<1.5$, blue full triangles $1.5|\eta|<2.5$ [7]. These values scaled by the ratios of charged particle pseudorapidity density in p+p collision at 2.76 TeV relative to 19.6 GeV and 200 GeV are represented by blue stars and full blue squares, respectively. The results of a further scaling of 19.6 and 200 GeV data by the ratio of the average number of collisions suffered by the wounded nucleons at 2.76 TeV and the corresponding values at 19.6 and 200 GeV as a function of N_{part} are represented by full green and red squares, respectively. Hatched zones correspond to the experimental uncertainties.

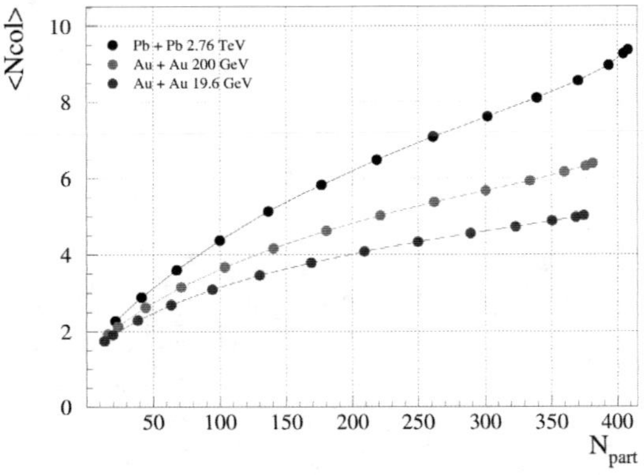

Figure 5. The average number of collisions suffered by a wounded nucleon in Au+Au collisions at 19.6 and 200 GeV and in Pb+Pb at 2.76 TeV as a function of N_{part}, blue, red and black full dots, respectively. The inelastic nucleon-nucleon cross section used in Glauber Monte Carlo calculation were taken from [35], see Fig.6

of N_{part}, blue, red and black full dots, respectively, estimated using the Glauber Monte Carlo approach. The inelastic nucleon-nucleon cross section used in Glauber calculation were taken from [35], see Fig.6. Based on this one could find the following scaling factors: $< N_{coll}^{2.76TeV} > / < N_{coll}^{0.2TeV} >$ and $< N_{coll}^{2.76TeV} > / < N_{coll}^{0.0196TeV} >$ as a function of N_{part}. If one multiplies by these factors the charged particle pseudorapidity densities at 19.6 and 200 GeV normalized in highly peripheral collisions as it was described above, the results represented by green and red full squares in Fig.4 are obtained. The hatched areas correspond to the

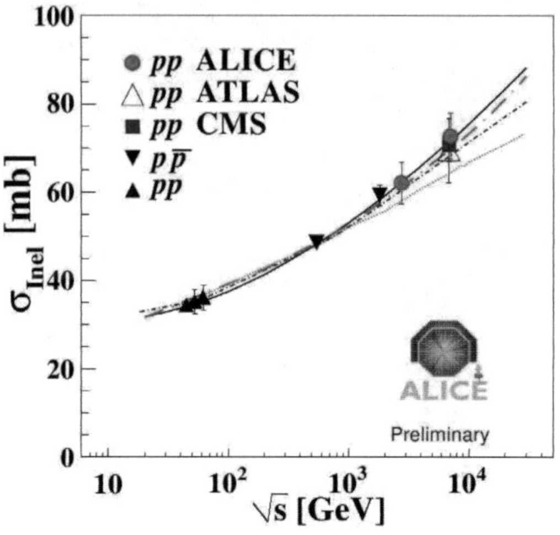

Figure 6. Inelastic cross section as a function of collision energy [35]

experimental uncertainties. A rather good agreement between the scaled centrality dependence of normalized charged particle pseudorapidity density at 19.6 and 200 GeV and recent results obtained at 2.76 TeV could be observed. The scaling is based on the ratio of charged particle pseudorapidity density in inelastic p+p collisions and the average number of collisions suffered by wounded nucleons at a given centrality at the corresponding energies.

4. Conclusions

The extrapolation of the fits based on normalized averaged total charged particles produced in $e^+ + e^-$ collisions up to $\sqrt{s}$ ~90 GeV at LHC energy overestimate the experimental results obtained in Pb+Pb central collisions at 2.76 TeV. The ratio between the results of the fits using the same parametrization for A + A relative to e^+e^+ shows an average difference of about 10%, a slight decrease of the ratio from 0.925 at $\sqrt{s_{NN}}$ ~20 GeV to 0.865 at $\sqrt{s_{NN}}$ ~2.76 TeV being observed.

The normalized charged particle pseudorapidity density as a function of centrality manifests a change from a constant behaviour at SPS and AGS energies through a slight increase with the centrality at RHIC energies to a significant centrality dependence at 2.76 LHC energy. An energy scaling of the centrality dependence of mid-rapidity charged particles pseudorapidity density, based on the ratio of charged particle pseudorapidity density in inelastic p+p collisions and the average number of collisions suffered by wounded nucleons at a given centrality at the corresponding energies is proposed and shown to be valid within the experimental uncertainties.

References

[1] Petrovici M 2012 *Relativistic Heavy Ion Collisions - Expectations, Experimental Facts and Future Plans http://www.theory.nipne.ro/predeal12/pdf/MPetrovici-Predeal12.pdf*

[2] Back B B et al PHOBOS Colaboration 2003 *Comparison of teh Total Charged-Particle Multiplicity in High-Energy Ion Collisions with* e^+e^- *and pp/p*$\bar{p}$ *Data arXiv* **2003** 0301017

[3] Basile M et al 1981 *Charged-Particle Multiplicities in pp Interactions and Comparison with* e^+e^- *Data Il Nuovo Cimento* **65** 400 and references therein

[4] Zheng Hong 1990 PhD Thesis *Charged Particle Multiplicities in Electron-Positron Annihilation at Center-of-Mass Energies from 50 to 61.4 GeV Department of Physics and Astronomy, University of Rochester, New York*

[5] Toia A et al ALICE Collaboration 2011 *Bulk Properties of Pb-Pb collisions at* $\sqrt{s_{NN}}$=2.76 *TeV measured by ALICE arXiv:nucl-ex* **1107.1973**

[6] Dalsgaard H H 2012 PhD Thesis *Pseudorapidity Densities in p+p and Pb+Pb collisions at LHC measured with the ALICE experiment Niels Bohr Institute*

[7] M. Guilbaud et al ALICE Collaboration 2012 *Charged-particle multiplicity in Pb-Pb at 2.76 TeV with the displaced vertex technique with ALICE at LHC Quark Matter*

[8] Klay J 2001 Thesis *Transverse Mass and rapidity spectra of protons and pions from Au+Au collisions at the AGS Universty of California* Davis CA

[9] Afanasiev S V et al 2002 *Energy dependence of pion and kaon production in central Pb+Pb collisions Phys. Rev. C* **66** 054902

[10] Mueller A H 1983 *Multiplicity and Hadron Distribution in QCD Jets Nucl. Phys. B* **213** 85

[11] Satz H 1975 *On the Phenomenology of Multihadron Production in* e^+e^- *Annihilation Lecture Notes in Physics* **56** 49

[12] Thome W et al 1977 *Charged Particle Multiplicity Distributions in p+p Collisions at ISR Energies Nucl. Phys. B* **129** 365

[13] Ahle L et al E802 Collaboration 1999 *Simultaneous multiplicity and forward energy characterization of particle spectra in Au+Au collisions at 11.6A GeV/c Phys. Rev. C* **59** 2173

[14] Ogilvie C A et al E866 and E917 Collaborations 1998 *Au+Au reactions at the AGS: Experiments E866 and E917 Nucl. Phys. A* **638** 57c

[15] Miskowiec D CERES Collaboration 2002 *Recent results from CERES Proceedings of the* 30th *International Workshop on Gross Properties of Nuclei and Nuclear Excitations: Ultrarelativistic heavy-ion collisions, Hirschegg, Austria* 238

[16] Ceretto F 1998 PhD thesis, *Charged hadron production in Pb-Au collisions at 158A GeV/c University of Heidelberg*

[17] Bearden I G et al BRAHMS Collaboration 2001 *Charged particle densities from Au+Au collisions at sNN=130 GeV Phys. Lett. B* **523** 227

[18] Bearden I G et al BRAHMS Collaboration 2002 *Pseudorapidity Distributions of Charged Particles from Au+Au Collisions at the Maximum RHIC Energy, sNN = 200GeV Phys. Rev. Lett.* **88** 202301

[19] Back B B et al PHOBOS Collaboration 2000 *Charged-Particle Multiplicity near Midrapidity in Central Au+Au Collisions at sNN = 56 and 130GeV Phys. Rev. Lett.* **85** 3100

[20] Back B B et al PHOBOS Collaboration 2002 *Centrality dependence of the charged particle multiplicity near midrapidity in Au+Au collisions at sNN=130 and 200 GeV Phys. Rev. C* **65** 061901(R)

[21] Baker M D et al PHOBOS Collaboration 2003 *Global observations from PHOBOS Nucl.Phys.A* **715** 65c

[22] Ullrich S et al STAR Collaboration 2003 *Experimental summary on global observables, hadron spectra and ratios Nucl. Phys. A* **715** 399c

[23] Adler S S et al PHENIX Collaboration 2005 *Systematic studies of the centrality and* $\sqrt{s_{NN}}$ *dependence of the* $dE_T/d\eta$ *and* $dN_{ch}/d\eta$ *in heavy ion collisions at midrapidity Phys. Rev. C* **71** 034908

[24] Aamodt K et al ALICE Collaboration 2011 *Centrality Dependence of the Charged-Particle Multiplicity Density at Midrapidity in Pb-Pb Collisions at* $\sqrt{s_{NN}} = 2.76$ *TeV Phys. Rev. Lett.* **106** 032301

[25] Glauber R J 1955 *Cross Sections in Deuterium at High Energies Phys. Rev.* **100** 242

[26] Franco V and Glauber R J 1966 *High-Energy Deuteron Cross Sections Phys. Rev.* **142** 1195

[27] Miller M I et al 2007 *Glauber Modeling in High Energy Nuclear Collisions Glauber Modeling in High Energy Nuclear Collisions arXiv:nucl-ex* 0701025

[28] Chatrchyan S et al CMS Collaboration 2011 *Dependence on pseudorapidity and on centrality of charged hadron production in PbPb collisions at* $\sqrt{s_{NN}} = 2.76$ *TeV Journal of High Energy Physics* **08** 141

[29] Aad G et al ATLAS Collaboration 2012 *Measurement of the centrality dependence of the charged particle pseudorapidity distribution in leadlead collisions at View the MathML source with the ATLAS detector Phys. Lett. B* **710** 363

[30] Alner G J et al UA5 Collaboration 1987 *UA5: A general study of proton-antiproton physics at* $\sqrt{s}=546$ *GeV Phys. Rep* **154** 247

[31] Albajar C et al UA1 Collaboration 2012 *A study of the general characteristics of proton-antiproton collisions at* $\sqrt{s}=0.2$ *to 0.9 TeV Phys. Lett. B* **710** 363

[32] Noucier R et al PHOBOS Collaboration 2004 *Pseudorapidity Distributions of charged particles in d+Au and p+p collisions at* $\sqrt{s}=200$ *GeV J.Phys.G* **30** S1133

[33] Khachatryan V et al CMS Collaboration 2010 *Transverse-momentum and pseudorapidity distributions of charged hadrons in pp collisions at* $\sqrt{s}=0.9$ *and 2.36 TeV J. High Energy Phys.* **02** 041

[34] Aamodt K et al ALICE Collaboration 2010 *Charged-particle multiplicity measurement in protonproton collisions at* $\sqrt{s} = 7$ *TeV with ALICE at LHC Eur. Phys. J. C* **68** 345

[35] Poghosyan M G et al ALICE Collaboration 2011 *Diffractive dissociation in proton-proton collisions at* $\sqrt{s}=0.9$ *TeV, 2.76 TeVand 7 TeV with ALICE at the LHC arXiv:hep-ex* 1109.4510

Int. Summer School for Advanced Studies 'Dynamics of open nuclear systems' (PREDEAL12) IOP Publishing
Journal of Physics: Conference Series **413** (2013) 012023 doi:10.1088/1742-6596/413/1/012023

Compact stars and the symmetry energy

Constana Providência[1], Rafael Cavagnoli[2], Debora P. Menezes[2], Prafulla K. Panda[1,3], Aziz Rabhi[1,4]

[1]Centro de Física Computacional, Department of Physics, University of Coimbra, P3004-516 Coimbra, Portugal
[2]Departamento de Física, CFM,Universidade Federal de Santa Catarina, Florianópolis - SC - CP. 476, CEP 88.040 - 900, Brazil
[3]Department of Physics, C.V. Raman College of Engneering, Vidya Nagar, Bhubaneswar-752054, India
[4]Laboratoire de Physique de la Matière Condensée, Faculté des Sciences de Tunis, Campus Universitaire, Le Belvédre-1060, Tunisia

E-mail: cp@fis.uc.pt

Abstract.
The effect of the symmetry energy on some properties of compact stars which contain strange degrees of freedom is discussed. Both the onset of hyperons or kaon condensation will be considered. The hyperon-meson couplings are chosen according to experimental values of the hyperon nuclear matter potentials and possible uncertainties are considered. It is shown that a softer symmetry energy affects the onset of strangeness, namely neutral (negatively charged) strange particles set on at larger (smaller) densities, and gives rise to a smaller strangeness fraction as a function of density. A softer symmetry energy will possibily give rise to maximum mass configurations with larger masses. Hyperon-meson couplings have a strong effect on the mass of the star. It is shown that, for stars with masses above 1 $M_\odot$, the radius of the star varies linearly with the symmetry energy slope L.

1. Introduction

The mesurement with high precision of the mass of the millisecond pulsar PSR J1614-2230, 1.97 ± 0.04 $M_\odot$ [1], puts strong constraints on the appearance of exotic degrees of freedom at high densities inside a compact star, namely strange degrees of freedom. In particular, calculations using microscopic approachs predict too small maximum masses if hyperons are included in the equation of state (EoS)[2], and not even the inclusion of hyperonic three-body forces seem to solve the problem [3].

The fraction of strangeness in a compact star is strongly dependent on the EoS that describes the hadronic matter: if the nucleonic EoS is very stiff the onset of strangeness occurs at smaller densities and the fraction of strangeness is larger, leading as a result to a softer EoS. Therefore, the fraction of the strangeness in a star is intrinsically related to some properties of the nucleonic EoS, namely its incompressibility at large densities, its density dependence of the symmetry energy and its effective mass.

In the present work we want to understand how sensitive are some properties of compact stars to the symmetry energy and its slope at saturation. We will discuss the strangeness content, the mass and radius of the maximum mass configuration and the onset of the direct

Urca process. The structure of the pasta phase and the crust-core transition are also influenced by the symmetry energy but this topic will not be discussed in the present contribution [4], [5].

Stars with a mass in the range $1.0\,M_\odot < M < 1.4\,M_\odot$ have a central density that goes from 1.5 ρ_0 to 2-3 ρ_0, and, therefore, it is of particular interest the study of these stars because we will be testing the equation of state at suprasaturation densities which may be attained in the near future at FAIR.

The work is performed in the relativistic mean field approximation and we consider different parametrizations of the non-linear Walecka model (NLWM) [6], which allow us to discuss the role of the density dependence of the symmetry energy on the star properties. We also present results obtained with the quark-meson-coupling (QMC) model [7], an effective nuclear model that takes into account the internal structure of the nucleon explicitly. Within the QMC model, matter at low densities and temperatures is a system of nucleons interacting through meson fields, with quarks and gluons confined within MIT bags. Finally, we discuss the possible onset of a kaon condensate.

2. Formalism

We describe hadronic matter within the framework of NLWM [6]. The nuclear interaction is described by the interchange of the σ, ω and ρ mesons. The Lagrangian density includes several non-linear terms in order to described adequately the saturation properties of nuclear matter.

In order to study the effect of the density dependence of the symmetry energy, a $\rho-\omega$ meson coupling term is introduced as in [8, 9]. The Lagrangian density reads

$$\begin{aligned}
\mathcal{L} &= \sum_{j=1}^{8} \bar{\psi}_j \Big[\gamma_\mu \left(i\partial^\mu - g_{\omega j}\,\omega^\mu - g_{\rho j}\,\vec{\tau}_j \,.\, \vec{\rho}^{\mu}\right) - m_j^* \Big] \psi_j + \sum_{l=1}^{2} \bar{\psi}_l \left(i\gamma_\mu \partial^\mu - M_l\right) \psi_l \\
&+ \frac{1}{2}\partial_\mu\sigma\partial^\mu\sigma - \frac{1}{2}m_\sigma^2\sigma^2 - \frac{1}{3!}k\sigma^3 - \frac{1}{4!}\lambda\sigma^4 - \frac{1}{4}\Omega_{\mu\nu}\,\Omega^{\mu\nu} + \frac{1}{2}m_\omega^2\,\omega_\mu\omega^\mu + \frac{1}{4!}\xi g_\omega^4(\omega_\mu\omega^\mu)^2 \\
&- \frac{1}{4}\vec{R}_{\mu\nu}\,.\,\vec{R}^{\mu\nu} + \frac{1}{2}m_\rho^2\,\vec{\rho}_\mu\,.\,\vec{\rho}^{\mu} + \Lambda_{\rm v}(g_\rho^2\,\vec{\rho}_\mu\,.\,\vec{\rho}^{\mu})(g_\omega^2\,\omega_\mu\omega^\mu)\,, \qquad (1)
\end{aligned}$$

where $m_j^* = m - g_{\sigma,j}\sigma$ is the effective mass of baryon j, $\Omega_{\mu\nu} = \partial_\mu\omega_\nu - \partial_\nu\omega_\mu$, $\vec{R}_{\mu\nu} = \partial_\mu\vec{\rho}_\nu - \partial_\nu\vec{\rho}_\mu - g_\rho\,(\vec{\rho}_\mu \times \vec{\rho}_\nu)$, g_{ij} are the coupling constants of mesons $i = \sigma, \omega, \rho$ with baryon j, m_i is the mass of meson i and l stands for the leptons e^- and μ^-. The couplings k, λ and ξ define the contribution of the non-linear σ and ω terms and $\vec{\tau}$ is the isospin operator.

The sum over j in (1) extends over the octet of lightest baryons $\{n, p, \Lambda, \Sigma^-, \Sigma^0, \Sigma^+, \Xi^-, \Xi^0\}$. In dense matter, due to the Pauli principle the nucleon Fermi energy increases and, if the Fermi energy of nucleons becomes larger than the hyperon masses, energy and pressure are lowered by conversion of some nucleons to hyperons. This softens the equation of state and has some direct consequences on the properties of compact stars as exemplified in Fig. 1: maximum star masses become smaller and neutrino fractions in neutrino trapped matter are larger.

Within the NLWM, we consider the following parametrizations, see Fig. 2: NL3 [10] and TM1 [11] both with a quite large symmetry energy and incompressibility at saturation and fitted in order to reproduce the ground state properties of both stable and unstable nuclei, TM1 has a quartic term on the ω-meson in order to reproduce Dirac-Brueckner results at high densities; FSU [8] and IUFSU [9], which were accurately calibrated to simultaneously describe the GMR in ^{90}Zr and ^{208}Pb, and the IVGDR in ^{208}Pb and still reproduce ground-state observables of stable and unstable nuclei. FSU is very soft at high densities. GM1 and GM3 [12] are generally used to describe stellar matter, and are softer than NL3, and NLρ [13], which has at high densities, a behavior between GM1 and GM3.

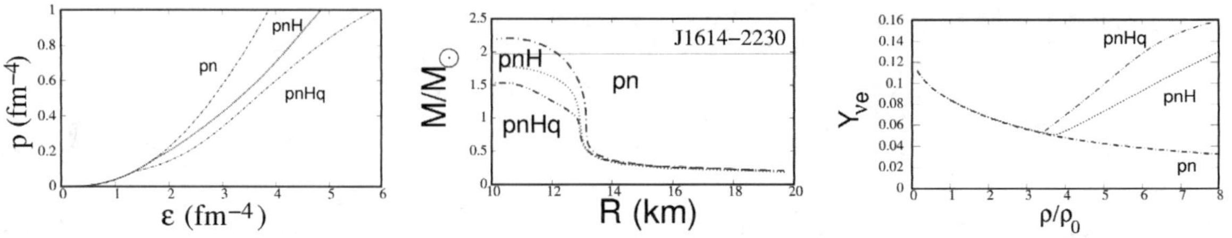

Figure 1. Left panel: Equation of state for stellar matter including only nucleonic degrees of freedom (pn), nucleonic and hyperonic degrees of freedom (pnH), nucleons, hyperons and quark degrees of freedom (pnHq); middle panel: the mass/radius curves of the corresponding star families; right panel: the effect of the presence of exotic degrees of freedom such as hyperons and quarks on the neutrino fraction of neutrino trapped stellar matter.

In general, the energy per particle of asymmetric nuclear matter characterized by the asymmetry parameter $\alpha = (\rho_n - \rho_p)/\rho$, where ρ_p, ρ_n and ρ are respectively the proton, the neutron and the total densities, may be expanded in powers of α

$$\frac{E}{A}(\rho,\alpha) = \epsilon_{SNM}(\rho) + \epsilon_{sym}(\rho)\alpha^2 + \mathcal{O}(4)\,,$$

where ϵ_{SNM} is the energy per particle of symmetric nuclear matter and ϵ_{sym} is the symmetry energy

$$\epsilon_{sym}(\rho) = \frac{1}{2}\frac{\partial^2 E/A}{\partial\alpha^2}\Big|_{\alpha=0} = E_{sym} + L\,x + \frac{K_{sym}}{2}\,x^2 + \frac{Q_{sym}}{6}\,x^3 + \mathcal{O}(4) \quad x = \frac{\rho-\rho_0}{3\rho_0}. \tag{2}$$

In the last expression, an expansion of ϵ_{sym} around the saturation density ρ_0 was done. E_{sym}, the symmetry energy at saturation, is constrained by experiments to be within $31 \le E_{sym} \le 35$ MeV [14]. The quantities L, K_{sym}, and Q_{sym} are related to its slope, curvature, and third derivative at saturation [4].

In the QMC model, the nucleon in nuclear medium is assumed to be a static spherical MIT bag in which quarks interact with the scalar (σ) and vector (ω, ρ) fields, and those are treated as classical fields in the mean field approximation [7, 15]. The quark field, ψ_{q_i}, inside the bag then satisfies the equation of motion:

$$\left[i\,\partial\!\!\!/ - (m_q^0 - g_\sigma^q) - g_\omega^q\,\omega\,\gamma^0 + \frac{1}{2}g_\rho^q\tau_z\rho_{03}\gamma^0\right]\,\psi_{q_i}(x) = 0\,, \quad q = u, d, s \tag{3}$$

where m_q^0 is the current quark mass, and g_σ^q, g_ω^q and g_ρ^q denote the quark-meson coupling constants. The effective mass of a nucleon bag at rest is taken to be equal to the energy of the bag, $M_i^* = E_i^{\rm bag}$. The equilibrium condition for the bag is obtained by minimizing the effective mass, M_i^* of baryon i with respect to the bag radius, see [7, 15]. The energy density and pressure of hadronic matter within QMC is equal to the one obtained for NLWM with the effective mass replaced by the energy of the bag, $E_i^{\rm bag}$, and excluding the non-linear σ and ω terms.

In order to study the effect of the isovector channel in the star properties we also consider a modified version of the IUFSU, TM1 and QMC parametrizations: we change g_ρ and Λ_v keeping the symmetry energy fixed at the density 0.1 fm^{-3} while the isoscalar channel remains unchanged. We generate a set of models that differ in their value of the symmetry energy and corresponding slope at saturation but have the same isoscalar properties.

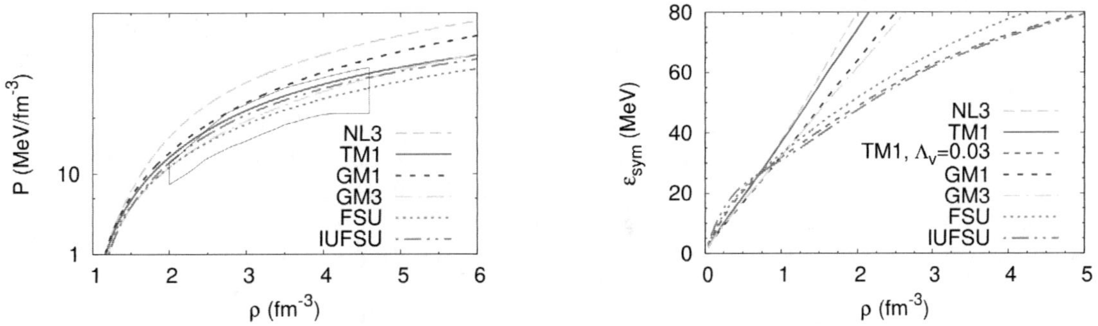

Figure 2. Left panel: Equation of state for symmetric nuclear matter and different NLWM models. The enclosed area represents experimental data according to Danielewicz *et al.*, [16]. Right panel: the corresponding symmetry energy

As extremes we take the parametrization IUFSU, with $L = 47$ MeV, and $L = 99$ MeV both within the experimental values obtained from isospin diffusion in heavy ion reactions [17]. Values of L below 47 MeV would give unacceptable EOS because they would predict that neutron matter is bound. In a second study we consider the TM1 parametrization with $L = 110$ MeV and a modified version of TM1 with $L = 55$ MeV corresponding to $\Lambda_v = 0.03$. For QMC we take values of Λ_v between 0 and 0.1 corresponding to $L = 95.6 - 39.$ MeV.

We consider different sets of hyperon-meson couplings. Within the coupling set A [18] the ω and ρ meson-hyperon coupling constants are obtained using the SU(6) symmetry: $\frac{1}{2}g_{\omega\Lambda} = \frac{1}{2}g_{\omega\Sigma} = g_{\omega\Xi} = \frac{1}{3}g_{\omega N}$, $\frac{1}{2}g_{\rho\Sigma} = g_{\rho\Xi} = g_{\rho N}$, $g_{\rho\Lambda} = 0$, where N means 'nucleon' ($g_{iN} \equiv g_i$). The coupling constants $\{g_{\sigma j}\}_{j=\Lambda,\Sigma,\Xi}$ of the hyperons with the scalar meson σ are constrained by the hypernuclear potentials in nuclear matter to be consistent with hypernuclear data [19]. We take $V_j = x_{\omega j} V_\omega - x_{\sigma j} V_\sigma$ where $x_{ij} \equiv g_{ij}/g_i$, $V_\omega \equiv g_\omega \omega_0$ and $V_\sigma \equiv g_\sigma \sigma_0$ are the nuclear potentials for symmetric nuclear matter at saturation with (see Ref. [19]), $V_\Lambda = -28$ MeV, $V_\Sigma = 30$ MeV , $V_\Xi = -18$ MeV. However, while the binding of the Λ in symmetric nuclear matter is well settled experimentally, the binding values of the Σ^- and Ξ^- still have a lot of uncertainties [20]. We, therefore, test also the effect of the cascade-meson couplings by taking $V_\Xi = -10$, 0, +18 MeV.

In QMC the couplings of the hyperons to the σ-meson do not need to be fixed because the effective masses of the hyperons are determined self-consistently at the bag level. We obtain $x_{\omega B}$ from the hyperon potentials in nuclear matter, $V_j = -(M_j^* - M_j) + x_{\omega j} g_\omega \omega_0$, and find $x_{\omega\Lambda} = 0.743$, $x_{\omega\Sigma} = 1.04$ and $x_{\omega\Xi} = 0.346$. $x_{\rho B} = 1$ for all the baryons [21].

In order to show how results are sensitive to the hyperon couplings we consider one of the coupling sets proposed in [12], set B, with x_σ=0.8 and equal for all the hyperons. The fraction x_ω is determined using $V_j = V_\Lambda = -28$ MeV for all the hyperons. For the hyperon-ρ-meson coupling we consider $x_\rho = x_\sigma$. This choice considers that the interaction of all hyperons in symmetric nuclear matter is attractive, and is restricted by acceptable maximum mass star configurations.

Strangeness may also occur in the interior of compact stars through kaon condensation. Kaplan and Nelson [22] suggested that the interaction of the K^- with the nuclear medium reduces its mass within chiral perturbation theory favoring a boson condensation in a zero momentum state which would replace electrons as the neutralizing agent in charge neutral matter. The existence of a kaon condensate has strong implications on the star properties, namely stronger neutrino fluxes or a late low mass black-hole formation are expected. Kaons

Journal of Physics: Conference Series **413** (2013) 012023 doi:10.1088/1742-6596/413/1/012023

may be included in a RMF theory through a kaon effective lagrangian density [23]

$$\mathcal{L}_K = \mathcal{D}^*_\mu K^* \mathcal{D}^\mu K - m^*_K K^* K, \tag{4}$$

with $\mathcal{D}_\mu = \partial_\mu + i g_{\omega K}\omega_\mu + \frac{i}{2} g_{\rho K}\ \vec{\tau} \cdot \mathbf{b}_\mu$, and $m^*_K = m_K - g_{\sigma K}\sigma$. The dispersion relation $\omega_{K^-} = m^*_K - g_{\omega K}\omega_0 - \frac{1}{2} g_{\rho K} b_{03}$ is obtained replacing $K = f_K e^{iEt}$ in the kaon equation of motion. The parameters of the kaon lagrangian density are the vacuum mass 497 MeV, the vector mesons coupling fixed using a simple quark model and isopsin counting rule $g_{\omega K} = \frac{1}{3} g_\omega$, $g_{\rho K} = g_\rho$, and the scalar coupling calculated from the optical potential of the kaon in symmetric nuclear matter $V_K = -g_{\sigma K}\sigma - g_{\omega K}\omega_0$. According to kaonic atom data $V_K(\rho_0) = -(50 - 200) MeV$ [20]. In the following we consider $V_K(\rho_0) = -125$ MeV, a value suggested by chiral models [24]. Charge neutrality in matter with a kaon condensate requires $\rho_p = \rho_e + \rho_\mu + \rho_K$, while β-equilibrium imposes $\omega_K = \mu_K = \mu_e$ and defines the kaon condensate amplitude. The chemical potential μ_e increases with density while the effective kaon mass m^*_K decreases. Eventually $\mu_K = \mu_e$, kaons become energetically favorable and condense. If a further increase of the density occurs, m^*_K decreases, thus μ_K and ρ_e decrease: since all kaons can condense in the lowest energy state (s wave), they become energetically more favorable than electrons as the neutralizing agent of positive charge. At finite temperature, thermal kaons should also be included.

3. Results

The stellar matter EoS is obtained after solving the meson field equations together with the chemical equilibrium and charge neutrality conditions. Chemical equilibrium is expressed in terms of the relations $\mu_B = \mu_n - q_B(\mu_e - \mu_{\nu_e})$ and $\mu_e - \mu_{\nu_e} = \mu_\mu - \mu_{\nu_\mu}$ where μ_i is the chemical potencial of baryon (B), neutron (n), electron (e), muon (μ) and neutrino (ν), q_B is the electric charge of baryon B, and the equilibrium is attained through the weak interaction. In neutrino free matter $\mu_{\nu_e} = \mu_{\nu_\mu} = 0$, and if trapped neutrinos are included we fix the fraction of electron leptons $Y_{Le} = (\rho_e + \rho_{\nu_e})/\rho$ and take the fraction of muon leptons equal to zero. The maximum value of Y_{Le} is 0.3–0.4. Charge neutrality is expressed by $\sum_{baryons} q_B \rho_B + \sum_{leptons} q_l \rho_l = 0$.

The effect of the symmetry energy on the strangeness fraction is seen in Figs. 3: the strangeness content is sensitive to the model and the meson-hyperon couplings. In general, the softer the nucleonic EoS the larger the strangeness onset density if the Λ is the first hyperon to set on, but the contrary will occur for the Σ^-, and the smaller the strangeness content. A large meson-hyperon vector coupling, as occurs in set B, hinders the formation of hyperons. However, we

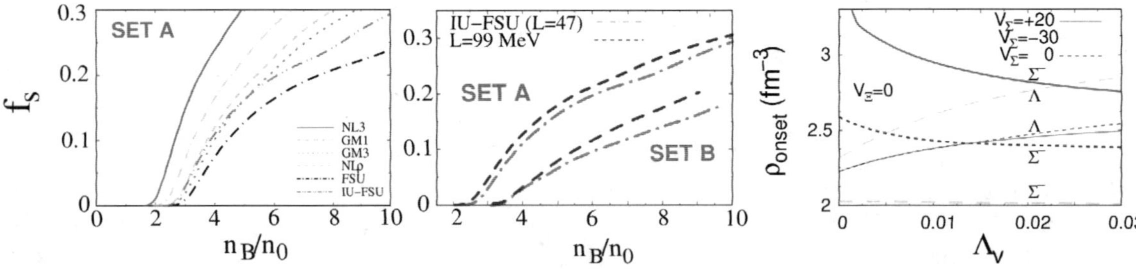

Figure 3. Strangeness fraction for different NLWM models with hyperon set A (left panel), and for the modified IUFSU models with $L = 47$ and 99 MeV, and with set A and set B for the hyperon-meson couplings (middle panel). Right panel: Onset of the Σ^- and the Λ hyperons as a function of the parameter Λ_v and with $V_\Lambda = -28$ MeV and $V_\Xi = 0$ MeV and several values of the Σ^- potential in symmetric nuclear matter at saturation.

should point out that when comparing different models we are comparing not only the effect of the density dependence of the symmetry energy but also the density dependence of the isoscalar channel, namely the incompressibility and effective mass. This explains why FSU, a softer EOS, has a smaller strangeness content than IUFSU, a model with a softer symmetry energy than FSU above saturation density. In the middle panel of Fig. 3 the strangeness content for the modified IUFSU models is shown for $L = 47$ MeV and 99 MeV: a smaller symmetry energy slope hinders the formation of hyperons because it gives rise to a softer EoS. The conditions for the onset of hyperons depend on the charge of the hyperon: Λ is the first hyperon to appear with set A and occurs at larger densities for a smaller slope L, on the contrary, with set B, Σ^- will occur first and at smaller densities for smaller values of L.

The influence of the symmetry energy on the onset density of the hyperons is seen in the right panel of Fig. 3: a smaller L implies a decrease (increase) of the onset density of the Σ^- (Λ). A convenient choice of the magnitude of the hyperon potentials may induce an interchange between the onset of Λ and Σ^- hyperons when the slope L increases. Large values of L (small values of Λ_v) favor the neutral hyperon.

Cooling of the star by neutrino emission can occur relatively fast if the direct Urca process, $n \to p + e^- + \bar{\nu}_e$, is allowed [25]. This occurs when the proton fraction exceeds a critical value $x_{\rm DU}$[25]. Cooling rates of neutron stars seem to indicate that this fast cooling process does not occur and, therefore, a constraint is set imposing that the direct Urca process is only allowed in stars with a mass larger than $1.5\,M_\odot$, or a less restrictive limit, $1.35\,M_\odot$ [26]. Since the onset of the direct Urca process is closely related with the density dependence of the symmetry energy, this constraint gives information on the isovector channel of the EOS.

In figures 4 the onset density of the nucleon direct Urca (DU) process is plotted as a function of the slope L for the IUFSU and modified versions in the left panel , and for NL3, GM1, GM3, NLρ, FSU and IUFSU models in the right panel. We conclude that the symmetry energy and the hyperon content affect the DU onset density according to a) for non-strange matter the larger the slope L the smaller the neutron-proton asymmetry above the saturation density and, therefore, the smaller the DU onset density. In fact, a larger slope L corresponds to a harder symmetry energy and, therefore, larger fractions of protons are favored; b) generally, for a low value of L the presence of hyperons decreases the onset density. This is always true if the first hyperon to occur is negatively charged because the proton fraction increases. However, if Λ is the first hyperon to set on, both the proton and the neutron fractions decrease and it is the net effect that defines the behavior.

Integrating the Tolman-Volkoff-Oppenheimer [27] equations we have obtained the mass-radius curves for the families of stars without hyperons and with hyperons with the meson-hyperon sets A and B. Both sets give similar results [28].

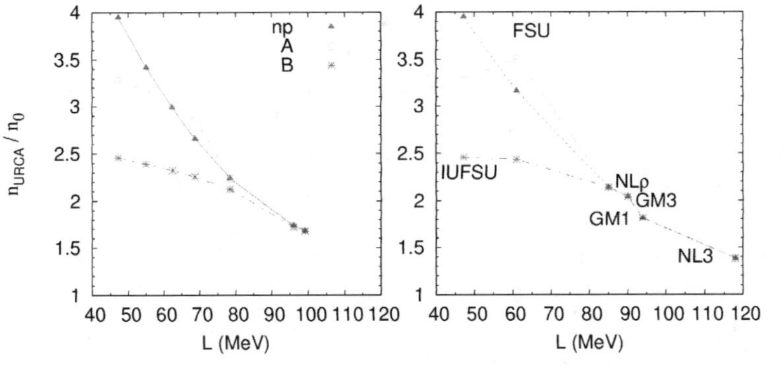

Figure 4. Onset of the direct Urca process in stellar matter for the modified IUFSU models (left panel) and NLWM (right panel), for non-hyperon matter (red triangles), hyperon coupling set A (green squares) and hyperon coupling set B (blue stars).

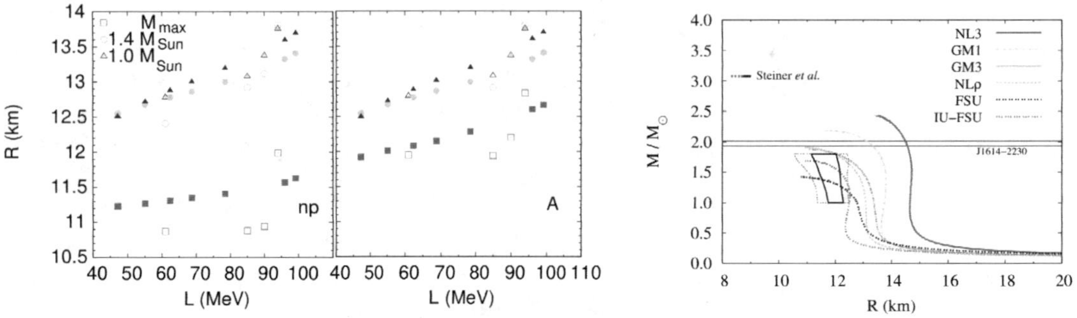

Figure 5. Star radius dependence on the slope of the symmetry energy for stars with maximum mass (squares), 1.4 $M_\odot$ (circles), 1 $M_\odot$ (triangles): non-hyperons (left panel), with hyperons and coupling A (middle panel). The empty symbols are for the set of different models considered. The full symbols are for the modified IUFSU models. Right panel: Mass/radius curves using set B for the hyperon couplings.

All the models except FSU are able to describe the mass of pulsar J1614-2230 [1] if hyperons are not included [28]. The star properties are determined by the underlying relativistic mean field (RMF) models: the harder models like NL3 and GM1 predict larger masses and radius, the softest EOS, FSU, has the smallest mass. It is also seen that the smaller the value of L the smaller the radius [28]. This last property is clearly seen in the left panel of Fig. 5 were the radius of maximum mass stars (squares), 1.4 $M_\odot$ stars (circles) and $1.0M_\odot$ stars (triangles) are plotted as a function of the symmetry energy slope for nucleonic stars. The full symbols are for the modified IUFSU models and the empty ones for the NL3, GM1, GM3, NLρ, FSU models. The modified IUFSU models show that the radius decreases when L. decreases. This reduction is larger for $1.0M_\odot$ stars The RMF models chosen also show the same trend, but, since the isoscalar properties differ among them the linear behavior is not present.

Including hyperons in the EOS makes the EOS softer at large densities and the mass of the maximum mass stars is smaller [29]. As an example, in the right panel of Fig. 5 the mass-radius curves obtained with hyperon-meson coupling set B is plotted. It is clearly seen that it is important to have correct couplings since the star masses are sensitive to the hyperon couplings. With set A only the NL3 model is able to describe the mass of the pulsar PSR J1316-2230 [1], while within set B NL3, GM1, GM3 and NLρ describe this star. The trend discussed above between the star radius and L is still present in these stars. We also conclude that the mass of the maximum mass configuration is quite insensitive to the symmetry energy slope [28], although one might expect that the strangeness content could affect the maximum mass configuration. As we will see, this conclusion depends on the hadronic model considered.

We next discuss the results obtained within the framework of the QMC model. In Fig. 6 we plot the QMC EoS with nucleons only, with nucleons and hyperons [15] and different values of the coupling parameter Λ_v. We also include the empirical EOS obtained by Steiner *et al.* [30]. The agreement of the calculated EOS with the empirical one when hyperons are included is defined by the hyperon-meson interaction and the Λ_v coupling, or, equivalently, by the symmetry energy. The QMC pn EOS satisfies the constraints. However, the inclusion of hyperons with the hyperon couplings obtained for the hyperon nuclear potentials taking $V_\Lambda = -28$ MeV, $V_\Sigma = 30$ MeV and $V_\Xi = -18$ MeV makes the EOS too soft. Increasing Λ_v makes the EOS harder bringing the EOS closer to the constraints defined by the empirical EOS. A larger Λ_v corresponds to a smaller symmetry energy at supersaturation densites, giving rise to a softer pn EOS at high densities and, therefore, hindering the onset of hyperons, and the drastic softening due to the

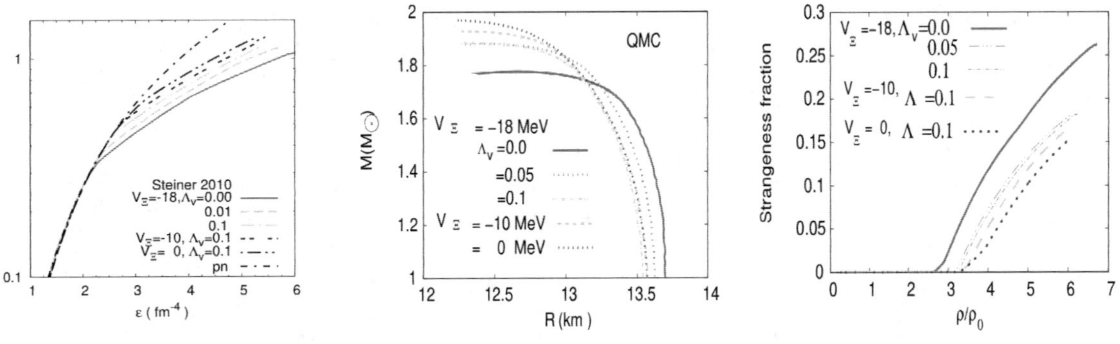

Figure 6. Equation of state for QMC and different values of V_Ξ and of the Λ_v coupling (left panel), corresponding mass/radius curve for the families of compact stars (middle panel), and strangeness content (right panel).

presence of these extra degrees of freedom. The effect of a less attractive V_Ξ potential is also clear: the EoS becomes harder because the onset of hyperons occurs at larger densities and a mechanism that hinders the formation of hyperons makes the EOS harder.

The mass/radius curve for stars with a mass larger than $1\,M_\odot$ are shown in the middle panel of Fig. 6. First let us discuss the effect of the symmetry energy and the hyperon couplings on the mass/radius curve. A larger Λ_v gives rise to a softer EOS and, therefore, a smaller radius. It is seen that when going from $\Lambda_v = 0$ to 0.1 the radius of stars with a mass $M = 1 - 1.5\,M_\odot$ decreases by ~ 0.3 Km. A similar effect was obtained with the NLWM [28, 21]. However, within the NLWM models the maximum mass did not depend so much on L, while in the framework of the QMC model there is a clear effect of almost $0.1 M_\odot$ if Λ_v increases from 0 to 0.1. This is mainly due to the smaller strangeness fraction inside the star as seen in the right panel of Fig. 6. The reduction of the attractiveness of V_Ξ has a similar effect on the maximum mass of the star, i.e., the mass increases $\sim 0.2\,M_\odot$ if V_Ξ increases from -18 to 0 MeV. There is still a large uncertainty on the coupling of hyperons to nuclear matter and, therefore, there is still room for a very massive star such as the pulsar J1614-2230, even including hyperons in the EOS.

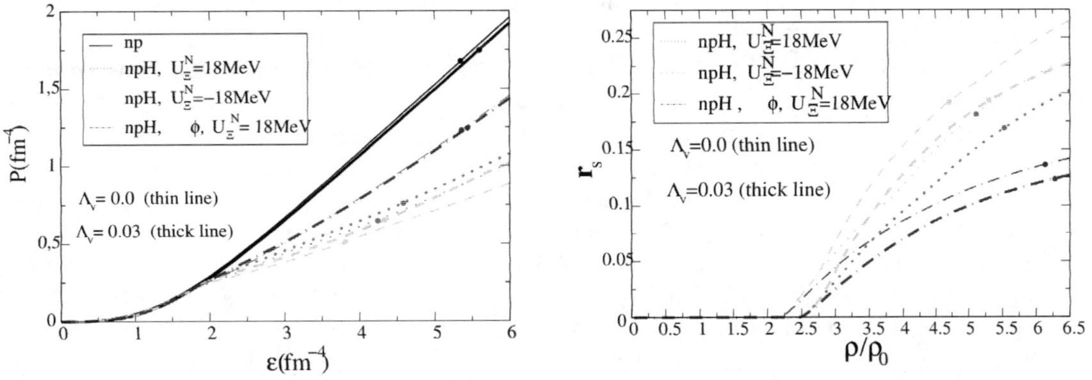

Figure 7. EoS and strangeness fraction for TM1 without and with hyperons (see text for discussion) the dots identify the maximum mass configuration.

In order to better understand the effect of the symmetry energy together with the uncertainty on the hyperon interaction on the strangeness content, the mass and radius of the stars, we consider the the parametrization TM1, a parametrization that satisfies the heavy-ion flow constraints for symmetric matter at 2-3ρ_0 [16] (see Fig. 2). TM1 has a value of $L = 110$ MeV at saturation density, a value which is presently considered too high, and, therefore, we also take a second parametrization changing the isovector channel of TM1 by making the symmetry energy slope softer at suprasaturation densities and choosing $L = 55$ MeV ($\Lambda_v = 0.03$). Using these two nucleonic EoS we test different hyperon interactions in nuclear matter. For the hyperon potentials in symmetric nuclear matter at saturation we take $V_\Lambda = -28$ MeV, $V_\Sigma = 30$ MeV and $V_\Xi = \pm 18$ MeV. The two values of V_Ξ take into account some uncertainty on the experimental data on this potential [20]. Finally, we also consider the inclusion of the strange vector meson ϕ. The strange scalar meson is not included because according to recent experimental $\Lambda - \Lambda$-hypernuclear data, the $\Lambda - \Lambda$ interaction is only weakly attractive [31]. The results of the calculations are shown in Fig. 7 and 8.

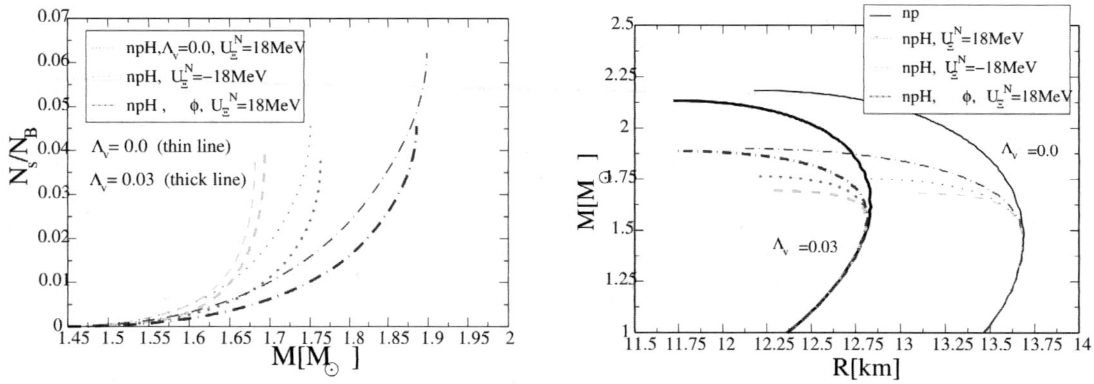

Figure 8. Left panel: The strangeness number over the total baryonic number, for $\Lambda_v = 0.0$ and 0.03 and different hyperon-meson interactions; right panel: the mass-radius curves for stars with a mass above $M_\odot$.

We conclude that a) a smaller L makes the nucleonic EoS softer giving rise to less massive stars with smaller radius as discussed in [32]; b) the inclusion of hyperons softens the EoS but if the ϕ meson is included the central densities attained in a hyperonic star may be as large as the one in a nucleonic star, and, although the strangeness fraction as a function of density is smaller, the total strangeness content in the star will be larger than when ϕ is not included; c) an attractive or repulsive optical potential of the hyperons, in this case the Ξ, clearly affects the fraction of strangeness as a function of density. A large fraction of strangeness at smaller densities makes the EoS too soft and smaller maximum masses and central densities are obtained; d) a softer symmetry energy hinders the appearance of hyperons, the onset is shifted to larger densities and at a given density the fraction of strangeness, with the appearance of the Λ, is smaller. As a result, although the initial nucleonic EoS is softer, if hyperons are included the EoS becomes harder and larger densities are attained at the center of the star. The total strangeness content of stars with the same mass is smaller; e) a consequence of the last comment is that the maximum mass configuration of a star with a softer symmetry energy maybe slightly larger 1.76 $M_\odot$ instead of 1.75 $M_\odot$ for $V_\Xi = 18$ MeV or 1.69 $M_\odot$instead of 1.68 $M_\odot$ for $V_\Xi = -18$ MeV. This effect was also observed in the calculation with the QMC model. This is not anymore true when the strange meson is included because it brings extra repulsion to the EoS, that is stronger the larger the hyperon density. Therefore an EoS with more strange content may become harder. In this schematic study the larger mass obtained for hyperonic stars was 1.89 $M_\odot$, below 1.97 $\pm$0.04 $M_\odot$ of the pulsar PSR J16142230. However, if a slightly harder at supra-saturation densities EoS had been used, and this is possible still within the heavy-ion flow data constraints [16], we would have obtained a larger mass.

The occurence of kaon condensation is another possible mechanism that gives rise to the appearance of strange degrees of freedom. In order to discuss the effect of the symmetry energy slope we consider again the TM1 parametrization and the TM1 modified parametrization with $\Lambda_v = 0.03$ for the nucleonic EoS and fix the kaon-meson coupling. The couplings are obtained from a kaon optical potential in symmetric nuclear matter at saturation $V_K = -125$ MeV, a value suggested by chiral models [24]. A more attractive potential lowers the onset density and softens the EoS as shown in [23], but we are not concerned with this point here. In the left panel of Fig. 9 we plot the fraction of particles in β-equilibrium matter with the possibility of kaon condensation for two EoS: the dash lines correspond to TM1 with a symmetry energy slope $L = 110$ MeV and the full lines to a modified TM1 model with $L = 55$ MeV. A larger L corresponds to a larger symmetry energy at suprasaturation densities and therefore, less neutrons and more protons and leptons. The kaon condensate replaces the letpons and a larger fraction of kaons occurs for a larger L. However, the onset of kaons occurs at lower densities if $L = 55$ MeV. In this case the coupling g_ρ is larger and ω_K decreases faster with density. Although a larger fraction of kaons is expected for the EoS with $L = 110$ MeV, this EoS is harder within the range of densities of interest, e.g. below the central density of the maximum mass configuration, therefore its central density is smaller and the star contains less exotic matter. These properties are reflected in the total strangeness content of the star (middle panel of Fig. 9) and the mass-radius star curves, right panel of Fig. 9. It is interesting to notice that with the present choice of parameters at the level of the nucleon EoS and the kaon interaction the mass of the millisecond pulsar PSR J16142230 would be obtained. The softer EoS with the smaller slope L gives rise to stars with smaller radius and with a larger total strangeness content and may favor the appearance of a larger content of exotic matter because larger central densities are possible and the onset density is smaller. A second conclusion is that the pulsar PSR J16142230 does not rule out kaon-condensation inside a compact star. The uncertainties existing at the level of the nucleon EoS at suprasaturation densities are enough to allow the appearance of exotic degrees of freedom.

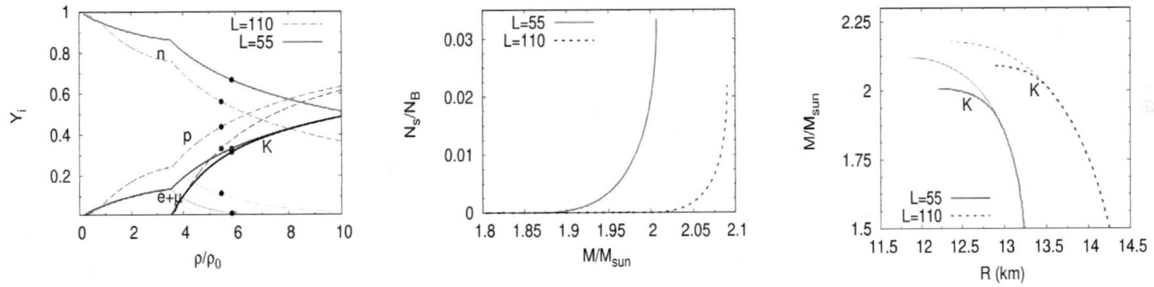

Figure 9. Left panel: fraction of particles in stellar matter with a kaon condensate. The full (dashed) lines correspond to $L = 110$ MeV ($L = 55$ MeV). The dots identify the maximum mass configurations; middle panel: the strangeness number over the baryonic number inside the two families of stars; right panel: the mass radius curves for stars with a mass above $M_\odot$.

4. Conclusions

The density dependence of the symmetry energy and the hyperon couplings have both a strong effect on the mass and radius of the star. We have tested different hyperon-meson parametrizations, using information from hypernuclei to fix the couplings. We took advantage of the fact that QMC predicts the hyperon effective masses without being necessary to fix the hyperon-σ couplings, in one of the parametrizations.

A softer symmetry energy gives rise to smaller stars. Also the hyperon fraction is affected: softer symmetry energy corresponds to a slower increase of the hyperon fraction with density [28, 21]. However, the onset of strangeness depends on the charge of the hyperons. Negatively charged hyperons or antikaons set on at smaller densities while neutral hyperons at larger densities for smaller values of L. The EoS of matter with less hyperons may become harder above hyperons onset, and larger central densities may be attained. The total strangeness fraction is generally smaller, but since larger central densities may occur, it may also happen that the total strangeness content is relatively larger. Generally a smaller strangeness content gives rise to larger maximum mass configurations, although this result depends on the model used to describe nucleonic matter and the hyperon interaction. This effect was clear in the QMC model, but for the NLWM models there was a dependence on the hyperon coupling. Even including hyperons in the QMC EOS we could explain the mass of the pulsar J1614-2230 if the cascade nuclear potential is set to be very little attractive. More data on hypernuclei is needed to constrain the hyperon-meson couplings.

The kaon condensation was studied. In this case the EoS with a softer symmetry energy has shown a total strangeness fraction much larger because much larger central densities were obtained. Also in this case using TM1 and a optical potential of K^- in nuclear matter at saturation equal to -125 MeV, of the order of the ones predicted by chiral models, a mass of the order ot 2 $M_\odot$ was obtained. A similar conclusion was taken in [33] with a different model. The present uncertainties on the kaon interaction in the medium and the EoS at high densities do not allow the exclusion of a kaon condensation in compact stars.

Acknowledgments

This work was partially supported by COMPETE/FEDER and FCT (Portugal) under the grants PTDC/FIS/113292/2009 and CERN/FP/116366/2010 and by CNPq (Brazil) and CAPES(Brazil)/FCT(Portugal) under project 232/09.

References

[1] Demorest P, Pennucci T, Ransom S, Roberts M and Hessels J 2010 *Nature* **467** 1081
[2] Baldo M, Burgio G F and Schulze H-J 2000 *Phys Rev* C **61** 055801; Vidaña I, Polls A, Ramos A, Engvik L and Hjorth-Jensen M 2000 *Phys Rev* C **62** 035801; Schulze H-J, Polls A, Ramos A and Vidaña I 2006 *Phys Rev* C **73** 058801
[3] Vidaña I, Logoteta D, Providência C, Polls A and Bombaci I 2011 *EPL* **94** 11002
[4] Vidaña I, Providência C, Polls A and Rios A 2009 *Phys Rev* C **80** 045806; Ducoin C, Margueron J, Providência C, and Vidaña I 2011 *Phys Rev* C **83** 045810
[5] Grill F, Providência C, Avancini S S 2012 *Phys Rev* C**85** 055808 ; Avancini S S Barros C C Jr, Brito L, Chiacchiera S, Menezes D P and Providência C *Phys Rev* C 2012 **85** 035806
[6] Boguta J and Bodmer A R 1997 *Nucl Phys* **A 292** 413
[7] Guichon P A M 1988 *Phys Lett* **B200** 235; Saito K and Thomas A W 1994 *Phys Lett* B **327** 9 ; Guichon P A M, K Saito, Rodionov E, and Thomas A W 1996 *Nucl Phys* **A601** 349
[8] Todd-Rutel B G and Piekarewicz J 2005 *Phys Rev Lett* **95** 122501
[9] Fattoyev F J, Horowitz C J , Piekarewicz J and Shen G 2010 *Phys Rev C* **82** 055803; (arxiv:10083030)
[10] Lalazissis G A, König J, Ring P 1997 *Phys Rev* C **55** 540
[11] Sugahara Y and Toki 1994 H *Nucl Phys* **A579** 557
[12] Glendenning N K and Moszkowski S A 1991 *Phys Rev Lett* **67** 2414
[13] Liu B, Greco V, Baran V, Colonna M and Di Toro M 2002 *Phys Rev* C **65** 045201
[14] Vretenar D, Niksic T, Ring P 2003 *Phys Rev* C**68** 024310

[15] Panda P K, Providência C and Menezes D P 2010 *Phys Rev* **C83** 045801
[16] Danielewicz P Lacey , R and Lynch W G 2002 *Science* **298** 1592
[17] Chen L W , Ko C M, and Li B A 2005 *Phys Rev Lett* **94** 032701
[18] Chiapparini M *et al* 2009 *Nucl Phys* A**826** 178
[19] Schaffner-Bielich J and Gal A 2000 *Phys Rev* C **62** 034311; Schaffner-Bielich J, Hanauske M, Stocker H and Greiner W 2002 *Phys Rev Lett***89** 171101; Friedman E and Gal A 2007 *Phys Rep* **452** 89
[20] Gal A *Prog Theor Phys Suppl* 2010 **186** 270 and references therein
[21] Panda P K, Santos A M S, Menezes P, Providência C 2012 *Phys Rev* C **85** 055802
[22] Kaplan D B andNelson A E 1986 *Phys Lett* B **175** 57
[23] Glendenning N K and Schaffner-Bielich J 1999 *Phys Rev* C **60** 025803
[24] Waas T and Weise W 1997 *Nucl Phys* A **625** 287; Weise W and Hartle R 2010 *Nucl Phys* A **835** 51
[25] Lattimer J M, Pethick C J, Prakash M , and Haensel P 1991 *Phys Rev Lett* **66** 2701
[26] Klähn T *et al* 2006 *Phys Rev* C **74** 035802
[27] Oppenheimer J R and Volkoff G M 1939 *Phys Rev* **55** 374; R C Tolman 1939 *Phys Rev* **55** 364
[28] Cavagnoli R, Providência C, Menezes D P 2011 *Phys Rev* C **84** 065810
[29] Prakash M, Bombaci I, Prakash M, Ellis P J, Lattimer JM and R Knorren 1997 *Phys Rep* **280** 1
[30] Steiner A W, Lattimer J M Brown , E F 2010 *Astrophys J* **722** 33
[31] Gal A and Millener D 2011 *Phys Lett* B**701** 342
[32] Horowitz C J and Piekarewicz J 2001 *Phys Rev* C **64** 062802
[33] Gupta N and Arumugam P 2012 *Phys Rev* C **85** 015804

Journal of Physics: Conference Series **413** (2013) 012024 doi:10.1088/1742-6596/413/1/012024

Selected problems in astrophysics of compact objects

Armen Sedrakian

Institute for Theoretical Physics, J. W. Goethe-University,
D-60438 Frankfurt-Main, Germany

E-mail: sedrakian@th.physik.uni-frankfurt.de

Abstract. I review three problems in astrophysics of compacts stars: (i) the phase diagram of warm pair-correlated nuclear matter at sub-saturation densities at finite isospin asymmtery; (ii) the Standard Model neutrino emission from superfluid phases in neutron stars within the Landau theory of Fermi (superfluid) liquids; (iii) the beyond Standard Model physics of axionic cooling of compact stars by the Cooper pair-breaking processes.

1. Introduction

The present lecture summarizes some of the recent work on three problems in the astrophysics of compact stars listed in the abstract. It attempts to provide sufficient perspective and detail within the limited span of pages as to be a useful introduction to these problems. The selection of the topics is largely motivated by the author's interests and covers only a small part of many facets of compact star physics.

Neutron stars are born in spectacular explosions of type-II supernovas [1]. The equation of state of matter at subnuclear densities and more generally its phase structure and composition are important ingredients of supernova physics, one key reason being their importance for the formation of neutrino signal at the neutrino-sphere and for transport of the energy that ultimately may generate a successful explosion. The temperature range relevant for subnuclear densities is several MeVs. In Sec. 2 we will discuss some recent progress in the understanding of the phase diagram of nuclear matter in this regime.

If a neutron star is formed in a supernova explosion, its further thermal evolution is determined by the neutrino emission from the interior of the star: neutrinos once produced in the star escape the stellar environment without further interactions [2, 3, 4]. The black-body radiation form the surface of the star in thermal equilibrium, while providing us with the information on the composition of matter and effective temperature of the emitting regions, is an unimportant cooling agent until the star is hundreds of thousands years old. The physics of neutrino emission from the interiors of neutron stars is discussed in Sec. 3 in the case of (non-exotic) baryonic matter.

Non-Standard Model particles emitted during the neutrino stage of cooling of a neutron star can drain sufficient energy as to "spoil" the Standard Model based cooling scenario. This may be used to place limits on the properties of exotic particles, such as the axions. Astrophysical bounds on the properties of axions, i.e., the mass and the coupling to the Standard Model particles, have been derived from stellar (red giant, white dwarf, supernova, and neutron star) physics, which are complementary to those obtained from cosmology and terrestrial laboratory searches [5]. Section 4 discusses the bounds on the axion properties derived from the cooling of neutron stars through inclusion of a new process of axion emission via the Cooper pair-breaking in superfluid baryonic matter [6].

2. Warm isospin asymmetrical nuclear matter

Nuclear matter at sub-saturation densities, i.e., at $\rho \leq 0.5\,\rho_0$, where $\rho_0 = 2.8 \times 10^{14}$g cm^{-3} is the nuclear saturation density, is a many-body system with well defined pair interactions, whose dominant attractive part is responsible for the formation of nuclear clusters and the Bardeen-Cooper-Schrieffer (BCS) type pair-condensate(s) of nucleons. In the supernova context the matter is at finite, but small compared to cold neutron stars, isospin asymmetry. Furthermore, low densities and relatively high (again compared to neutron stars) temperatures allow for the existence of a substantial amount of light clusters, notably deuterons, tritons, He^3 nuclei and alpha particles [7, 8, 9, 10, 11, 12, 13].

Because the unbound nucleons and deuterons are the dominant component of the matter at all relevant densities it is useful to consider only two-body correlations first. In the extreme low density limit such matter can be considered as a mixture of quasi-free nucleons and deuterons, where the only effect of the interaction is to renormalize the mass of the constituents. The deuterons being bosons may also condense and from a Bose-Einstein condensate (BEC) even without interactions if, of course, the temperature is sufficiently low [14, 15, 16, 17, 18, 19]. Imagine now increasing density while keeping the temperature of the system constant, such that the degeneracy of the system is effectively increased. As the density $\rho \to \rho_0$ the abundances of deuterons are reduced because of the Pauli-blocking of the phase space available to nucleons [16, 20]. However, the emergence of the Fermi-surface of nucleons (at finite isospin - two Fermi surfaces of neutrons and of protons) leads to BCS type correlations and formation of macroscopic coherent state - the condensate of nucleons. The interaction channel responsible for this pairing phenomenon is the $^3S_1$-$^3D_1$ partial wave, i.e., the same interaction channel which binds the deuteron. Thus, it has been conjectured that the nuclear matter may undergo a BCS-BEC phase transition, first in the context of intermediate energy heavy-ion collisions [14, 15] and more recently in the context of supernovas [8, 21].

The theoretical framework for the description of the BCS-BEC transition was developed by Nozières and Schmitt-Rink in condensed matter physics [22]. Isospin asymmetry, induced by weak interactions in stellar environments and expected in exotic nuclei, disrupts isoscalar neutron-proton (np) pairing, since the mismatch in the Fermi surfaces of protons and neutrons suppresses the pairing correlations [23]. The standard Nozières-Schmitt-Rink theory of the BCS-BEC crossover must also be modified, such that the low-density asymptotic state becomes a gaseous mixture of neutrons and deuterons [24].

Two relevant energy scales for the problem domain under study are provided by the shift $\delta\mu = (\mu_n - \mu_p)/2$ in the chemical potentials μ_n and μ_p of neurons and protons from their common value μ_0 and the pairing gap Δ_0 in the SD channel at $\delta\mu = 0$. With increasing isospin symmetry, i.e., as $\delta\mu$ increases from zero to values of order Δ_0, a sequence of unconventional phases may emerge. One of these is a neutron-proton condensate whose Cooper pairs have non-zero center-of-mass (CM) momentum [25, 26]. This phase is the analogue of the Larkin-Ovchinnikov-Fulde-Ferrell (LOFF) phase in electronic superconductors [27, 28]. Another possibility is phase separation into superconducting and normal components, proposed in the context of cold atomic gases [29]. At large isospin asymmetry, where SD pairing is strongly suppressed, a BCS-BEC crossover may also occur in the isotriplet $^1S_0$ pairing channel. The ideas of unconventional SD pairing and the BCS-BEC crossover in a model of isospin-asymmetric nuclear matter where combined recently in Ref. [21]. A phase diagram for superfluid nuclear matter over wide ranges of density, temperature, and isospin asymmetry was constructed. A self-consistent set of equations, which includes the gap equation and the expressions for the densities of constituents (neutrons and protons) was solved allowing for the ordinary BCS state, its low-density asymptotic counterpart BEC state and two phases that owe their existence to the isospin asymmetry: the phase with a moving condensate (LOFF phase) and the phase where the normal fluid and superfluid break down into separate domains.

The phase diagram found in Ref. [21] is shown in Fig 1. The results for the BCS phase and BCS-BEC crossover are consistent with the earlier studies: one observes a smooth crossover to an asymptotic state corresponding to a mixture of a deuteron Bose condensate and a gas of excess neutrons. The transition from BCS to BEC is established according the following criteria: (i) The average chemical potential $\bar{\mu}$ changes its sign from positive to negative values, (ii) the coherence length of a Cooper pair becomes comparable to the interparticle distance, i.e., $\xi \sim d \sim \rho^{-1/3}$ as conditions change from $\xi \gg d$ to $\xi \ll d$.

The nuclear LOFF phase arises as a result of the energetic advantage of translational symmetry breaking by the condensate, in which pairs acquire a non-zero CM momentum $\vec{Q}$. At constant asymmetry, a temperature increase shifts the gap maximum and the free-energy minimum of the LOFF phase toward small Q, and at sufficiently high temperature and small asymmetry the BCS state is favored over the LOFF phase. This behavior is well understood in terms of the phase-space overlap of the Fermi surfaces of neutrons and protons, which (at finite asymmetry) increases with temperature and the momentum Q of the Cooper pairs.

Thus, as the temperature increases, we expect a restoration of the BCS phase

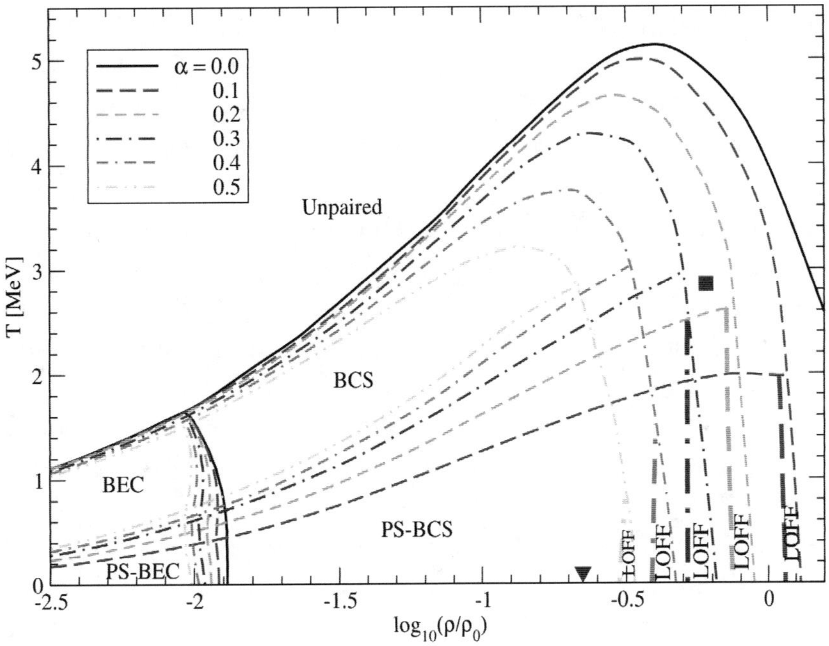

Figure 1. Phase diagram of dilute nuclear matter in the temperature-density plane for several isospin asymmetries α (from Ref. [21]). Included are four phases: unpaired phase, BCS (BEC) phase, LOFF phase, and PS-BCS (PS-BEC) phase. For each asymmetry there are two tri-critical points, one of which is always a Lifshitz point [30]. For special values of asymmetry these two points degenerate into single tetra-critical point at $\log(\rho/\rho_0) = -0.22$ and $T = 2.85$ MeV (shown by a square dot) for $\alpha_4 = 0.255$. The LOFF phase disappears at the point $\log(\rho/\rho_0) = -0.65$ and $T = 0$ (shown by a triangle) for $\alpha = 0.62$.

and of the translational symmetry in the superfluid. Obviously, the same restoration occurs when the isospin asymmetry is small enough. The superfluid phase with phase separation (PS) has the symmetrical BCS phase as one of its components. The temperature dependence of this phase is well established within BCS theory. The second component, which accommodates the neutron excess, is a normal Fermi liquid whose low-temperature thermodynamics is controlled by the excitations in the narrow strip of width $T/\epsilon_{F,n/p}$ around the Fermi surfaces of neutrons and protons.

The transition to the BEC regime of strongly-coupled neutron-proton pairs, which are asymptotically identical with deuterons, occurs at low densities. As already well established, in the case of neutron-proton pairing the criteria for the BCS-BEC transition are fulfilled, i.e., $\bar{\mu}$ changes sign and the mean distance between the pairs becomes larger than the coherence length of the superfluid.

We now turn to the question of how the BCS-BEC crossover is affected by the existence of nuclear LOFF and PS phases at non-zero isospin asymmetries, and conversely how these phases evolve in the strongly-coupled regime if the density of the system is decreased. Four different phases of matter are present in the phase diagram

Journal of Physics: Conference Series **413** (2013) 012024 doi:10.1088/1742-6596/413/1/012024

(Fig. 1). (i) The unpaired phase is always the ground state of matter at sufficiently high temperatures $T > T_{c0}$, where $T_{c0}(\rho)$ is the critical temperature of the superfluid phase transition at $\alpha = 0$. (ii) The LOFF phase is the ground state in a narrow temperature-density strip at low temperatures and high densities. (iii) The PS phase appears at low temperatures and low densities, while the isospin-asymmetric BCS phase is the ground state for all densities at intermediate temperatures. In the extreme low-density and strong-coupling regime the BCS superfluid phases have two counterparts: the BCS phase evolves into the BEC phase of deuterons, whereas the PS-BCS phase evolves into the PS-BEC phase, in which the superfluid fraction of matter is a BEC of deuterons. The superfluid-unpaired phase transitions and the phase transitions between the superfluid phases are of second order (thin solid lines in Fig. 1), with the exception of the PS-BCS to LOFF transition, which is of first order (thick solid lines in Fig. 1). The BCS-BEC transition and the PS-BCS to PS-BEC transition are smooth crossovers. At non-zero isospin asymmetry the phase diagram features two tri-critical points where the simpler pairwise phase coexistence terminates and three different phases coexist.

The extreme low-density region of the phase diagram features two crossovers. At intermediate temperatures one recovers the well-known BCS-BEC crossover, where the neutron-proton BCS condensate transforms smoothly into a BEC gas of deuterons with some excess of neutrons. The new ingredient of our phase diagram is the second crossover at low temperatures, where the heterogeneous superfluid phase is replaced by a heterogeneous mixture of a phase containing a deuteron condensate and a phase containing neutron-rich unpaired nuclear matter.

What would be the effect of adding somewhat heavier clusters, notably ^{3}He, ^{3}H and ^{4}He, to the phases discussed above? At low densities statistical equilibrium suggests that the heavier clusters may "eat up" some of the phase-space, therefore their main effect would be to reduce the phase space available to pair-correlated particles, which would eventually lead to some shifts in the phase diagram without qualitative modifications of its structure or topology. The physics might not be as trivial at high densities, for we know that the asymptotically dense phase of matter does not contain clusters. The transition to the homogenous phase can be attributed to the Pauli-blocking of the phase space that can be used to form three and four body bound states. In other words, the rise of the Fermi surface suppresses the formation of three and four-body bound states. It remains an open problem to tackle the interplay of the clusters and pair-correlations on the ultimate composition of matter at subnuclear densities. As stated in the introduction, the answer is not merely of academic interest. The neutrinos (and more generally leptons) transport energy in the supernova processes while moving through this environment; its composition, many-body effects, etc remain the key unknowns in the accurate description of the neutrino transport.

3. Neutrino pair-breaking processes

Next we trasport ourselves in time several weeks past the supernova explosion and the formation of a neutron star. During this period of time the star has cooled down to the temperature of superfluid phase transition of baryons, which is roughly in the range $T_c \simeq 0.5$ MeV. At this stage the crust of the star cools predominantly via the electron (e^-) bremsstrahlung process: $e^- + A \to e^- + A + \nu + \bar{\nu}$, where A stands for a nucleus and ν and $\bar{\nu}$ for the neutrino and antineutrino [31]. The surface of the star cools by emitting thermal soft X-rays (photons). Both processes are not important sinks of energy up until the star's age $t \geq 10^4 - 10^5$ yr [2, 3, 4]. The exact value of the transition to the crust plus surface cooling depends on a number of factors, which we will not discuss. While it was widely accepted from the beginning that the superfluidity will suppress the neutrino radiation processes on baryons in the core, once it becomes superfluid, the less trivial pair-breaking processes where initially neglected, although the rates of these processes, computed at one-loop, where available since 1976 [32, 33, 34]. The Standard Model weak neutral interactions proceed via vector and axial vector interactions. Thus, the neutrino emissivity (phase space integrated rate of neutrino emission) is mainly determined by the response functions of matter to weak vector and axial-vector currents. These can be computed within the finite temperature Green's functions technique in general. Specific applications within the real-time Schwinger-Keldysh formalism can be found in Refs. [33, 35].

Before discussing these response functions, we pause to recall the many-body developments of the formal theory. The polarization tensor in superfluid system involves a re-summation of infinite series of particle-hole diagrams. The methods for doing so were developed in the case of symmetrical nuclear matter, with the purpose of applications to finite nuclei in Ref. [36]. This work is one of the first applications of the Landau Fermi liquid theory (which was designed to describe the properties of non-superfluid liquid He^3) to superfluid systems. The diagrammatic methods of re-summation of particle-hole ladders were developed even earlier [37, 38]. While the theory of superfluid Fermi liquids was generalized further by Leggett [39] to finite temperatures (in the context of condensed matter theories), Leggett did so by computing directly the response functions and not the vertex functions. In nuclear physics, the theory took a turn that focused the entire subsequent work on finite systems (i.e. nuclei), which were treated in the equation of motion formalism for second-quantized operators.

The importance of the vertex corrections for the case of the vector current interactions was first pointed out in Ref. [40], where the authors implemented a polarization tensor adopted from condensed matter work (derived not quite in the regime needed for neutron stars). A number of subsequent works derived directly the finite temperature vertex functions and polarization tensors for neutron and proton components of the core of the star [41, 42, 43, 44]. The current consensus is that the vector current emissivity is suppressed compared to the one-loop results by a factor $(v_f/c)^4$, where $v_f/c \sim 0.1$ is the Fermi velocity of the baryons in units of the speed of light. At the same time it was established that the axial vector interactions are adequately represented by the bare vertices and one-loop polarization tensors. Explicitly, the axial vector emissivity given by [34, 42]

$$\epsilon_\nu = \frac{24G_F^2 g_A^2}{105\pi^3}\nu(0){v_f}^2T^7 I_\nu, \quad I_\nu = z^7\int_1^\infty dy\frac{y^5}{\sqrt{y^2-1}}f_F\left(zy\right)^2, \tag{1}$$

where $G_F = 1.166 \times 10^{-5}$ GeV^{-2} is the Fermi weak coupling constant, $g_A = 1.25$ is (unquenched) axial vector coupling constant, $\nu(0)$ is the density of states at the Fermi surface, T is the temperature, $z = \Delta(T)/T$ is the ratio of the pairing gap to the temperature, and $f_F(x) = [1 + \exp(x)]^{-1}$ is the Fermi distribution function. The emissivity of the axial vector current is of the order $(v_f/c)^2$ and, therefore, is the dominant channel of the energy loss.

4. Axion emission from compact stars

The strong sector of the Standard Model may feature a CP violating interaction, which arises due to a topological interaction term in the QCD Lagrangian [45]

$$\mathscr{L}_\theta = \frac{g^2\theta}{32\pi^2}F^a_{\mu\nu}\tilde{F}^{\mu\nu a}, \tag{2}$$

where $F^a_{\mu\nu} = \partial_\mu A_\nu - \partial_\nu A_\mu + gf^{abc}A_{\mu b}A_{\nu c}$ is the gluon field strength tensor, $\tilde{F}^a_{\mu\nu} = \epsilon_{\mu\nu\lambda\rho}F^{\lambda\rho a}/2$, f^{abc} are the structure constants of $SU(3)$ group, θ is the parameter which parametrizes the non-perturbative vacuum states of QCD $|\theta\rangle = \sum_n \exp(-in\theta)|n\rangle$, where n is the winding number characterizing each distinct state of QCD, which is not connected to another by any gauge transformation. The QCD action changes by 2π under the shift $\theta \to \theta + 2\pi$, i.e., θ is a periodic function with a period of 2π. In presence of quarks the physical parameter is not θ, but $\bar{\theta} = \theta + \arg\det m_q$, where m_q is the matrix of quark masses. Experimentally, the upper bound on the value of this parameter is $\bar{\theta} \le 10^{-10}$, which is based on the measurements of the electric dipole moment of neutron $d_n < 6.3 \cdot 10^{-26} e$ cm [46]. The smallness of $\bar{\theta}$ is the strong CP problem: the Standard Model does not provide any explanation on why this number should not be of order unity.

An elegant solution to the strong CP problem is provided by the Peccei-Quinn mechanism [47, 48, 49]. This solution amounts to introducing a global $U(1)_{PQ}$ symmetry, which adds an additional anomaly term to the QCD action proportional to the axion field a. This term acts as a potential for the axion field and gives rise to an expectation value of the axion field $\langle a\rangle \sim -\bar{\theta}$. The physical axion field is then $a - \langle a\rangle$, so that the undesirable θ term in the action is replaced by the physical axion field. The axion is the Nambu-Goldstone boson of the Peccei-Quinn $U(1)_{PQ}$ symmetry breaking [48, 49], and its effective Lagrangian has the form

$$\mathscr{L}_a = -\frac{1}{2}\partial_\mu a\partial^\mu a + \mathscr{L}_{int}(\partial_\mu a, \psi), \tag{3}$$

where the second term describes the coupling of the axion to fermion fields (ψ) of the Standard Model.

Figure 2. The two diagrams contributing to the polarization tensor of baryonic matter, which defines the axion emissivity. The "normal" baryon propagators for particles (holes) are shown by single-arrowed lines directed from left to right (right to left). The double arrowed lines correspond to the "anomalous" propagators with two incoming or outgoing arrows. The horizontal dashed lines represent the axion a.

There are ongoing experimental searches for the axion. Cosmology and astrophysics provide strong complementary constraints. Because axions can be effectively produced in the interiors of stars they act as an additional sink of energy. The requirements that the energy loss from a star is consistent with the astrophysical observations place lower bounds on the coupling of axions to the standard model particles, and hence on the Peccei-Quinn symmetry breaking scale [5, 50]. The latter limit translates into an upper limit on the axion mass. Such arguments have been applied to the physics of supernova explosions [51, 52, 53, 54, 55] and white dwarfs [56]. In the case of supernova explosions the dominant energy loss process is the emission of an axion in the nucleon (n) bremsstrahlung $n + n \to n + n + a$. The same process was considered earlier by Iwamoto as a cooling mechanism for mature neutron stars, i.e., neutron stars with core temperature in the range $10^8 - 10^9$ K [57]. The implications of the axion emission by the modified Urca and nucleon bremsstrahlung, as calculated in Ref. [57], were briefly studied in Ref. [58]. However, as discussed in the previous section the pair-breaking processes are the dominant energy loss channels for temperatures below the critical temperature. Therefore, to obtain bounds on the properties of axions from the cooling of compact stars, the rate of the axion cooling is required. This was computed recently in Ref. [6] in the case of S-wave superfluid.

We now briefly outline this calculation. The energy radiated per unit time in axions (axion emissivity) is given by the phase-space integral over the probability of the emission process

$$\epsilon_a = -f_a^{-2} \int \frac{d^3q}{(2\pi)^3 2\omega}\, \omega g_B(\omega)\kappa_a, \quad \kappa_a = q_\mu q_\nu \Im\Pi_a^{\mu\nu}(q), \tag{4}$$

where q and ω are the axion momentum and energy, $\Pi_a^{\mu\nu}(q)$ is the polarization tensor of the superfluid baryonic matter, $g_B(\omega)$ is the Bose distribution function, f_a is the axion-baryon coupling strength. The polarization tensor of a superfluid obtains contributions from four distinct diagrams that can be formed from the normal and anomalous propagators with four distinct effective vertices [41, 42, 43, 44]. However, for the axial vector perturbations the vertices are not renormalized in the medium and, therefore, one proceeds with the bare vertices, in which case the number of the distinct contributions to the polarization tensor reduces to a sum of two admissible bare loops (see Fig. 2). The axion emissivity obtained from Eq. (4) is given by

$$\epsilon_a = \frac{8}{3\pi} f_a^{-2}\, \nu(0)\, v_f^2\, T^5\, I_a, \quad I_a = z^5 \int_1^\infty dy\, \frac{y^3}{\sqrt{y^2-1}} f_F\,(zy)^2\,. \tag{5}$$

The T^5 scaling of the emissivity is understood as follows. The integration over the phase space of neutrons carries a power of T, since for degenerate neutrons the phase-space integrals are confined to a narrow strip around the Fermi surface of thickness T. The axion is emitted thermally and being relativistic contributes a factor T^3 to the emissivity. The one power of T from the energy of the axion and the inverse one power of T from the energy conserving delta function cancel. The transition matrix element is proportional to the combinations of u_p and v_p amplitudes, which are dimensionless, but contain *implicit* temperature dependence due to the temperature dependence of the gap function. This dependence is not manifest in Eq. (5), i.e., was absorbed in the definition of the integral I_a. Thus, the explicit temperature dependence of the axion emission rate Eq. (5) is T^5. As discussed in Sec. 3 at temperatures of order the critical temperature $T_c \simeq 10^9$ K the superfluid cools primarily by emission of neutrinos via the pair-breaking processes driven by the axial-vector currents (we continue to assume that potential fast cooling via direct Urca processes is prohibited). In a first approximation one may require that the axion luminosity does not exceed the neutrino luminosity, i.e.,

$$\frac{\epsilon_a}{\epsilon_\nu} = \frac{10\pi^2}{f_a^2 G_F^2 g_A^2 \zeta_A} \frac{I_a}{I_\nu} \simeq \frac{59.2}{f_a^2 G_F^2 \Delta(T)^2}\, r(z) < 1, \tag{6}$$

where $r(z) \equiv z^2(I_a/I_\nu)$. Not far from the critical temperature $\Delta(T) \simeq 3.06 T_c\sqrt{1-T/T_c}$, which translates into $z = 3.06\, t^{-1}\sqrt{1-t}$, where $t = T/T_c$. Numerical evaluations of the integrals provides the following values $r(0.5) = 0.07$, $r(1) = 0.26$, $r(2) = 0.6$ and asymptotically $r(z) \to 1$ for $z \gg 1$. Noting that $r(z) \le 1$, one finally obtains

$$f_a > 5.92 \times 10^9\, \mathrm{GeV} \left[\frac{0.1\ \mathrm{MeV}}{\Delta(T)}\right], \tag{7}$$

which translates into an upper bound on the axion mass

$$m_a = 0.62 \times 10^{-3}\, \mathrm{eV} \left(\frac{10^{10}\mathrm{GeV}}{f_a}\right) \le 1.05 \times 10^{-3}\, \mathrm{eV} \left[\frac{\Delta(T)}{0.1\ \mathrm{MeV}}\right]. \tag{8}$$

The bound Eq. (7) can be written in terms of the critical temperature by noting that $\Delta(T) \simeq T_c$ in the temperature range $0.5 \le t < 1$ of interest.

The neutrino cooling era of compact stars, which spans the time-period $t \le 10^4 - 10^5$ yr after their birth in supernova explosions is a sensitive probe of the particle physics of their interiors. If one assumes that there are no rapid channels of cooling in neutron stars, i.e., deconfined quarks, above Urca threshold fractions of protons or hyperons (all of which lead to a rapid Urca cooling), then neutron stars cool primarily by neutrino emission in Cooper pair-breaking processes in baryonic superfluids. If, however, axions exist in Nature, the neutron stars must cool via axion emission in Cooper pair-breaking processes, whose axion emission rate scales as T^5. This scaling differs from the T^7 scaling of the counterpart neutrino processes. The difference arises from the different phase spaces required for the pairs of neutrinos and the axion and is independent of the baryonic polarization tensor. Note also that the rate of axion emission from a P-wave superfluid will differ from the S-wave rate, derived above, by a factor $O(1)$ and, therefore, will not change quantitatively the obtained bounds on the axion parameters.

Similar bounds to those quoted above were obtained previously by Iwamoto [57] ($f/10^{10}$GeV > 0.3) from a comparison of the rates of axion bremsstrahlung and modified Urca neutrino emission by mature neutron stars, and by Umeda et al [58] ($f/10^{10}$GeV $> 0.1 - 0.2$) from fits of cooling simulations to the PSR 0656+14 data‡. The lower bound on f_a derived in Ref. [6] is somewhat larger than the one that follows form the requirement that the axions do not "drain" too much energy from supernova process so that it fails [5, 51, 52, 53, 54, 55]. Note the dependence of the bound Eq. (7) on the pairing gap. This is an ingredient, which originates from superfluidity of cold neutron stars, that does not appear in other bounds on axion parameters. Because, the magnitude and density dependence of the gap are not well-known (for a review see Ref. [59]) there remains enough room for speculations on the impact of the axions on the cooling of compact stars.

5. Perspectives

Although the equation of state and composition of the dilute and warm nuclear matter have been studied for several decades, there remain a number of unsettled issues related to the complex many-body character of this system. The interplay of the pair-correlations and clustering is one of the challenging problems. As we have seen the phase structure of the matter at finite isospin becomes more complicated due to the emergence of novel superconducting phases. It is very likely that along with the deuteron condensate and its BCS counterpart a BEC of alpha particles will coexist with these phases. Thus, one needs to work out the physics of a mixtures of several superfluids phases in this context. How these new features will affect the neutrino transport in supernovas remains an open question.

Neutron star cooling offers a new playground for the studies of beyond Standard Model physics. The example presented above shows that with microscopic rates of the axion emission from superfluid phases at hand one can place useful bounds on the axion properties. Axions offer a unique channel of rapid energy loss by medium to low mass compact stars, where the central densities are below the direct Urca thresholds and/or below the density of deconfinement into quark matter. Thus, the long-standing paradigm that the light and medium mass neutron stars cannot cool rapidly will not hold should the axions exist in Nature and couple to matter strong enough to cool a compact star faster than the Standard Model neutrinos.

‡ Note that Eq. (9) of this work is incorrect, which could be the reason why the limits on the axion's mass reported in their Figs. 2-4 are by an order of magnitude larger compared to those quoted in Ref. [6].

Acknowledgements

I gratefully acknowledge the discussions and collaboration with J. W. Clark, X.-G. Huang, J. Keller, and M. Stein on the topics presented in this lecture.

References

[1] Burrows A arXiv:1210.4921 [astro-ph.SR]
[2] Yakovlev D G, Gnedin O Y, Kaminker A D and Potekhin A Y 2008 AIP Conf. Proc. **983**, 379
[3] Page D, in *Neutron Stars and Pulsars*, Astrophys. Space Sci. Library, vol. **357** ed. W. Becker, (Springer-Verlag, 2009) 247
[4] Sedrakian A 2007 *Prog. Part. Nucl. Phys.* **58** 168
[5] Raffelt G G 2008 *Lect. Notes Phys.* **741** 51
[6] Keller J and Sedrakian A arXiv:1205.6940 [astro-ph.CO]
[7] Sumiyoshi K and Röpke G 2008 *Phys. Rev.* C **77** 055804
[8] Heckel S, Schneider P P and Sedrakian A 2009 *Phys. Rev.* C **80** 015805
[9] Typel S, Röpke G, Klahn T, Blaschke D and Wolter H H 2010 *Phys. Rev.* C **81** 015803
[10] Hempel M and Schaffner-Bielich J 2010 *Nucl. Phys.* A **837** 210
[11] Raduta A R and Gulminelli F 2010 *Phys. Rev.* C **82** 065801
[12] Oertel M, Fantina A F and Novak J 2012 *Phys. Rev.* C **85** 055806
[13] Gulminelli F and Raduta A R 2012 *Phys. Rev.* C **85** 025803
[14] Alm T, Friman B L, Röpke G and Schulz H 1993 *Nucl. Phys.* A **551** 45
[15] Baldo M, Lombardo U and Schuck P 1995 *Phys. Rev.* C **52** 975
[16] Sedrakian A and Clark J W 2006 *Phys. Rev.* C **73** 035803
[17] Mao S, Huang X and Zhuang P 2009 *Phys. Rev.* C **79** 034304.
[18] Huang X G 2010 Phys. Rev. C **81** 034007
[19] Jin M, Urban U and Schuck P 2010 *Phys. Rev.* C **82** 024911
[20] Röpke G, Bastian N U, Blaschke D, Klahn T, Typel S and Wolter H H, arXiv:1209.0212 [nucl-th].
[21] Stein M, Huang X G, Sedrakian A and Clark J W, arXiv:1208.0123 [nucl-th].
[22] Nozieres P and Schmitt-Rink S 1985 *J. Low. Temp. Phys.* **59** 195
[23] Sedrakian A and Lombardo U 2000 *Phys. Rev. Lett.* **84** 602
[24] Lombardo U, Nozieres P, Schuck P, Schulze J J and Sedrakian A 2001 *Phys. Rev.* C **64** 064314
[25] Sedrakian A 2001 Phys. Rev. C **63** 025801
[26] Müther H and Sedrakian A 2003 *Phys. Rev.* C **67** 015802
[27] Larkin A I and Ovchinnikov Y N. 1965 *Zh. Eksp. Teor. Fiz.* **47** 762
[28] Fulde P and Ferrell R A 1965 *Phys. Rev.* **135** 550
[29] Bedaque P F, Caldas H and Rupak G 2003 *Phys. Rev. Lett.* **91** 247002
[30] Chaikin P and Lubensky T, *Principles of Condensed Matter Physics* (Cambridge University Press, 1995).
[31] Shapiro S L and Tuekolsky S A, *Black Holes, White Dwarfs, and Neutron Stars* (John. Wiley & Sons, New York, 1983).
[32] Flowers E, Ruderman M, Sutherland P 1976 *Astrophys. J.* **205** 541
[33] Voskresensky D N, Senatorov A V 1987 *Sov. J. Nucl. Phys.* **45** 657; 1987 [Yad. Fiz. **45** 411].
[34] Kaminker A D, Haensel P, Yakovlev D G 1999 *Astron. Astrophys.* **345** L14-L16
[35] Sedrakian A, Dieperink A 1999 *Phys. Lett.* B **463** 145-152; 2000 *Phys. Rev.* **D62** 083002
[36] Larkin A I and Migdal A B, Sov. Phys. 1963 JETP **17** 1146; Migdal A B, *Theory of Finite Fermi Systems and Applications to Atomic Nuclei* (Interscience, London, 1967).
[37] Abrikosov A A, Gorkov L P 1959 *Sov. Phys. JETP* **8**, 1090 (1959); *Sov. Phys. JETP* **9** 220
[38] Abrikosov A A, Gorkov L P, and Dzyaloshinski I E, *Methods of quantum field theory in statistical physics*, (Dover, New York, 1975).

[39] Leggett A J 1966 *Phys. Rev.* **147** 119
[40] Leinson L B and Pérez A 2006 *Phys. Lett.* B **638** 114
[41] Sedrakian A, Müther H, Schuck P 2007 *Phys. Rev.* C **76** 055805
[42] Kolomeitsev E E, Voskresensky D N 2008 *Phys. Rev.* C **77** 065808; 2010 *Phys. Rev.* C **81** 065801
[43] Steiner A W, Reddy S 2009 *Phys. Rev.* C **79** 015802
[44] Sedrakian A 2012 *Phys. Rev.* C **86** 025803
[45] Hooft G 1976 *Phys. Rev. Lett.* **37** 8
[46] Baker C A *et al.* 2006 *Phys. Rev. Lett.* **97** 131801
[47] Peccei R D and Quinn H R 1977 *Phys. Rev. Lett.* **38** 1440
[48] Weinberg S 1978 *Phys. Rev. Lett.* **40** 223
[49] Wilczek F 1978 *Phys. Rev. Lett.* **40** 279
[50] Raffelt G G, Redondo J and Maira N V 2011 *Phys. Rev.* D **84** 103008
[51] Brinkmann R P and Turner M S 1988 *Phys. Rev.* D **38** 2338
[52] Burrows A, Turner M S and Brinkmann R P 1989 *Phys. Rev.* D **39** 1020
[53] Raffelt G and Seckel D 1995 *Phys. Rev.* D **52** 1780
[54] Janka H T, Keil W, Raffelt G and Seckel D 1996 *Phys. Rev. Lett.* **76** 2621
[55] Hanhart C, Phillips D R and Reddy S 2001 *Phys. Lett.* B **499** 9
[56] Corsico A H, Althaus L G, Bertolami M M M, Romero A D, Garcia-Berro E, Isern J and Kepler S O 2012 *Mon. Not. R. Astron. Soc.*, **424** 2792
[57] Iwamoto N 1984 *Phys. Rev. Lett.* **53** 1198
[58] Umeda H, Iwamoto N, Tsuruta S, Qin L and Nomoto K, in Proceedings of the International Conference on Neutron Stars and Pulsars, edited by N. Shibazaki et al., Frontiers science series no. 24, (Universal Academy Press, 1998) 213 [astro-ph/9806337]
[59] Sedrakian A and Clark J W, in *"Pairing in Fermionic Systems: Basic Concepts and Modern Applications"*, edited by Sedrakian A, Clark J W and Alford M, (World Scientific, Singapore, 2006) 135 [arXiv:nucl-th/0607028].

Nuclear astrophysics with exotic nuclei and rare ion beams

Livius Trache
Cyclotron Institute, Texas A&M University, College Station, TX 77843, USA, and
"Horia Hulubei" National Institute for Physics and Nuclear Engineering, Bucharest-Magurele, Romania

E-mail: livius_trache@tamu.edu, or livius.trache@nipne.ro

Abstract. Nuclear astrophysics has become a major motivation for nuclear physics research in the latest few decades. The quests to understand grand scale cosmic phenomena, the origin of elements and isotopes, the sources of energy in stars, were advanced by studies at the microscopic scale of nuclei. Advances in the production, separation and acceleration of unstable nuclei lead not only to new knowledge in the structure of nuclei and nuclear matter, but also have revolutionized nuclear physics for astrophysics. I will review some of the many contributions that nuclear astrophysics made to our fundamental knowledge, and then will describe a few indirect methods used in nuclear astrophysics using radioactive beams, concentrating on those used by the groups I work with.

1. Introduction

I need to start not merely by thanking the organizers for inviting me here, but also by "justifying" my presence here at this school that celebrates the 80th year of life of prof. Aurel Sandulescu! He was, and is, a theoretician in nuclear physics! And I am an experimentalist! Most of the speakers before me are reputed theoreticians, his collaborators, or experimentalists working in fields where prof. Sandulescu made important contributions. And I am neither! However, he was my diploma advisor, most probably the only experimentalist to claim this qualification! He may or may not remember, but I do remember very well: I was for 3 months in Dubna for my diplomawork, in the Laboratory for Nuclear Reactions (today the Flerov Laboratory), where he was at the time deputy director. We had long conversations at the time: I was learning "deep inelastic collisions" (DIC) from their discoverer, dr. V.V. Volkov; he was "munching" at the time the idea of cluster radioactivity. I said munching, and I believe it is not the wrong word, because they were mostly physical images prof. Sandulescu was using to try to figure out the phenomena, not equations or existing models. And the idea of the double nuclear system used at the time in DIC, a dynamical nuclear system at temperature $T\neq0$, was probably encouraging his ideas about cluster decay from $T=0$ systems. I wish him good health and a long, productive life!

Going into the subject of my lecture: after attending the lectures of the first days I realized that I have to rethink and retool my own presentation! I am the first, and the only one for some time, talking about *nuclear astrophysics* (NA). Even though the actual subject is better called '*nuclear physics for astrophysics*', a brief introduction in nuclear astrophysics is in order. I will start by making a few general considerations about the contributions of nuclear astrophysics to fundamental science, to our understanding of the Universe in general. Contributions made in the last 100, or 60 years, but more so in the last few decades. Given the many students in the audience, I find that going directly into the details of the subject to which the title refers specifically would be counter-productive! These should also be good starting points for discussions with "the adults" in the audience. Second, I will introduce the domain we call nuclear physics for astrophysics (NPA), including its specific vocabulary and main problems (very low energies on nuclear scale, very low cross sections and reactions involving in many cases unstable nuclei, far from stability). Third, I will go into the subject of the indirect methods for nuclear astrophysics using rare ion beams (RIB). I will insist on a few only and will use mostly examples from I studies I was participating.

The above was the structure of the lecture. For this paper I will retain some of the points made in the general, introductory discussion. I will treat only briefly the second part, as those can be found in many books, or lectures at other conferences. And I will review the third part, with only one example for each of the methods treated. References to which I send throughout should be good reading to cover some of the details missing here, for those students wanting to go deeper into the subject.

2. Nuclear Astrophysics

As essentially a fundamental science, in addition to so many practical applications that it brought us, Nuclear Physics (NP) was from its beginnings taking a front place in the human endeavour of understanding of the Universe. Our Universe! It was and continues to be part in understanding its composition, its dynamics, its origins and history, and possibly, its future. Some of these advances were made through its branch which we call *nuclear astrophysics*. Here is a short, non-exhaustive, list of the important successes of nuclear astrophysics:

- **Nuclear physics for astrophysics** – is increasingly motivation for NP research:
  - We know that nuclei are the fuel of the stars
  - Origin of chemical elements: nucleosynthesis = a large series of nuclear reactions & elemental/isotopic abundances are indelible fingerprints of cosmic processes. We need better nuclear data to have convincing quantitative descriptions of various scenarios.
- **Big successes of NA**:
  - **BBN** (Big Bang Nucleosynthesis) – is a quantitative, parameter free theory explaining the formation of the lightest elements. Alternatively, we should say that BBN theory was the first to determine that fundamental parameter of the standard model which is the baryon-to-photon ratio η. CMB (Cosmic Microwave Background) studies later confirmed the value and decreased the error bar! BBN theory lead also to the demonstration that there are 3 types of neutrinos, before the 3rd type was discovered.
  - **Heavier elements were created in stars** through a number of complex processes.
  - **Solar reactions are basically understood** (pp-chains, CNO cycle, solar neutrinos, neutrino oscillations…)
  - Nucleosynthesis is an **on-going** process!
  - We (quasi-) understand **novae, XRB, neutron stars** …,
  - but not the super-novae – mechanisms, quantitative description, etc…

We study our Universe through observations, but also through experiments in the laboratory. It is actually considered that cosmology went from the realm of philosophy and speculation into that of science when physicists started to use nuclear physics data to model the genesis of chemical elements (Bethe and Critchfield, 1938 [1]; Alpher, Bethe and Gamow, 1948 [2]) and compare their quantitative predictions with the observations. Since then, many and fundamental advances were made, a large and rich spectrum of new astrophysical observations was added to our knowledge, and for their interpretation more detailed nuclear and particle data were necessary. Isotopic abundances, available from astronomical observations, are unique fingerprints of the evolution of stars. Sir A. Edington was the first to suggest that nuclei only can hold the key to the production of solar energy. In the 1920s and

'30s the hypothesis was advanced that nuclear reactions are the source of the solar energy, the very source that made and makes possible life on Earth, our life. It was only possible to explain the origin of the solar energy when nuclear processes started to be understood in the 1930s. The detailed mechanisms of this energy production could only be described, in part yet, much later with the advance of *nuclear physics for astrophysics*, or *nuclear astrophysics* (NA). Because nuclear reactions could not happen at the measured temperature of the solar spectrum! One can say that only in the early '70s the existence of the nuclear reactions was proved by the detection of solar neutrinos originating from the much hotter interior of the Sun. This was a joint achievement of *nuclear astrophysics, nuclear chemistry* and *astroparticle physics.*

Nuclei are the fuel of the stars! And all chemical elements in the Universe as we know it were produced in processes that we call generically nucleosynthesis. Nucleosynthesis occurred in various stages of the evolution of the Universe, in various places and in different types of events: Big Bang Nucleosynthesis (BBN) or later stellar evolution, far away or around us, explosive or steady burning. And we have firm evidence collected in the latter decades that nucleosynthesis happens today, even in our own galaxy, close to where we live. We also know today that the nuclear processes occurring in stars are not only the source of energy for cosmic processes, but also that nucleosynthesis gives us unique and indelible fingerprints of these processes. Many nucleosynthesis scenarios exist today. Some were formulated for some time, beginning with the seminal works by Burbidge, Burbidge, Fowler & Hoyle, 1957 [3] and independently by Cameron, 1957 [4]: Big Bang Nucleosynthesis (BBN), Inhomogeneous Big Bang Nucleosynthesis (IBBN), the s-process, the r-process, the rp-process, etc., and some are newer proposals. The possibilities to check the detailed predictions of specific models occurred only recently, with the availability of more and better astrophysical observations, of more nuclear data, of advances in understanding the dynamics of non-equilibrium processes, and of increased computing power. It turns out that an important component of all these nucleosynthesis model calculations is represented by the data for the nuclear processes involved. Only good nuclear physics data permit to make definite, quantitative predictions that can be checked against the ever increasing observational data sought and obtained by astrophysicists. This is the object of the nuclear physics for astrophysics, a subject that is most often called *nuclear astrophysics*. It does not deal with the specificities of the dynamics of different stellar processes, but only with the nuclear reactions involved, in particular with how we obtain these data from direct or indirect measurements. However, more recently the modeling of stellar processes and the dynamics of stars came closer and closer to the realm of interest of nuclear physicists and there is increased synergy of the two fields.

There are thousands of nuclear reactions and nuclear processes that occur in stars. Some are very important, some are less important and some are irrelevant in one type of process, while becoming important in another, depending on the conditions of the particular process: composition, densities and temperatures involved. There are also many nucleosynthesis processes, and our knowledge about them differs.

It is an important success of physics in general that we can describe now the **primordial abundances** (in BBN) over ten orders of magnitude. This description is parameter free after the baryon-to-photon ratio was determined independently and quite exactly η_{WMAP}=6.19(15)*10^{-10} from the measurement of the Cosmic Microwave Background using 7-year WMAP data. Only the abundance of ^{7}Li is not exactly matching the observations and remains "the Li puzzle of BBN" (see Fig. 1). It is not clear now if this is due to the existing nuclear reaction data, to the list of 11 reactions important being incomplete, or is due to observational problems.

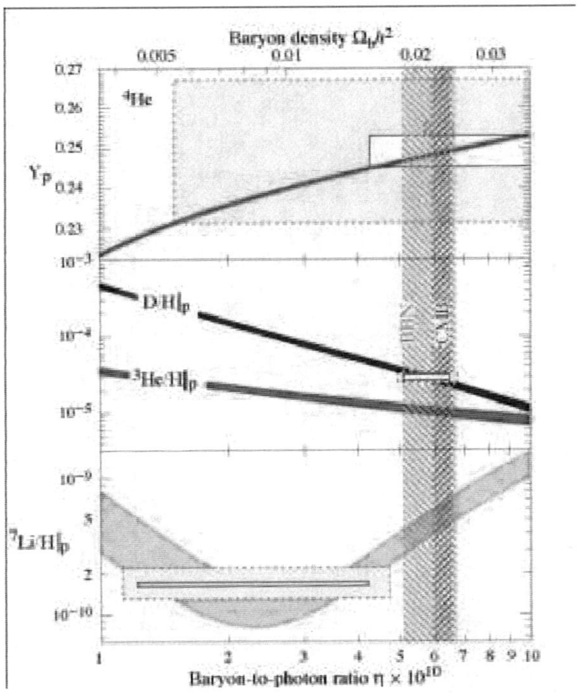

Figure 1 The abundances of ^{4}He, D, ^{3}He, and ^{7}Li as predicted by the standard Big Bang Nucleosynthesis, as function of the baryon-to-photon ratio η. The vertical bands show the η values given by the CMB and by the BBN. Boxes indicate the observed light isotopes abundances ($\pm 1\sigma$, and $\pm 2\ \sigma$ errors). From Ref. [5].

Closer to home, we have a good understanding of how our **Sun** works. Nuclear astrophysics measurements provide currently good data for most of the reactions important in Sun: those in the pp-I and pp-II chains responsible for most of the energy production and those in the CNO cycle. And for the $^3He(^4He,\gamma)^7Be$ and $^7Be(p,\gamma)^8B$ reactions at the end of the pp-III chain (Figure 2, right), reactions crucial for the evaluation of the solar neutrino production. However, the cross sections accuracies of around 5% called for by the current Standard Solar Model are not met in all cases. Much progress was made lately through the work of the underground facility LUNA at Gran Sasso National Laboratory, in Italy, where for the first time cross sections were measured into the energies in the Gamow peak. See for these the lecture of M. Junker at the recent Sinaia Carpathian School [6]. The "solar neutrino puzzle"= the discrepancy between the number of solar neutrinos produced and measured on Earth, lead to the discovery of neutrino oscillations, and implicitly to the proof for neutrino mass, and all current revolution in neutrino physics (see, e.g [7]).

Jointly nuclear astrophysics and observational astrophysics have also proven that **nucleosynthesis is an on-going process** in the Universe: it happened at various evolution stages in the past, but is still happening now. This very important concept has been proven by the gamma-ray space-based telescopes like COMPTEL and INTEGRAL, through the identification of characteristic gamma-rays emitted following the β-decay of long-lived isotopes like ^{26}Al ($T_{1/2}$=0.7 My) or ^{60}Fe ($T_{1/2}$=1.5 My), or not so long-lived ones, like ^{44}Ti ($T_{1/2}$=60 y) or ^{22}Na ($T_{1/2}$=2.6 y). The detection of gamma-rays originating from ^{26}Al, with a lifetime considerably shorter than that of the Universe, or of that of our Galaxy, was the first proof that nucleosynthesis is an on-going process. *(Note: this is a common statement! However, we should not forget that He was first identified in Sun's spectra, and*

Journal of Physics: Conference Series **413** (2013) 012025 doi:10.1088/1742-6596/413/1/012025

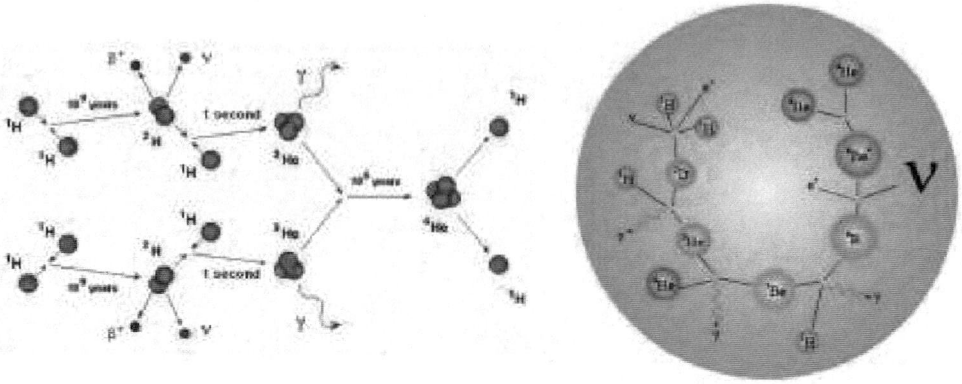

Figure 2 The reactions in the pp-chains taking place in Sun and producing most of the energy (pp-I left) and most of the observed neutrinos (pp-III, right).

the element Tc (Z=43), which does not have a stable element, was also first identified in stellar spectra). Figure x of Ref. [8] presents schematically the nuclear process involved and a sky map of the measured distribution of ^{26}Al sources. The distributions of sources can give information not only about the location of the nucleosynthesis sites, but also of the dynamics of mixing of the matter in the galaxy by measuring distributions for sources of various lifetimes. However, we do not have yet a precise and quantitative understanding of the nuclear processes leading to the production of these isotopes. Nor of the transport dynamics of the matter ejected from the underlying cosmic processes and more nuclear physics data are needed.

We can presently describe relatively well **H- and He-burning** in some environments like novae and X-Ray Bursters (XRB) and we have models for various types of **supernovae** (SN), more or less successful, but we do not know major things, for example the cosmic environments of the **s- and r-processes** [8]. We should say here that these processes account, each, for the production of about 50% of the chemical elements heavier than Fe, essential for life and our own existence. The origin of these heavy elements is considered one of the greatest unanswered questions of contemporary physics. The least we know today about the formation of heavier elements through the repeated absorption of neutrons at high neutron densities and high temperatures, the so called r-process. It is not clear what the exact path of these reactions is because we do not know key elements like the lifetimes of very neutron-rich nuclei or their neutron absorption cross sections. And for sure we do not know the exact location of the neutron dripline for medium and heavy elements. As this is dominantly a fast chain of reactions, followed by decays, it may not be needed to know all reactions precisely, but currently we have very limited knowledge even about the crucial ones at the waiting points at N=82 and N=126 shell closures. For many of the reactions involved the uncertainties are a few orders of magnitude! Therefore, much more work is needed before we fully understand and describe stellar nucleosynthesis and it is to be expected that the new facilities will bear answers to some of the above questions and to new ones that will appear.

3. Introduction to Nuclear Physics for Astrophysics

A number of particularities occur when we talk about nuclear data needed to describe reactions taking place in stars or in stellar environments. *Cross sections* are needed, but for practical reasons, in cases where barrier penetrations are important, it is helpful to introduce the *astrophysical S-factors*. In fact, what the nucleosynthesis modelists are using are *reaction rates*, averaging cross sections' contributions over the whole energy ranges in gases at the appropriate energies. The weighing with the Maxwell-distributions lead to the so called *Gamow peak*, which specifies for what energies we would actually need to know/measure reactions cross sections. In reactions we can have *direct* and *resonant contributions*, etc… Discussions of these notions and their precise definitions can be found in books (like [9], e.g.) and in some previous lectures [10]. I will not repeat them in print, here.

4. Indirect methods for nuclear astrophysics with RIBs

The use of indirect methods in nuclear astrophysics is prompted by the difficulties that one encounters in attempting to make direct nuclear astrophysics measurements. Direct experiments mean trying to measure exactly the reactions that happen in stars, in those exact conditions (targets and projectiles, energies, charge states, etc…). The main difficulties arise because:

- Stars are cold! Compared with the energies typical in our nuclear physics laboratories, the energies of the partners in stars are very small (10s-100s keV) and the corresponding cross sections, in particular when charged particles are involved, are very small, therefore difficult to measure.
- In stars many reaction partners are unstable nuclei, and some are so short-lived that even with the recent advances in the rare isotope production they are not available, or not easily available, for the exact projectile-target combination at the energies they have in stars.

We have to resort to indirect methods. Several such methods are known in literature, some dedicated and labelled as such, some not. All these experiments are done at laboratory energies (1-10-100 MeV/nucleon) to extract nuclear structure information. This nuclear structure information is then used for nuclear astrophysics, that is, to evaluate reaction cross sections at low energies (10s-100s keV) and the resulting reaction rates at appropriate stellar temperatures. There are two steps here where theoretical calculations occur, and these calculations need to be seriously tested, well parameterized if necessary, using a large variety of data. For this, the use of good quality data with stable beams is still crucial. I want to stress this, because even if common sense, it is too many times overlooked and neglected.

In this lecture I will present three of these indirect methods:

1. One-nucleon transfer reactions (the ANC method)
2. Breakup reactions at intermediate energies
3. Decay spectroscopy. Beta-delayed gamma and proton decays.

In all three cases they are being used to evaluate reaction rates for radiative proton capture, with the difference that the first two are applied to find the continuum (non-resonant) component of the reaction cross sections, while the latter is used for resonant capture.

This being a school, I will not attempt below to be exhaustive in the description of the methods, but rather to be illustrative. I will also prefer to use relevant cases as illustrations, not necessarily 'newest' data. All of the examples will be from work done in the group I am working at Texas A&M University, even though many groups in the world have by now accepted these methods and are using them.

4.1 One-nucleon transfer reactions (the ANC method)

A direct transfer reaction is characterized by the rearrangement of only a few nucleons during a fast process. From the early days of nuclear physics, nucleon transfer reactions were the tool to study the single-particle degrees of freedom of nuclei and were crucial in establishing our current understanding of the structure of nuclei. Typically, spectra of final states and angular distributions were measured. Due to the direct character of the interaction, the tool of choice for the description of transfer reactions was the Born Approximation, either in the Plane Wave (PWBA), or the Distorted Wave (DWBA) form. By comparing the shape of the measured angular distributions with DWBA, the quantum numbers *nlj* of the single-particle orbitals involved could be determined, and by comparing the absolute values of experimental cross sections with those calculated, the spectroscopic factors S_{nlj} were found for the states populated. The spectroscopic factor is proportional to the "probability" that a many-body system is found in a given configuration. In the case we are talking about, single particle orbitals *nlj*, the classical definition (from Macfarlane and French, 1960 to Bohr and Mottelson, 1969 etc...) relates the spectroscopic factors to the occupation number for the *nlj* orbital in question. One nuclear state may present several spectroscopic factors: e.g. the ground state (g.s.) of ^{8}B has $S(p_{3/2})$, $S(p_{1/2})$... related to the probability that the last proton is bound around the g.s. of the ^{7}Be core in a $1p_{3/2}$, or a $1p_{1/2}$ orbital. The determination of spectroscopic factors from one-nucleon transfer reactions was and is crucial in building our current understanding of the fermionic degrees of freedom in nuclei and their coupling to other types of excitations. The Asymptotic Normalization Coefficient (ANC) method is an indirect NA method introduced by our group more than a decade ago to determine astrophysical S-factors for the non-resonant component of radiative proton capture at low energies (tens or hundreds of keV) from one-proton transfer reactions involving complex nuclei at laboratory energies (about 10 MeV/u) [10]. The method was explained in detail in many previous publications, I summarize the main ideas below.

We can choose peripheral proton transfer reactions to extract the ANCs, which can be used to evaluate (p,γ) cross sections important in different types of H-burning processes. The idea behind it is that in peripheral processes it is sufficient to know the overlap integral at large distances, and this is given by a known Whittaker function times a normalization coefficient C_{nlj}, to be determined by experiment. Figure also stresses the importance of having good and reliable optical model potentials (OMP) to make the DWBA calculations, a problem I will not discuss here.

The technique was used in several experiments of this type; I will mention one of the latest studies, on the ^{12}N(p,γ)^{13}O proton capture reaction at stellar energies. It uses the proton transfer reaction ^{14}N(^{12}N,^{13}O)^{13}C with a ^{12}N beam at 12 MeV/u [11]. Figure 3 below, also the image of a slide shown during the lecture, summarizes the whole process. Going from bottom left, clockwise: we have measured the elastic scattering and the one-proton transfer using a ^{12}N beam produced and separated with the MARS spectrometer [12] at Texas A&M University. The elastic scattering data (lower left corner) were used to determine the OMP needed in the DWBA calculations for transfer. The ANC for the system ^{13}O→^{12}N+p was extracted from the transfer data (top left) after which was used to evaluate the non-resonant component of the astrophysical S-factor for the radiative proton capture ^{12}N(p,γ)^{13}O and the corresponding reaction rate as a function of stellar temperature (top right). Finally, the astrophysical consequences are shown in a plot (bottom right) which shows the region of density-temperature where the capture process competes with its competitor (β-decay), in first stars. For comparison, the curves from literature before our data were measured are shown. There is a big change from the original estimates (dashed curves) based on theoretical estimates only.

A variation of the ANC method uses one-neutron transfer reactions to obtain information about the mirror nuclei, for example studying the ^{13}C(^{7}Li,^{8}Li)^{12}C reaction to determine the ANC for ^{8}Li which we then translate into the corresponding structure information (the proton ANC) for its mirror ^{8}B and from there $S_{17}(0)$ for the reaction important in the neutrino production in Sun ^{7}Be(p,γ)^{8}B [4]. We did this using the mirror symmetry of these nuclei: the similarity of their wave functions, expressed best by the identity of the neutron and proton spectroscopic factors for the same *nlj* orbital in the two nuclei $S_p(nlj)=S_n(nlj)$ (of course, the radial wave functions are not identical!). The experiment using these concepts and the results were published in Ref. [13].

I mentioned before that in order to extract data, either the spectroscopic factors, or the ANCs, the experiments have to be compared with calculations, and in the above conditions, the knowledge of the optical potentials is crucial. We established a procedure based on double folding, starting from an effective nucleon-nucleon interaction we call JLM. Florin Carstoiu of Bucharest was instrumental in this work. I will not insist on all these here, but I send you to literature [14].

4.2 Breakup reactions at intermediate energies

Work done in the last decade in several laboratories has demonstrated that one-nucleon removal reactions (or breakup reactions) can be a good and reliable spectroscopic tool. In a typical experiment a loosely bound projectile at energies above the Fermi energy impinges on a target and loses one nucleon. The momentum distributions (parallel and/or transversal) of the remaining core measured after reaction give information about the momentum distribution of the removed nucleon in the wave function of the ground state of the projectile.

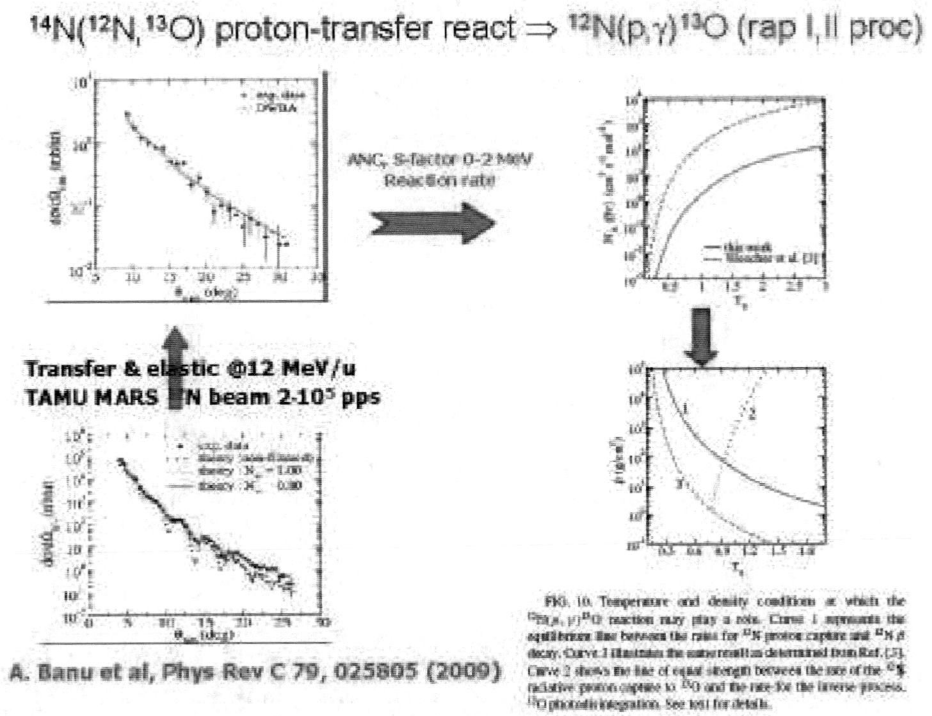

Figure 3. Summary of how elastic and one-proton transfer data measured with secondary RIB (left side) are transformed in nuclear astrophysics information (right side).

The shape of the momentum distributions allows determining the quantum numbers *nlj* of the s.p. wave function (unambiguously only *l* is determined; shell model systematics are needed for the others). It was shown in Ref. 15 that on a large range of projectile energies breakup reactions are peripheral and, therefore, the breakup cross sections can be used to extract asymptotic normalization coefficients. For this to be true, we need, again, careful and reliable reaction model calculations. They need to reproduce all available data from such measurements if they are to be believed. This is a very important point, which I stressed in the lecture. The method to use breakup reaction for nuclear astrophysics was first applied in [15,16] to the breakup of ^{8}B to determine $S_{17}(0)$. All available breakup data, on targets from C to Pb and at energies from 27 MeV/u to 1400 MeV/u were used to determine the ANC for $^{8}B \rightarrow {}^{7}Be+p$. Different reaction models and different nucleon-nucleon effective interactions were used. Consistent ANCs values were obtained, with an overall uncertainty estimated at about 10%. This is a very good agreement, a fact that validates both the $S_{17}(0)$ adopted in the neutrino production calculations pertinent to what was called the "solar neutrino puzzle" before the neutrino oscillations were demonstrated, and the validity of indirect methods in NA.

Another example is the breakup of ^{23}Al and ^{24}Si at intermediate energies [17,18]. The first is a good example as it takes a case where several configurations contribute to make the ground state of the projectile. The participating configurations were disentangled using the detection of gamma-rays from the de-excitation of the remaining core after a proton is removed from the projectile moving at 50-60 MeV/nucleon. It also shows how it is important to combine the results of this nuclear breakup reaction to evaluate the continuum contribution to the reaction rate with those of Coulomb breakup of the same projectile to evaluate the contribution of the resonant part. It is treated in the paper by A. Banu *et al.* and I refer the reader to it [17].

4.3 Decay spectroscopy. Beta-delayed gamma and proton decays.

Among the indirect methods, a large class is the spectroscopy of resonances, in general (transfer reactions, gamma-ray in-beam spectroscopy, decay spectroscopy, etc...). These resonances are meta-stable states in the compound nuclear system produced in reaction as an intermediate step. To evaluate the corresponding contributions to the reaction rates (for narrow, isolated resonances) it is sufficient to determine the location of the resonances (E_r) and their resonance strengths ($\omega\gamma$):

$$\langle \sigma \upsilon \rangle_{res} = \left(\frac{2\pi}{\mu kT} \right)^{3/2} \hbar^2 \omega\gamma \exp\left(-\frac{E_r}{kT} \right)$$

These may be obtained by studying the spectroscopic properties of the corresponding meta-stable states, populated through another, more convenient method. The decay spectroscopy is one such method: instead of measuring radiative proton capture (p,γ) one can study the inverse of its first step, the proton decay of the same state. The states populated by beta-delayed proton decay: in the same compound nucleus, states above the proton threshold are populated by β-decay, and then they decay emitting a proton. This happens if the selection rules for (p,γ) and βp allow for the population of the same states (energy and spin-parity selection rules). One can determine that way the energy of the resonance, determine or restrict the spins and parity and determine the branching ratios. This simple connection is schematically presented in figure 1 of Ref. [ ] for the case of the ^{22}Na(p,γ)^{23}Mg radiative proton capture: we aim at populating and study states in the ^{23}Mg daughter nucleus following the β-decay of ^{23}Al. The selection rules allow that: s-wave radiative capture involves $J^{\pi}=5/2^+$ and $7/2^+$ states; beta-decay populates predominantly positive parity states with spins 3/2, 5/2 and 7/2. Figure 1, a slide from the actual lecture, underlines that we need to locate the resonances and determine their properties (spin and parity and partial widths). Similar situations for the other two proton capture in our list, which we study through the decay of ^{27}P and ^{31}Cl, respectively. I will skip these in favor of sending the reader to the recent papers describing these experiments, the equipment and experimental methods involved, and their results [20,21].

5 Conclusions

After a brief review of nuclear astrophysics most important contributions to our understanding of the Universe, I showed that in many cases we need to use indirect methods to obtain data leading to the evaluation of stellar reaction rates. Three methods involving rare ion beams are described: one-nucleon transfer (the ANC method), nuclear breakup at intermediate energies and decay spectroscopy.

Acknowledgements
All of the examples used in this paper are based on work done along the years with my colleagues from the Cyclotron Institute at Texas A&M University: R.E. Tribble, A.M. Mukhamedzhanov, A. Banu (now at J. Madison University, Harrisonburg, VA, USA), B. Roeder, M. McCleskey, E. Simmons, and A. Spiridon. F. Carstoiu of IFIN-HH Bucharest very beneficially collaborated with us over many years on the reaction theory of subjects touched here. E. Pollacco (CEA/IRFU Saclay)

collaborated on the beta-delayed proton-decay spectroscopy. I thank them all. The original articles or planned publications are cited throughout this paper. The work presented in this paper was supported by U.S. Department of Energy under Grant No. DE-FG03-93ER40773.

References

[1] Crichfield C L and Bethe H 1938 Phys. Rev. **248** 862(L)
[2] Alpher R Bethe H and Gamov G 1948, Phys. Rev. **73** 803
[3] Burbidge E M Fowler B and Hoyle F 1957, *Rev Mod. Phys.* **29** 547
[4] Cameron A C W 1957 *Pub. Astron. Soc. Pac.* **6** 201
[5] Particle Data Group (http://pdg.lbl.gov/2010)
[6] Bahcall J N and Pinsonneault M H 2004 *Phys. Rev. Lett.* **92** 121301
[7] Diehl R et al. 2006 *Nature* **439**,45 doi:10.1038/nature04364
[8] Kratz K –L 2008 in "Exotic Nuclei and Nuclear/Particle Astrophysics (II)", Proc. CSSP07, Sinaia, Romania, eds. Trache L and Stoica S, AIP Conf Proceedings, vol. **972** Melville, New York, p. 298
[9] Rolfs C and Rodney W S 1988 *Cauldrons in the Cosmos*, University of Chicago Press
[10] Mukhamedzhanov A M Gagliardi C A and Tribble R E 2001 *Phys. Rev. C* **63** 024612
[11] Banu A et al. 2009 *Phys. Rev. C* **79** 025805
[12] Tribble R E et al. 1991 *Nucl. Instr. Meth. A* **285** 441
[13] Adelberger et al. 1998 *Rev. Mod. Phys.* **70** 1265; ibidem 2010 **83** 195
[14] Trache L et al. 2003 *Phys. Rev. C* **67**, 062801(R)
[15] Trache L et al. 2000 *Phys. Rev. C* **61** 024612; Carstoiu F et al. 2004 *Phys. Rev. C* **70** 054610
[16] Trache L Carstoiu F Gagliardi C A and Tribble R E 2001 *Phys. Rev. Lett.* **87** 271102
[17] Trache L Carstoiu F Gagliardi C A and Tribble R E 2004 *Phys.Rev. C* **69** 032802(R)
[18] Banu A et al. 2010 *Phys. Rev. C* **84** 015803
[19] Banu A et al. 2012 Phys. Rev. C **86** 015806
[20] Saastamoinen A et al. 2011 *Phys. Rev. C* **83** 045808
[21] Saastamoinen A et al. 2012 *Nuclear Instruments and Methods A* **698** in press

Journal of Physics: Conference Series **413** (2013) 012026 doi:10.1088/1742-6596/413/1/012026

Nuclei in Dense Matter and Equation of State

Stefan Typel

GSI Helmholtzzentrum für Schwerionenforschung, Planckstraße 1, D-64291 Darmstadt, Germany

E-mail: s.typel@gsi.de

Abstract. Correlations are an essential feature of interacting many-body systems such as nuclear and stellar matter. They cause the formation of clusters, i.e. nuclei and scattering resonances, changing the chemical composition and affecting the thermodynamical properties of the system. The equation of state has to be known in a wide range of density, temperature and isospin asymmetry for astrophysical simulations of core-collapse supernovae and compact stars. In a generalized relativistic density functional approach, the properties of dense stellar matter can be described with the correct limits at low densities, the model-independent virial equation of state, and at high densities using a quasiparticle mean-field approach. Effects of correlations and changes of the cluster properties in dilute matter can be studied experimentally in heavy-ion collisions.

1. Introduction

The final stage in the evolution of a massive star ($M_{\rm star} \gtrsim 8 M_{\rm sun}$) is reached when the growing iron core collapses under its own gravitational attraction since nuclear fusion reactions can no longer provide energy and pressure to stabilize the system. A violent supernovae explosion is launched and the outer shells of the star are ejected leaving a neutron star or a black hole as a remnant [1]. The equation of state (EoS) of dense stellar matter determines the dynamical evolution of such a core-collapse supernova, the properties of the (proto-)neutron star [2] and the thermodynamic conditions and chemical composition of the expanding matter where nucleosynthesis processes are presumed to create heavy elements.

The thermodynamic conditions of the matter are set by three parameters: the mass density ϱ (or baryon number density n), the temperature T and the electron fraction Y_e (or neutron-proton asymmetry $\beta = 1 - 2Y_e$). In supernovae simulations these quantities cover large ranges with typical values of $10^{-9} \lesssim \varrho/\varrho_{\rm sat} \lesssim 10$ (with the nuclear saturations density $\varrho_{\rm sat} \approx 2.5 \cdot 10^{14} {\rm g/cm}^3$), 0.1 MeV $\lesssim k_B T \lesssim 50$ MeV and $0 \leq Y_e \lesssim 0.6$. At densities below $\varrho_{\rm sat}$, the relevant degrees of freedom in stellar matter are nucleons, nuclei, electrons and thermal photons. An equation of state of dense matter in thermodynamical and chemical equilibrium assuming local charge neutrality can be used in astrophysical simulations because the timescale of the nuclear reactions is much shorter than that of the dynamical evolution of the system. Neutrinos are not included in the EoS because they are not in equilibrium with the matter and therefore treated explicitly in supernovae simulations.

A large variety of EoS can be found in the literature, from simple parametrizations to sophisticated models, often focusing an specific aspects and properties of the matter. For many years, there were only few global EoS available that were used in astrophysical simulations covering the relevant parameter ranges, e.g. [3, 4, 5], These conventional EoS tables consider only those particles that are also incorporated in astrophysical simulations, i.e. nucleons, electrons, photons, ^{4}He and a representative heavy nucleus. Recently, several new and extended models were developed [6, 7, 8, 9, 10, 11] including the full distribution of nuclei or more exotic particles such as hyperons or quarks.

At densites around and above nuclear saturation density, nuclear matter is expected to be uniform and mean-field models with nucleons as degrees of freedom are very sucessful. A quasiparticle approach is used to take the effects of the interaction into account. At lower densities and temperatures below approximately 15 MeV, inhomogeneities will develop with a coexistence of gas and liquid phases on a macroscopic scale. This feature is the origin of the "liquid-gas" first-order phase transition in nuclear matter, a fictitious system where the long-range Coulomb interaction is neglected. However, in realistic stellar matter with Coulomb interaction and electrons under the condition of charge neutrality, low- and high-density regions are separated by various shapes of the interfaces on a microscopic scale that are often described by the so-called "pasta phases". Solid phases with crystal/lattice structures develop at very low temperatures, e.g. in the outer crust of neutron stars. At very low densities and sufficiently high temperature, light clusters (^{2}H, ^{3}H, ^{3}He, ^{4}He) have to be considered in the EoS model as explicit two-, three-, ..., many-body correlations.

The theoretical description of dense matter at nuclear saturation density and (much) below is the focus of this contribution, in particular the formation of clusters due to correlations and their dissolution with increasing density. Different concepts and approaches to describe dense matter with clusters at low densities are discussed in section 2. A specific approach introducing a generalized relativistic density functional (gRDF) is presented in section 3. The model takes the effects of few- and many-body correlations explicitly into account with the correct low- and high-density limits. The change of the properties of light and heavy nuclei due to the medium are discussed in section 4. Conditions from the low-density limit of the model are examined in section 5. In section 6, the abundances of light clusters and the effects of two-nucleon scattering correlations are considered. The behavior of symmetry energy with cluster formation is studied in section 7. Finally, experimental approaches to test the model predictions are mentioned. Details of the material presented in this contribution and further references can be found in references [12, 13, 14]. Natural units are used in order to simplify the equations, i.e. $\hbar = c = k_B = 1$.

2. Theoretical approaches

The chemical composition and thermodynamic properties of dilute matter vary substantially with density, temperature and neutron-proton asymmetry. Theoretical models of the system can be constructed by starting from two different points of view.

A mixture of nucleons and nuclei in chemical equilibrium is considered in the *chemical picture*. The properties of the particles composed of nucleons are assumed to be independent of the medium. Correlation effects can be taken into account by specifying the interaction individually for all pairs of constituents. A major problem is the description of the cluster dissolution at high densities. A simple geometric concept such as the excluded-volume mechanism can be introduced for this purpose.

Only nucleons are considered as fundamental constituents in the *physical picture*. Correlations are consequences of the nucleon-nucleon interaction. It is the only and essential ingredient in this type of approach leading to the formation of bound states and scattering resonances. In order to treat two-, three-, ... many-body correlations inside the medium, appropriate theoretical methods have to be applied.

In the following, several models are presented that combine both pictures to various degrees considering different effects of the nuclear interaction.

2.1. Low-density models

2.1.1. Nuclear statistical equilibrium model The simplest NSE model assumes an ideal mixture of nucleons (p, n) and nuclei (X) with mass numbers $A = N + Z$ in chemical equilibrium $Zp + Nn \Leftrightarrow {}^A_Z X$ neglecting mutual interactions. Hence the nonrelativistic chemical potentials μ_i of the particles are given by $\mu_X = Z\mu_p + N\mu_n + B_X$ with the binding energies $B_X > 0$ of the nuclei X with mass $m_X = Zm_p + Nm_n - B_X$. For nonrelativistic kinematics and Maxwell-Boltzmann statistics, the grand canonical potential assumes the form

$$\Omega(T, V, \mu_i) = -TV \sum_i \frac{g_i}{\lambda_i^3} \exp\left(\frac{\mu_i}{T}\right) \tag{1}$$

with thermal wavelenghts $\lambda_i = \sqrt{2\pi/(m_i T)}$ and degeneracy factors g_i by summing over protons, neutrons and all nuclei. Excited states of nuclei are usually taken into account by introducing temperature dependent degeneracy factors $g_i(T)$. All thermodynamical properties, e.g. the equations of state, such as $pV = NT$ and $E = 3NT/2$ with $N = V\sum_i n_i$, for a mixture of ideal gases can be derived from Eq. (1). In particular, the individual particle number densities are given by $n_i = -\left.\partial\Omega/\partial\mu_i\right|_{T,V,\mu_{j\neq i}} = g_i \lambda_i^{-3} \exp\left(\mu_i/T\right)$. The densities of the particles species are connected by the law of mass action that is well known in chemistry. The simple NSE model cannot describe the dissolution of nuclei with increasing density. Their abundance can be suppressed with increasing density using an excluded-volume mechanism, see, e.g., reference [6].

2.1.2. Virial equation of state The VEoS can incorporate two-, three-, ..., many-body correlations due to the interaction between the constituents using an expansion of the grand canonical partition function in powers of the particle fugacities $z_i = \exp\left(\mu_i/T\right)$. This approximation is only valid for small $z_i \approx n_i \lambda_i^3 \ll 1$, i.e. for low densities and not too low temperatures. The grand canonial potential in the virial expansion is given by

$$\Omega(\mu_i, T, V) = -TV \left(\sum_i b_i \frac{z_i}{\lambda_i^3} + \sum_{ij} b_{ij} \frac{z_i z_j}{\lambda_i^{3/2} \lambda_j^{3/2}} + \dots \right) \tag{2}$$

with the dimensionless cluster (virial) coefficients $b_i = g_i$, b_{ij}, ... The second, third, ... virial coefficients vanish for independent particles without interaction and the NSE result is recovered.

The second virial coefficient b_{ij} encodes the effect of correlations between particles i and j. In classical mechanics it depends directly on the two-body interaction potential V_{ij}. G.E. Beth and E. Uhlenbeck derived the quantum mechanical generalization of b_{ij} [15, 16] by replacing the integration over phase space by a summation over quantum states. The second virial coefficient

$$b_{ij}(T) = \frac{1 + \delta_{ij}}{2} \left(\frac{m_i + m_j}{\sqrt{m_i m_j}} \right)^{3/2} \int dE \, D_{ij}(E) \exp\left(-\frac{E}{T}\right) \pm \delta_{ij} g_i 2^{-5/2} \, . \tag{3}$$

depends on the quantity

$$D_{ij}(E) = \sum_k g_k^{(ij)} \delta\left(E - E_k^{(ij)}\right) + \sum_l \frac{g_l^{(ij)}}{\pi} \frac{d\delta_l^{(ij)}}{dE} \tag{4}$$

that can be seen as the difference of the level densities between the correlated and uncorrelated two-body system. There are contributions from bound states at energies $E_k^{(ij)} < 0$ and scattering states with phase shifts $\delta_l^{(ij)}$ in channels k and l, respectively. The last term in equation (3) is a quantum statistical correction with positive (negative) sign if i and j are identical bosons (fermions).

The low-density behavior of the EoS can be established model-independently [17] since the binding energy of the deuteron and the scattering phase shifts are known experimentally in the two-nucleon system. Thus, the VEoS can serve as a benchmark at low densities but the dissolution of clusters with increasing density cannot be accounted for.

2.1.3. Quantum statistical model In the QS or generalized Beth-Uhlenbeck approach, medium effects on the properties of nuclei in matter can be included. The approach was formulated using thermodynamic Green's functions [18]. The EoS of an interacting many-body system is derived by expressing the total nucleon number density with the help of the spectral functions $A(j,E)$ for single-particle states $j \equiv (\vec{p}_j, \sigma_j, \tau_j)$. The spectral function depends on the complex valued self-energy that contains the information on the interaction. After expanding $A(j,E)$ for small imaginary parts of the self-energy, the total density splits into two parts $n = n_{\text{free}} + 2n_{\text{corr}}$. The first contribution $n_{\text{free}} = \sum_j f_{FD}[e(j)]$ with the Fermi-Dirac distribution function f_{FD} is the density of free quasiparticles with quasiparticle energies $e(j)$. The correlation density

$$n_{\text{corr}} = \sum_k g_k^{(2)} \sum_{\vec{P},P>P_{\text{Mott}}} f_{BE}(E_{\text{cont}} - B_k) + \sum_l g_l^{(2)} \sum_{\vec{P}} \int \frac{dE}{\pi} 2\sin^2\delta_l \frac{d\delta_l}{dE} f_{BE}(E_{\text{cont}} + E) \quad (5)$$

with the Bose-Einstein distribution function f_{BE} receives contributions from bound states and from scattering states. The form of the correlation density resembles that of the second virial coefficient (3) in the VEoS. However, the properties of the two-body states in the QS model depend on their c.m. momentum $\vec{P}$ with respect to the medium. There are no two-body bound states for P below the Mott momentum P_{Mott} due to the action of the Pauli principle that blocks states by the medium. The energy of the scattering states with zero relative momentum defines the continuum edge E_{cont}. The binding energies B_k and the in-medium scattering phase shifts δ_l depend on the medium properties and $\vec{P}$. They have to be determined from the in-medium T matrix in general. The additional $2\sin^2\delta_l$ factor in the continuum contribution reduces the strength of the explicit scattering correlations since the self-energies contain a part of the correlation effect due to the interaction.

The medium-dependent shift of the nuclear binding energies can be calculated by solving the appropriate in-medium Schrödinger equation for light nuclei with mass number $A \leq 4$ with realistic nucleon-nucleon potentials or with nucleon self-energies taken from phenomenological mean-field models. See subsection 4.1 for results. A different approach is used for heavier nuclei ($A > 4$), see subsection 4.2.

2.2. Intermediate/high-density models

There are two major classes of theoretical approaches that describe nuclear matter successfully near nuclear saturation density. Phenomenological approaches use effective interactions, e.g. nonrelativistic Hartree-Fock models with the Skyrme or Gogny interaction or relativistic mean-field models with meson exchange. Results depend on a small number of parameters that are usually determined by fits to properties of finite nuclei. "Ab-initio" approaches employ realistic nucleon-nucleon interactions that were fitted to two- (and three-) nucleon bound and scattering states, see, e.g., (non-)relativistic Brueckner-Hartree-Fock calculations. The interaction is given from the outset and the quality of the results is determined by the sophistication of the many-body method. Unfortunately, the models have some limitations, e.g., often they can be applied only to uniform nuclear matter.

Phenomenological mean-field models are the main choice in order to obtain quantitatively reasonable results for an EoS for astrophysical applications in a wide range of temperature, density and neutron-proton asymmetry. Relativistic approaches are the preferred choice, as they avoid, e.g., the problem of a superluminal speed of sound. Despite their virtues, explicit correlations are not considered in mean-field models with only nucleons as basic constituents. The formation and dissolution of nuclei is not incorporated into the description.

3. Generalized relativistic density functional

Nucleons are considered as quasiparticles in conventional relativistic mean-field (RMF) with self-energies that depend on the medium properties and take into account the effects of the interaction. For astrophysical applications when the EoS of charge neutral stellar matter is needed and the Coulomb interaction has to be included, electrons (and myons) can be added. The nuclear/electromagnetic interaction is modeled by an exchange of mesons/photons that couple minimally to the nucleons. The mesons/photon are generally treated as classical fields.

Additional degrees of freedom are introduced in a generalized relativistic density functional (gRDF) approach by including also nuclei as constituents and considering correlations in an effective way. The Lagrangian density of a RMF model was extended in reference [12] by introducing light nuclei (^{2}H, ^{3}H, ^{3}He and ^{4}He) as spin 1, 1/2 and 0 fields. They interact minimally with the meson and photon fields like the nucleons, however, with rescaled couplings. In addition, it is assumed that the binding energies of the nuclei depend on the medium properties. This dependence enters through functions of the temperature and vector meson fields in the particular approach of reference [12].

In reference [13], nucleon-nucleon scattering correlations, i.e. the nn, np and pp scattering states, were added to the model. The continuum correlation in a particular channel is represented by a single state with an effective "resonance" energy, that depends on the medium properties like the binding energy of nuclei, and a temperature dependent degeneracy factor, similar as excited states of nuclei in NSE models.

Heavy nuclei with mass number $A > 4$ can be considered in a similar way as light nuclei. See subsection 4.2 concering the medium dependence of their binding energy shifts.

For an quantitative description of nuclear matter and finite nuclei, it is not suffcient to assume minimal nucleon-meson couplings with constant strength Γ_{im} between particles i and mesons m in RMF models. A medium dependence of the effective nuclear interaction has to be considered in the model. The traditional approach uses nonlinear self-interactions of the mesons in the Lagrangian density with a polynomial dependence on various combinations of the meson fields. A more recent and more flexible approach is motivated by results of Dirac-Brueckner calculations of nuclear matter [19]. Here the couplings Γ_{im} depend on the density of the medium with a well controlled asymptotic behavior. The gRDF model assumes such density dependent couplings.

The parameters in phenomenological RMF models are determined in fits to properties of nuclei and nuclear matter. Their number is rather small (≈ 10). When clusters are introduced in the gRDF approach, additional parameters appear that determine the medium dependence of the binding energies. EoS models can be tested by using constraints from nuclear physics and astrophysics, see, e.g., reference [20].

In the gRDF approach a density functional serves as the starting point to derive all equations and thermodynamical quantities. In the particular case of reference [12], the grand canonical thermodynamical potential

$$\Omega = \int d^3r\, \omega_g(T, \mu_i, \sigma, \delta, \omega_0, \rho_0, \vec{\nabla}\sigma, \vec{\nabla}\delta, \vec{\nabla}\omega_0, \vec{\nabla}\rho_0, \vec{\nabla}A) \tag{6}$$

with the potential density ω_g is chosen. The temperature T, the chemical potentials μ_i of all particles i, the meson and photon fields σ, δ, ω_0, ρ_0, A and their derivatives are the independent variables. There are contributions from nucleons, bound states of nuclei and two-nucleon scattering states in the baryonic sector. The particle masses are given by

$m_i = N_i m_n + Z_i m_p - (1-\delta_{in})(1-\delta_{ip})B_i^{(\mathrm{vac})}$ with the vacuum binding energy $B_i^{(\mathrm{vac})} \geq 0$ of the composite systems ($B_i^{(\mathrm{vac})} = 0$ for the scattering states). The energy of a baryon or lepton with momentm $\vec{k}$ has the relativistic form $E_i = \sqrt{k^2 + (m_i - S_i)^2} + V_i$ with scalar potentials S_i and vector potentials V_i that depend on the meson fields σ, δ, ω_0, ρ_0, the photon field A and the couplings Γ_{im}. The scalar self-energy contains medium dependent binding energy shifts. In the vecter self-energy, a rearrangment contribution appears due to the density dependence of the couplings. Neglecting antiparticle and boson condensate contributions, the potential density in equation (6) assumes the form

$$\begin{aligned}
\omega_g &= \sum_i \mp g_i T \int \frac{d^3k}{(2\pi)^3} \ln\left[1 \pm \exp\left(-\frac{E_i - \tilde{\mu}_i}{T}\right)\right] - \frac{1}{2}\left(m_\omega^2 \omega_0^2 + m_\rho^2 \rho_0^2 - m_\sigma^2 \sigma^2 - m_\delta^2 \delta^2\right) \quad (7) \\
&+ \frac{1}{2}\left(\vec{\nabla}\omega_0 \cdot \vec{\nabla}\omega_0 + \vec{\nabla}\rho_0 \cdot \vec{\nabla}\rho_0 - \vec{\nabla}\sigma \cdot \vec{\nabla}\sigma - \vec{\nabla}\delta \cdot \vec{\nabla}\delta + \vec{\nabla}A_0 \cdot \vec{\nabla}A_0\right) \\
&+ \left(\Gamma'_\omega \omega_0 n_\omega + \Gamma'_\rho \rho_0 n_\rho - \Gamma'_\sigma \sigma n_\sigma - \Gamma'_\delta \delta n_\delta\right)(n_n + n_p)
\end{aligned}$$

where the upper (lower) signs applies to fermions (bosons). The source densities in equation (7) are determined by $n_k = \sum_i g_{ik} n_i$ for $k = \omega, \rho$ and $n_k = \sum_i g_{ik} n_i^s$ for $k = \sigma, \delta$, respectively, with scaling factors g_{ik}, vector densities n_i and scalar densities n_i^s of the individual constituents. The relativistic chemical potentials of the baryonic particles are given by $\tilde{\mu}_i = N_i \tilde{\mu}_n + Z_i \tilde{\mu}_p = \mu_i + m_i$. Hence, there are only two independent chemical potentials due to the condition of chemical equilibrium. The degeneracy factors g_i can depend on the temperature T. Note that n_i and n_i^s have to be considered as functions of the independent variables of Ω. The field equations, derived from ω_g, include rearrangement contributions that ensure the thermodynamic consistency of the model. The set of coupled field equations has to be solved self-consistently for given parameters T, n and Y_e.

4. Medium effects on nuclear binding energies

The medium dependence of the nuclear binding energies is the essential feature of the gRDF approach that causes the reduction of the cluster abundances with increasing density. The shift originates mainly from action of the Pauli principle suppressing the possibility of a cluster formation.

4.1. Light nuclei

For light clusters ($A \leq 4$), results of the QS model are adopted for the medium dependent shifts of the binding energies. In the gRDF approach, they are represented as functions of temperature and an effective density, see reference [12] for details, in particular figure 1. With increasing medium density, the binding energy of a cluster decreases smoothly from its experimental value in the vacuum. Depending on the medium temperature, it finally crosses zero, indicating that the cluster becomes unbound. In reference [13] different extrapolations to high densities were explored. See reference [21] for more recent parametrizations of the energy shifts, including the momentum dependence.

4.2. Heavy nuclei

The binding energy shifts of heavy nuclei are calculated using the relativistic density functional considering nucleons and electrons. For this purpose, the field equations are solved self-consistently in a spherical Wigner-Seitz cell employing an extended Thomas-Fermi approximation with inhomogeneous density distribution of nucleons and electrons under the condition of a vanishing total charge. A heavy cluster is formed in the center of the cell that is surrounded by a low-density gas of nucleons, see figure 11 of reference [14]. Electrons form

Journal of Physics: Conference Series **413** (2013) 012026 doi:10.1088/1742-6596/413/1/012026

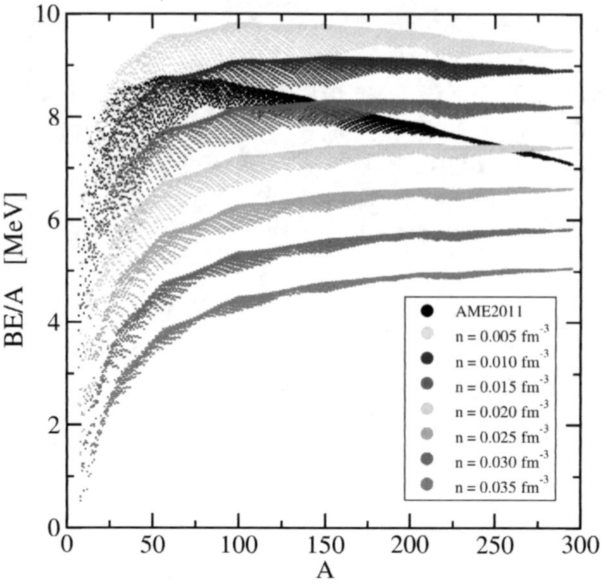

Figure 1. Binding energy per nucleon of nuclei with mass number A in vacuum (AME2011 [22]) and in the medium at temperature $T = 0$ MeV for various densities n. Figure from reference [14].

a highly degenerate Fermi gas resulting in an practically uniform distribution inside the cell. They screen the Coulomb potential of the protons. When light clusters are considered in the calculation in addition to nucleons and electrons, an enhancement of the light cluster abundancies at the surface of the heavy cluster is observed.

When the energies of the Wigner-Seiz cell calculation with uniform and nonuniform density distributions are compared, the binding energy of a nucleus at a certain density of the medium can be determined since it represents the energy gain due to the correlations. The extended Thomas-Fermi approximation does not take into account shell effects and it is more reasonable to extract only relative shifts of the binding energies from the calculations than to extract absolute binding energies in the medium.

In figure 1, experimental binding energies per nucleon [22] as a function of the mass number A are depicted in by black points. The maximum near $A = 60$ and shell effects close to doubly magic nuclei are clearly visible. Taking into account the the binding energy shifts in the medium, the A dependence clearly changes. The screening of the Coulomb field by the electrons causes a stronger binding of the nuclei at low medium densities. At higher medium densities, the binding energies of the nuclei reduce substantially. Nuclei with low mass numbers dissolve at lower medium densities than clusters with larger mass numbers.

In the so-called single-nucleus approximation (SNA), only one representative nucleus is considered in traditional EoS tables for astrophysical applications, see, e.g., Refs. [3, 4]. The full distribution of nuclei is covered only in more recent NSE-type models [6, 7]. The latter approaches usually account for the Coulomb screening effect but not for the reduction of the nuclear binding energies at higher medium densities.

5. Low-density limit

The VEoS represents the correct description of the matter properties at low densities and finite temperatures. The gRDF model in constructed such that the model-independent low-density results of the VEoS are reproduced. Comparing the fugacity expansions of the grand thermodynamical potential Ω in both models, consistency relations can be derived to determine the effective resonance energies and degeneracy factors of the nucleon-nucleon scattering states that both depend on T. Due to relativistic kinematics, there are already corrections to the

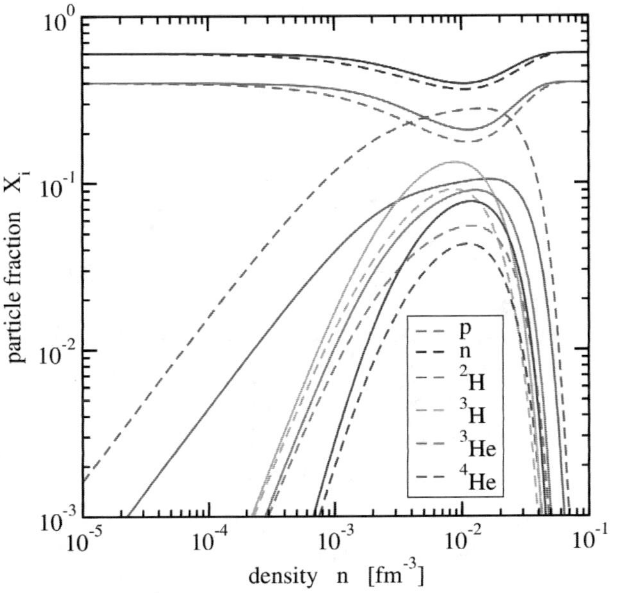

Figure 2. Particle fractions $X_i = n_i/n$ in nuclear matter with temperature $T = 10$ MeV and total proton fraction $Y_p = 0.4$ as a function of the total nucleon density n in the gRDF approach without heavy clusters. Calculation without (dashed lines) and with (full lines) nucleon-nucleon scattering correlations. Heavy clusters with mass number $A > 4$ are not considered in this calculation. Figure adapted from reference [14].

first order coefficient b_i in equation (6) of the VEoS. The effective resonance energies increase smoothly with temperature as depicted in figure 2 of reference [13]. At low temperatures, the contribution of the s-wave channel dominates and analytical results for the effective resonance energies can be obtained using the effective-range expansion of the phase shifts. Consistency relations for the nn, pp and np channels are found from a comparison of the second virial coefficients in the VEoS and gRDF approaches. The occurrence of a term that depends on the meson couplings Γ_k at zero density through the coefficients $C_\pm = C_\omega - C_\sigma \pm C_\rho \mp C_\delta$ with $C_k = [\Gamma_k(0)]^2/m_k^2$ is the important feature in these relations. The consistency relations serve to define the effective degeneracy factors for the effective two-nucleon scattering states. Again, a smooth dependence on the temperature is found as depicted in figure 3 of reference [13].

In the zero-temperature limit, two conditions are found that connect the differences $C_\omega - C_\sigma$ and $C_\rho - C_\delta$ with the scattering lengths of the four relevant nucleon-nucleon s-wave scattering channels, see references [13, 14] for details.

In pure neutron matter, there is no bound state and two-nucleon correlations appear only in the nn scattering channel. In figure 5 of reference [13] the internal energy per neutron E/N is depicted as a function of the total neutron density in different approximations. The gRDF model with the constribution of the effective nn scattering correlation perfectly reproduces the VEoS at low densities.

6. Light clusters and continuum correlations

The composition of matter changes substantially when the parameters T, n and Y_p are varied. In figure 2 the evolution of the particle mass number fractions $X_i = A_i n_i/n$ is depicted as a function of the total nucleon density n for $T = 10$ MeV and $Y_p = 0.4$ for the gRDF model using the parametrization DD2 [12]. At very low densities, nucleons dominate the composition of matter. There is a tiny fraction of deuterons and the abundance of heavier clusters can be neglected. Contributions of clusters to the composition become more important with increasing density and the abundance of free nucleons is reduced. The total cluster fraction reaches a maximum at approximately 1/10 of the nuclear saturation density. A further increase of the medium density causes the reduction of the cluster fraction and finally they disappear. Hence, the gRDF model successfully describes the dissolution of clusters in matter and the Mott effect is observed.

Journal of Physics: Conference Series **413** (2013) 012026 doi:10.1088/1742-6596/413/1/012026

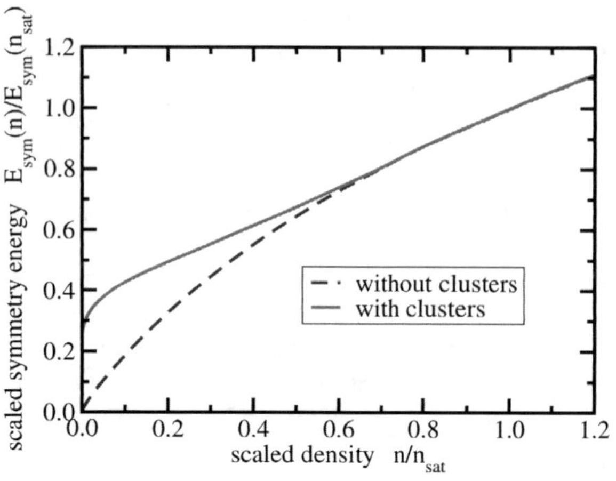

Figure 3. Scaled symmetry energy of nuclear matter at zero temperature as function of the scaled density in the gRDF model with parametrization DD2 without (blue dashed line) and with (red full line) cluster formation taken into account. Figure adapted from reference [14].

Two-nucleon scattering correlations substantially affect the cluster fractions. The number of deuteron-like correlations is reduced and the remaining particle fractions are redistributed. Note that the particle abundancies in the gRDF model represent those of quasiparticles. Some part of the correlation strength is accounted for by the finite self-energies of the quasiparticles and the size of explicit correlations is reduced.

7. Symmetry energy

The symmetry energy $E_{\text{sym}}(n)$ of nuclear matter is currently investigated with much experimental and theoretical effort. See, e.g., references [23, 24] for constraints and the importance in nuclear physics and in astrophysical applications. The appearance of clusters strongly modifies the density dependence of E_{sym} (including Coulomb contributions) at low densities, as depicted in figure 3 for matter at zero temperature. Conventional mean-field models of uniform matter without cluster degrees of freedom predict that $E_{\text{sym}}(n)$ approaches zero in the limit of vanishing density n. With correlations and cluster formation the symmetry energies rises as compared to that in mean-field models without correlations. For vanishing temperature it even approaches a finite value in the limit of zero density. The effects will be less pronounced at higher temperatures, but possibly still detectable in experiments.

The analysis of heavy-ion collisions can help to extract the symmetry energy and to study the properties of fragments in dilute matter by determining the thermodynamical properties of the expanding system. Recent results indicate an increase of the symmetry energy as predicted by models with correlations and cluster formation, see references [25, 26, 27, 28, 29] for details.

8. Conclusions

Correlations strongly affect the thermodynamic properties and composition of an interacting many-body system such as nuclear matter. Clusters can form and dissolve with changing density and temperature of the medium. This aspect has to be considered in theoretical models for the EoS of dense matter with possible consequences in astrophysical applications. The effects can be studied also experimentally in heavy-ion collisions.

In this contribution, a generalized relativistic density functional approach was presented that interpolates between the correct low-density limit, the VEoS, and the high-density quasiparticle description. Nucleons, two-nucleon continuum states and nuclei are considered as degrees of freedom. They are treated as quasiparticles with medium-dependent properties. The description of light clusters was examined in some some detail, see references [12, 13]. At high densities, the dissolution of the composite systems is observed. The approach of the gRDF model can be extended to include heavy clusters in the future. The final aim for the application of the gRDF model is to provide extensive EoS tables of dense matter.

Acknowledgments
The methods and results presented in this contribution were developed in close cooperation with Gerd Röpke (Universität Rostock), David Blaschke (Uniwersytet Wrocławski), Thomas Klähn (Uniwersytet Wrocławski), Hermann Wolter (Ludwig Maximilians-Universität München) and Maria Voskresenskaya (GSI Darmstadt). The authors thanks them for the fruitful collaboration. This work was supported in part by the DFG cluster of excellence "Origin and Structure of the Universe", by CompStar, a Research Networking Program of the European Science Foundation (ESF), by the Helmholtz International Center for FAIR within the framework of the LOEWE program launched by the state of Hesse via the Technical University Darmstadt and by the Helmholtz Association (HGF) through the Nuclear Astrophysics Virtual Institute (VH-VI-417).

References

[1] Janka H-Th, Langanke K, Marek A, Martinez-Pinedo G and Mueller B 2007 *Phys. Rep.* **442** 38
[2] Glendenning N K 2000 *Compact Stars: Nuclear Physics, Particle Physics, and General Relativity* (New York: Springer)
[3] Lattimer J M and Swesty F D 1991 *Nucl. Phys.* A **535** 331
[4] Shen H, Toki H, Oyamatsu K and Sumiyoshi K 1998 *Nucl. Phys.* A **637** 435
[5] Shen H, Toki H, Oyamatsu K and Sumiyoshi K 1998 *Prog. Theor. Phys.* **100** 1013
[6] Hempel M and Schaffner-Bielich J 2010 *Nucl. Phys.* A **837** 210
[7] Botvina A S and Mishustin I N 2010 *Nucl. Phys.* A **843** 98
[8] Shen G, Horowitz C J and Teige S 2010 *Phys. Rev.* C **82** 015806
[9] Shen H, Toki H, Oyamatsu K and Sumiyoshi K 2011 *Ap. J. Suppl. Series* **197**, 20
[10] Shen G, Horowitz C J and Teige S 2010 *Phys. Rev.* C **83** 035802
[11] Shen G, Horowitz C J and O'Connor E 2011 *Phys. Rev.* C **83** 065808
[12] Typel S, Röpke G, Klähn T, Blaschke D, and Wolter H H 2010 *Phys. Rev.* C **81** 015803
[13] Voskresenskaya M D and Typel S 2012 *Nucl. Phys.* A **887** 42
[14] Typel S 2012 Clusters in Nuclear Matter and the Equation of State for Astrophysical Applications *Proc. Int. Workshop XII Hadron Physics (April 22 - 27, 2012, Bento Gonçalves, RS, Brazil)* ed M V T Machado (Melville, NY: American Institute of Physics) to be published
[15] Beth G E and Uhlenbeck E 1936 *Physica* **3** 729
[16] Beth G E and Uhlenbeck E 1937 *Physica* **4** 915
[17] Horowitz C J and Schwenk A 2006 *Nucl. Phys.* A **776** 55
[18] Schmidt M, Röpke G and Schulz H 1990 *Ann. Phys. (N.Y.)* **202** 57
[19] Typel S and Wolter H H 1999 *Nucl. Phys.* A **656** 331
[20] Klähn T, *et al.* 2006 *Phys. Rev.* C **74** 035802
[21] Röpke G 2011 *Nucl. Phys.* A **867** 66
[22] Audi G and Meng W 2011 private communication
[23] Tsang M B *et al.* 2012 Constraints on the symmetry energy and neutron skins from experiments and theory *Preprint* arXiv:1204.0466
[24] Lattimer J M and Lim Y 2012 Constraining the Symmetry Parameters of the Nuclear Interaction *Preprint* arXiv:1203.4286
[25] Kowalski A *et al.* 2007 *Phys. Rev.* C **75** 014601
[26] Natowitz J B *et al.* 2010 *Phys. Rev. Lett.* **104** 202501
[27] Hagel K *et al.* 2012 *Phys. Rev. Lett.* **108** 062702
[28] Qin L *et al.* 2012 *Phys. Rev. Lett.* **108** 172701
[29] Wada R *et al.* 2012 *Phys. Rev.* C **85** 064618

Probing the shell structure of heaviest nuclei with α-decay data

Andreea-Ioana Budaca[1] and Ion Silisteanu

Horia Hulubei National Institute of Physics and Nuclear Engineering, Bucharest-Magurele, RO-077125, Romania, EU

E-mail: [1] `abudaca@theory.nipne.ro`

Abstract. A standard many-body theory of alpha decay is developed using shell model clustering and resonance scattering amplitudes. Formulas are obtained for the α half-lives and applied extensively to known superheavy nuclei. These formulas are combined with measured decay energy to investigate the structure of the decaying states. The work includes formal considerations, as well as practical computational methods, based on self-consistent models for the nuclear structure and reaction dynamics.

1. Introduction

For the determination of the α-formation amplitude we employ here the microscopic shell model wave functions. This method has been initiated for the harmonic oscillator single-particle wave functions [1] and extended to Woods-Saxon wave functions [2]. We use two kinds of formation amplitudes: one-body resonance amplitude that results in Breit procedure [3] and the shell model formation amplitude given on the basis of shell model single-particle wave functions [4]. In case of the α-decay of a single resonance state k into a set n of channels, the partial half-live is: $T_n^k = ln2 \cdot \hbar/\Gamma_n^k$, the partial decay width being expressed [5] as:

$$\Gamma_n^k = 2\pi \left| \frac{\int_{r_{min}}^{r_{max}} I_n^k(r) u_n^0(r) dr}{\int_{r_{min}}^{r_{max}} I_n^k(r) u_n^k(r) dr} \right|^2 , \tag{1}$$

where $I_n^k(r)$ is the particle (cluster) formation amplitude (FA)

2. Methods

Resonance formation amplitude (RFA) The RFA was defined [5] by means only of the eigenvalues and eigenfunctions. The asymptotic overlap integral or the RFA

$$I_n^k(r \to \infty) \longrightarrow I_n^{res}(r) = u_n^0(r) \simeq \bar{G}_n(r), \tag{2}$$

is represented by the normalized irregular Coulomb w.f. $\bar{G}_n$.

Shell model α-formation amplitude (SM) Following [1] we use in Eq.(2) the w.f. $\mid \Psi_k^{SM}(r_i)\rangle = \det \parallel \psi_{nlj}^{SM}(r_i) \parallel$, i=1,A and $\mid \Phi_D^{SM} >= \det \parallel \psi_{nlj}^{SM}(\eta_{1i}) \parallel$, i=1,A-4 , and the α-particle wave function is chosen as: $\Phi_p(\eta_2) = (2/(1/2)!)^{3/2}(\beta/\pi)^{9/4}(\rho_1^2 + \rho_2^2 + \rho_3^2)(4\pi)^{-3/2}\chi_{00}(s_1 s_2)\chi_{00}(s_3 s_4)$, where the α-particle oscillator parameter is $\beta = 0.484\ fm^{-2}$, χ_{00} is the singlet spin function and the internal spatial coordinates $\rho_1 = (r_1 - r_2)/\sqrt{2}$, $\rho_2 = (r_3 - r_4)/\sqrt{2}$, $\rho_3 = (r_1 + r_2 - r_3 - r_4)/2$, are connected to the individual coordinates r_i of four nucleons. Replacing these functions in Eq.(2) we get

$$I_n^{k[SM]}(r) = r\langle \Psi_k^{SM}(r_i) \mid \mathcal{A}\{[\Phi_D^{SM}(\eta_1)\Phi_p(\eta_2)Y_{lm}(\hat{r})]_n\}\rangle. \tag{3}$$

We used s.p. configurations to α-half-lives ($\rightarrow$ $\log T_\alpha^{SM}$) Z=102-120$\rightarrow$p (s.p. proton states): $1i_{13/2}, 2f_{7/2}, 2f_{5/2}, 3p_{3/2}, 3p_{1/2}$; N=150-178$\rightarrow$n (s.p. neutron states): $2g_{9/2}, 2g_{7/2}, 3d_{5/2}, 3d_{3/2}, 4s_{1/2}$.

Empirical approximations:

Viola-Seaborg formula [6] (**VS**): $\log T_\alpha(s) = (aZ_d + b)Q_\alpha^{-1/2} + (cZ_d + d) + h_{e-o}$, with the parameters a, b, c, d [8] and even-odd corrections h_{e-o} for different (Z, N) parities.

Brown formula [7] (**B**): $\log T_\alpha(s) = 9.54 Z_d^{(0.6)} \cdot Q_\alpha^{-1/2} - 51.37 + h_{e-o}$, with the numerical constants being determined from the fit of 119 data points (T_α, Q_α), Z_d=74-106, only for even-even nuclei.

The effective decay energy includes screening corrections $E_{sc} = 6.53 \cdot 10^{-5} \cdot Z_d^{7/5} - 8.0 \cdot 10^{-5} \cdot Z_d^{2/5}$: $Q_\alpha(MeV) = (A/(A-4)) \cdot E_\alpha^{exp} + E_{sc}$. T_α^{exp} and E_α^{exp} are taken from Refs. [9–12] and therein references.

3. Results

The earliest systematics of α-decay lifetimes of naturally emitters from actinides region was obtained by Geiger and Nuttal in 1911 plotting the experimental values of $logT_\alpha$ vs. $Q_\alpha^{-1/2}$. Fig.1a shows a version of this plot for the g.s.-g.s. α half-lives of known SHN calculated for the simplest case with "one body" approximation (Eq.(5)). We see that all values are distributed on parallel lines for different Z_d values. Inspired by later formulations as Viola-Seaborg or Brown we plot $logT_\alpha$ vs. $Z_d Q_\alpha^{-1/2}$ (Fig.1b). The order of Z_d lines is now reversed suggesting a possible interpolation between Fig.1a and Fig.1b. After several representations we found that for $\log T_\alpha$ vs. $(Z_d^{0.6} Q_\alpha^{-1/2})$ the data fall on a single line and have the minimum rms value (Fig.1c). Although, this dependence does not have a physical interpretation, it came out as well numerically from the WKB calculations performed by Brown in [7].

Further we see that the presently known data points for SHN (Fig. 3) also fall on a nearly straight universal line and it is parallel with the Brown fit line. From the fit of the calculated values of $logT_\alpha^{res} + h_{e-o}$ (s) and experimental ones $logT_\alpha^{exp}$ (s) as functions $Z_d^{0.6} Q_\alpha^{-1/2}$, we obtain two simple formulas for α half-live:

$$logT_\alpha^{res}(\mathrm{s}) = 9.93(Z_d^{0.6} Q_\alpha^{-1/2}) - 53.83, \quad rms^{res} = 0.37, \tag{4}$$
$$logT_\alpha{}^{exp}(\mathrm{s}) = 9.69(Z_d^{0.6} Q_\alpha^{-1/2}) - 51.37, \quad rms^{exp} = 0.52. \tag{5}$$

Eqs.(7,8) can be written in a generalized form as: $logT_\alpha = AZ_d^{0.6} Q_\alpha^{-1/2} + B$ where the parameters A and B are determined by a fitting procedure, like in the case of the empirical formulas presented above.

In the case of the SHN the growth of the available data amount has become so considerable as to demand a reexamination of regularities in α-decay properties [5]. In order to do this, we plotted in Figs.4-9 the Q_α and $\log T_\alpha$ values as function of the neutron number for even-Z and respectively odd-Z isotopic chains. Large gaps in Q_α are observed in Fig.4 when passing from $Z = 110$ to 108, and from $Z = 116$ to 114 regions. This is the direct confirmation of the shell closures at $Z = 108$, $N = 162$ and $Z = 114$, $N = 184$. Similar behavior of Q_α can be seen in Fig.5 when passing from $Z = 109$ to 107. The slope of Q_α vs. N changes significantly for the nuclides with $Z = 108$ and 110. Such an effect might be due to the transition from deformed to spherical shapes in successive α-decays, in agreement with the results of Ref.[11]. The trends of T_α^{exp} and T_α^{SM} vs. N remain practically the same, excepting the heaviest isotopes of $Z = 111 - 115$. These discrepancies can be due to the measurements errors All the SHN found up to now are believed to be well deformed. The measured E_α^{exp}-values and α half-lives Figs.4-9 confirm the special stability of the neutron deformed shell at $N = 162$ ($Z = 108$), and the the spherical shell in the proximity of $N = 184$ ($Z = 114$). The four magic character of the

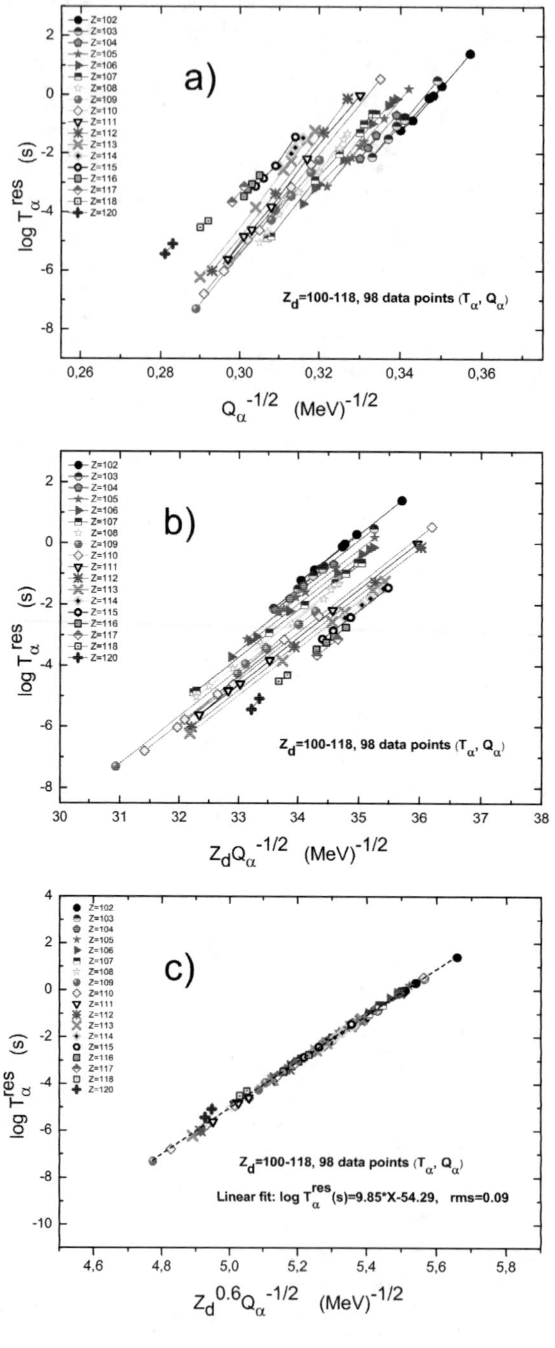

Figure 1. Calculated values for $\log T_\alpha^{res}$ with Eq.(1), I_n^k from Eq.(5). Q_α is the effective decay energy and Z_d is the charge number of the daughter nucleus.

systems $\alpha + ({}^{270}\text{Hs}, {}^{298}114)$ manifests trough exceptionally large Q_α-values (Figs.4,5) and short α half-lives (Figs.6-9). These indicate not only the major influence and importance of magicity for the α-decay channels, but also high probabilities in the formation and transmission probabilities.

Since the SHN are joined in long α-decay chains, it is possible to interrelate on the energy content basis most different nuclear species and to determine regions of nuclear stability higher than in neighboring regions where the nucleons are less bound. The study of nuclei passing from spherical to deformed shapes in the course of their successive α-decays provides valuable information about the influence of significant structure changes on the nuclear decay properties of SHE.

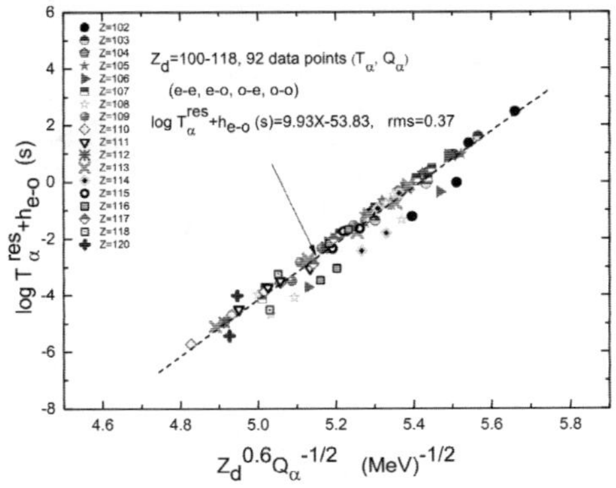

Figure 2. Calculated values (Eq.(1), I_n^k from Eq.(5)) of $\log T_\alpha^{res}$ (s), with even-odd corrections h_{e-o} added, are plotted vs. $Z_d^{0.6} Q_\alpha^{-1/2}$.

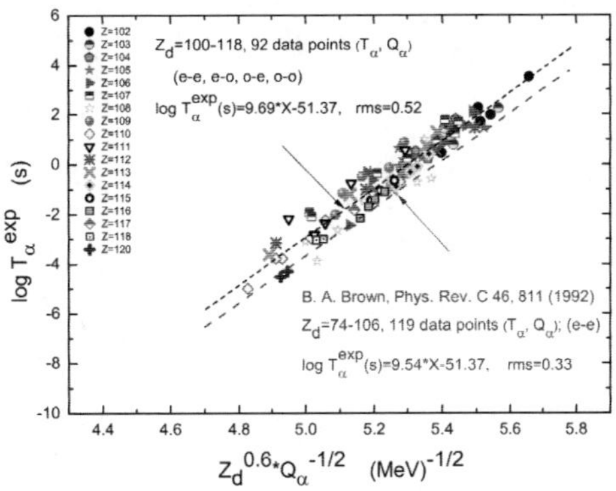

Figure 3. The values of $\log T_\alpha^{exp}$ (s) are plotted *vs.* $Z_d^{0.6} Q_\alpha^{-1/2}$. The upper line represent the best fit to the T_α^{exp} values, while the lower one is the result of the linear fit from Ref.[7].

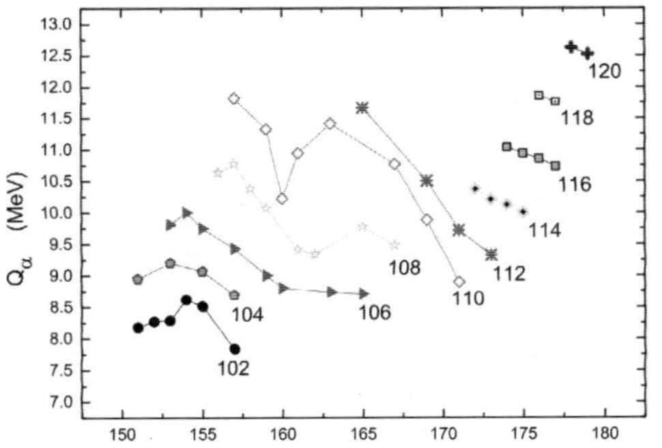

Figure 4. Experimental values for Q_α (MeV) are plotted vs. neutron number for isotopes of even-Z elements $Z = 102 - 120$.

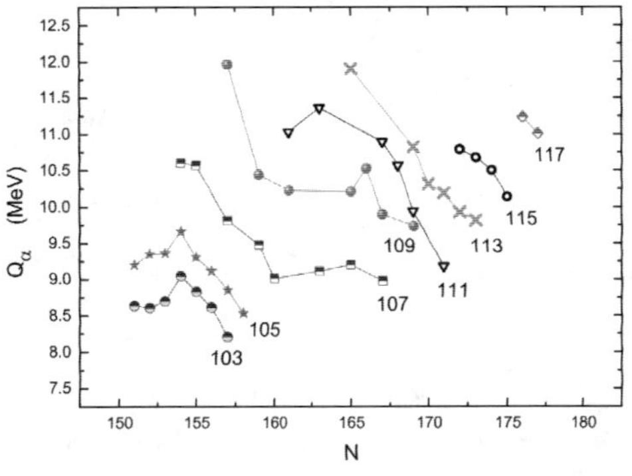

Figure 5. Calculated values for Q_α (MeV) are plotted vs. neutron number for isotopes of odd-Z elements $Z = 103 - 117$.

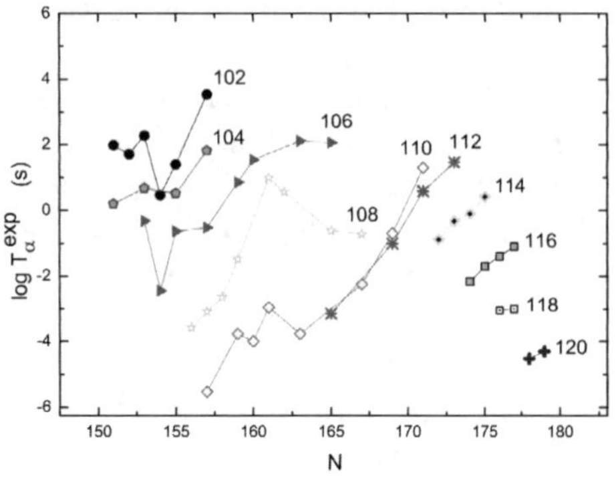

Figure 6. Experimental values for $logT_\alpha^{exp}$ (s) are plotted vs. neutron number for isotopes of even-Z elements $Z = 102 - 120$.

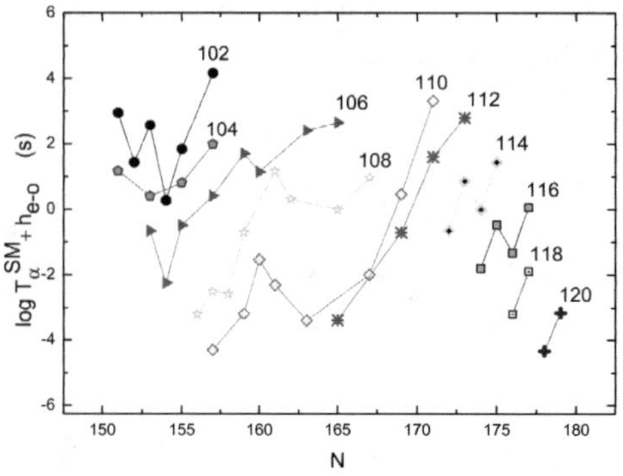

Figure 7. Calculated values for $logT_\alpha^{SM}$ (s), with even-odd corrections h_{e-o} added, are plotted vs. neutron number for isotopes of even-Z elements $Z = 102 - 120$

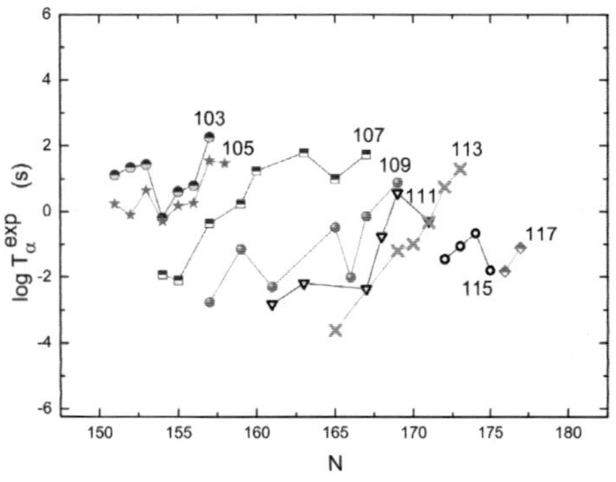

Figure 8. Experimental values for $logT_\alpha^{exp}$ (s) are plotted vs. neutron number for isotopes of odd-Z elements $Z = 103 - 117$.

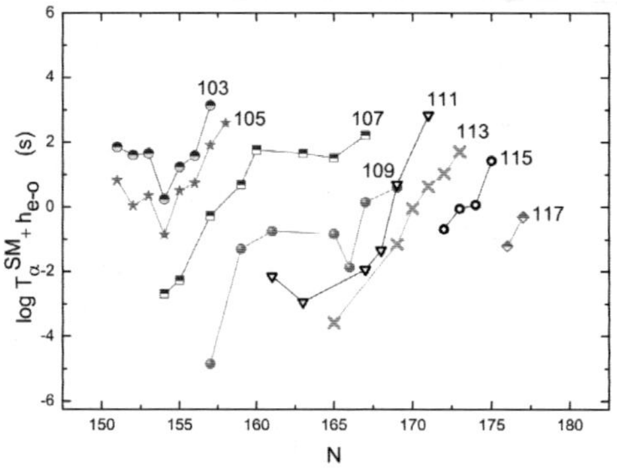

Figure 9. Calculated values for $logT_\alpha^{SM}$ (s), with even-odd corrections h_{e-o} added, are plotted vs. neutron number for isotopes of odd-Z elements $Z = 103 - 117$.

This work was supported from Contract: PN II 55/05.10.2011 and PN II 68/05.10.2011.

References

[1] Mang H J 1960 *Phys. Rev.* **119** 1069
[2] Săndulescu A, Silişteanu I and Wuensch R 1978 *Nucl. Phys.* A **305** 205
[3] Breit G 1959 *Encyclopedia of Physics XLI* S. Flugge (Springer-Verlag, Berlin)
[4] Bender M, Rutz K, Reinhard P-G *et al.* 1999 *Phys. Rev.* C **60** 034304
[5] Silişteanu I and Budaca A I 2012 *At. Data Nucl. Data Tab.* DOI:10.1016/j.adt.2011.12.007
[6] Viola V E and Seaborg G T 1966 *J. Inorg. Nucl. Chem.* **28** 741
[7] Brown B A 1992 *Phys. Rev.* C **46** 811
[8] Sobiczewski A, Patyk Z and Cwiok S 1989 *Phys. Lett.* B **224** 1
[9] Oganessian Y T, Abdulin F S, Bailey P D *et al.* 2010 *Phys. Rev. Lett.* **104** 0142502
[10] Oganessian Y T 2006 *Pure. Appl. Chem.* **78** 889
[11] Oganessian Y T, Utyonkov V K, Dmitriev S N *et al.* 2005 *Phys. Rev.* C **72** 034611
[12] Hofmann S 2012 *Rom. J. Phys.* **57**, 733

A semi-microscopic approach to the backbending phenomena in even-even nuclei

A A Raduta[1,2] and R Budaca[1]

[1]Horia Hulubei National Institute of Physics and Nuclear Engineering (IFIN-HH), PO Box MG-6, R-077125 Bucharest-Magurele, Romania
[2]Academy of Romanian Scientists, 54 Splaiul Independentei, Bucharest 050094, Romania

E-mail: raduta@nipne.ro

Abstract. The phenomenon of backbending in rare earth nuclei is described by means of the band hybridization mechanism. The hybridization of two rotational bands is performed by treating a model Hamiltonian associated with a set of interacting particles moving in a deformed mean field and coupled to a phenomenological core described in terms of quadrupole boson operators, in a product space of angular momentum projected states. The model states for the ground and S bands are obtained by angular momentum projection of a deformed product state with an axially symmetric coherent boson state standing for the core factor function and a BCS type single-particle factor function defining the nature of the band. The yrast band energies are defined by the lowest eigenvalues of the model Hamiltonian in an orthogonal basis constructed by diagonalizing the overlap matrix of the ground and S band states, which are not mutually orthogonal. The formalism is positively tested on six deformed even-even nuclei known to exhibit a backbending behavior in their moment of inertia.

1. Introduction

The backbending phenomenon is known to be the result of the crossing between the ground band and another band with a larger moment of inertia known as S(tockholm) band [1]. The S band is built on a broken pair with aligned individual high angular momentum. Such that in the rare earth region the first backbending is caused by the breaking of a neutron pair from the intruder orbital $i_{13/2}$, while the second one is due to the subsequent breaking of a proton $h_{11/2}$ pair. The first backbending is described in this work by means of a semi-microscopic model based on the bands hybridization mechanism. The physical system to be treated consists of a set of particles from the intruder orbital where the pair breaking occurs and a phenomenological deformed core defined by the remaining nucleons. The motion of the $i_{13/2}$ intruder neutrons is described by a BCS state, while the collective core by the Coherent State Model (CSM) [2]. The core deformation generates through the mutual coupling a deformed mean field for the single nucleon motion. However, the system is considered in the laboratory reference frame where the rotation invariance holds and thereby the angular momentum projected states should be constructed based on the intrinsic wave function defined above. Due to the deformation of the particle-core space the projected states are not mutually orthogonal. From this basis a new orthogonal basis is constructed which mixes the original projected states. The lowest eigenvalues of the model Hamiltonian in the orthogonal basis define the yrast band. Six even-even nuclei from the rare earth region with $N = 90 - 94$, which exhibit a backbending behavior are quantitatively treated within the proposed formalism.

2. Theoretical framework

A particle-core Hamiltonian is treated in a space defined by a set of spherical projected product functions:

$$\Psi \equiv \psi_f \psi_c, \tag{1}$$

where ψ_f stands for the fermion factor state corresponding to the intruder particles and is chosen to be of BCS type, while the collective factor state ψ_c describing the core, is a coherent state for the quadrupole bosons $b^\dagger_{20}$:

$$\psi_c = e^{d(b^\dagger_{20} - b_{20})}|0\rangle_b. \tag{2}$$

$|0\rangle_b$ is the quadrupole boson vacuum state whereas d is a real parameter which actually simulates the nuclear deformation. The fermion factor function specifies the nature of the rotational band. Thus, the ground band (g band), where all neutrons are paired, is given by a deformed BCS state associated to the set of neutrons placed in the intruder deformed states. On the other hand, the S band, which crosses the g band and produces the backbending, is build upon a broken pair of particles whose states are no longer connected by a time reversal transformation. The time reversal symmetry breaking, in the present formalism, is achieved by applying the operator $J_+\alpha^\dagger_{jk}\alpha^\dagger_{j-k}$ on the mentioned deformed BCS state with $\alpha^\dagger_{jk}$ being the creation operator for a quasiparticle in the single-particle state $|jk\rangle$.

The projected particle-core functions of the g and S band are obtained by applying the Hill-Wheeler projection operator on the product state (1) with a specific fermionic factor state and can be written in the following form:

$$\Psi^{(1)}_{JM} = \mathcal{N}^{(1)}_J P^J_{M0}|BCS\rangle_d \psi_c = \mathcal{N}^{(1)}_J \sum_{J_f J_c} \frac{C^{J_f J_c J}_{0\ 0\ 0}}{\mathcal{N}^{BCS}_{J_f} N^{(g)}_{J_c}} \left[\Phi^{BCS}_{J_f} \phi^{(g)}_{J_c}\right]_{JM}, \tag{3}$$

$$\begin{aligned}\Psi^{(2)}_{JM;1}(jk) &= \mathcal{N}^{(2)}_{J1}(jk) P^J_{M1}\left[J_+\alpha^\dagger_{jk}\alpha^\dagger_{j-k}|BCS\rangle_d\right]\psi_c \\ &= \mathcal{N}^{(2)}_{J1}(jk) \sum_{J_f J_c} \frac{C^{J_f J_c J}_{1\ 0\ 1}}{\mathcal{N}^{jk}_{J_f 1} N^{(g)}_{J_c}} \left[\Phi^{jk}_{J_f 1} \phi^{(g)}_{J_c}\right]_{JM},\end{aligned} \tag{4}$$

with the corresponding normalization factors

$$\left[\mathcal{N}^{(1)}_J\right]^{-2} = \sum_{J_f J_c} \left(\frac{C^{J_f J_c J}_{0\ 0\ 0}}{\mathcal{N}^{BCS}_{J_f} N^{(g)}_{J_c}}\right)^2, \quad \left[\mathcal{N}^{(2)}_{J1}(jk)\right]^{-2} = \sum_{J_f J_c} \left(\frac{C^{J_f J_c J}_{1\ 0\ 1}}{\mathcal{N}^{jk}_{J_f 1} N^{(g)}_{J_c}}\right)^2. \tag{5}$$

$\phi^{(g)}_{J_c}$ is the core projected state (2) with the norm equal to unity, while $\Phi^{BCS}_{J_f M_f}$ and $\Phi^{jk}_{J_f 1; M_f}$ from (3) and (4) are the angular momentum projected fermionic factor states corresponding to the $0qp$ and $2qp$ BCS states, respectively:

$$\Phi^{BCS}_{J_f M_f} = \mathcal{N}^{BCS}_{J_f} P^{J_f}_{M_f 0}|BCS\rangle_d, \quad \Phi^{jk}_{J_f 1; M_f} = \mathcal{N}^{jk}_{J_f 1} P^{J_f}_{M_f 1} J_+ \alpha^\dagger_{jk}\alpha^\dagger_{j-k}|BCS\rangle_d. \tag{6}$$

The norms of these states are calculated using the recipe given in Ref.[3], while the norm $N^{(g)}_{J_c}$ of the projected coherent boson state (2) was presented by A. A. Raduta in Ref.[2] in connection to CSM. It is worth to mention that the summations from Eqs.(3) and (4) over J_f are restricted by $J_f \leqslant 24$. This upper limit value is provided by the group theory [4] and represents the maximal angular momentum that can be realized in the $j = 13/2$ configuration, when the Pauli principle is obeyed.

The model Hamiltonian describing the particles and core interacting system is a sum of three terms:

$$H = H_c + H_p + H_{pc}. \tag{7}$$

The core Hamiltonian H_c is represented by the quadrupole boson number operator:

$$H_c = \hbar\omega_b \sum_{\mu} b^{\dagger}_{2\mu} b_{2\mu}, \tag{8}$$

while H_p is associated to a set of particles from an intruder shell model orbital j which interact between themselves through a pairing force:

$$H_p = (\varepsilon_{nlj} - \lambda) \sum_{m=all} c^{\dagger}_{nljm} c_{nljm} - \frac{G}{4} P^{\dagger}_j P_j, \tag{9}$$

where $P^{\dagger}_j(P_j)$ are creation(annihilation) operators of a Cooper pair in the intruder orbital j. The particles and core subsystems interact with each other by a quadrupole-quadrupole (qQ) and a spin-spin interaction

$$H_{pc} = H_{qQ} + H_C = -A_C \sum_{\mu,m,m'} \langle nljm|r^2 Y_{2\mu}|nljm'\rangle c^{\dagger}_{nljm} c_{nljm'} \left[b^{\dagger}_{2-\mu}(-)^{\mu} + b_{2\mu}\right] + C\vec{J}_c \cdot \vec{J}_f. \tag{10}$$

The particle-core interaction induces a mutual deformation for both the quadrupole bosons and the single-nucleon mean field. The later one can be written as

$$H_{mf} = \sum_{m=all} \varepsilon_{nljm} c^{\dagger}_{nljm} c_{nljm}, \tag{11}$$

with energies given in the first order of perturbation by:

$$\varepsilon_{nljm} = \varepsilon_{nlj} - 4dX_C(2n+3) C^{j\,2\,j}_{\frac{1}{2}\,0\,\frac{1}{2}} C^{j\,2\,j}_{m\,0\,m}, \quad \text{where} \quad X_C = \frac{\hbar}{8M\omega_0}\sqrt{\frac{5}{\pi}} A_C. \tag{12}$$

Treating the sum $H_{mf} + H_{pair}$ through the BCS formalism, one obtains the occupation probabilities V and U, the energy gap Δ, the quasiparticle energies and the Fermi level energy, λ. With these quantities, the single-particle factor of the projected particle-core product basis is completely determined. Due to the dangerous graphs vanishing the matrix elements between total projected states of the single-particle term $H_{mf} + H_{pair}$ and of the qQ interaction, written in the quasiparticle representation, acquire very compact expressions due.

The matrix elements of the remaining Hamiltonian terms are easily obtained by using the tensorial form of the projected states (3) and (4). For the harmonic boson Hamiltonian we make use of its matrix elements on projected coherent states given in Ref.[5], while the spin-spin interaction matrix elements are obtained from the basic properties of the angular momentum operators. In this way the energy spectra of the g and S bands are completely determined by the average of the total Hamiltonian on the corresponding projected state, (3) or (4). In order to achieve the hybridization of these bands, one must diagonalize the total Hamiltonian in a basis defined by the projected states (3) and (4), which unfortunately are not orthogonal. However, using the eigenvalues $\alpha_m(J, jk)$ and the eigenvectors $V_{nm}(J, jk)$ of the overlap matrix defined by the non-orthogonal states, one can construct the functions [6]:

$$\Phi^{JM}_m(jk) = \frac{1}{\sqrt{\alpha_m(J, jk)}} \left[\Psi^{(1)}_{JM} V_{1m}(J, jk) + \Psi^{(2)}_{JM;1}(jk) V_{2m}(J, jk)\right], \quad m = 1, 2, \tag{13}$$

which are mutually orthogonal. Diagonalizing the total Hamiltonian in this orthogonal basis, for each angular momentum J one obtains a set of two hybridization energies, with the lowest ones defining the yrast band.

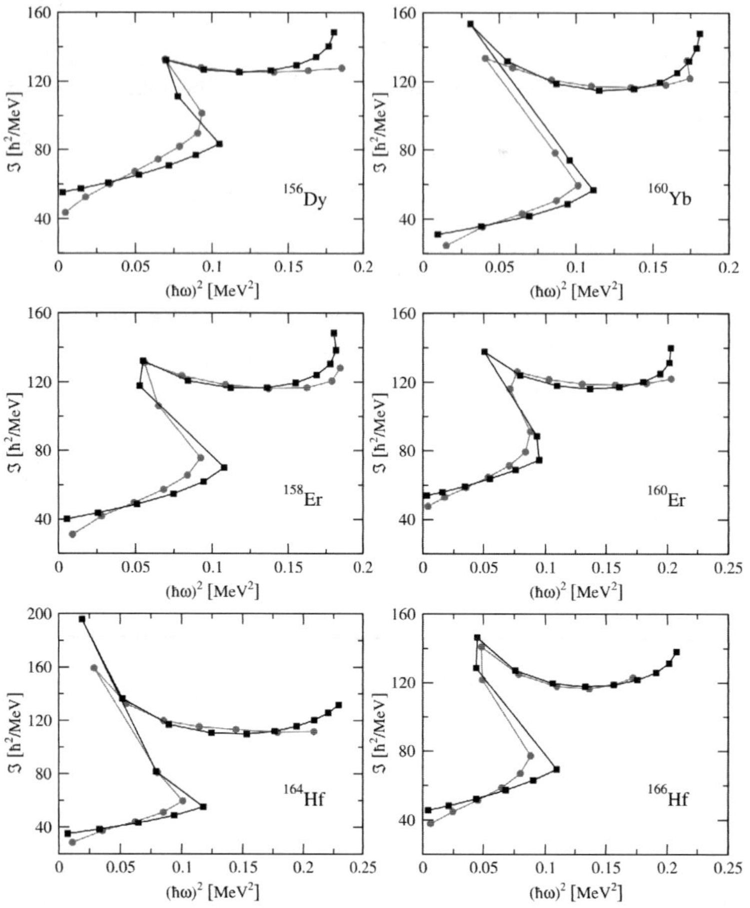

Figure 1. (color online)Backbending plots for ^{156}Dy, ^{160}Yb, ^{158,160}Er and ^{164,166}Hf isotopes comparing theory (black squares) with experiment (red circles). Experimental data are taken from [7, 8, 9, 10, 11].

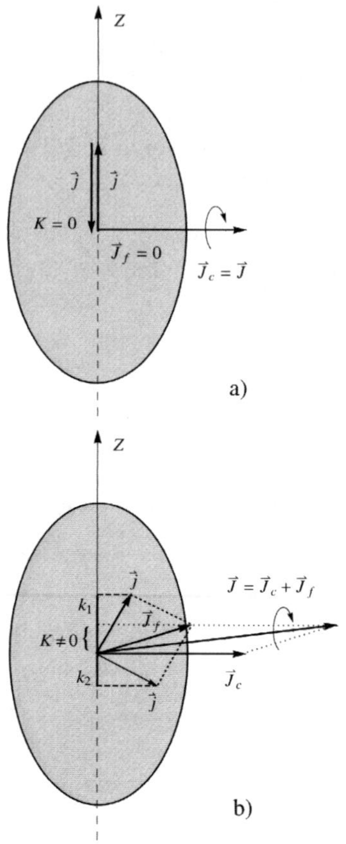

Figure 2. The coupling scheme for g band (a) and for S band (b).

3. Numerical application

The nuclei ^{156}Dy, ^{160}Yb, ^{158,160}Er and ^{164,166}Hf are known to be good backbenders. These nuclei are treated within the formalism described in the previous section. The model involves five free parameters, with four of them, namely, the pairing constant G, the qQ and spin-spin interaction strengths X_C and C, and the boson frequency $\hbar\omega_b$ being the structure coefficients of the model Hamiltonian. The remaining parameter is the deformation d which defines the collective factor state. The manner in which these parameters are fixed is detailed in Ref.[3]. Solving the BCS equations for an extended subset of 23 single-particle states comprising all possible levels which might interact with the $i_{13/2}$ intruder states, one obtains the gap parameter Δ, the Fermi level energy, the quasiparticle energies and consequently the occupation probabilities U and V. The input data for the BCS equations are the pairing constant G and the mean field single-particle energies defined by Eq.(12) and which depend linearly on the deformation parameter d. Collecting the BCS quantities only for the $i_{13/2}$ neutron orbital and using the values of the model parameters found in Ref.[3], one calculates the system energy by diagonalizing the total Hamiltonian in the orthogonal basis of states (13). The lowest hybridization energies for each angular momentum J defines the yrast band. It is worth mentioning that besides the description of the energy spectra, the fitted parameters also reproduce the observed sequence of the single-particle levels with a deformation d being in the range of values determined in Refs.[12, 13].

The theoretical results are compared with experimental data in Fig.1 by means of backbending plots. The backbending plot is a graphical representation of the moment of inertia dependence on the angular frequency squared. From Fig.1 it is obvious that in all cases the general zigzag behavior is reproduced quite well. An especially good agreement is found for moderate spin states situated right in the backbending region. The theoretical yrast energies associated to the backbending plots from Fig.1 have the *r.m.s.* deviations of about 30 keV from the corresponding experimental energies. The angular momentum where the backbending starts represent the band crossing point. From Fig.1 one can see that for Er isotopes and ^{166}Hf the band crossing takes place at $J = 12$, while for ^{164}Hf and ^{160}Yb at $J = 10$, and for ^{156}Dy at $J = 14$. At the band crossing point there is a transition of the yrast band from $0qp$ to $2qp$ states caused by the breaking of a neutron intruder pair. The mechanism of pair breaking and the alignment of the involved angular momenta is visualized in Fig.2. At low values of the core angular momentum the nucleon angular momenta are anti-aligned along the symmetry axis of the mean-field ($\vec{J}_f = 0$) and the whole angular momentum is due to the core alone ($\vec{J}_c = \vec{J}$) which is perpendicular to the symmetry axis (Fig.2(a)). The alignment of the particle angular momenta to the core angular momentum is shown in Fig.2(b). The full alignment between the core and fermion angular momenta is not possible due to the fact that the total K for the two neutrons is equal to unity.

4. Conclusions

A semi-microscopic formalism was applied to describe the moment of inertia anomalies at low spins in even-even rare earth nuclei. The theoretical yrast energies obtained as the lowest eigenvalues of a particle-core Hamiltonian in an orthogonal basis constructed from the projected product states are used to calculate the backbending plots for six nuclei. The experimental backbending curves are fairly well reproduced by the numerical results of the proposed formalism. The shape and sharpness of the backbendings are in a good agreement with experiment. The first energy levels before the backbending are determined by the $0qp$ projected states and after a critical angular momentum, where the bands g and S cross each other, the states become of a $2qp$ nature. The advantage of the present formalism over other models consists in the fact that it also provides a qualitative explanation of the combined effect of the pair breaking mechanism and rotational alignment of the angular momenta implied in the system, on the backbending behavior of the moment of inertia. Concluding, one can say that the model proposed is able to describe quantitatively the main features of the first backbending in rare earth nuclei, revealing in the same time new features regarding its mechanism.

References

[1] Johnson A, Reyde H and Sztarkier J 1971 *Phys. Lett.* B **34** 605
[2] Raduta A A, Ceausescu V, Gheorghe A and Dreizler R M 1982 *Nucl. Phys.* A **381** 253
[3] Raduta A A and Budaca R 2012 *Phys. Rev.* C **84** 044323
[4] Hamermesh N 1962 *Group Theory and its Application to Physical Problems* (New York: Dover Publications)
[5] Raduta A A, Lo Iudice N and Ursu I I 1995 *Nucl. Phys.* A **584** 84
[6] Raduta A A, Raduta C M and Faessler A 2009 *Phys. Rev.* C **80** 044327
[7] Reich C W 2003 *Nucl. Data Sheets* **99** 753
[8] Reich C W 1996 *Nucl. Data Sheets* **78** 547
[9] Helmer R G 2004 *Nucl. Data Sheets* **101** 325
[10] Balraj Singh 2001 *Nucl. Data Sheets* **93** 243
[11] Shurshikov E N and Timofeeva N V 1992 *Nucl. Data Sheets* **67** 45
[12] Raduta A A, Budaca R and Faessler A 2010 *J. Phys.* G: *Nucl. Part. Phys.* **37** 085108
[13] Raduta A A, Budaca R and Faessler A 2012 *Ann. Phys.* (NY) **327** 671

Towards a new solvable model for the even-even triaxial nuclei

A. A. Raduta[1,2], P. Buganu[1]

[1)] Department of Theoretical Physics, Institute of Physics and Nuclear Engineering, P.O.B. MG-6, RO-077125, Romania

[2)]Academy of Romanian Scientists, 54 Splaiul Independentei, Bucharest 050094, Romania

E-mail: raduta@nipne.ro,buganu@theory.nipne.ro

Abstract. The even-even triaxial nuclei are described by amending the Bohr-Mottelson Hamiltonian with an energy potential consisting of two terms: a sextic oscillator with centrifugal barrier in the β variable and a periodic function in the γ variable. After the variable separation is performed, the β equation is quasi-exactly solved, while the γ equation is satisfied by the Mathieu function. The reduced E2 transition probabilities are determined using an anharmonic transition operator. The formalism is conventionally called the Sextic and Mathieu Approach (SMA). Numerical applications concerned seven non-axial nuclei: ^{188}Os, ^{190}Os, ^{192}Os, ^{228}Th, ^{230}Th, ^{182}W and ^{180}Hf. SMA results are compared with the experimental data as well as with those yielded by the Coherent State Model (CSM).

1. Introduction

The field of the nuclear shape phase transitions received a considerable attention when it was noticed that the critical points may be described by differential equations which are exactly solvable. In some of the situations the solutions of these equations achieve the irreducible representations of a certain symmetry group. For example the E(5) symmetry [1] describes the critical point for the transition U(5)→O(6) while the one associated to the transition U(5)→SU(3) is not yet known and thereby referred to as X(5) [2]. Since the proposal for the two critical points showed up, on that matter a huge number of papers were published, many of them being reviewed in Refs. [3, 4]. In the mean time other two critical symmetries were proposed, namely Y(5) [5] and Z(5) [6], for the axial-triaxial shape phase transition and for the prolate-oblate shape phase transition, respectively. All these critical point symmetries are analytical solutions of the Bohr-Mottelson Hamiltonian [7, 8] amended with a potential which depends on both β and γ variables.

Here, we present a new solution of the Bohr-Mottelson Hamiltonian equation which seems to be suitable for the description of the even-even triaxial nuclei having an axial deformation close to $\gamma_0 = 30^0$. In order, to improve the agreement between the theoretical predictions and the corresponding experimental data, we used for the β variable a sextic oscillator with centrifugal barrier potential. The β equation with sextic potential is quasi-exactly solvable, which means that is still exactly solvable but for a finite number of states. For the γ equation, choosing a periodic potential with a minimum in $\gamma_0 = 30^0$ and avoiding the approximations made in the previous models, we obtained as solutions the Mathieu functions which are periodic in the interval $[0, 2\pi]$. In this way, the hermiticity of the γ Hamiltonian in respect with the integration measure $|\sin 3\gamma|d\gamma$ is preserved. The reduced E2 transition probabilities were determined using a transition operator written in the intrinsic reference frame and which contains two terms, a harmonic part and an anharmonic part, respectively. The formalism developed in this way was conventionally called the Sextic and Mathieu Approach (SMA). More details about the SMA and its numerical applications may be found in the Refs. [9, 10, 11]. Here, we present only its main ingredients and some numerical examples. The SMA results were also compared with those yielded by the Coherent State Model (CSM) [12]. Numerical analysis of the two models results suggested a possible relationship between the two approaches. The connection between SMA and CSM was analytically established in Refs. [10, 11], where the SMA equations were obtained through a semi-classical treatment of the CSM Hamiltonian.

The description of the results presented in this communication is organized as follows. In Section II, the main ingredients of the SMA model are given, while in Section III an illustrative example of ^{192}Os is discussed. In Section IV we show how the SMA equations were obtained from the CSM Hamiltonian and finally, in Section IV, the main conclusions are summarized.

2. The Sextic and Mathieu Approach

In order to describe the critical nuclei of the prolate-oblate shape phase transition, we invoke the Bohr-Mottelson Hamiltonian with a potential depending on both the β and γ variables:

$$H\psi(\beta,\gamma,\Omega) = E\psi(\beta,\gamma,\Omega), \text{ where,} \tag{1}$$

$$H = -\frac{\hbar^2}{2B}\left[\frac{1}{\beta^4}\frac{\partial}{\partial\beta}\beta^4\frac{\partial}{\partial\beta} + \frac{1}{\beta^2\sin 3\gamma}\frac{\partial}{\partial\gamma}\sin 3\gamma\frac{\partial}{\partial\gamma} - \frac{1}{4\beta^2}\sum_{k=1}^{3}\frac{Q_k^2}{\sin^2(\gamma-\frac{2}{3}\pi k)}\right] + V(\beta,\gamma),$$
$$V(\beta,\gamma) = V_1(\beta) + \frac{V_2(\gamma)}{\beta^2}. \tag{2}$$

Here, β, γ and Ω are the intrinsic deformation variables and the Euler angles, respectively, while with Q_k are denoted the angular momentum projections in the intrinsic reference frame. After the separation of variables [9], the following equations are obtained:

$$\left[-\frac{1}{\beta^4}\frac{\partial}{\partial\beta}\beta^4\frac{\partial}{\partial\beta} + \frac{L(L+1)}{\beta^2} + v_1(\beta)\right]f(\beta) = \varepsilon_\beta f(\beta), \tag{3}$$

$$\left[-\frac{1}{\sin 3\gamma}\frac{\partial}{\partial\gamma}\sin 3\gamma\frac{\partial}{\partial\gamma} - \frac{3}{4}R^2 + \left(10L(L+1) - \frac{39}{4}R^2\right)\left(\gamma-\frac{\pi}{6}\right)^2 + v_2(\gamma)\right]\phi(\gamma) = \tilde{\varepsilon}_\gamma\phi(\gamma), \tag{4}$$

where the following notations are used:

$$v_1(\beta) = \frac{2B}{\hbar^2}V_1(\beta), \quad v_2(\gamma) = \frac{2B}{\hbar^2}V_2(\gamma), \quad \varepsilon_\beta = \frac{2B}{\hbar^2}E_\beta, \quad \tilde{\varepsilon}_\gamma = \langle\beta^2\rangle\frac{2B}{\hbar^2}E_\gamma. \tag{5}$$

Eq. (3) is reduced to the sextic oscillator equation if we change the function $f(\beta) = \beta^{-2}\varphi(\beta)$ and take the β potential in the form:

$$v_1^{\mp}(\beta) = (b^2 - 4ac^{\pm})\beta^2 + 2ab\beta^4 + a^2\beta^6 + u_0^{\pm} \;\; c^{\pm} = \frac{L}{2} + \frac{5}{4} + M, \;\; M = 0,1,2,..., \tag{6}$$

where, $c^{\pm}$ and $u_0^{\pm}$ are constants with the signs + and − for L even and L odd, respectively. The solutions of Eq. (3), with the above specified potential are

$$\varphi^{(M)}_{n_\beta,L}(\beta) = N_{n_\beta,L}P^{(M)}_{n_\beta,L}(\beta^2)\beta^{L+1}e^{-\frac{a}{4}\beta^4-\frac{b}{2}\beta^2}, \;\; n_\beta = 0,1,2,...M, \tag{7}$$

where $N_{n_\beta,L}$ are the normalization factor, while $P^{(M)}_{n_\beta,L}(\beta^2)$ are polynomials in x^2 of n_β order. The corresponding excitation energies are:

$$E_\beta(n_\beta, L) = \frac{\hbar^2}{2B}\left[b(2L+3) + \lambda^{(M)}_{n_\beta}(L) + u_0^\pm\right], \quad n_\beta = 0, 1, 2, ..., M, \tag{8}$$

where $\lambda^{(M)}_{n_\beta} = \varepsilon_\beta - u_0^\pm - 4bs$ is the eigenvalue of the equation:

$$\left[-\left(\frac{\partial^2}{\partial\beta^2} + \frac{4s-1}{\beta}\frac{\partial}{\partial\beta}\right) + 2b\beta\frac{\partial}{\partial\beta} + 2a\beta^2\left(\beta\frac{\partial}{\partial\beta} - 2M\right)\right] P^{(M)}_{n_\beta,L}(\beta^2) = \lambda^{(M)}_{n_\beta} P^{(M)}_{n_\beta,L}(\beta^2). \tag{9}$$

Considering the periodic γ potential, $v_2(\gamma) = \mu\cos^2 3\gamma$ (4), which exhibits a minimum in $\gamma_0 = \pi/6$ and then changing the function $\phi(\gamma) = \mathcal{M}(3\gamma)/\sqrt{|\sin 3\gamma|}$ one arrives at a differential equation for the Mathieu functions. The expression for the excitation energy of the γ equation is:

$$E_\gamma(n_\gamma, L, R) = \frac{\hbar^2}{2B}\frac{1}{\langle\beta^2\rangle}\left[9a_{n_\gamma}(L,R) + 18q(L,R) - \frac{3}{4}R^2 - \frac{5}{2}\right], \quad n_\gamma = 0, 1, 2, ... \tag{10}$$

The Mathieu functions have the advantage that are periodic and defined on a bounded interval $[0, 2\pi]$, preserving the hermiticity of the initial γ Hamiltonian in respect with the integration measure $|\sin 3\gamma|d\gamma$. The total energy of the system is obtain by adding the contributions coming from both β and γ equations.

Using the Rose's convention [15], the reduced E2 transition probabilities are determined by:

$$B(E2, L_i \to L_f) = |\langle L_i||T_2^{(E2)}||L_f\rangle|^2, \tag{11}$$

where the transition operator has the following expression:

$$\begin{aligned} T^{(E2)}_{2\mu} &= t_1\beta\left[\cos\left(\gamma - \frac{2\pi}{3}\right)D^2_{\mu 0} + \frac{1}{\sqrt{2}}\sin\left(\gamma - \frac{2\pi}{3}\right)(D^2_{\mu 2} + D^2_{\mu,-2})\right] + \\ & t_2\sqrt{\frac{2}{7}}\beta^2\left[-\cos\left(2\gamma - \frac{4\pi}{3}\right)D^2_{\mu 0} + \frac{1}{\sqrt{2}}\sin\left(2\gamma - \frac{4\pi}{3}\right)(D^2_{\mu 2} + D^2_{\mu,-2})\right]. \end{aligned} \tag{12}$$

3. Numerical results

SMA was successfully applied for several non-axial nuclei: ^{188}Os, ^{190}Os, ^{192}Os, ^{228}Th, ^{230}Th, ^{182}W and ^{180}Hf [9, 10, 11]. For the sake of saving the space here we shall present only the case of ^{192}Os. This nucleus is a good candidate for a triaxial deformation close to $\gamma_0 = 30^0$. Indeed, its equilibrium value predicted by Leander [16] is $\gamma_0 = 25^0$. On the other hand one signature of the triaxial rigid rotor is $\Delta E = |E_{2_1^+} + E_{2_2^+} - E_{3_1^+}| = 0$. In the chosen case this equality is obeyed with a good accuracy having $\Delta E = 5$keV. Moreover, its γ band has a pronounced staggering behavior (Fig.1). As seen in Fig. 1 and Fig. 2, the staggering and the spectrum of ^{192}Os are quite well described by SMA. The agreement with the data is appraised by the value of the *r.m.s.* value associated to the predicted energy deviation, which in the mentioned case amounts of 16 keV. The calculated E2 properties regard the intraband transitions of the ground and γ bands, as well as the interband $\gamma \to g$. Results are compared with the corresponding experimental data in Table 1. Comparing the SMA results with those of CSM, we noticed that they are quantitatively close to each other, although the two approaches have apparently different grounds. The connection between SMA and CSM was explained in details in Refs. [10, 11].

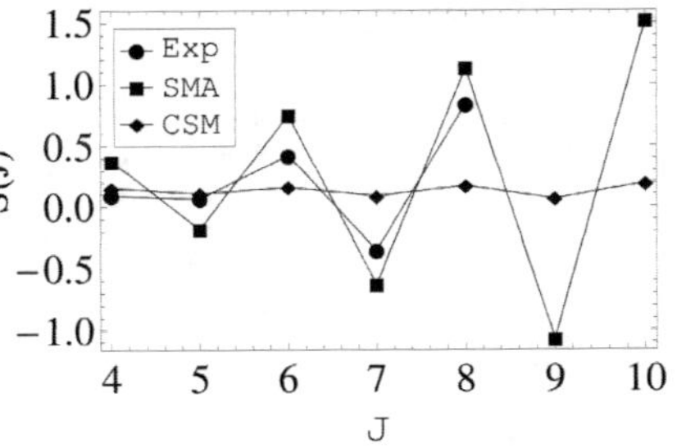

Figure 1. The staggering behavior of ^{192}Os (Exp) compared with the SMA and CSM predictions.

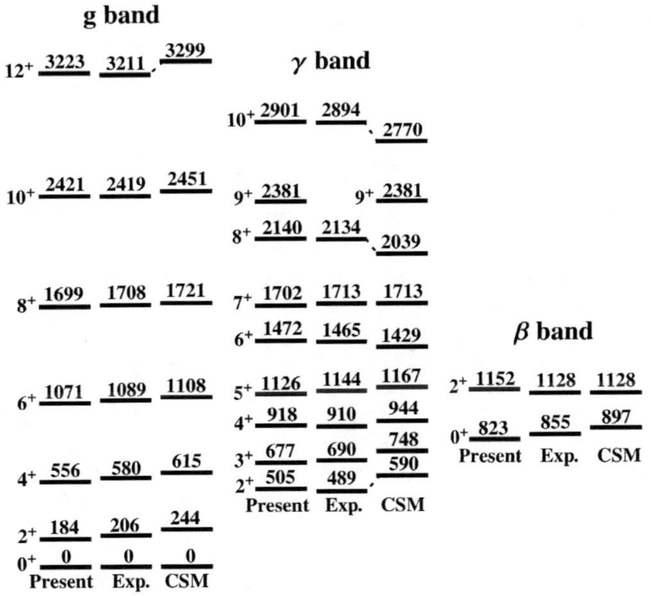

Figure 2. The experimental spectrum of ^{192}Os [17, 18] for the first three bands, given in keV units, is compared with the results yielded by the SMA (Present) and CSM.

Table 1. Some B(E2) values for ^{192}Os obtained with SMA and CSM, are compared with the corresponding experimental data [17].

$B(E2)[e^2b^2]$	Exp	SMA	CSM
$2_g^+ \to 0_g^+$	0.424	0.424	0.236
$4_g^+ \to 2_g^+$	0.497	0.632	0.449
$6_g^+ \to 4_g^+$	0.660	0.858	0.611
$8_g^+ \to 6_g^+$	0.754	1.030	0.754
$10_g^+ \to 8_g^+$	0.688	1.175	0.887
$4_\gamma^+ \to 2_\gamma^+$	0.298	0.261	0.277
$6_\gamma^+ \to 4_\gamma^+$	0.336	0.352	0.595
$8_\gamma^+ \to 6_\gamma^+$	0.314	0.549	0.814
$2_\gamma^+ \to 0_g^+$	0.037	0.006	0.192
$2_\gamma^+ \to 2_g^+$	0.303	0.303	0.055
$2_\gamma^+ \to 4_g^+$	0.024	0.000	0.000
$4_\gamma^+ \to 2_g^+$	0.002	0.004	0.274
$4_\gamma^+ \to 4_g^+$	0.203	0.068	0.137
$4_\gamma^+ \to 6_g^+$	0.018	0.000	0.000
$6_\gamma^+ \to 4_g^+$	0.000	0.002	0.357
$6_\gamma^+ \to 6_g^+$	0.171	0.042	0.171

4. The connection between the SMA and CSM formalisms

Aiming at obtaining the SMA equations starting with CSM, first the CSM boson Hamiltonian is dequantized using a two parameter coherent state function as a trial variational function:

$$\mathcal{H} = \langle\psi|H_{\rm CSM}|\psi\rangle,\ \ |\psi\rangle = \exp\left[z_0 b_0^\dagger + z_2 b_2^\dagger + z_{-2} b_{-2}^\dagger - z_0^* b_0 - z_2^* b_2 - z_{-2}^* b_{-2}\right]|0\rangle, \tag{13}$$

Journal of Physics: Conference Series **413** (2013) 012029 doi:10.1088/1742-6596/413/1/012029

The basic property of the coherent state is:

$$b_\mu|\psi\rangle = (\delta_{\mu,0}z_0 + \delta_{\mu,2}z_2 + \delta_{\mu,-2}z_{-2})\,|\psi\rangle. \tag{14}$$

New coordinates which bring the classical equations to the Hamilton canonical form are desirable. Such coordinates are obtained by the restriction $z_2 = z_{-2}$ and the transformation $q_0 = \sqrt{2}\mathrm{Re}[z_0]$, $p_0 = \sqrt{2}\mathrm{Im}[z_0]$, $q_2 = 2\mathrm{Re}[z_2]$, $p_2 = 2\mathrm{Im}[z_2]$. In the expression of the classical Hamilton energy function, the terms which couple the coordinate with momentum and also those which are not quadratic in momenta are neglected. The resulting Hamiltonian is quantized in polar coordinates which results in obtaining:

$$\hat{H} = -\left(11A_1 + 3A_2 + A_1 d^2 + \frac{3}{70}d^4 A_3\right)\left(\frac{1}{r}\frac{\partial}{\partial r} + \frac{\partial^2}{\partial r^2} + \frac{1}{r^2}\frac{\partial^2}{\partial \gamma^2}\right) + V_1(r) + V_2(\gamma), \tag{15}$$

$$V_1(r) = \left[11A_1 + 3A_2 - \frac{3d^2}{4}A_1 - \frac{3A_3}{70}d^4\right]r^2 + \left[\frac{A_1}{4} + \frac{9A_3}{280}d^2\right]r^4 + \frac{A_3}{280}r^6,\ V_2(\gamma) = \frac{A_3 r_0^6}{280}\cos 6\gamma. \tag{16}$$

Finally, the SMA equations are obtained after the separation of variables $r = k\beta$ and γ through Taylor expansions around the equilibrium values, β_0 and γ_0.

5. Conclusions

A new formalism, called the Sextic and Mathieu Approach (SMA), aimed at describing the even-even triaxial nuclei was proposed. The β equation is that of a sextic oscillator with centrifugal barrier, while that of the γ variable is reduced to a Mathieu equation. The comparison between the SMA results and experimental data of seven nuclei ^{188,190,192}Os, ^{228,230}Th, ^{180}Hf and ^{182}W showed a good agreement. Also, a connection between the SMA and CSM formalism was pointed out. The closeness of the SMA and CSM results was explained by obtaining the SMA equations through a semi-classical treatment of the CSM Hamiltonian. With this demonstration, the SMA potentials get a theoretical support in contrast with their intuitive choice when the SMA equations emerge from the BohrMottelson Hamiltonian.

References

[1] Iachello F 2000, *Phys. Rev. Lett.* **85** 3580-83.
[2] Iachello F 2001, *Phys. Rev. Lett.* **87** 052502.
[3] Fortunato L 2005, *Eur. J. Phys.* A **26 s01**, 1-30.
[4] Cejnar P, Jolie J and Casten R F 2010, *Rev. Mod. Phys.* **82** 2155-212.
[5] Iachello F 2003, *Phys. Rev. Lett.* **91** 132502.
[6] Bonatsos D, Lenis D, Petrellis D, Terziev P A 2004,*Phys. Lett.* B **588**, 172-9.
[7] Bohr A 1952, *Mat. Fys. Medd. Dan. Vid. Selsk.* **26** no.14;
[8] Bohr A and Mottelson B. 1953, *Mat. Fys. Medd. Dan. Vid. Selsk.* **27** no. 16.
[9] Raduta A A and Buganu P 2011, *Phys. Rev.* C **83**, 034313.
[10] Buganu P, Raduta A A and Faessler A 2012, *J. Phys. G: Nucl. Part. Phys.* **39**, 025103.
[11] Buganu P and Raduta A A 2012, *Rom. Jour. Phys.* **57**, 1103-12.
[12] Raduta A A, Ceausescu V, Gheorghe A and Dreizler R M 1981, *Phys. Lett.* B **99**, 444-8; 1982, *Nucl. Phys.* A **381**, 253-76.
[13] Raduta A A 2004, *Recent Res. Devel. Nuclear Phys.*,1-70, ISBN:81-7895-124-X.
[14] Wilets L and Jean M 1956, *Phys. Rev.* **102**, 788.
[15] Rose M E 1957, *Elementary Theory of Angular Momentum* (Wiley, New York).
[16] Leander G 1976, *Nucl. Phys.* A **273**, 286-300.
[17] Wu C Y *et al.*, 1996, *Nucl. Phys.* A **607**, 178-234.
[18] Baglin Coral M 1998, *Nuclear Data Sheets* **84**, 717-900.

Study of the neutron rich sulfure isotope ^{43}S through intermediate energy Coulomb excitation

S Calinescu[1,9], L Cáceres[2], S Grévy [3], D Sohler[4], M Stanoiu[1], F Negoita[1], C Borcea[1], R Borcea[1], M Bowry[5], W Catford[5], Z Dombradi[4], S Franchoo[6], R Gillibert[10], JC Thomas[2], I Kuti[4], S Lukyanov[7], A Lepailleur[2] , J Mrazek[8], M Niikura[6], Z Podolyak[5], C Petrone[1,9], Y Penionzhkevich[7], T Roger[2] , F Rotaru[1], O Sorlin[2], I Stefan[6], Z Vajta[4], E Wilson[5]

[1]Horia Hulubei National Institute for Physics and Nuclear Engineering, P.O. Box MG-6, 077125 Bucharest-Magurele, Romania

[2]Grand Accélérateur National d'Ions Lours (GANIL), CEA/DSM-CNRS/IN2P3, Caen, France

[3] CENBG, CNRS/IN2P3, Université de Bordeaux 1, Gradignan, France

[4]Institute of Nuclear Research (ATOMKI), Debrecen, Hungary

[5]Department of physics, University of Surrey, Guildford GU2 7HX. U.K.

[6]IPN, Orsay, France

[7]FLNR, JINR, Dubna, Russia

[8]NPI, Rez, Czech Republic

[9]Faculty of Physics, University of Bucharest, Romania

[10]Université de Caen Basse Normandie, Caen, France

E-mail address: stefana.calinescu@tandem.nipne.ro

Abstract. The reduced transition probability B(E2: $3/2^- \rightarrow 7/2^-_2$)has been measured in ^{43}S using Coulomb excitation at intermediate energy. The nucleus of interest was produced by fragmentation of a ^{48}Ca beam at GANIL. The reaction products were separated in the LISE spectrometer. After Coulomb-excitation of ^{43}S in a ^{208}Pb target, the γ rays emitted in-flight were detected by 64 BaF_2 detectors of the Château de Cristal array. The preliminary value deduced for the reduced transition probability B(E2: $3/2^- \rightarrow 7/2^-_2$) is in agreement with the predictions of the shell model calculations and supports a prolate-spherical shape coexistence in the ^{43}S nucleus.

1. Motivation

The structure of atomic nuclei near the line of beta stability can be understood in terms of the shell model assuming a modified harmonic oscillator nuclear potential with a spin-orbit interaction. One of the greatest successes of this model was the reproduction of the magic numbers at 2, 8, 20, 28, 50, 82 and 126 observed in the stable nuclei. Nuclei with magic numbers have a closed shell configuration, and are characterized by a large 2^+_1 excitation energy and low B(E2:$2^+_1 \rightarrow 0^+_1$) reduced transition probability. The universality of the properties of nuclei with neutron magic has been questioned in the last decades as a result of the new experimental observations provided by the radioactive beam facilities.

At N=28, there is a large drop of the excitation energy of the first 2^+ state from ^{48}Ca to ^{42}Si (from 3832 keV to 770 keV) revealing a strong change in their structure, which has been ascribed to a reduction of the N=28 shell gap combined with the degeneracy of the proton $s_{1/2}$ and $d_{3/2}$ orbitals. In order to prove the reduction of the shell gap, it is necessary to understand the evolution of the single particle energy levels involved in this region of the nuclear chart and to determine where the deformed intruder configurations obeying deformation become the ground state of the N = 28 isotones.

Recent results on the structure of the N=28 sulfur isotone ^{44}S [1, 2, 3] were interpreted as indications of shape coexistence at low excitation energy in this nucleus. Different theoretical approaches provide different interpretations for the type of shape coexistence: prolate - spherical shape coexistence is predicted in the shell model approach, while mean-field calculations consider a wide configuration mixing [4].

Spectroscopic information on the N=27 nucleus ^{43}S is of special interest, as it can be used to deduce information on the shape of the nucleus in different neutron configurations and on the predicted competition between spherical and deformed configurations.

First experimental results regarding the structure of ^{43}S were obtained by Ibbotson et al. [5] in a Coulomb excitation experiment, which revealed the presence of a level at around 940 keV excitation energy. From the excitation cross section Ibbotson deduced 175(69) e^2fm^4 for the reduced transition probability of this transition. In addition, an isomeric state at 319 keV [6] with a lifetime of 475(48) ns and a reduced transition probability to the g.s. of 352 e^2fm^4 was extracted. This value suggest an E2 transition type between the isomeric state and the ground state. Magnetic moment measurements by Gaudefroy et al. [7] confirmed a spin and parity of $7/2^-$ for the isomeric state and a spin and parity of $3/2^-$ was assigned to the ground state. Within the shell model framework, these spectroscopic data are interpreted as the result of the progressive reduction of the N = 28 shell gap and the increase of correlation energy going from stability toward exotic nuclei.

The present experiment aimed to re-measure the reduced transition probability B(E2: $3/2^- \rightarrow 7/2^-_2$). This observable provides direct information on the deformation of the states. Additionally, we expected to populate higher spin states which will give access to additional information regarding the structure of ^{43}S.

2. Experimental set-up.

A schematic picture of the experimental set-up is shown in Figure 1. The nuclei of interest were produced by fragmentation of a $^{48}Ca^{19+}$ primary beam of 60 MeV/u energy with an average intensity of ~4 μA impinging on a 145 μm Be target. The produced fragments were separated by means of the Bρ-ΔE-Bρ method [8] in the LISE3 spectrometer, and their energies were determined in terms of their magnetic rigidity values. At the final focal plane of the spectrometer two position sensitive detectors were placed in order to reconstruct the trajectories of the incoming ions [9]. These detectors were also used for the Time-of-Flight (TOF) measurement with respect to the cyclotron radio-frequency. Afterwards, the selected and identified nuclei were excited in the Coulomb field of a lead target of 200 $mg.cm^{-2}$ thickness.

After passing the lead target, the nuclei that suffered Coulomb excitation were scattered into a 500 μm thick round shape Double Sided Silicon Strip Detector (DSSSD) [10], which had a central hole of 3 cm diameter. It consisted of four quadrants with 16 annular strips of 1.9 mm width and 2 mm pitch on the front side and 8 radial strips at 3.4^0 pitch on the back side. The inter-strip distance was 100 μm. This detector was mounted at a distance of 41.2 cm from the secondary target. In this way the DSSSD covered angles from 1.5 to 6.5^0 in the laboratory frame. This angular covering ensures that nuclei up to the grazing angle of the reaction are detected. The scattered ions were identified in an event-by-event basis by the combined measurement of ΔE and the TOF in the DSSSD (Figure 2). The DSSSD was followed by a residual energy detector, which consisted of four non-segmented silicon quadrants of 1500 μm thickness, which ensured that all particles that emerge from the DSSSD were stopped in its active area.

The nuclei that suffered Coulomb excitation, de-excited in flight by emission of γ-rays. 65 hexagonal BaF_2 crystals of the Château de Cristal were placed in a close geometry surrounding the secondary target. The γ-ray efficiency was determined to be 31% at 1274 keV, allowing for performing a coincidence measurement of the prompt γ-radiation and the ion identified in the DSSSD. The energy resolution was 11% and a time resolution of ~ 1 ns was obtained.

Nuclei that passed through the central hole of the DSSSD-PAD telescope were identified in an ionization chamber through the measurement of their energy loss, and came to rest in a plastic scintillator mounted at 0 degrees downstream. Two HP Ge detectors were placed surrounding the implantation setup in order to detect the delayed gamma transitions corresponding to the decay of the isomeric states populated in the fragmentation reaction at the entrance of the spectrometer and survived the flight path through LISE.

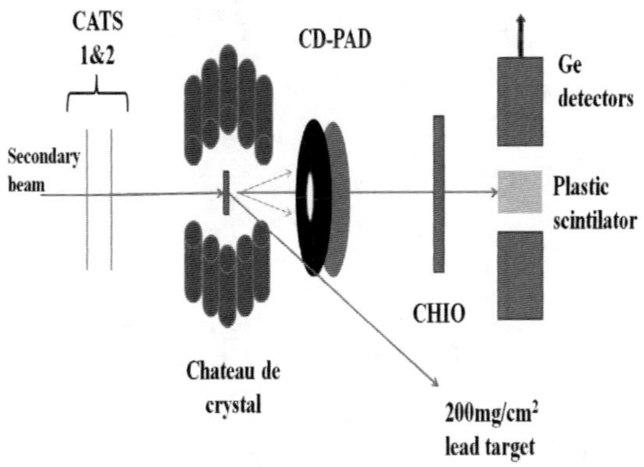

Figure 1: Schematic picture of the experimental Set-up

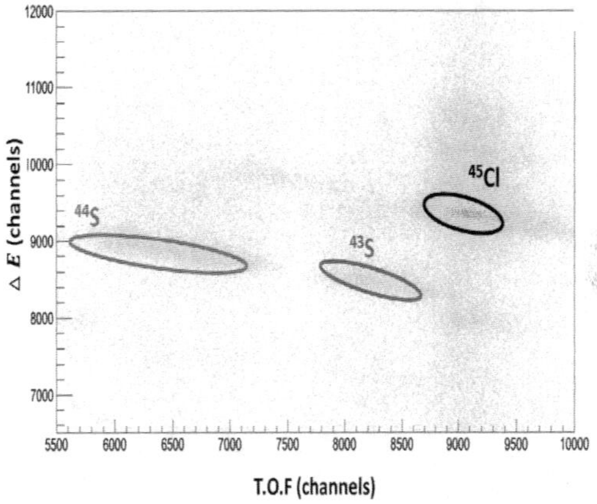

Figure 2: ΔE in one quadrant of the DSSSD vs. TOF identification plot

3. Preliminary results

The first nucleus investigated was the ^{44}Ca due to the high probability of its production and its good selection by the LISE spectrometer. In addition, the reduced transition probability B ($E2:0^+_1 \rightarrow 2^+_1$) is well known in this nucleus [11], thus it can be used for normalization in the relative determination of other B (E_2) values. Moreover, the transition of 1157 keV from ^{44}Ca was also used to determine the effective angle of each of the Château de Cristal detectors, necessary to have an accurate Doppler correction. In Figure 3, the singles spectrum obtained without (a) and with (b) Doppler correction for a ^{44}Ca beam is shown. These spectra have been obtained by gating on the ^{44}Ca ions identified in the DSSSD detectors and requiring that the time difference between detected ion and the γ-ray emission is maximum 5 ns. In addition, only events with γ-ray multiplicity one were taken into account.

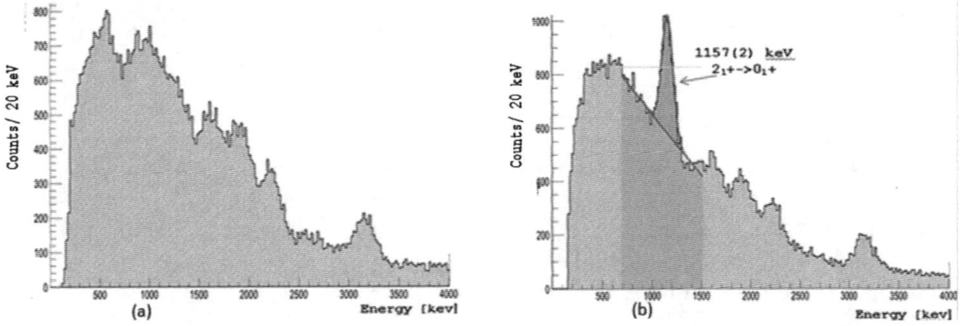

Figure 3: ^{44}Ca spectrum from the whole Chateau de Crystal.
(a) without Doppler correction, (b) Doppler corrected spectrum

Similarly to the ^{44}Ca spectrum, the spectrum associated with the Coulomb excitation of ^{43}S was also extracted and is presented in Fig. 4. Two γ-rays resulting from $7/2^-_1 \rightarrow 3/2^-$ and $7/2^-_2 \rightarrow 3/2^-$ transitions with the energies of 319 keV and 920 keV can be clearly seen.

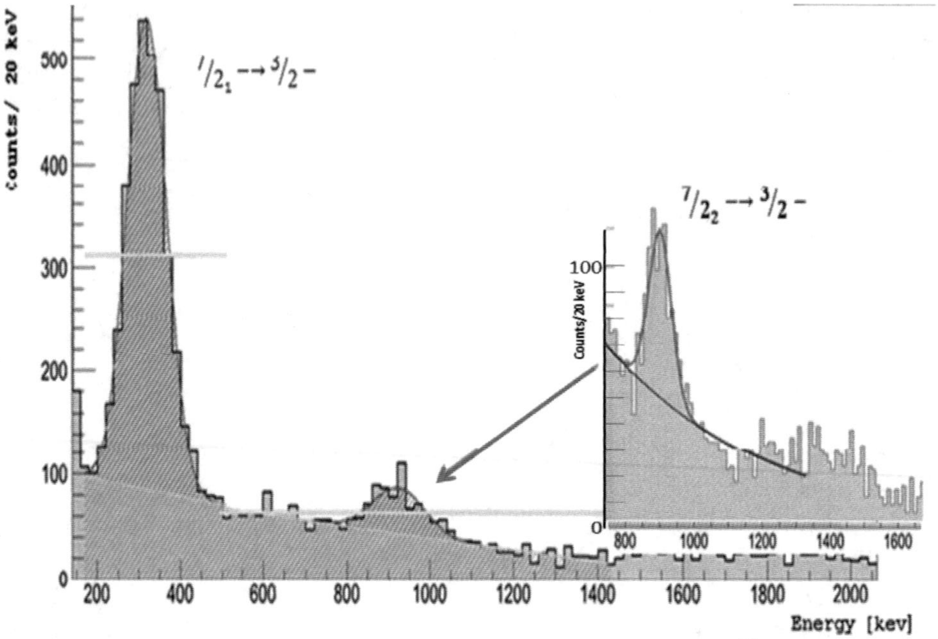

Figure 4: Coulomb excitation singles spectrum of ^{43}S

The number of counts in the 926 keV peak were determined by fitting a Gaussian plus an exponential background function the single spectra. Using the known reduced transition probability $B(E2:0^+_1 \rightarrow 2^+_1)$ from ^{44}Ca, a preliminary value of 257(27) e^2fm^4 was obtained for the reduced transition probability $B(E2:7/2^-_2 \rightarrow 3/2^-)$ in ^{43}S.

Following work will consist of implementation of the addback algorithm between the BaF_2 detectors, which will reduce the background in the γ-ray spectrum and increasing the statistics for the transitions of interest. Therefore, a more accurate value of the B(E2) reduced transition probability will be extracted.

The large yield of the 319 keV transition in the prompt γ-ray spectrum is under study in order to establish if its decay is compatible with previous measurements.

4. Conclusions

The preliminary value for the transition probability $B(E2: 3/2^- \rightarrow 7/2^-_2)$ of 257(27) e^2fm^4 has been deduced. This value is larger than the previous one obtained by Ibbotson et al. [5] and further support the presence of deformation in ^{43}S. Further work will focus on the implementation of the addback routine to reduce the uncertainties of the B(E2) values and on a detailed analysis of the origin of the observed prompt 319 keV transition.

5. Acknowledgments

The author S. Calinescu is grateful for the financial support from the European Social Fond through POSDRU 107/1.5/S/82514 Project

This work was supported by a grant of the Romanian National Authority for Scientific Research, CNCS – UEFISCDI, project number PN-II-RU-TE-2011-3-0051, the Hungarian Scientific Research Fund OTKA (contract number K100835), the TÁMOP-4.2.2/B-10/1-2010-0024 project and the Bolyai János Foundation.

The work of C. Petrone was possible with the financial support of the Sectorial Operational Program for Human Resources Development 2007-2013, co-financed by the European Social Fund, under the project number POSDRU/107/1.5/S/80765.

We are grateful for the technical support provided to us by staff at the GANIL facility.

References

[1] Force C et al., 2010, *Phys. Rev. Lett.* **105**, 102501
[2] Santiago-Gonzales D et al., 2011, *Phys. Rev.* C **83**, 061305(R)
[3] Cáceres L et al., 2012, *Phys. Rev.* C **85**, 024311
[4] Rodriguez T R et al., 2011, *Phys. Rev.* C 84, 051307(R)
[5] Ibbotson R et al., 1999, *Phys. Rev.* C **59**
[6] Sarazin F et al., 2000, *Phys. Rev. Lett.* **84** 5062
[7] Gaudefroy L et al., 2009, *Phys. Rev. Lett.* **102**
[8] Dufour J P et al., 1986, *Nucl. Instr. Methods* A **248**, 267
[9] Ottini S et al., 1999, *NIM* A **431**
[10] Ostrowski A N et al., 2002, *NIM*, A **480**
[11] Cline D, Horoshko R N and Towsley C W, 1973, *N. Phys* .A **204**, 574

Int. Summer School for Advanced Studies 'Dynamics of open nuclear systems' (PREDEAL12) IOP Publishing
Journal of Physics: Conference Series **413** (2013) 012031 doi:10.1088/1742-6596/413/1/012031

Model dependence of elliptic flow differences

M D Cozma

IFIN-HH, Reactorului 30, 077125 Măgurele-Bucharest, Romania

E-mail: `cozma@niham.nipne.ro`

Abstract. An isospin dependent version of the QMD transport model is used to study the influence of the isovector part of the equation of state of nuclear matter on observables that can be measured in heavy-ion collisions at intermediate energy. The model dependence of neutron-proton elliptic flow difference is studied for AuAu collisions at an incident energy of 400 MeV per nucleon. It is found that the sensitivity to microscopical nucleon-nucleon cross-sections, momentum dependence of the optical potential, compressibility modulus of nuclear matter and width of nucleon wave function are moderate compared to the dependence on the stiffness of the isospin asymmetric part of the equation of state. It is concluded that neutron-proton elliptic flow difference is a suitable observable for setting constraints on the supra-saturation density dependence of symmetry energy.

1. Introduction

One of the remaining open questions in nuclear physics is the equation of state (EoS) of isospin asymmetric nuclear matter (asy-EoS), *i.e.* the density dependence of the symmetry energy (SE). Its precise knowledge is mandatory for a proper understanding of nuclear structure of rare isotopes, dynamics and products of heavy-ion collisions, and most importantly for astrophysical processes such as neutron star cooling and supernovae explosions [1, 2]. Intermediate energy nuclear reactions involving radioactive beams have allowed, by studying the thickness of neutron skins, deformation, binding energies and isospin diffusion to constrain the density dependence of SE at densities below saturation (ρ_0) [3, 4]. Existing theoretical models describing its density dependence generally agree with each other in this density regime, but their predictions start to diverge well before regions with densities $\rho \geq 2\rho_0$ are reached [2].

Nuclear matter at supra-saturation densities is created in the laboratory in the process of collision of heavy nuclei. Several observables that can be measured in such reactions have been determined to bear information on the behavior of the SE above ρ_0: the neutron/proton ratio of squeezed out nucleons [5], light cluster emission [6], π^-/π^+ multiplicity ratio in central collisions [7, 8], double neutron to proton ratios of nucleon emission from isospin-asymmetric but mass-symmetric reactions [9], elliptic flow ratios [10] and others. The FOPI experimental data for the π^-/π^+ ratio [11] have been used to set constraints on the suprasaturation density behavior of SE by various authors with contradicting results: Xiao *et al.* [7] made use of the IBUU transport model supplemented by the isovector momentum dependent Gogny inspired parametrization of SE [12] to point toward a soft asy-EoS, while the study of Feng *et al.* [8], which employed IQMD and a power-law parametrization of the symmetry energy $S(\rho) = S_0\,(\rho/\rho_0)^\gamma$, favors a stiff SE. The study of elliptic flow ratios (npEFR) of Ref. [10] favors, by making use of the power-law parametrization of SE, an asy-EoS dependence on density above saturation point close to a linear one: γ=0.9±0.4.

Elliptic flows of protons and neutrons cannot be used separately to constrain the isovector part of the equation of state above saturation point due to their sizable dependence on the values of transport model parameters, that are either inaccurately determined or do not represent measurable quantities, like in-medium nucleon-nucleon (NN) cross-sections, compressibility modulus of nuclear matter and width of nucleon wave function [13]. A precise knowledge of the dependence on these parameters would allow the elimination of the most uncertain ones leaving one with a set of observables that bear no or almost no model dependence. In practice, one is forced to make assumptions that are verified or disproved by employing a definite transport model. For example, neutron proton elliptic flow ratios (npEFR), studied in Ref. [10], require a scenario in which elliptic flows scale linearly with model parameters. In this study we consider the scenario in which the derivatives of elliptic flows of neutrons and protons with respect to model parameters are assumed to be equal, *i.e.* we study the model dependence of neutron-proton elliptic flow difference (npEFD). Differences in the constraints on the density dependence of SE extracted from npEFR and npEFD can be viewed as due to residual model dependence and can be used to estimate the accuracy of such an approach.

2. The model

The quantum molecular dynamics (QMD) model [20, 21] is employed to simulate heavy-ion collisions. It has been updated by adding explicit density dependence of the microscopical nucleon-nucleon cross-sections and building in an isospin dependent EoS [13]. The same model has been previously used to study dilepton emission in heavy-ion collisions [14, 15, 16], stiffness of the equation of state of symmetric nuclear matter [17] and various in-medium effects relevant for the dynamics of heavy-ion collisions [18, 19]. The ingredients of the model most relevant for the current study are presented in the following.

One of the key ingredients of the transport models are the microscopic elastic and inelastic cross-sections. For the case of elastic collisions theoretically [22, 23] and experimentally driven [24, 25] studies hint for important in-medium modifications. For the case of total cross-sections the following density dependent parametrizations are being used [22, 23]

$$\begin{aligned}\sigma_{np}(E_{lab},\rho) &= [31.5+0.092\times(20.2-E_{lab}^{0.53})^{2.9}]\times\frac{1.0+0.0034E_{lab}^{1.51}\,\rho^2}{1.0+21.55\,\rho^{1.34}} \\ \sigma_{pp}(E_{lab},\rho) &= [23.5+0.256\times(18.501-E_{lab}^{0.52})^{3.1}]\times\frac{1.0+0.1667E_{lab}^{1.05}\,\rho^3}{1.0+9.704\,\rho^{1.2}}\end{aligned} \quad (1)$$

for energies below the pion production threshold; here E_{lab} is the incident kinetic energy in MeV, while the density ρ is expressed in fm^{-3}. Values of the elementary cross-sections of reactions involving at least one excited baryon (either Δ or N^*) are supposed to remain unmodified in a dense nuclear medium. Alternatively the Cugnon parametrization of vacuum NN cross-sections [26] can be used. The only sizable difference between the two mentioned parametrizations occurs in the neutron-proton channel at incident energies below 100 MeV.

The dependence of in-medium nucleon-nucleon cross-sections on isospin asymmetry is introduced by making use of the assumption that NN matrix elements retain their vacuum analytical expressions in the medium, leading to a isospin asymmetry dependence indirectly through the expressions of the effective nucleon masses as functions of β [27]

$$\sigma_{N_1N_2}(\rho,\beta) = \sigma_{N_1N_2}(\rho,\beta=0)\,\frac{m_{N_1}(\rho,\beta)\,m_{N_2}(\rho,\beta)}{m_{N_1}(\rho,\beta=0)\,m_{N_2}(\rho,\beta=0)}\,. \quad (2)$$

The above expression is used with minimal changes above the pion production threshold to describe both the density and isospin asymmetry dependence of NN cross-sections by considering

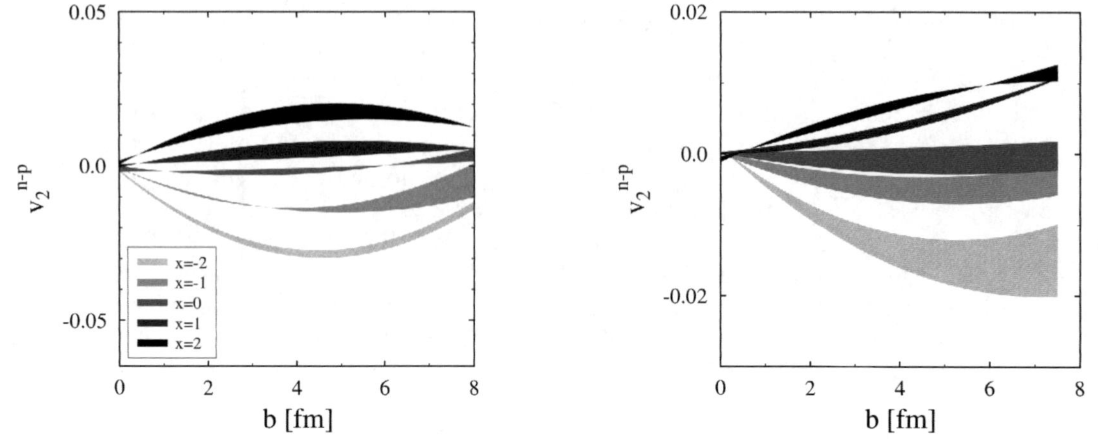

Figure 1. *(Left)* Sensitivity of npEFD to changes in the compressibility modulus. For the isoscalar part of EoS the Skyrme type parametrization has been used. The width of the bands corresponds to a variation of K from K=210 MeV (soft EoS) to K=380 MeV (stiff EoS). *(Right)* Variations in the values of the npEFD due to different parametrizations of the momentum dependent part of the isoscalar EoS. Different bands correspond, in both panels, to various stiffnesses for asy-EoS.

a dependence of nucleon mass on both parameters. The in-medium nucleon mass is derived from the momentum dependent part of the mean field potential by making use of the expression: $m_N(\rho,\beta)/m_N(\rho=0,\beta=0) = 1/(1 + m_N(\rho=0,\beta=0)/p \times dU/dp)$.

To describe the mean-field motion of nucleons a parametrization of the EoS developed from the Gogny effective interaction is used [12]. It contains a isospin dependent part which also exhibits an explicit momentum dependence. It reads

$$\begin{aligned} U(\rho,\beta,p,\tau,x) &= A_u(x)\frac{\rho_{\tau'}}{\rho_0} + A_l(x)\frac{\rho_\tau}{\rho_0} + B(\rho/\rho_0)^\sigma(1-x\beta^2) - 8\tau x\frac{B}{\sigma+1}\frac{\rho^{\sigma-1}}{\rho_0^\sigma}\beta\rho_{\tau'} \\ &+\frac{2C_{\tau\tau}}{\rho_0}\int d^3p'\,\frac{f_\tau(\vec{r},\vec{p'})}{1+(\vec{p}-\vec{p'})^2/\Lambda^2} + \frac{2C_{\tau\tau'}}{\rho_0}\int d^3p'\,\frac{f_{\tau'}(\vec{r},\vec{p'})}{1+(\vec{p}-\vec{p'})^2/\Lambda^2} \end{aligned} \quad (3)$$

where the parameters $A_u(x), A_l(x), B, C_{\tau,\tau'}, C_{\tau,\tau}$ and Λ have been obtained by fitting the momentum-dependence of $U(\rho,\beta,p,\tau,x)$ to the one predicted by the Gogny potential and requiring that the properties of nuclear matter at saturation density are reproduced. The compressibility modulus K is set to be a soft one (210 MeV) and the value of energy symmetry at ρ_0 is set to 30 MeV. The parameter x can be adjusted to mimic the density dependence of the symmetry energy of various microscopical theoretical models and has been varied in the current study between -2 and 2. The value $x = -2$ corresponds to a super-stiff asy-EoS, $x = 2$ to a super-soft asy-EoS, while setting $x = 0$ the momentum independent part of SE is omitted. The parameters used in the calculations can be found in [12].

3. Elliptic flow difference

The azimuthal distribution of protons (or neutrons) resulted in heavy-ion collisions can be approximately described by $dN/d\phi = (N/(2\pi)\,[\,1 + v_1\cos\phi + 2\,v_2\cos 2\phi\,]$, v_1 and v_2 being called the sidewards and elliptic flow (EF) parameters respectively. The elliptic flow can by extracted from simulated or experimental data by computing the following average over the respective particle specie in the final state $v_2 = (1/N)\sum_{i=1,N}({p_x^i}^2 - {p_y^i}^2)/{p_T^i}^2$. The neutron-proton elliptic flow difference can readily be obtained: $v_2^{n-p} = v_2^n - v_2^p$. The results presented in this Section are restricted to Au+Au collisions at an incident energy of 400 MeV per nucleon.

The most reliable extraction of the compressibility modulus K of nuclear matter has been made possible by studying the multiplicity ratio of K^+ production in heavy (Au+Au) over light (C+C) nuclei at incident energies close to 1 AGeV pointing towards a soft EoS. At lower incident energies the situation is not as clear: an even softer EoS is favored while the study of sidewards flow by the FOPI collaboration favors a soft or hard EoS of state depending on system size. In view of this we have investigated the sensitivity of EF to changes from a soft (K=210 MeV) to a hard (K=380 MeV) EoS. Results are shown in left panel of Fig. 1 for different choices of the parameter x determining the stiffness of asy-EoS. A clear separation of results corresponding to different values of the asy-EoS stiffness is observed. The kinematic constraints adopted here are compatible with those of the FOPI collaboration [28].

The importance and impact of momentum dependent interactions on heavy-ion observables is well known [29], yet the momentum dependence of the nucleon optical potential is not fully agreed upon. We study its impact on npEFD by employing two popular parametrizations: the Gogny inspired one (Eq. 3) and U_{mdi}=-0.054+0.00158 $\cdot \ln^2[500\,(\vec{p}_1 - \vec{p}_2)^2 + 1]$ [GeV] [30]. They differ mainly in the high momentum region where the Aichelin parametrization [30] becomes repulsive while the former stays attractive at all energies. In the right panel of Fig. 1 changes in the predicted values for npEFD are studied when switching between the two parametrizations of the optical potential, while preserving the saturation properties of nuclear matter. It is observed that while for a soft asy-EoS scenario the impact of different parametrizations of the optical potential is negligible the situation changes considerably once an as asy-stiff scenario is used. This is in agreement with the results of Ref. [31].

An additional indetermination of theoretical estimates is brought in by in-medium values of the nucleon-nucleon cross-sections and the value of the nucleon wave-packet width L [13]. For the former we employ two different vacuum parametrizations: Cugnon and Li-Machleidt and several scenarios for the density and asymmetry dependence of second (details in Ref. [13]). The width of the nucleon wave function is usually set in literature to $2L^2$=4 fm^2 for light systems, while for heavy systems an increase to the value $2L^2$=8 fm^2 is found necessary in order to generate nuclei with stable static properties (*e.g.* rms). The sensitivity of npEFR to variations in these two model parameters are plotted in Figure 2, in the left and right plots for NN cross-sections and nucleon wave-function width respectively. The variations amount to about 20% of the splitting of npEFR between a super-stiff and super-soft asy-EoS in each case.

To conclude, we have studied the sensitivity of npEFR to different model parameters like microscopic NN cross-sections, compressibility modulus of nuclear matter, optical potential and width of nucleon functions as compared to the sensitivity to the stiffness of the asy-EoS. We have found that summed together (in quadrature) the indetermination of npEFR due to inaccurately known model parameters amounts to about 40% of the splitting of the same observable between the super-stiff and super-soft asy-EoS scenarios. This would allow, by comparing with the experimental npEFD data of the FOPI-LAND [32] or the ASY-EOS [33] Collaborations, to set constraints on the density dependence of symmetry energy above saturation point with a very limited model bias.

4. Acknowledgments

M.D.C. would like to thank the organizers of the summer school "Dynamics of Open Nuclear Systems" 2012 for the invitation and generous financial support and W. Trautmann and for extensive discussions on the topic. M.D.C. acknowledges financial support from PN09370103 grant of the Romanian Ministry of Education and Research.

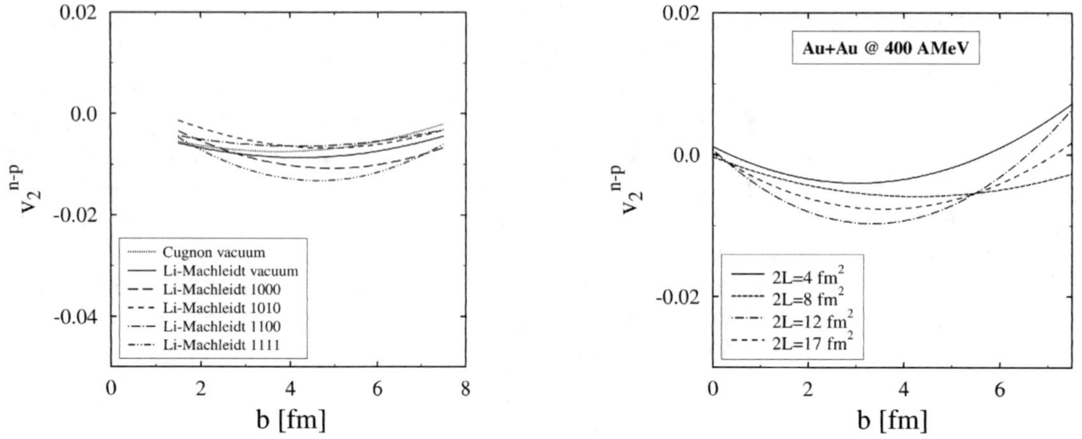

Figure 2. Sensitivity of the npEFD due to changes in the parametrization for the microscopic NN cross-sections (left) and the width of the nucleon waves function (right).

References

[1] Baran V, Colonna M, Greco V and Di Toro M 2005 *Phys. Rep.* **410** 335
[2] Li B-A, Chen K W and Ko C M 2008 *Phys. Rep.* **464** 113
[3] Li B-A, Ko C M and Bauer W 1998 *Int. J. Mod. Phys.* E **7** 147
[4] Chen L W, Ko C M and Li B-A 2005 *Phys. Rev. Lett.* **94** 032701
[5] Yong G-C, Li B-A and Chen L-W 2007 *Phys. Lett.* B **650** 344
[6] Chen L-W, Ko C M and Li B-A 2003 *Phys. Rev.* C **68** 017601
[7] Xiao Z, Li B-A, Chen L-W, Yong G-C and Zhang M 2009 *Phys. Rev. Lett.* **102** 062502
[8] Feng Z-Q and Jin G-M 2010 *Phys. Lett.* B **683** 140
[9] Li Q, Li Z and Stöcker H 2006 *Phys. Rev.* C **73** 051602
[10] Russotto P, Wu P Z, Zoric M, Chartier M, Leifels Y, Lemmon R C, Li Q *et al.* 2011 *Phys. Lett.* B **697** 471
[11] Reisdorf W*et al.* [FOPI Collaboration] 2007 *Nucl. Phys.* A **781** 459
[12] Das C B, Das Gupta S, Gale C and Li B-A 2003 *Phys. Rev.* C **67** 034611
[13] Cozma M D 2011 *Phys. Lett.* B **700** 139
[14] Shekhter K, Fuchs C, Faessler A, Krivoruchenko M and Martemyanov B 2003 *Phys. Rev.* C **68** 014904
[15] Cozma M D, Fuchs C, Santini E and Fassler A 2006 *Phys. Lett.* B **640** 170
[16] Santini E, Cozma M D, Faessler A, Fuchs C *et al.* 2008 *Phys. Rev.* C **78** 034910
[17] Fuchs C, Faessler A, Zabrodin E and Zheng Y-M 2001 *Phys. Rev. Lett.* **86** 1974
[18] Fuchs C, Essler P, Gaitanos T and Wolter H H 1997 *Nucl. Phys.* A **626** 987
[19] Uma Maheswari V S, Fuchs C, Faessler A, Sehn L, Kosov D S and Wang Z 1998 *Nucl. Phys.* A **628** 669
[20] Khoa D T *et al.* 1992 *Nucl. Phys.* A **548** 102
[21] Uma Maheswari W S *et al.* 1998 *Nucl. Phys.* A **628** 669
[22] Li G Q and Machleidt R 1993 *Phys. Rev.* C **48** 1702
[23] Li G Q and Machleidt R 1994 *Phys. Rev.* C **49** 566
[24] Li Q, Li Z, Soff S, Bleicher M and Stöcker H 2006 *J. Phys.* G **32** 407
[25] Gaitanos T, Fuchs C and Wolter H H 2005 *Phys. Lett.* B **609** 241
[26] Cugnon J, Mizutani T and Vandermuelen J 1981 *Nucl. Phys.* A **352** 505
[27] Li B-A and Chen L-W 2005 *Phys. Rev.* C **72** 064611
[28] Andronic A *et al.* 2001 *Nucl. Phys.* A **679** 765
[29] Aichelin J, Rosenhauer A, Peilert G, Stoecker H and Greiner W 1987 *Phys. Rev. Lett.* **58** 1926
[30] Aichelin J 1991 *Phys. Rep.* **202** 233
[31] Zhang L, Gao Y, Du Y, Zuo G H and Yong G C 2012 *Eur. Phys. Jour.* A **48** 30
[32] Leifels Y *et al.* [FOPI Collaboration] 1993 *Phys. Rev. Lett.* **71** 963
Lambrecht D *et al.* [FOPI-LAND Collaboration] 1994 *Z. Phys.* A **350** 115
[33] Russotto P *et al.* [ASY-EOS Collaboration] arXiv:1209.5961 [nucl-ex].

Int. Summer School for Advanced Studies 'Dynamics of open nuclear systems' (PREDEAL12) IOP Publishing
Journal of Physics: Conference Series **413** (2013) 012032 doi:10.1088/1742-6596/413/1/012032

Systematic study of the isovector pairing effect on the moment of inertia of proton-rich nuclei in the region $30 \leq Z \leq 40$

I Ami [1,2], M Fellah [2,3], N Benhamouda [2] and N H Allal [2,3]

[1] Département de Physique, Faculté des Sciences, Université M'hamed Bougara de Boumerdes. Campus Sud-35000, Boumerdès, Algeria.
[2] Laboratoire de Physique Théorique, Faculté de Physique, USTHB, BP 32, El Alia, 16111 Bab Ezzouar, Algiers, Algeria.
[3] Centre de Recherche Nucléaire d'Alger, 2 Bd. Frantz Fanon, BP. 399 Alger-Gare, Algiers, Algeria.

Abstract. A systematic study of the isovector neutron-proton (np) pairing effect on the moment of inertia is performed at zero temperature. This study is based on a recently established expression obtained using the framework of the quantum perturbation theory and the Inglis cranking method, at the limit when the temperature is nil.

We considered even–even proton-rich nuclei such as $30 \leq Z \leq 40$ and $N - Z = 0, 2, 4$ using the single-particle energies and eigen-states of a deformed Woods-Saxon mean-field. The obtained results are compared to their homologues of the conventional BCS theory (i. e. when only the pairing between like-particles is considered).

1. Introduction

The nuclear moment of inertia plays a very important role in the description of rotating nuclei. Indeed, this observable being sensitive to the deformation, its measurement indirectly provides significant information about the shape of the nuclei. The study of the moment of inertia has been the subject of several works taking into account the pairing correlations between like-particles at zero and finite temperature [1, 2, 3, 4]. It has been also calculated at zero and finite temperature with inclusion of isovector pairing correlations [5, 6, 7, 8, 9].
The aim of the present work is to carry out a systematic study of the isovector pairing effect on the moment of inertia of proton-rich nuclei in the region $30 \leq Z \leq 40$.

2. Formalism

Let us consider a heated nucleus constituted by N neutrons and Z protons. The symmetry axis being Oz, the system is cranked around the Ox axis of a rotating frame. The grand-partition function is given by [8, 9]:

$$Z = Tr\left\{\exp(-\beta[H - \sum_t \lambda_t N_t - \hbar\omega J_x])\right\} \tag{1}$$

where β is the inverse of the temperature T and H is the Hamiltonian of the system given, in the second quantization and isospin formalism, in the isovector pairing case, by [10]:

$$H = \sum_{\nu>0,t} \varepsilon_{\nu t}\left(a^{\dagger}_{\nu t}a_{\nu t} + a^{\dagger}_{\tilde{\nu}t}a_{\tilde{\nu}t}\right) - \sum_{tt'} G_{tt'} \sum_{\nu\mu>0}\left(a^{\dagger}_{\nu t}a^{\dagger}_{\tilde{\nu}t'}a_{\tilde{\mu}t'}a_{\mu t} + a^{\dagger}_{\nu t}a^{\dagger}_{\tilde{\nu}t'}a_{\tilde{\mu}t}a_{\mu t'}\right) \tag{2}$$

The subscript t corresponds to the isospin component ($t = n, p$), $a^{\dagger}_{\nu t}$ and $a_{\nu t}$ respectively represent the creation and annihilation operators of the particle in the state $|\nu t\rangle$, of energy $\varepsilon_{\nu t}$; $|\tilde{\nu}t\rangle$ is the time-reverse of $|\nu t\rangle$, $G_{tt'}$ characterizes the pairing-strength. The neutron and proton are supposed to occupy the same energy levels. In all that follows, it is assumed that the single-particle energies are independent from the temperature [11]. $\lambda_t (t = n, p)$ are the Fermi level energies and N_t are the particle-number operators given by:

$$N_t = \sum_{\nu>0}\left(a^{\dagger}_{\nu t}a_{\nu t} + a^{\dagger}_{\tilde{\nu}t}a_{\tilde{\nu}t}\right) \qquad t = n, p \tag{3}$$

ω is the rotation frequency and J_x is the projection over the Ox axis of the angular momentum. The usual Inglis [12] expression of the energy may be easily generalized in order to include the temperature effects as well as the np pairing correlations. The energy E is then given by [4]:

$$E = \left(-\frac{\partial \ln Z}{\partial \beta}\right)_{\lambda_t \beta = cte} \tag{4}$$

The expansion to the second order in ω of expression (4) is given by:

$$E \simeq E_0 - \omega^2\hbar^2 \int_0^{\beta} \langle J_x(\beta)\, J_x(\chi)\rangle_0\, d\chi \tag{5}$$

where

$$J_x(\kappa) = e^{\kappa H'} J_x e^{-\kappa H'} \quad and \quad \langle J_x(\beta)\, J_x(\chi)\rangle = \frac{Tre^{-\beta H'} J_x(\beta)\, J_x(\chi)}{Tre^{-\beta H'}} \qquad \kappa = \beta, \chi \tag{6}$$

$J_x(\kappa)$ is the Heisenberg transform of J_x and the thermal average in Eq. (5) is evaluated using the grand-canonical ensemble associated to the Hamiltonian without rotation H' given by:

$$H' = H - \sum_{t=n,p} \lambda_t N_t \tag{7}$$

This thermal average value may easily be determined using the quasiparticle representation. In the latter, the auxiliary Hamiltonian has been approximately diagonalized by means of the Feynman path integral technique and using the Hubbard-Stratonovich transformation [10]. In fact, the diagonal form of the Hamiltonian H' becomes:

$$H' = \sum_{\nu>0,\ \tau=1,2} E_{\nu\tau}\left(\alpha^{\dagger}_{\nu\tau}\alpha_{\nu\tau} - \alpha_{\tilde{\nu}\tau}\alpha^{\dagger}_{\tilde{\nu}\tau}\right) + \sum_{\nu>0,t} \widetilde{\varepsilon}_{\nu t} \tag{8}$$

where we set

$$\widetilde{\varepsilon}_{\nu t} = (\varepsilon_{\nu t} - \lambda_t) - (G_{tt} + G_{np})/2 \qquad t = p, n \tag{9}$$

$\alpha^{\dagger}_{\nu\tau}$ (respectively $\alpha_{\nu\tau}$) is the creation (respectively annihilation) operator of a quasiparticle of $\tau\,(\tau = 1, 2)$ type, given by the generalized Bogoliubov-Valatin transformation [10]:

$$\alpha^{\dagger}_{\nu\tau} = \sum_{\nu>0,\ t} \left(u_{\nu\tau t} a^{\dagger}_{\nu t} + v_{\nu\tau t} a_{\tilde{\nu} t} \right) \qquad \tau = 1, 2 \tag{10}$$

The corresponding energies $E_{\nu\tau}$ are defined in Ref. [10]. The perpendicular moment of inertia, is then defined by [8, 9]:

$$\Im^{np}_{\perp} = 2\hbar^2 \int_0^{\beta} \langle J_x(\beta)\, J_x(\chi) \rangle_0 \, d\chi \tag{11}$$

One has, after some algebra,

$$\begin{aligned} \Im^{np}_{\perp} &= \hbar^2 \sum_{\nu\mu\tau\tau' tt'} \langle \nu t \left| J_x \right| \mu t \rangle \langle \mu t' \left| J_x \right| \nu t' \rangle \\ &\left\{ \left[v_{\nu\tau t} u_{\mu\tau' t} - v_{\mu\tau' t} u_{\nu\tau t} \right] \left[v_{\nu\tau t'} u_{\mu\tau' t'} - v_{\mu\tau' t'} u_{\nu\tau t'} \right] \left(\frac{\tanh\left(\frac{\beta E_{\nu\tau}}{2}\right) + \tanh\left(\frac{\beta E_{\mu\tau'}}{2}\right)}{E_{\nu\tau} + E_{\mu\tau'}} \right) \right. \\ &\left. + \left[u_{\nu\tau t} u_{\mu\tau' t} + v_{\nu\tau t} v_{\mu\tau' t} \right] \left[u_{\nu\tau t'} u_{\mu\tau' t'} + v_{\nu\tau t'} v_{\mu\tau' t'} \right] \left(\frac{\tanh\left(\frac{\beta E_{\mu\tau'}}{2}\right) - \tanh\left(\frac{\beta E_{\nu\tau}}{2}\right)}{E_{\mu\tau'} - E_{\nu\tau}} \right) \right\} \end{aligned} \tag{12}$$

If the np pairing effects vanish, expression (12) reduces to the sum of the moments of inertia of the neutron and proton systems considered separately in the framework of the pairing between like-particles approach [4, 13, 14].

3. Numerical results-Discussion

The previously described formalism has been numerically applied using the single-particle energies and eigen-states of a deformed Woods-Saxon mean-field [15]. We used the parameters of Ref. [16] with a maximum number of shells $N_{max} = 12$. The ground state deformations are those of the Möller table [17]. We considered even-even nuclei such as $N - Z = 0, 2, 4$. Indeed, in such nuclei, it has been shown that np pairing effects are not negligible [18, 19]. In the present work, we considered only nuclei with a deformed ground-state such as $30 \leq Z \leq 40$. The pairing-strength values G_{pp}, G_{nn} and G_{np} have been deduced from the Δ_{pp}, Δ_{nn} and Δ_{np} values at zero temperature. The latter are obtained using the odd-even mass differences [20].
It appears that the inclusion of the np pairing effect clearly modifies the values of the moment of inertia obtained within the framework of usual BCS theory. Indeed, the relative discrepancy between the previsions of the present model and those of the BCS one varies from one nucleus to another and is about 15.57% on average for all the studied nuclei. In addition, the relative discrepancy between the theoretical values and the experimental data for $N = Z$ nuclei is on average 28.60% and 23.15% respectively with and without inclusion of the np pairing. With regard to other nuclei, this discrepancy is, with and without inclusion of the np pairing, respectively 27.11% and 37.73% for nuclei such as $(N - Z) = 2$ and 39.57% and 45.56% for nuclei such as $(N - Z) = 4$.
It is thus necessary to take into account this type of correlations in the evaluation of the moment of inertia of proton rich nuclei like those considered in this work.

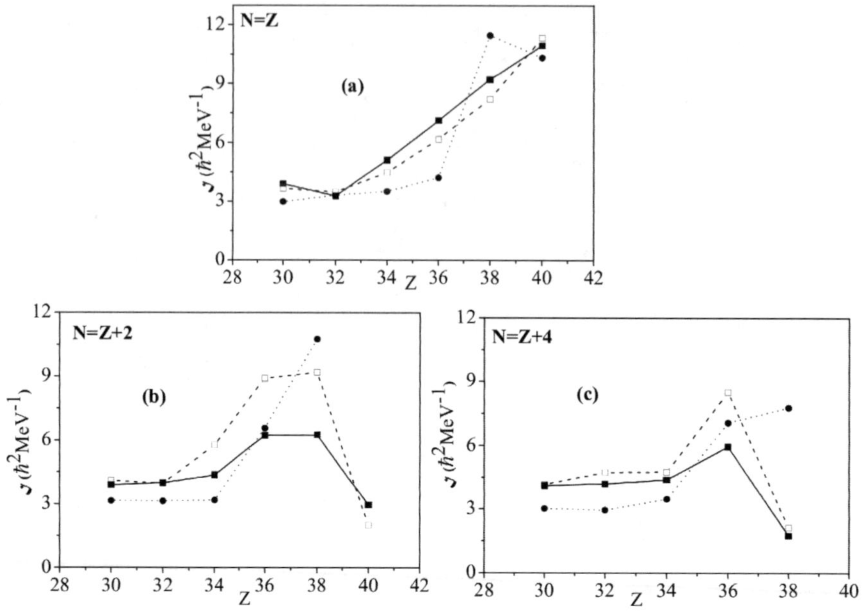

Figure 1. Variation of the moment of inertia as a function of the proton number Z, for nuclei such as $N = Z$ **(a)**, $N - Z = 2$ **(b)** and $N - Z = 4$ **(c)** and whose experimental values are known; with (–■–) and without (–□–) inclusion of np pairing. (– • –) represent the experimental values [21, 22].

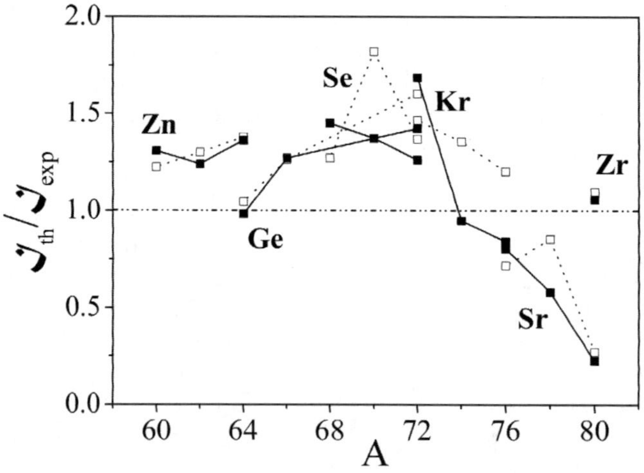

Figure 2. Points (–■–) are our results and (–□–) those of the BCS approximation .

Table 1. Moment of inertia of even-even proton-rich nuclei evaluated without (column 2) and with (column 3) inclusion of np pairing. The last column deals with the relative discrepancy.

Nucleus	$\Im_{FTBCS}(\hbar^2 MeV^{-1})$	$\Im_{\perp}^{np}(\hbar^2 MeV^{-1})$	$\left\| \frac{\Im_{\perp}^{np} - \Im_{FTBCS}}{\Im_{FTBCS}} \right\|$ %
^{60}Zn	3.66	3.91	6.80
^{62}Zn	4.09	3.90	4.78
^{64}Zn	4.17	4.12	1.14
^{64}Ge	3.48	3.27	5.96
^{66}Ge	3.96	3.98	0.71
^{68}Ge	4.73	4.20	11.23
^{68}Se	4.47	5.10	13.93
^{70}Se	5.78	4.36	24.52
^{72}Se	4.76	4.39	7.68
^{72}Kr	6.19	7.12	15.15
^{74}Kr	8.91	6.23	30.06
^{76}Kr	8.51	5.95	30.09
^{76}Sr	8.23	9.21	12.03
^{78}Sr	9.19	6.25	32.00
^{80}Sr	2.12	1.76	16.60
^{80}Zr	11.35	10.96	3.48
^{82}Zr	2.00	2.97	48.50

[1] Allal N. H and Fellah M, 1991 *Phys. Rev.* C **43** 2648.
[2] Hasegawa M and Tazaki S, 1993 *Phys. Rev.* C **47** 188.
[3] Zeng J. Y, Jin T. H and Zhao Z. J, 1994 *Phys. Rev.* C **50** 1388.
[4] Allal N. H, Fellah M, Benhamouda N and Oudih M. R, 2008 *Phys.Rev.* C **77** 054310.
[5] Mokhtari D, Ami I, Fellah M and Allal N.H, 2008 *Int. J. Mod. Phys.* E **17** 655.
[6] Gerceklioğlu M and Calik A. E, 2005 *Acta Phys. Slo.* **55** 197.
[7] Calik E, Deniz C and Gercelioglu M, 2009 *Pramana-J. Phys.* **73** 847.
[8] Ami I, Fellah M, Benhamouda N, Allal N. H and Belabbas M, 2012 *J. Phys.: Conf. Series* **338** 012019.
[9] Ami I, Fellah M, Allal N.H, Benhamouda N, Belabbas M and Oudih M.R, 2011 *Int. J. Mod. Phys.* E **20** 1947.
[10] Fellah M, Allal N.H, Belabbas M, Oudih M.R and Benhamouda N, 2007 *Phys. Rev.* C **76** 047306.
[11] Brack M and Quentin P, 1981 *Nucl. Phys.* A **361** 35.
[12] Inglis D. R, 1954 *Phys. Rev.* **96** 1059; 1955 *Phys. Rev.***97** 701.
[13] Brack M et al., 1972 *Rev. Mod. Phys.* **44** 320.
[14] Brack M et al., 1974 *Nucl. Phys.* A **234** 185.
[15] Woods R. D and Saxon D. S, 1954 *Phys. Rev.* **95** 577.
[16] Allal N.H and Fellah M, 1994 *Phys. Rev.* C **50** 1404.
[17] Moller P, Nix J. R, Myers W. D and Swiatecki W. J, 1995 *Atomic Data and Nuclear Data Tables* **59** 185.
[18] Civitarese O, Reboiro O, 1997 *Phys. Rev.* C **56** 1179.
[19] Civitarese O, Reboiro M and Vogel P, 1997 *Phys. Rev.* C **56** 1840.
[20] Simkovic F, Moustakidis Ch. C, Pacearescu L and Faessler A, 2003 *Phys. Rev.* C **68** 054319.
[21] Firestone R. B and Shirley V. S, *Table of isotopes* (John and Wiley & Sons, New York,1999).
[22] Fischer S.M, Lister C. J and Balamuth D. P, 2003 *Phys. Rev.* C **67** 064318.

Journal of Physics: Conference Series **413** (2013) 012033 doi:10.1088/1742-6596/413/1/012033

Runge-Kutta methods of special form

L. Gr. Ixaru

"Horia Hulubei" National Institute of Physics and Nuclear Engineering,
Department of Theoretical Physics,
P.O.Box MG-6, Bucharest, Romania
e-mail: ixaru@theory.nipne.ro
and
Academy of Romanian Scientists,
54 Splaiul Independenţei, 050094, Bucharest, Romania

E-mail: ixaru@theory.nipne.ro

Abstract. Many problems in physics and engineering require the solution of stiff systems of differential equations and only A-stable numerical methods are appropriate for this purpose. The theory covering the methods in current use, which are methods with constant coefficients, says that A-stable methods of order greater than 2 are necessarily implicit, but in practice this asks for time consuming iteration procedures. We present versions of explicit Runge-Kutta methods of a different form: their coefficients are no more constant but equation dependent. Due to this peculiarity the new versions are A-stable, such that the iteration process is avoided. We also give a numerical illustration.

1. Introduction

Quite often the theoretical approach of problems in physics and engineering requires the numerical solution of differential equations. The literature on such methods is quite vast and the way of selecting the version which is best suited for the given problem is not always so simple. The selection is typically restricted on the class of the most popular methods, and the Runge Kutta methods belong to this. Still, the methods in the Runge Kutta (RK) family are not equivalent for performances. Some are convenient when the accuracy is an issue while others act better for stiff problems. In this paper we present some recently formulated versions which offer a reasonable ballance between the two.
To fix the ideas, the one-step problem to solve is

$$y' = f(x, y), \ x \in [x_n, x_{n+1} = x_n + h], \ y(x_n) = y_n, \tag{1}$$

and the general algorithm of an s-stage RK method is (see [1], [2])

$$y_{n+1} = y_n + h \sum_{i=1}^{s} b_i f(x_n + c_i h, Y_i) \,,$$

where

$$Y_i = y_n + h \sum_{j=1}^{s} a_{ij} f(x_n + c_j h, Y_j), \qquad i = 1, 2, ..., s.$$

This allows computing y_{n+1} in terms of the input y_n by the formula written in the first line (so called final stage), where the values of Y_i are as resulting from the set of formulae in the second line (internal stages). The coefficients are usually collected in the Butcher array

$$\begin{array}{c|cccc} c_1 & a_{11} & a_{12} & \cdots & a_{1s} \\ c_2 & a_{21} & a_{22} & \cdots & a_{2s} \\ \cdot & \cdot & \cdot & \cdots & \cdot \\ c_s & a_{s1} & a_{s2} & \cdots & a_{ss} \\ \hline & b_1 & b_2 & \cdots & b_s \end{array}$$

If at least one a_{ij} with $j \geq i$, $i = 1, 2, \ldots, s$, is nonvanishing, then the method is called implicit. Otherwise it is explicit, and if only $a_{i,i-1}$, $i = 2, 3, \ldots, s$, are nonvanishing the method is called *diagonally* explicit.

Examples

1. The wellknown three-stage Kutta method is an explicit method. It has the Butcher array

$$\begin{array}{c|ccc} 0 & 0 & 0 & 0 \\ 1/2 & 1/2 & 0 & 0 \\ 1 & -1 & 2 & 0 \\ \hline & 1/6 & 2/3 & 1/6 \end{array}$$

and its order is three.
2. Also of order three is the three-stage diagonally explicit version

$$\begin{array}{c|ccc} 0 & 0 & 0 & 0 \\ c_2 & a_{21} & 0 & 0 \\ c_3 & 0 & a_{32} & 0 \\ \hline & b_1 & b_2 & b_3 \end{array}$$

if c_2 and c_3 are correlated, $c_3 = 3c_2(1 - c_2)$, and

$$a_{21} = c_2,\ a_{32} = c_3,\ b_2 = \frac{3c_2 - 1}{6c_2^2},\ b_3 = -\frac{1}{18c_2^2(c_2 - 1)},\ b_1 = 1 - b_2 - b_3,$$

see [2].
3. The standard four-stage Runge-Kutta method (abbreviated RK4) is diagonally explicit. Its Butcher array is

$$\begin{array}{c|cccc} 0 & 0 & 0 & 0 & 0 \\ 1/2 & 1/2 & 0 & 0 & 0 \\ 1/2 & 0 & 1/2 & 0 & 0 \\ 1 & 0 & 0 & 1 & 0 \\ \hline & 1/6 & 2/3 & 2/3 & 1/6 \end{array}$$

and its order is four. As a matter of fact, if the number of stages is not larger than four, then diagonally explicit versions exist such that their order is equal with the number of stages.

The main element of attractivity with the explicit methods is that they are easy to use: the computation of each new Y_i requires only data available from the previous stages. This is in contrast with the implicit methods where all Y_i are coupled and therefore their computation implies solving a set of nonlinear equations. An iteration procedure is then required, whose success depends on how good are the starting data. Since for methods with constant coefficients only implicit versions can hopefully solve stiff problems, the need for iterations cannot be avoided, and this is a serious difficulty when solving such problems.

2. New versions

What is special with these versions is that their coefficients are no longer constant. They are *equation dependent*. Versions of this type have been obtained recently in [3] and [4] by means of the exponential fitting formalism. Technical details are not given here, but we list the main results.

2.1. Two-stage explicit method

The reference Butcher array is:

$$\begin{array}{c|cc} 0 & 0 & 0 \\ c_2 & a_{21} & 0 \\ \hline & b_1 & b_2 \end{array}$$

With $c_2 \in (0,1]$ as a free parameter the coefficients are:

(i) Standard version:

$$a_{21} = c_2, \; b_1 = 1 - 1/2c_2, \; b_2 = 1 - b_1. \tag{2}$$

(ii) New version:

$$a_{21} = c_2, \; b_1 = 1 - 1/[2c_2(1 - c_2 M_2/2)]\,, \quad b_2 = 1 - b_1\,, \tag{3}$$

where $M_2 = hf_y(x_n + c_2 h, y)|_{y=Y_2}$.

Remarks

1. Both versions have order two for all c_2. However, if $c_2 = 2/3$ the order of new version becomes three.
2. Due to the presence of the equation dependent M_2 the coefficients b_1 and b_2 are also equation dependent and thus they must be updated at each step.
3. For systems of differential equations M_2 becomes h times the Jacobian matrix. Also, b_1 and b_2 become matrices.

2.1.1. Stability A-stability is a necessary condition for a method in order to be successful on stiff problems. The way how a method is checked for stability is based on a simple test equation. This is

$$y' = \lambda y, \; y \geq 0, \;\; y(0) = y_0 \tag{4}$$

with the exact solution $y(x) = y_0 \exp(\lambda x)$. If $Re(\lambda) < 0$ the solution tends to zero when $x \to \infty$ and the same must hold true also for the numerical solution. Since $f(x,y) = \lambda\, y$, the first and second internal stages give $Y_1 = y_n$ and $Y_2 = (1 + c_2\nu)y_n$, respectively, where $\nu = \lambda\, h$, such that the final stage results in $y_{n+1} = R(\nu)y_n$ where $R(\nu) = (1 + \nu + b_2 c_2 \nu^2)$. The solution propagated numerically from 0 up to $x_N = Nh$ is $y_N = R^N(\nu)y_0$ and the condition that $y_N \to 0$ if $N \to \infty$ is satisfied iff $|R(\nu)| < 1$ for all ν with negative real part. A method with this property for $R(\nu)$ is called A-stable.

With the parameters of the new method, and by taking into account the fact that $M_2 = \nu$, the stability function $R(\nu)$ reads

$$R_f(\nu) = \frac{2 + (2 - c_2)\nu + (1 - c_2)\nu^2}{2 - c_2\nu}.$$

When $c_2 = 1$ this becomes

$$R_f(\nu) = \frac{2 + \nu}{2 - \nu},$$

a form which clearly satisfies the mentioned condition. The new version with $c_2 = 1$ is therefore an A-stable method.

2.2. Three-stage diagonally explicit method
The reference Butcher array is now as in Example 2 above. A family of versions with equation dependent coefficients was developed in [4] but two versions are of particular interest for practice. These are:
1. $c_2 = 1/2$, $c_3 = 1$, $b^{den} = 12 - 4M_2 - 2M_3 + M_3M_2$ and

$$b_1 = \frac{2 - 3M_2}{b^{den}},\ b_2 = \frac{8 - 2M_3 + M_3M_2}{b^{den}},\ b_3 = \frac{2 - M_2}{b^{den}}, \tag{5}$$

2. $c_2 = 1$, $c_3 = 1/2$, $b^{den} = 12 - 2M_2 - 4M_3 + M_3M_2$ and

$$b_1 = \frac{2 - 2M_2 + M_3 + M_3M_2}{b^{den}},\ b_2 = \frac{2 - 5M_3}{b^{den}},\ b_3 = \frac{8}{b^{den}}. \tag{6}$$

where $M_i(x) = h\, f_y(x + c_ih, y)|_{y=Y_i(x)}$. In both versions $a_{21} = c_2$ and $a_{32} = c_3$.
As before, for systems of equations M_i and b_i become matrices. The formulae are the same: the matrix products have to be carried out in the written order and the division by b^{den} must be replaced by multiplication on the right by matrix ${b^{den}}^{-1}$.

The special importance of these versions is realated to two properties:
(a) they are of order 4, that is higher by one unit than the classical three-stage methods with the same Butcher array;
(b) they are A-stable, with one and the same stability function,

$$R_f(\nu) = \frac{12 + 6\nu + \nu^2}{12 - 6\nu + \nu^2}. \tag{7}$$

A general remark. Any discussion on the properties of a numerical method tacitly accepts that the method is sharply defined. Mentioning this is redundant for methods with constant coefficients but we have to be careful in the case of methods with equation dependent coefficients. In fact, in the latter case the existence of the method depends on how accurate is the computation of the coefficients. Seen from this perspective, situations may appear when the accurate determination of these parameters becomes a problem. Think for example on eqs.(5) and (6) which, for systems of equations, require inversion of matrix b^{den}; if this is singular the inversion is impossible and then the method is not defined.

3. Numerical illustration

The system of two differential equations

$$y^{1'} = (10\lambda + 9)y^1 - 10(\lambda + 1)y^2,\ y^{2'} = -9(\lambda + 1)y^1 - (9\lambda + 10)y^2,$$
$$x \in [0,\, 5],\ y^1(0) = y_0^1,\ y^2(0) = y_0^2,$$

has the exact solution:

$$y^1(x) = 10(y_0^1 - y_0^2)e^{\lambda x} + (-9y_0^1 + 10y_0^2)e^{-x},\ y^2(x) = 9(y_0^1 - y_0^2)e^{\lambda x} + (-9y_0^1 + 10y_0^2)e^{-x}.$$

For reasonable sized values of the initial conditions, and when λ is negative and big, the solution is largely dominated in magnitude by the component with e^{-x}. Yet, the presence of the component with $e^{\lambda x}$ seriously affects the numerical solution if the problem is solved by a method with low stability properties.
To illustrate this phenomenon we take the initial conditions $y_0^1 = y_0^2 = 1$, for which the exact solution is independent of λ,

$$y^1(x) = y^2(x) = e^{-x}.$$

Table 1. Absolute error $err = y^1_{exact}(5) - y^1_{comput.}(5)$ for the test equation.

h	$RK4$	New method, eq.(5)
1/ 2	8.51(+57)	-4.41(-04)
1/ 4	-7.82(+83)	-2.72(-05)
1/ 8	4.11(+216)	-1.70(-06)
1/ 16	-NaN	-1.06(-07)
1/ 32	-NaN	-6.62(-09)
1/ 64	-NaN	-4.12(-10)
1/128	-NaN	-2.55(-11)
1/256	-9.84(-12)	-1.13(-12)

We use $\lambda = -600$.
We compare two methods: standard four-stage method RK4 from example 3, with low stability properties, and the new three-stage method with $c_2 = 1/2$, eq.(5), which is A-stable. The errors in the values at the end point $x = 5$ of the component y^1 of the solution are collected on Table 1 for a set of step sizes. The errors in y^2 are identical. We see that the first method gives reliable results only for $h = 1/256$, which corresponds to $\nu = \lambda h \approx -2.34$. This is quite normal because this method is known to be stable only if ν is bigger than approximately -2.8. For contrast, no stability problem exists for the new method. Its order four is also confirmed by the data in Table 1.

In conclusion, replacement of the standard methods, based on constant coefficients, by versions with equation dependent coefficients leads to an improvement of the quality of the results in two respects:
(a) the accuracy is increased: the order is typically increased by one unit;
(b) there is a massive improvement for stability: the presented versions are A-stable.
This recommends the new versions for being included in the short list of potential solvers for differential equations.

Acknowledgments

I want to express my gratitude to Prof. A. Sandulescu for gently guiding my first steps in scientific research in the mid-sixties. The investigations presented in this paper benefited partially from financial support by the project PN-II-ID-PCE-2011-3-0092 of the Romanian Ministry of Education and Research.

References

[1] Butcher J C 2008 *Numerical Methods for Ordinary Differential Equations* (Chichester:Wiley)
[2] Lambert J D 1991 *Numerical methods for ordinary differential systems: The initial value problem* (Chichester:Wiley)
[3] D'Ambrosio R, Ixaru L Gr and Paternoster B 2011 *Comput. Phys. Commun.* **182** 322
[4] Ixaru L Gr 2012 *Comput. Phys. Commun.* **183** 63

Int. Summer School for Advanced Studies 'Dynamics of open nuclear systems' (PREDEAL12) IOP Publishing
Journal of Physics: Conference Series **413** (2013) 012034 doi:10.1088/1742-6596/413/1/012034

MAGNETIC ROTATION (MR) BAND-CROSSING AT HIGH SPIN STATES : Role of nucleons in this crossing in N = 78 odd-Z isotones.

Suresh Kumar[1], Ritika Garg[1] and Samit Kumar Mandal[1]

[1] Department of Physics and Astrophysics, University of Delhi, Delhi - 110007, India

E-mail: sursvmk123@gmail.com

Abstract. Magnetic-dipole rotational (MR) bands were discovered about 15 years ago, and have been explained using shears mechanism. The theoretical understanding of these bands has been provided using tilted axis cranking (TAC). At present , magnetic rotation has been seen in whole nuclear landscape and about 180 bands in 80 nuclides has been observed in mass regions A=20, 60, 80,110, 135 and 195 respectively. The crossing of these bands (ΔI=1) is very much similar to normal band crossing (ΔI=2) and already exhibited in different mass regions. We have observed new MR bands and their crossing in the A = 130 mass region in ^{135}La, ^{137}Pr and ^{139}Pm nuclei. The systematic evolution of this phenomenon in N=78 odd-Z isotones leads us to understand the role on nucleons in this crossing .

1. Introduction

The observation of regular rotational-like bands of strongly enhanced M1 transitions in the near spherical Pb region was very surprising[1]. To explain the observation of rotational bands in near spherical nuclei, a new type of nuclear excitation called 'magnetic rotation (MR)' was discovered. In this rotation, the total angular momentum is generated by step-by-step alignment of the particle and hole spins in the direction of total angular momentum. Theoretically TAC has been used extensively to interpret the bands. MR phenomenon has been explored in several mass regions like A = 20, 60, 80, 110, 135 and 195 [2, 3, 4, 5]. Some of the features of MR bands are:

(i) Typically observed in near-spherical nuclei.

(ii) Strong intraband M1 transitions with weak or absent E2 crossover transitions.

(iii) The M1 transitions have a reduced transition probability B(M1) $\sim$ 2-10 $\mu_N{}^2$.

(iv) Decrease in B(M1) and B(M1)/B(E2) values with increasing angular momentum.

In the rare-earth region (A $\approx$ 135), several nuclei have been investigated to explore MR bands. For these nuclei, the proton $g_{7/2}$, $d_{5/2}$, $h_{11/2}$ spherical single states are originating from the particle orbitals and neutron $h_{11/2}$ spherical states from hole orbitals. In the present experimental study, ^{135}La was studied for MR phenomenon. The observed ΔI=1 bands were compared with other odd-mass N=78 isotones namely ^{137}Pr, ^{139}Pm and ^{141}Eu. Apart from the observation of MR bands, band crossing between the MR bands has been observed for ^{135}La. A systematic comparision of band crossings across N = 78 isotones has been done.

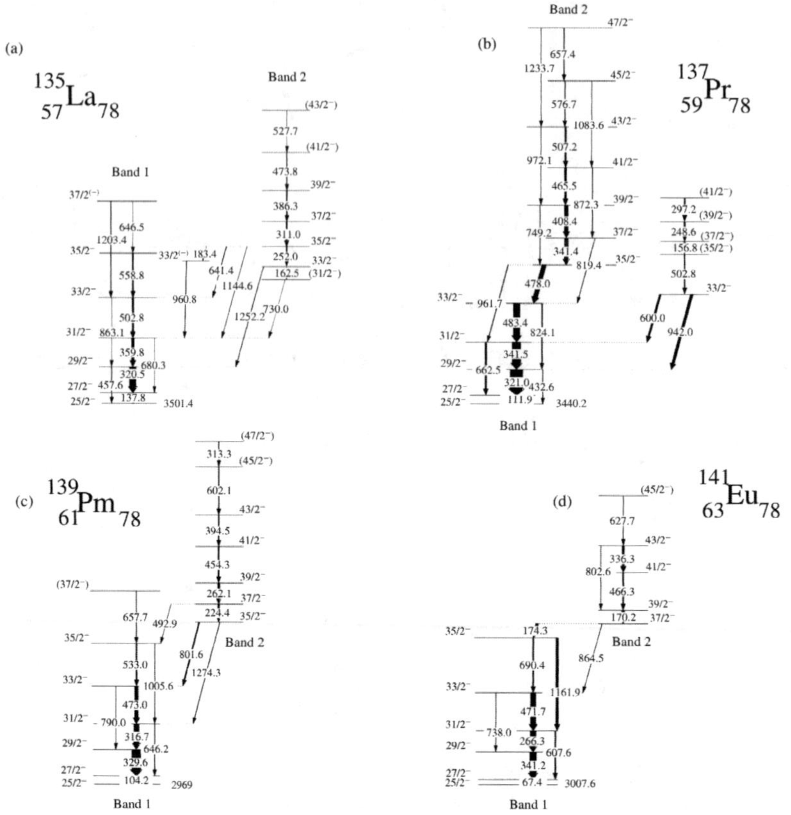

Figure 1. Partial level schemes of some N = 78 isotones [6, 7, 8, 9].

Bandcrossing between rotational bands arises due to rotation alignment of certain pairs of nucleons under the influence of coriolis force. Alignment gain Δi, crossing spin and band crossing frequency $\hbar\omega$ characterise a band crossing. The rotational frequency at which band crossing occurs is called the band crossing frequency ($\hbar\omega$).

2. Experimental details

High spin states in ^{135}La were populated by the reaction ^{128}Te(^{11}B,4n)^{135}La using a ^{11}B beam of 50.5 MeV from the Pelletron Linac Facility at Tata Institute of Fundamental Research (TIFR), Mumbai. The target consisted of 1.02 mg/cm^2 ^{128}Te on 4 mg/cm^2 ^{197}Au backing. The gamma rays were detected using INGA [10, 11]. In the current experiment, the array consisted of 16 Compton-suppressed clover detectors arranged in spherical geometry with 3, 2, 2, 4, 2 and 3 number of clover detectors placed at 157°, 140°, 115°, 90°, 65° and 40° with respect to the beam direction respectively. The detectors are placed at a distance of 25 cm from the target. The online data was collected via XIA based triggerless Digital Data AcQuisition (DDAQ) system [10]. A total of about 3.9 $\times$ 10^9 two and higher fold events were recorded. The data were sorted using Multi pARameter time-stamped based COincidence Search program (MARCOS) to generate double and triple gamma coincidence matrices. The RADWARE software package was used for analysis of these matrices [12].

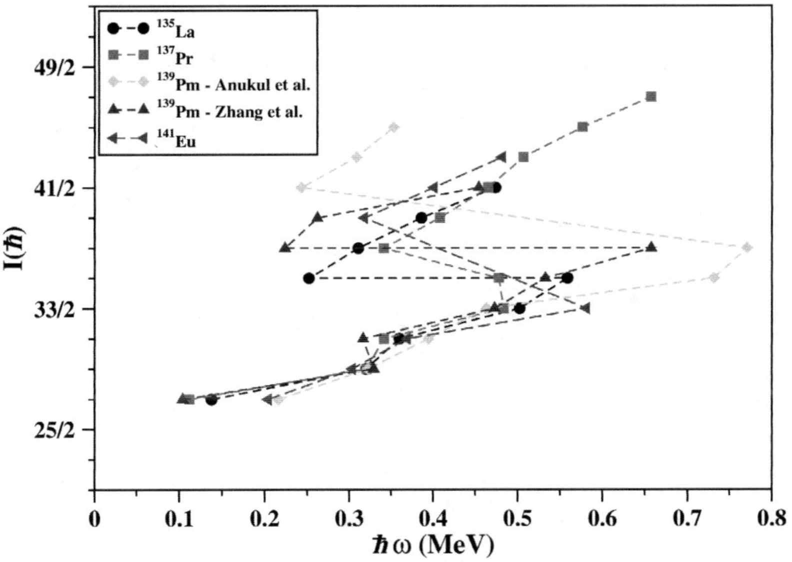

Figure 2. Plot of $I(\hbar)$ vs $\hbar\omega$ of MR bands in ^{135}La, ^{137}Pr, ^{139}Pm and ^{141}Eu .

Table 1. Configurations, deformation parameter ε_2 and triaxiality parameter γ obtained from TAC calculations across some N = 78 isotones.

Nucleus	3qp, Band 1 in Fig. 1			5qp, Band 2 in Fig. 1		
	Configuration	ε_2	γ	Configuration	ε_2	γ
^{135}La	$\pi(h_{11/2})^1 \otimes \nu(h_{11/2})^{-2}$	0.12	58°	$\pi(h_{11/2})^1(g_{7/2}/d_{5/2})^2 \otimes \nu(h_{11/2})^{-2}$	0.116	60°
^{137}Pr	$\pi(h_{11/2})^1 \otimes \nu(h_{11/2})^{-2}$	0.135	58°	$\pi(h_{11/2})^1(g_{7/2})^2 \otimes \nu(h_{11/2})^{-2}$	0.126	58°
^{139}Pm	$\pi(h_{11/2})^1 \otimes \nu(h_{11/2})^{-2}$	0.124	52°	$\pi(h_{11/2})^1 \otimes \nu(h_{11/2})^{-4}$	0.11	60°
^{141}Eu	$\pi(h_{11/2})^1 \otimes \nu(h_{11/2})^{-2}$	–	-	$\pi(h_{11/2})^1 \otimes \nu(h_{11/2})^{-4}$	-	

3. Discussion

In Fig. 1, the partial level schemes for different nuclei are shown for the interested part of the nuclear structure. It is very much clear that all the nuclei have ΔI=1 bands which can be interpreted as arising from 3qp and 5qp configurations using Tilted Axis Cranking(TAC) calculations [6, 7, 13, 14]. The 3qp configuration is built on the $\pi(h_{11/2})^1 \otimes \nu(h_{11/2})^{-2}$ configuration. 5qp configuration can be understood by considering different alignments of proton/neutron pair as given in Table 1. For ^{135}La and ^{137}Pr, $g_{7/2}/d_{5/2}$ proton pair particles decouple and align their angular momentum along the axis of rotation to produce a 5qp configuration at a particular rotational frequency. At this band crossing frequency, a band crossing occurs between the 3qp and 5qp configuration. However, for ^{139}Pm, band crossing occurs due to the alignment of neutron pair holes. Thus all the nuclei exhibit ΔI=1 band crossing from 3qp to 5qp configuration. The experimental features of 5qp configuration suggest different configurations for ^{135}La, ^{137}Pr nuclei as compared to ^{139}Pm, ^{141}Eu nuclei. In Fig. 2 is given a plot of angular momentum/spin I($\hbar$) as a function of rotational frequency which is taken as $\hbar\omega = E_\gamma$ for dipole transitions. All the nuclei show band crossing at $\hbar\omega \approx 0.40$ MeV. The behaviour of I vs $\hbar\omega$ plot is very similar for all nuclei for lower bands before band crossing.

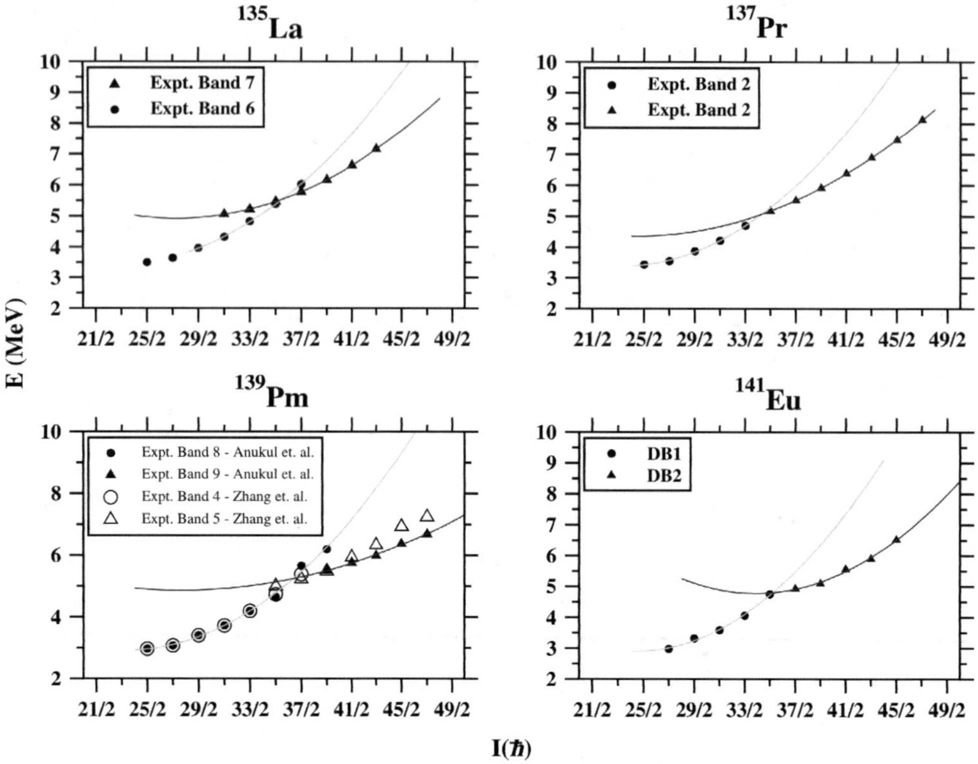

Figure 3. Experimental band crossings shown by the behaviour of excitation energy E(MeV) vs spin $I(\hbar)$. The fitted lines are guide to see band crossing (E vs I fit).

However, after the band crossing ^{135}La,^{137}Pr nuclei behave in similar way as compared to ^{139}Pm, ^{141}Eu nuclei which show similar features. Also Δi in ^{135}La, ^{137}Pr is $3.5\hbar$ as compared to $4.5\hbar$ for ^{139}Pm, ^{141}Eu nuclei. The systematic comparision for excitation energy E(MeV) vs spin I($\hbar$) provides the information of crossing spin. It is clear from Fig. 3 that band crossing took place between $33/2\ \hbar$ to $37/2\ \hbar$ ($\approx 35/2\ \hbar$). The trends of E vs I also show different crossing behaviour for ^{135}La, ^{137}Pr as compared to ^{139}Pm, ^{141}Eu.

Electromagnetic properties for ΔI=1 bands, which play an important role in the understanding of nuclear structure of these bands, can be obtained from the absolute values of the reduced M1 and E2 transition probabilities, $B(M1)$ and $B(E2)$ respectively. The B(M1) values are sensitive to the details of the nucleon configuration, coupling scheme of the nucleons angular momenta and their contribution to net perpendicular magnetic moment. Lifetime measurements are not carried in all the nuclei of interest. In such a situation, the B(M1)/B(E2) ratios are very useful to get insight into the details of nuclear structure. Therefore, the B(M1)/B(E2) values have been compared for the ΔI=1 bands for different nuclei and plotted in Fig. 4. The $B(M1)/B(E2)$ behaviour again supports the argument made above for the change in configuration as one moves from ^{137}Pr to ^{139}Pm. Apart from this, very interesting observation can be seen for inter-mixing between two bands and the role played by pair of nucleons responsible for band crossing by observing $B(M1)/B(E2)$ values as function of angular momentum. It can be seen from the Fig. 4 that the behaviour of B(M1)/B(E2) vs I is very similar in ^{135}La, ^{137}Pr than^{139}Pm, ^{141}Eu nuclei, and hence justifies the effect of nucleon pair on the band crossing of ΔI=1 bands.

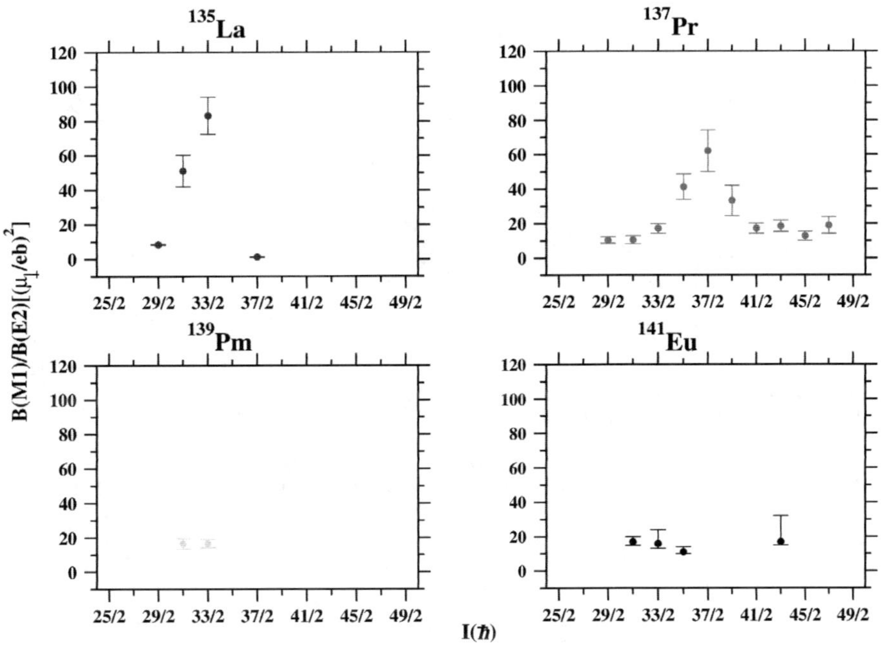

Figure 4. Variation of $B(M1)/B(E2)$ as a function of $I(\hbar)$ for the negative parity MR bands.

The MR band crossing has been explored systematically for N = 78 odd-Z isotones. The influence of nucleon pair alignment in such band crossing provide the information of role of nucleons. The systematic study of different experimental properties of band crossing gives insight on the participating aligning nucleons. In future, lifetime measurements will be neccesary to provide information on B(M1) transition strengths and hence the role of nucleons in MR band crossing.

Acknowledgements

The authors gratefully acknowledge the support provided by the staff of TIFR-BARC and IUAC pelletron facility, the INGA collaboration and the Department of Science and Technology, India, for providing funds for INGA project. Financial support from the UGC-DAE-Kolkata and CSIR, India at Department of Physics and Astrophysics, University of Delhi is acknowledged.

References

[1] Hubel H 2005 *Prog. Part. Nucl. Phys.* **54** 1–69
[2] Peng J, Yao J M, Zhang S Q and Meng J 2010 *Chin. Phys. Lett.* **27** 122101–4
[3] Torres D A *et al* 2008 *Phys. Rev.* C **78** 054318–24
[4] Steppenbeck D *et al* 2012 *Phys. Rev.* C **85** 044316–23
[5] Jain A, Jain A K and Singh B 2000 *Atomic Data and Nuclear Data Tables* **74** 283–331
[6] Garg R *et al* 2012 *Phys. Rev.* C submitted
[7] Agarwal P *et al* 2007 *Phys. Rev.* C **76** 024321
[8] Zhang N T *et al* 2011 *Phys. Rev.* C **84** 057302–4
[9] Podsvirova E O *et al* 2004 *Eur. Phys. J.* A **21** 1–6
[10] Palit R *et al* 2012 *Nucl. Instrum. Methods* A **680** 90–6
[11] Palit R 2011 *AIP Conf. Proc.* **1336** 573–75
[12] Radford D C 1995 *Nucl. Instrum. Methods* A **361** 297–305
[13] Dhal A *et al* 2009 *Phys. Rev.* C **80** 014320
[14] Marcinkowska Z *et al* 2003 *Acta Phys. Pol.* B **34** 2319–22

Int. Summer School for Advanced Studies 'Dynamics of open nuclear systems' (PREDEAL12) IOP Publishing
Journal of Physics: Conference Series **413** (2013) 012035 doi:10.1088/1742-6596/413/1/012035

Dissipation for Z-channels in fission, alpha-decay and cluster emission

Nafiseh Shayan Shakib[1], Mohammad Farhad Rahimi[2] and Mohammad Mahdi Firoozabadi[1]

[1] Physics Faculty, University of Birjand, P.O. Box 615/97175, Birjand, Iran
[2] Physics Faculty, Ferdowsi University of Mashhad, Azadi Square, Post Code: 9177948974, Mashhad, Iran

E-mail: nafiseh.shakib@gmail.com

Abstract. The dissipated energy in fission, cluster and alpha decay was estimated microscopically by using the time dependent pairing equations. The single particle level schemes were determined in the framework of the superasymmetric two center shell model. A strong dependence of the dissipated energy as function of the mass asymmetry is evidenced. The dissipated energy is much lower for alpha decay and cluster emission than for fission.

1. Introduction

As mentioned earlier in Ref. [1], the motion of any real macroscopic physical system is not only managed by conservative forces, but also by frictional ones. Therefore, in dynamical treatments we need to take into account the dissipative forces that produce an irreversible flow of energy or angular momentum from the macroscopic degrees of freedom to the intrinsic ones. The concept of nuclear friction can be employed if we use collective coordinates and microscopic ones. A such ensemble can be found in the case when the Schrodinger equation is solved for microscopic potentials that vary in time. In nuclear physics, an interesting case could be that of nucleons that are moving in the Nilsson potential.

In the following, we will use the time dependent pairing equations [2, 3] to determine the dissipated energy for the fission, for the alpha decay and for the ^{14}C emission of ^{236}U. The dissipated energy will be determined for predefined families of shapes that start from one spherical nucleus and ends as two separated fragments. A nuclear shape parametrization given by two intersected spheres of different radii will be used. The variations of the single particle energies is obtained with the Nilsson two center shell model [4, 5, 6]. Only the two degrees of freedom, namely, the elongation and to the mass asymmetry are used. Such calculations were realized in the past only for three fission channels [7]. We will investigate systematically the dissipated energy for all even-even partitions in the Z-channels.

2. The model

In our investigation, the time dependent pairing equations is used to evaluate the dissipation. The calculations are based on solutions given by the superasymmetric two center shell model.

2.1. The two center shell model

Investigating the fission in a wide range of mass asymmetries, the two-center shell models allow a continuous description of the changes of nuclear shapes from one initial nucleus to two final fragments. It is a generalization of the one center Nilsson model. The two-center shell-model Hamiltonian is obtained from the two-center oscillator V_{2c} by adding the spin orbit interaction term V_{ls} and a correcting term V_{l^2} [5].

$$H = T + V_{2c} + V_{ls} + V_{l^2} \tag{1}$$

where T is the kinetic energy. The general expressions of these terms can be found in Refs. [4, 6]. The Nilsson parameters coupling constants are taken from previous works [8].

2.2. Dissipated energy

The microscopic description of the dynamics of a many nucleon system has been extensively investigated with the time dependent Hartree-Fork method [11]. However, in this method, the residual interactions are neglected. That leads to the unpleasant feature that a system even moving infinitely slowly could not end up in its ground state. In order to avoid all these difficulties one could use the pairing residual interaction. We shall start from the variational principle taking the following energy functional

$$\delta L = \delta < \varphi | H - i\hbar\frac{\partial}{\partial t} - \lambda N | \varphi >, \tag{2}$$

and we assume the many-body state as a BCS seniority zero wave function

$$|\varphi(t) >= \prod_l (u_l(t) + v_l(t) a_l^+ a_{\bar{l}}^+)|0 > . \tag{3}$$

where $|0>$ denotes the vacuum state. a_l^+ and $a_{\bar{l}}^+$ are creation operators of the coupled states l and $\bar{l}$, and u_l and v_l are vacancy and occupation amplitudes, respectively. H is the time dependent many-body Haminltonian with pairing residual interactions

$$H(t) = \sum_{k>0} \epsilon_k(t)(a_k^+ a_k + a_{\bar{k}}^+ a_{\bar{k}}) - G \sum_{k,l>0} a_k^+ a_{\bar{k}}^+ a_l a_{\bar{l}}. \tag{4}$$

and the particle number operator is N. By performing the variation of the Lagrangian in a way similar as in [9, 10], the next system of coupled differential equations are obtained.

$$i\hbar\dot{\rho}_l = \kappa_l \Delta^* - \kappa_l^* \Delta, \tag{5}$$

$$i\hbar\dot{\kappa}_l = (2\rho_l - 1)\Delta + 2\kappa_l(\epsilon_l(t) - \lambda(t)) - 2G\rho_l\kappa_l. \tag{6}$$

where $\Delta = G\sum_k \kappa_k$ is the pairing gap, $\kappa_k = u_k v_k$ are pairing moment components, $\rho_k = |v_k|^2$ are single-particle densities. In our formula, the sum over pairs generally runs over index k.

In the frame of our model, the difference

$$E = E' - E_0 \tag{7}$$

behaves as a dissipated energy [2]. Here E' is the energy of the system obtained within solutions of the time dependent pairing equations and E_0 is the BCS energy. These equations were recentely used in the study of fission [12, 13, 14, 15]

Int. Summer School for Advanced Studies 'Dynamics of open nuclear systems' (PREDEAL12) IOP Publishing
Journal of Physics: Conference Series **413** (2013) 012035 doi:10.1088/1742-6596/413/1/012035

3. Results

Three nuclear decay modes are treated in an unified manner. A nuclear shape parametrization given by two intersected spheres of different radii is used. We need to know the variation of the mass asymmetry as function of the elongation for each nuclear decay mode. Information about this dependence can be found in Refs. [16, 17] where it was shown that in the case of very large mass asymmetries, that is, for alpha decay and cluster emission, the radius of the light fragment must be approximately constant $R_2 = 1.16 \times A_2^{1/3}$ fm. In the case of fission the light fragment emerges with a radius that gives approximately a constant emitted volume. A and A_2 denote the mass numbers of the parent and of the light fission fragment, respectively.

Figure 1. Familly nuclear shapes, from the spherical configuration up to α emission. The distances between the center of the fragments in fm are marked on the plot.

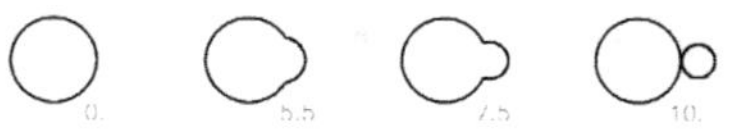

Figure 2. Same as Fig. 1 for the ^{14}C emission.

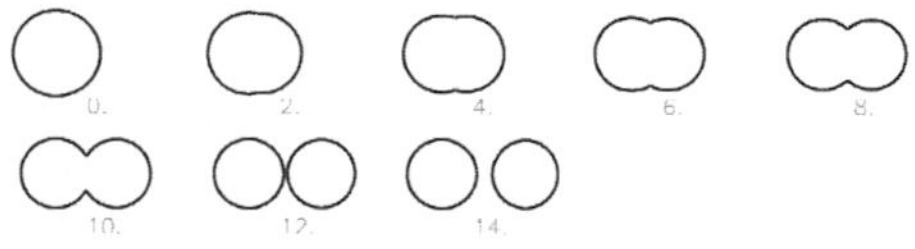

Figure 3. Same as Fig. 1 for fission.

In the case of the alpha decay and of the ^{14}C emission, the shapes are displayed in Figs. 1 and 2, respectively. The emitted fragment starts to appear at an elongation $R = 1.16 \times ((A - A_2)^{1/3} - A_2^{1/3})$ fm. A different situation appears in the case of fission for which the shapes are plotted in Fig. 3. In this last case, a deviation from the spherical shape can be observed from the beginning of the process, that is $R \approx 0$. The neutron and proton level schemes are obtained with the superasymmetric two-center shell model improved in Ref. [6]. The fact that the nucleus begins to deforms at larger values of R is reflected by the variations of single particle levels schemes in the case of alpha decay and carbon emission as evidenced in Figs. 4 and 5. In the case of fission, a different situation appears, as displayed in Fig. 6. In the tree mentioned cases, the level scheme behave as a Nilsson diagram for prolate deformation.

Having the level schemes for the three decay modes it is possible to compute the dissipation during the disintegrations with the time dependent pairing equations. The velocity of the internuclear distance was fixed at $\partial R/\partial t = 10^6$ m/s, that gives approximately 10^{-21} s as the time to tunnel the barrier [6]. This is considered as a characteristic time for scission.

Several Z-channels were tested in the case of ^{236}U fission. We used the partitions that give a maximum fission yield as given by existing compilations [18]. In this context we simulated the even-even comlementary fragments like:(^{118}Pd, ^{118}Pd) ,(^{114}Ru, ^{122}Cd), (^{106}Mo, ^{130}Sn), (^{100}Zr, ^{132}Te), (^{96}Sr, ^{140}Xe), (^{92}Kr, ^{144}Ba), (^{86}Se, ^{150}Ce), (^{82}Ge, ^{154}Nd), (^{78}Zn, ^{158}Sm), (^{72}Ni, ^{164}Gd) and the last ones (^{68}Fe, ^{168}Dy).

In Fig. 7, the total dissipated energy in fission as function of of the charge number Z of the light fragment and the elongation R is represented. It is important to note that the dissipated energy at scission surpasses in general 7 MeV. For alpha decay and cluster emission,

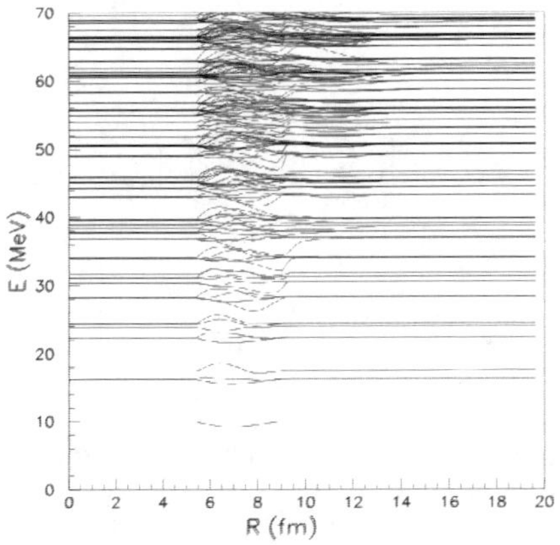

Figure 4. Neutron level scheme for α emission from ^{236}U.

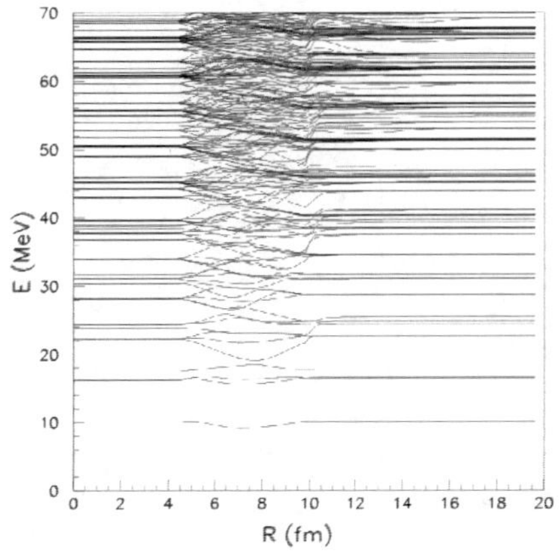

Figure 5. Neutron level scheme for ^{14}C emission from ^{236}U.

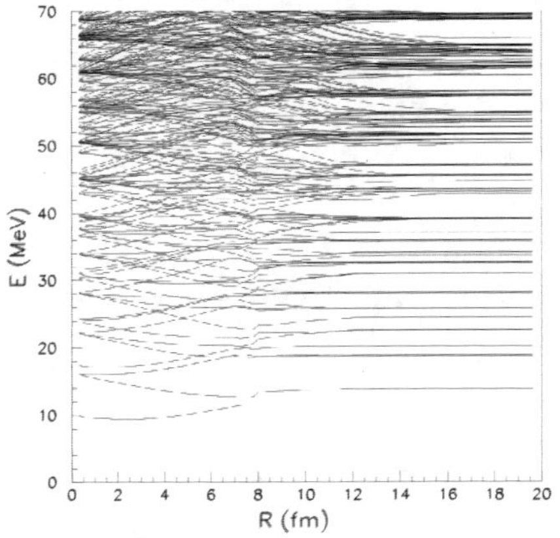

Figure 6. Neutron level scheme for the reaction $^{236}\text{U} \longmapsto {}^{86}_{34}\text{Se} + {}^{150}_{58}\text{Ce}$

the dissipated energies as function of the elongation are represented in Figs. 8 and 9. At scission the total dissipated energy amounts to 3 MeV for α decay and it is lower than 4.5 MeV for cluster emission. The dissipated energy reflect differences in the structure of these three decay modes.

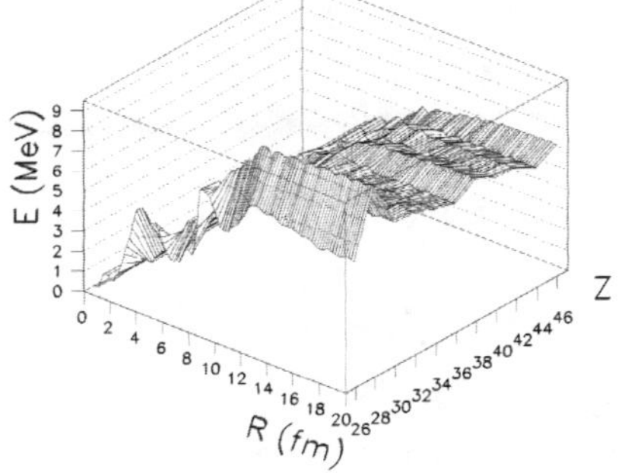

Figure 7. Dissipated energy as function of elongation R and the charge number Z of the light fragment in fission.

The superasymmetric fission proceeds through adiabatic states, without dissipation, while in the fission at low energy a considerable amount of intrinsic excitation is produced. Therefore, the alpha decay and the cluster emission can be considered as cold rearrangement processes [19, 20, 21, 22]. It is a confirmation of cold rearrangement hypothesis used in the prediction of

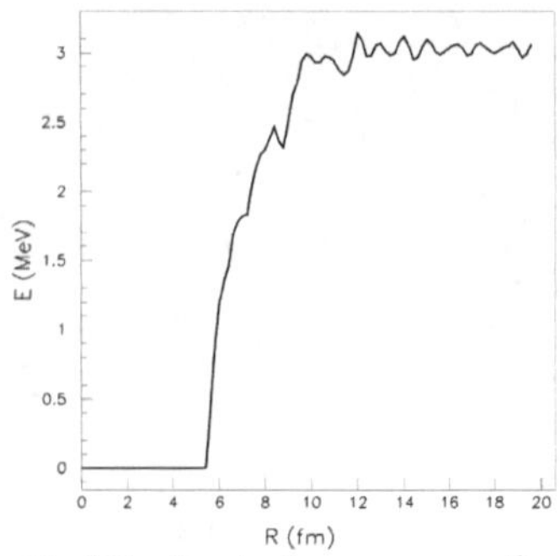

Figure 8. Dissipated energy as function of elongation in α-emission.

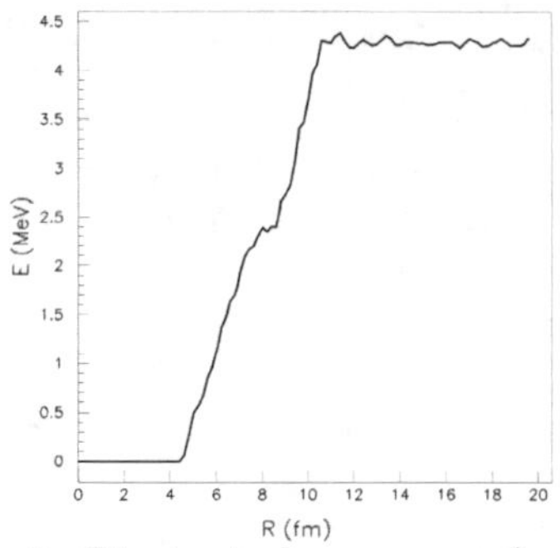

Figure 9. Dissipated energy as function of elongation in ^{14}C emission.

cluster decay [23]. The differences in dissipated energy have their origin in the variations of the single particle energies. As observed in the level schemes, the degree of rearrangement of the levels is much lower for superasymmetric fission than for fission.

Our study treats the alpha decay within fission models as in Ref. [24]. In general this decay modes is analysed with preformation models [25, 26] or liquid drop ones [27].

Acknowledgments

One of us (N.S.S.) wishes to express her love and gratitude to her beloved families especially to her mother; for their understanding and endless love, through the duration of her studies.

References

[1] Hasse R W 1978 *Rep. Progr. Phys.* **41** 1027
[2] Koonin S E and Nix J R 1976 *Phys. Rev.* C **13** 209
[3] Blocki J and Flocard H 1976 *Nucl. Phys. A* **273** 45
[4] Mirea M 1996 *Phys. Rev.* C **54** 302
[5] Poenaru D N, Ivascu M S 2000 *Particle Emission from Nuclei*(CRC Press) vol 1 p 113-173
[6] Mirea M 1998 *Phys. Rev.* C **57** 2484
[7] Mirea M, Tassan-Got L, Stephan C and Bacri C O 2004 *Nucl. Phys.* A **735** 21
[8] Maruhn J and Greiner W 1972 *Z. Phys.* A **780** 431
[9] Mirea M 2008 *Phys. Rev.* C **78** 044618
[10] Mirea M 2003 *Mod. Phys. Lett.* A **18** 1809
[11] Dirac P A M 1930 *Proc. Cambridge Phil. Roy. Soc.* **26** 376
[12] Avez B, Simenel C and Chomaz Ph 2008 *Phys Rev* C **78** 044318
[13] Ebata S et al. 2010 *Phys. Rev.* C **82** 034306
[14] Mirea M 2012 *Int. J. Mod. Phys.* E **21** 1250035
[15] Mirea M 2011 *Phys. Rev.* C **83** 054608
[16] Mirea M, Poenaru D N and Greiner W 1994 *Z. Phys* A **349** 39
[17] Mirea M, Poenaru D N and Greiner W 1992 *Nuovo Cimento* A **105** 571
[18] Wahl A C 1988 *At. Data and Nucl. Data Tabl.* **39** 1
[19] Mirea M, Sandulescu A and Delion D S 2012 *Eur. Phys. J.* A **48** 86
[20] Mirea M, Sandulescu A and Delion D S 2011 *Nucl. Phys.* A **870-871** 23
[21] Mirea M 2012 *Rom. J. Phys.* **57** 372
[22] Mirea M, Sandulescu A and Delion D S 2011 *Proc. Rom. Acad. Series A* **12** 203
[23] Sandulescu A and Greiner W 1977 *J. Phys.* G **3** L189
[24] Mirea M 2001 *Phys. Rev.* C **63** 034603
[25] Silisteanu I, Budaca A I 2012 *Rom. J. Phys* **57** 493
[26] Silisteanu I, Rizea A, Ciobanu B I and Neacsu A 2008 *Rom. J. Phys.* **53** 1191
[27] Mirzaei V and Miri-Hakimabad H 2012 *Rom. Rep. Phys.* **64** 50

Int. Summer School for Advanced Studies 'Dynamics of open nuclear systems' (PREDEAL12) IOP Publishing
Journal of Physics: Conference Series **413** (2013) 012036 doi:10.1088/1742-6596/413/1/012036

Low energy isomeric levels of nuclei near N=40

C Petrone[1,2], J M Daugas[3], M Stanoiu[1], F Negoita[1], G Simpson[4], C Borcea[1], R Borcea[1], L Caceres[5], S Calinescu[1], R Chevrier[3], L Gaudefroy[3], G Georgiev[6], G Gey[4], C Plaisir[3], F Rotaru[1], O Sorlin[5], J C Thomas[5]

[1] Horia Hulubei National Institute for Physics and Nuclear Engineering, P.O. Box MG-6, 077125 Bucharest-Magurele, Romania

[2] Faculty of Physics, University of Bucharest - P.O. Box MG 11, RO 77125, Bucharest-Magurele, Romania

[3] CEA, DAM, DIF, F-91297 Arpajon, France

[4] ILL, 38042 Grenoble Cedex, France

[5] Grand Accélérateur National d'Ions Lours (GANIL), CEA/DSM-CNRS/IN2P3, Caen, France

[6] CSNSM, CNRS/IN2P3, 91405 Orsay-Campus, France

E-mail: cristina.petrone@tandem.nipne.ro

Abstract. We present the results of an experiment in which the structure of neutron-rich nuclei located in the vicinity of N=40 was studied. The importance of our results comes from the fact that knowing the behaviour of the neutron $g_{9/2}$ orbital with increasing number of neutrons is one of the key points in defining the structure of these nuclei at low excitation energy. The nuclei of interest were produced by fragmentation of a ^{86}Kr beam at 60MeV/u on a thick Be target at GANIL (France). Preliminary results on ^{75}Cu and ^{78}Ga isomers will be presented together with tentative spin and parity assignments.

1. Introduction

One of the compelling questions in nuclear physics is the evolution of single-particle states around shell closures when moving into a region of nuclei with unusual proton-to neutron ratio. In particular the nuclear structure and decay properties of neutron-rich nuclei in the vicinity of the magic nuclei ^{68}Ni have been intensively investigated over the last decade [1]. The comparison between shell-model calculations using different nucleon-nucleon interaction with the existing experimental results has shown the importance of tensor force effects in understanding the structure of nuclei around Z~28 and N>40. The importance of the monopole term of the tensor force was pointed out by Otsuka et al.[2,3] in understanding the evolution of the nuclear structure in this region of the nuclear chart. The lowering of the $\pi 1f_{5/2}$ orbital was predicted, while the energy of the $\pi 1f_{7/2}$ increases when the neutrons start filling the $g_{9/2}$ orbital. The same theoretical estimation predicted the cross over between $\pi\ 2p_{3/2}$ and $\pi 1f_{5/2}$ around N=45, given the repulsive character of the tensor force. This was experimentally proven by Flanagan and Mane using laser spectroscopy techniques for copper and gallium isotopic chains [4, 5].

2. Experimental Set-up

In order to investigate the shell evolution near ^{68}Ni towards the doubly magic ^{78}Ni , neutron-rich nuclei were produced by the fragmentation reaction of a primary beam of ^{86}Kr at the incident energy of 60.4MeV/A on a Be target. The cocktail beam was separated in flight using the LISE2000 spectrometer by means of the energy loss and magnetic rigidity. Isotope identification was performed by combined measurements of the energy loss, magnetic rigidity and Time-of-Flight (TOF) in an event-by-event basis. At the focal plane, the ions were implanted in a 75 μm kapton foil inclined at 70^0. Behind this foil a position sensitive Si detector was used in order to reject the events associated with the fragments that were not stopped in the foil. Two high-purity Ge detectors and a LEPS surrounded the implantation set-up in a very compact geometry. In this paper we present the preliminary results regarding the nuclear structure of ^{75}Cu and ^{78}Ga.

3.Results

3.1 ^{75}Cu

This nucleus was previously studied by Daugas *et al.*[6], showing the existence of two isomeric states. In the case of this previous experiment, γ-γ coincidences studies were not possible due to low statistics, the two observed gammas were being placed in coincidence by means of time spectra, which for the low isomeric state clearly indicates a feeding from the above state. In this experiment due to higher statistics it was possible to construct γ-γ coincidence matrix. The result by gating with the two transitions can be seen in figure 1. It can be observed that the two gammas are not in coincidence, the peaks present in the spectra correspond to random coincidences associated to the Compton scattering of 137.5 and 82keV transition, from to the decay of ^{72}Cu, which was the main contaminant in the identification matrix.

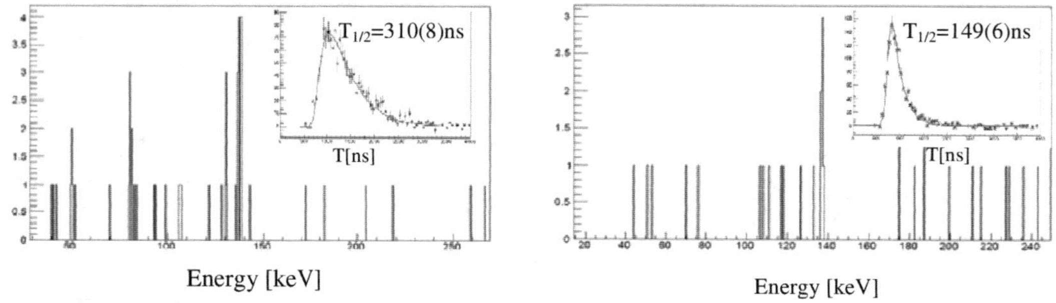

Figure 1. Gamma coincidence spectra gated on 61.7(2) keV (left) and 66.0(2) keV (right). In the inset the decay curves of both transitions are shown.

The lifetimes associated to the two observed transition were measured to be 310(8) ns for the 61.7(2) keV and 149(6) ns, for the decay curve of the 66.0(2) keV, respectively. From fitting the decay of low-lying isomeric state with a sum of double-exponential function convoluted with a Gaussian function (representing the detector response function), it has been proven that this transition is populated mainly from the decay of the isomeric state above. These lifetimes translate into the following reduced transition probabilities (assuming pure transitions multipolarity) : B(M1)=1.0(4)10^{-4} W.u, B(E2)=32(11) W.u for the 66.0(2) keV transition, and B(M1)=2.43(4)10^{-4}W.u, B(E2)=221(1) W.u for the 61.7(2) keV transition, respectively.

The known spin and parity of the ground state and given the low energy state systematics for the odd copper isotopic chain give rise to two possible scenario for this two low-lying states observed (scenario A and scenario B, figure 2).

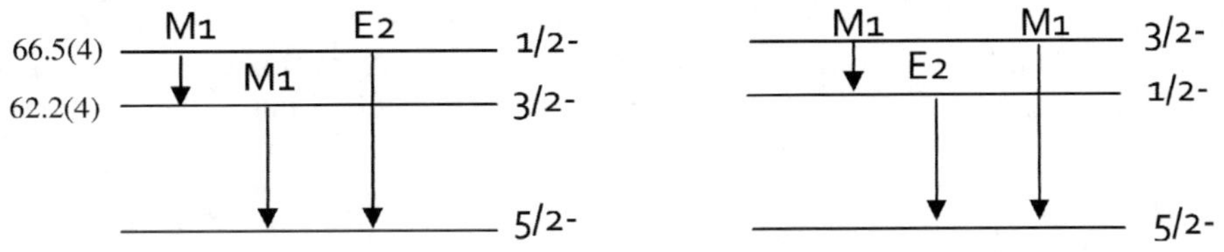

Figure 2: Proposed experimental spin assignment scenario A and B

The interpretation of the obtained values for the reduced transition probabilities and their implication on the nuclear structure of this nucleus is still in progress by means of large scale shell model calculation.

3.2 ^{78}Ga

Previous studies performed on this nucleus indicate the existence of an isomeric state of 499.2(4) keV of 110(6) ns, which decays via three transitions towards lower excited states [7]. More recent studies performed at ISOLDE by Mane using laser spectroscopy assigned a 2^- as spin and parity for the ground state of this nucleus [5]. The delayed gamma ray-spectrum associated to this nucleus is presented in Figure 3, the existence it can be seen of a new transition of 211.3(5) keV, unobserved so far.

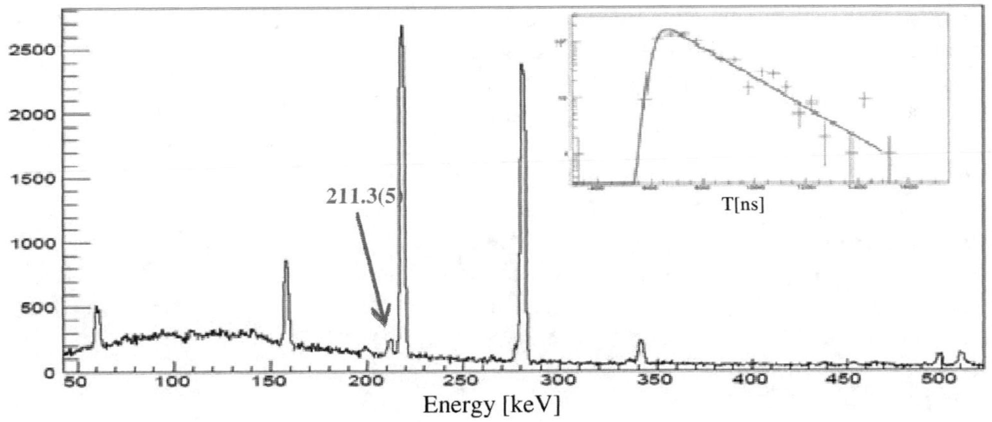

Figure 3: Delay gamma ray spectrum associated to ^{78}Ga. Inset: decay curve of new observed 211(4) keV transition.

γ-γ coincidence matrix conditioned by the decay of ^{78}Ga and gated by the known transition of this nucleus showed the fact that this new transition is in coincidence with the first excited state of 280.4(2) keV, which implies the existence of a new state of 491.8(3) keV, below the isomeric state. The obtain half-life of 110(5) ns for this state indicated the fact that this transition is also populate in the decay of the isomeric state and the existence of a transition of 6.6 (3) keV between the two. The comparison between the calculated reduced transition probabilities and systematics for this mass region near the shell closure Z=28, gives the most probable transition multipolarity for the observed gamma rays: M2 for the 498.9(8)keV, a transition type E2 for the 218.4(2) and 157.5(2) , while for the 6.6(3)keV a multipolarity of E2 is proposed.

4. Summary

A new level scheme was proposed for ^{75}Cu based on γ-γ coincidence results, as for ^{78}Ga a new observed gamma transition has been seen observed from the decay of the known isomeric state. Based on this new observation and knowing the ground state spin and parity, a new level scheme is proposed. The implications of those new results in understanding the structure of studied nuclei are in progress by means of large scale shell model calculation.

5. Acknowledgments

We are grateful to the GANIL staff for the technical support provided. The work of C. Petrone was possible with the financial support of the Sectorial Operational Program for Human Resources Development 2007-2013, co-financed by the European Social Fund, under the project number POSDRU/107/1.5/S/80765. This work was supported by a grant of the Romanian National Authority for Scientific Research, CNCS – UEFISCDI, project number PN-II-ID-PCE-2011-3-0487.
The author S. Calinescu is grateful for the financial support from the European Social Fond through POSDRU 107/1.5/S/82514 Project.

References

[1] Broda R , Fornal B, Krolas W, Pawlat T, Bazzacco D, Lunardi S, Rossi-Alvarez C, Menegazzo R, de Angelis G, Bednarczyk P, Rico J, De Acuna D, Daly P J, Mayer R H, Sferrazza M, Grawe H, Maier K H and Schubart R 1995 *Phys.Rev.Lett.* **74**

[2] Otsuka T, Suzuki T, Fujimoto R, Grawe H and Akaishi Y 2005 *Phys. Rev. Lett.* **95** 232502

[3] Otsuka T, Suzuki T, Honma M, Utsuno Y, Tsunoda N, Tsukiyama K and Hjorth-Jensen M 2010 *Phys. Rev. Lett.* **104** 012501

[4] Kosner U, Stone N J, Flanagan K T, Rikovska Stone J, Fedosseev V N, Kratz K L, Marsh B A, Materna T, Mathieu L, Molkanov P L, Seliverstov M D, Serot O, Sjodin A M and Volkov Yu M 2009 *Phys. Rev. Lett.* **103** 142501

[5] Mane E, Cheal B, Billowes J, Bissell M L, Blaum K, Charlwood F C, Flanagan K T, Forest D H, Geppert Ch, Kowalska M, Krieger A, Kramer J, Moore I D, Neugart R, Neyens G, Nortershauser W, Rajabali M M, Sanchez R, Schug M, Stroke H H, Vingerhoets P, Yordanov D T and Zakova M et al. 2011 *Phys. Rev.* C **84** 024303

[6] Daugas J M, Faul T, Grawe H, Pfutzner M, Grywacz R, Lewitowicz M, Achouri N L, Angelique J C, Baiborodin D, Bentida R, Beraud R, Brocea C, Bingham C R, Catford W N, Emsallem A, de France G, Grzywacz K L, Lemmon R C, Lopez Jimm]enez M J, de Oliveira Santos F, Regan P H, Rykaczewski K, Sauvestre J E, Sawicka M, Stanoiu M, Sieja K and Nowacki F 2010 *Phys. Rev.* C **81** 034304

[7] J.M. Daugas 1999 *PhD thesis* Caen

The structure of tin isotopes with a global optimized effective interaction

Chong Qi
Royal Institute of Technology (KTH), AlbaNova University Center, 10691 Stockholm, Sweden
E-mail: chongq@kth.se

Abstract. A systematic shell model calculation on the structure of light tin isotopes is presented with a monopole-optimized effective interaction. We started with the realistic CD-Bonn potential. The unknown single-particle energies and the $T = 1$ monopole interactions are determined by fitting to the binding energies of the yrast states of nuclei $^{102-132}$Sn. The mean deviation from experimental data is around 120 keV. In particular, we analyzed the origin of the spin inversion between ^{101}Sn and ^{103}Sn.

In this contribution I present a systematic study on the structure of light tin isotopes. We apply the so-called nuclear shell model. It should be mentioned that the term "shell model" is somewhat confusing since it is more often referred to the independent particle model proposed by Mayer and Jensen. The present shell model may be more precisely described as a full configuration interaction approach. It constructs the wave function as a linear expansion of all possible anti-symmetric Slater determinants within a given model space (Hibert space). In nuclear physics the model space is usually defined by taking into account the single-particle orbits between two adjacent magic numbers. On the first glance one may say that this truncation is a rather coarse approximation. But it works astonishingly well in most nuclei that have been studied. The present work is based on a paper we published recently [1].

The light tin isotopes between shell closures $N = 50$ and 82 are the longest chain that can be reached by contemporary shell model calculations. It is well established that the excitation energies of the first 2^+ states in even-A tin isotopes possess an almost constant value [2], which can be explained from the simple generalized seniority scheme [3]. A realistic description of these nuclei requires a knowledge of the effective interaction between the valence nucleons [4]. Shell model calculations have been very successful in explaining the structure and decay properties of light nuclei below ^{100}Sn (see, e.g., Refs. [4, 5, 6, 7] and references therein). The key to these calculations is a proper description of the monopole channel of the effective interaction which determines its bulk properties. Extensive shell-model calculations suggest that the realistic interaction can give a satisfactory description of the multipole part but not the monopole channel, which may be due to the lack of three-body forces [8, 9]. Thus we are motivated to fine-tune the monopole part of the realistic interaction by fitting to available experimental data in tin isotopes. We expect that the refined effective Hamiltonian will give a better understanding of the structure of trans-tin nuclei [1, 10, 11, 12, 13].

We assumed the doubly-magic nucleus ^{100}Sn as the core. For the model space we choose the neutron and proton orbitals between the shell closures $N = Z = 50$ and 82, comprising $0g_{7/2}$, $1d_{5/2}$, $1d_{3/2}$, $2s_{1/2}$ and $0h_{11/2}$. We also assume isospin symmetry in the effective Hamiltonian.

The monopole interaction is defined as the angular-momentum-weighted average value of the diagonal matrix elements $\langle j_\alpha j_\beta|V|j_\alpha j_\beta\rangle_{JT}$ for a given set of j_α, j_β and T. For the chosen model space there are 15 $T = 1$ monopole terms. The single-particle energies are assumed to be the same for all nuclei within the model space. They are given relative to the neutron $0g_{7/2}$ state. The energy of the $1d_{5/2}$ is taken as $\varepsilon(1d_{5/2}) = 0.172$ MeV [14]. The energies of other states are adjusted to fit the experimental binding energies of tin isotopes. The starting point of our calculation is the realistic CD-Bonn nucleon-nucleon potential which was renormalized using the perturbative G-matrix approach [4]. To minimize the χ^2 function we apply a Monte Carlo global optimization (MC) method which we developed recently. The advantage of the MC method is that no information on the derivatives is required. This is very convenient when other observables (e.g., $B(E2)$ values) are included in the fitting. As a comparison, the singular value decomposition (SVD) approach is also employed in the fitting process. We have 20 variables in total. The fitting was carried out in three steps. In the first step we only consider 131 states in the nuclei $^{102-112}$Sn and $^{120-132}$Sn. The nuclei $^{113-114}$Sn and $^{118-119}$Sn are considered in the second step to further fine-tune the effective interaction. The three isotopes $^{115-117}$Sn are added to the calculation in the last step [1]. Diagonalizations are done within the so-called M-scheme with a parallel shell model program we developed a few years ago [15].

As tests, calculations are done with 10 sets of random monopole Hamiltonians. They are generated by the Monte Carlo sampling approach and are optimized by fitting to the 131 states mentioned above. We found that in all cases one can get convergence with the MC approach within ten iterations, as seen in the left panel of Fig. 1 where one set of these calculations is presented. In the right panel of Fig. 1 the same calculations are done starting from the realistic CD-Bonn potential.

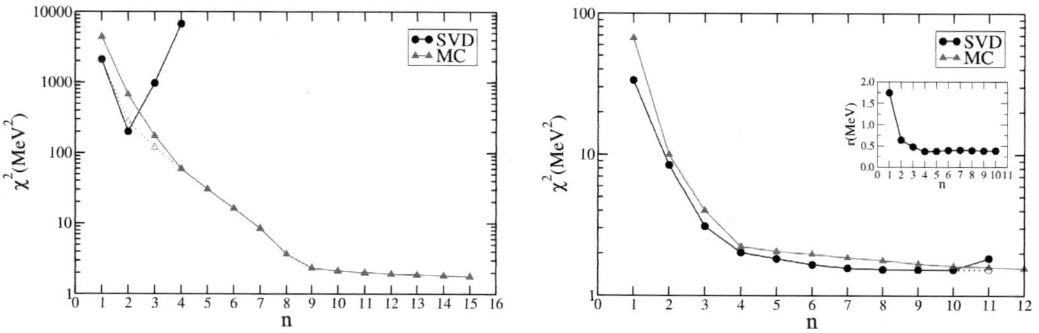

Figure 1. Left: Convergence of the mean-square deviations χ^2 within the SVD and MC approaches for calculations with a random monopole Hamiltonian; Right: Same as the left but for calculations with the CD-Bonn potential [1].

The monopole Hamiltonians we derived in Fig. 1 are slightly refined by including into the fitting the nuclei $^{113-119}$Sn. We include a total number of 157 states in the fitting [16, 17]. After around 15 iterations both calculations give a mean-square deviation $\chi^2 \sim 2.35$ MeV. It means that these states can be reproduced within an average deviation of about 123 keV. The monopole Hamiltonians optimized with the SVD and MC approaches are similar to each other. The largest uncertainties of the optimized monopole Hamiltonian are related to the single-particle energies. The values predicted by the MC approach are $\varepsilon_{1d_{3/2}} = 5.013 \pm 3.10$, $\varepsilon_{2s_{1/2}} = 0.369 \pm 2.63$ and $\varepsilon_{0h_{11/2}} = 3.249 \pm 0.83$ MeV.

In Fig. 2 we plotted the deviation from experimental data for calculations with the two effective Hamiltonians. The difference between the calculations is practically negligible. The largest deviation from experiments, denoted as r in Fig. 1, is seen at $N = 15$ where the ground

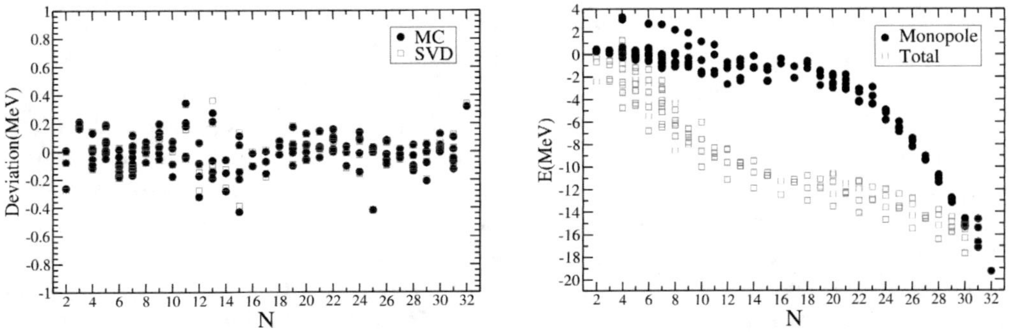

Figure 2. Left: Differences between experimental and calculated binding energies; Right: The calculated shell model energies (solid), $E^{SM} =< H >$, and contributions from monopole Hamiltonian $< H_m >$ (open).

state energy of ^{115}Sn is under-estimated by around 420 keV. In the following only calculations done with the MC optimized effective Hamiltonian are presented for simplicity. The calculated shell model energies, $E^{SM} =< H >$, of the selected states in tin isotopes are plotted in the right panel of Fig. 2. The contributions from the monopole Hamiltonian are also presented for comparison. The comparion between experimental and calculated binding energies of the ground states in Sn isotopes is plotted in Fig. 3.

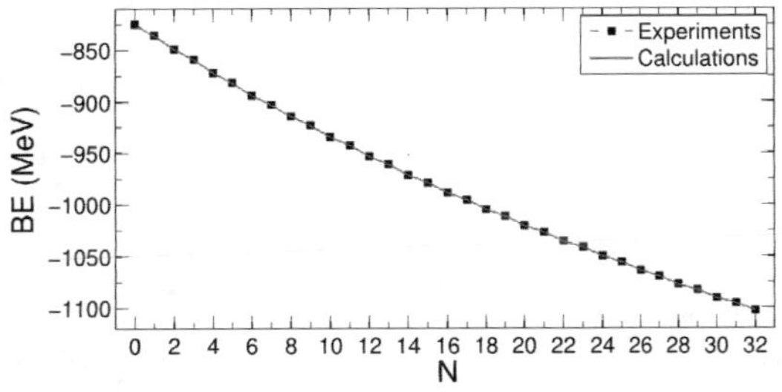

Figure 3. Experimental [16] and calculated binding energies of the ground states in Sn isotopes as a function of valence neutron number.

The excitation energies of the 126 excited states can be reproduced within an average deviation of about 150 keV. The largest difference is seen in the $3/2_1^+$ state in ^{115}Sn mentioned above, where the experimental datum is under-estimated by about 540 keV. Extensive experimental efforts are made recently to explore higher-seniority states built on the 10^+ states in Sn isotopes. The calculated excitation energies of the $11/2^-$, $1/2^+$, $5/2^+$ and $7/2^+$ one-neutron-hole states in ^{131}Sn, relative to the $3/2^+$ state, are -0.022, 0.475, 1.720 and 2.521 MeV, respectively.

Detailed systematic calculations on the spectra and decay properties of tin isotopes as well as the list of the two-body matrix elements will be presented in Ref. [18]. We have done a preliminary optimization of the $T = 0$ monopole interaction by fitting to the binding energies of Sb, Te and I isotopes around the $N = 50$ and 82 shell closures. This will also be available in Ref. [18].

The spins of the ground state and first excited state in ^{103}Sn are $I = 5/2$ and $7/2$, respectively, which are reversed with respect to those in ^{101}Sn [14]. Ref. [14] suggested that the inversion is dominated by orbital-dependent pairing correlations, namely the strength of the pairing matrix elements $\langle 0g^2_{7/2}|V|0g^2_{7/2}\rangle_{J=0}$ is much larger than that of $\langle 1d^2_{5/2}|V|1d^2_{5/2}\rangle_{J=0}$. This produces strong additional binding for the $J^\pi = 5/2^+$ state in ^{103}Sn, which eventually becomes the ground state. The effect of other interaction terms on the spin inversion was not considered in Ref. [14]. In Fig. 4 we analyze the contribution from different components of the effective Hamiltonian to the spin inversion between ^{101}Sn and ^{103}Sn. For that purpose the energies of the first $J^\pi = 5/2^+$ and $7/2^+$ states are calculated with a limited Hamiltonian H' containing the single-particle terms and two-body matrix elements with $J \leq J_{\text{max}}$ only, where J_{max} denotes the maximal spin value of the two-body matrix elements to be considered. Two different types of calculations are presented in the figure. The solid symbols correspond to the results calculated by diagonalizing the Hamiltonian H'. The expectation values of such a Hamiltonian with respect to the corresponding wave functions $|\Psi_I\rangle$ of the full Hamiltonian H, $\langle\Psi_I|H'|\Psi_I\rangle$, are plotted as open symbols. It is thus seen from Fig. 4 that both calculations give similar results concerning the order of the $5/2^+$ and $7/2^+$ states. Calculations with the pairing matrix elements only (i.e., $J_{\text{max}} = 0$) show that the pairing terms, in particular the element $\langle 0g^2_{7/2}|V|0g^2_{7/2}\rangle_{J=0}$ mentioned above, can significantly reduce the gap between the two states but were not strong enough to induce the inversion. A sudden switch is seen when the $J = 6$ two-body matrix elements are considered.

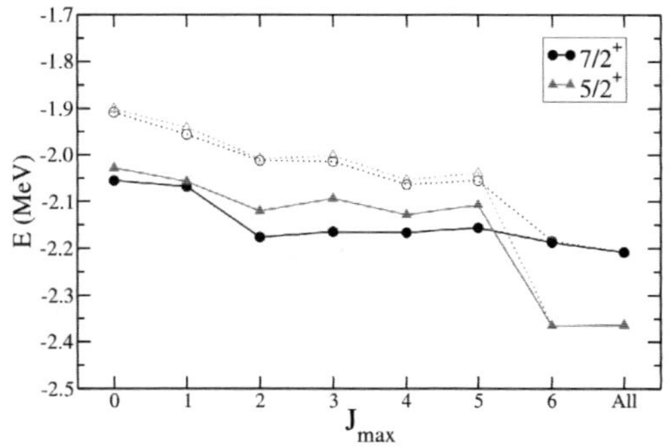

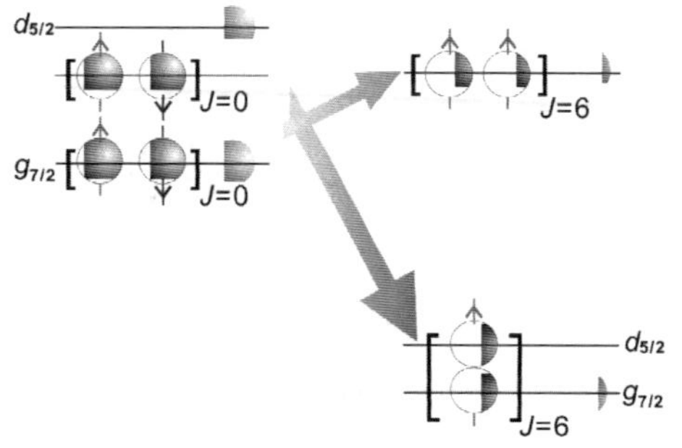

Figure 4. Left: The calculated shell-model energies of the first $5/2^+$ and $7/2^+$ states in ^{103}Sn with a Hamiltonian H' containing the single-particle terms and two-body matrix elements with $J \leq J_{\text{max}}$; Right: A schematic plot on the effect of configurations with high J values.

It can be seen from Fig. 4 that in both calculations the exact results are also approached by including terms with $J \leq 6$ only. This is expected since the low-lying states of light tin isotopes mainly occupy the nearly degenerate orbitals $0g_{7/2}$ and $1d_{5/2}$ for which the maximal spin is $J = 6$. Among the $J = 6$ two-body matrix elements the ones that contribute most to the spin inversion are the repulsive matrix element $\langle 0g^2_{7/2}|V|0g^2_{7/2}\rangle_{J=6}$ and the strongly attractive one $\langle 0g_{7/2}1d_{5/2}|V|0g_{7/2}1d_{5/2}\rangle_{J=6}$. This can be understood by considering the structure of the wave functions of the two states. In the $5/2^+$ ground state in ^{103}Sn, the leading component is $|(0g^2_{7/2})_{J=0}1d_{5/2}\rangle_I$ [14]. Its overlap with the total wave function is calculated to be $|\langle(0g^2_{7/2})_{J=0}1d_{5/2}|\Psi\rangle_I| = 0.86$. One can also construct a three-body state starting from the pair $|0g_{7/2}1d_{5/2}\rangle_{J=6}$. The overlap between the state thus constructed and the total wave function is $|(0g_{7/2}1d_{5/2})_{J=6}0g_{7/2}|\Psi\rangle_I = 0.75$. The $\langle 0g_{7/2}1d_{5/2}|V|0g_{7/2}1d_{5/2}\rangle_{J=6}$ term induces a significant additional binding for the $J^\pi = 5/2^+$ state in ^{103}Sn, as is illustrated in the right panel of Fig. 4. It should be mentioned that states generated by the two couplings $|(0g^2_{7/2})_{J=0}1d_{5/2}\rangle$ and $(0g_{7/2}1d_{5/2})_{J=6}0g_{7/2}\rangle_I$ are not perpendicular to each other. Their overlap is quite large, $\langle(0g^2_{7/2})_{J=0}1d_{5/2}|(0g_{7/2}1d_{5/2})_{J=6}0g_{7/2}\rangle_I = 0.74$. This can be evaluated analytically [19, 20, 21, 22].

The overlaps of the total wave function of the first $7/2^+$ state ^{103}Sn with its leading components are calculated to be $|\langle (1d^2_{5/2})_{J=0} 0g_{7/2}|\Psi\rangle_I| = 0.65$, $|\langle (0g^2_{7/2})_J 0g_{7/2}|\Psi\rangle_I| = 0.62$ and $|\langle (0g_{7/2}1d_{5/2})_{J=6} 1d_{5/2}|\Psi\rangle_I|$=0.57. From a shell-model point of view, the couplings $|(0g^2_{7/2})_{J=0}0g_{7/2}\rangle_I$ and $|(0g^2_{7/2})_{J=6}0g_{7/2}\rangle_I$ generate exactly the same three-particle state. All interaction terms $\langle 0g^2_{7/2}|V|0g^2_{7/2}\rangle_J$ contribute to the total energy of the state [19]. It may be interesting to mention that the effect of the maximally aligned pair in single-j systems was discussed in Refs. [19, 23, 24].

In summary, the structure properties of light tin isotopes are calculated with a global optimized effective interaction. The unknown single-particle energies of the orbitals $1d_{3/2}$, $2s_{1/2}$ and $0h_{11/2}$ and the monopole interactions are refined by fitting to experimental binding energies. A total number of 157 states in $^{102-132}$Sn are considered in the fitting. The binding energies of these states can be reproduced within an average deviation of about 120 keV. The largest deviation is around 400 keV which is seen in the nucleus ^{115}Sn. With the effective Hamiltonian thus derived we analyze the origin of the spin inversion between the $7/2^+$ and $5/2^+$ states in ^{103}Sn and heavier odd tin isotopes in order to explore the possible influence of different interaction channels. We thus find that both the $J = 0$ pairing and the maximally aligned $J = 6$ two-body matrix elements produce strong additional binding for the $5/2^+$ states. The non-diagonal pairing matrix elements play an essential role in inducing the mixing of different configurations in the wave functions of these states.

Acknowledgment

We thank T. Bäck, B. Cederwall, R. Liotta and R. Wyss for discussions. This work has been supported by the Swedish Research Council (VR) under grant No. 621-2010-4723. We also acknowledges the computational support from the Swedish National Infrastructure for Computing (SNIC) at PDC and NSC.

References

[1] Qi C, Xu Z X 2012 *Phys. Rev. C* (in press); arXiv: 1207.5056
[2] Grawe H and Lewitowicz M 2001 *Nucl. Phys. A* **693** 116
[3] Talmi I 1971 *Nucl. Phys. A* **172** 1
[4] Hjorth-Jensen M, Kuo T T S, Osnes E 1995 *Phys. Rep.* **261** 125
[5] Poves A and Zuker A 1981 *Phys. Rep.* **70** 235
[6] Honma M, Otsuka T, Mizusaki T and Hjorth-Jensen M 2009 *Phys. Rev. C* **80** 064323
[7] Qi C, Xu F R 2008 *Nucl. Phys. A* **800** 47; Qi C, Xu F R 2008 *Nucl. Phys. A* **814** 48
[8] Zuker A P 2003 *Phys. Rev. Lett.* **90** 042502
[9] Otsuka T *et al* 2010 *Phys. Rev. Lett.* **105** 032501
[10] Procter M G *et al* 2011 *Phys. Lett. B* **704** 118
[11] Bäck T *et al* 2011 *Phys. Rev. C* **84** 041306(R)
[12] Procter M G *et al* 2012 *Phys. Rev. C* **86** 034308
[13] Bäck T *et al* 2012 *Phys. Scr.* **T150** 014003.
[14] Darby I *et al 2010 Phys. Rev. Lett.* **105** 162502
[15] Qi C and Xu F R 2008 *Chin. Phys. C* **32** (S2) 112
[16] Audi G, Wapstra A H and Thibault C 2003 *Nucl. Phys. A* **729** 337
[17] http://www.nndc.bnl.gov/nudat2/
[18] http://www.nuclear.kth.se/cqi/sn100
[19] Qi C 2010 *Phys. Rev. C* **81** 034318
[20] Qi C 2011 *Phys. Rev. C* **83** 014307
[21] Qi C, Wang X B, Xu Z X, Liotta R J, Wyss R, and Xu F R 2010 *Phys. Rev. C* **82** 014304
[22] Qi C, Xu Z X, Liotta R J 2012 Nucl. Phys. A **884-885** 21
[23] Qi C, Blomqvist J, Bäck T, Cederwall B, Johnson A, Liotta R J, and Wyss R 2011 *Phys. Rev. C* **84** 021301(R)
[24] Xu Z X, Qi C, Blomqvist J, Liotta R J, Wyss R 2012 *Nucl. Phys. A* **877** 51

Open Quantum Systems of Particles and Principle 2 of Thermodynamics

Eliade Stefanescu[1]

[1] Center of Advanced Studies in Physics
at the Institute of Mathematics "Simion Stoilow" of the Romanian Academy,
21 Calea Grivitei, 010702 Bucharest, Romania

E-mail: `eliadestefanescu@yahoo.fr`

Abstract. We consider quantum master equations for a system of Fermions and an electromagnetic field, and apply these equations to a superradiant semiconductor structure converting environmental heat into coherent electromagnetic energy. For a non-irradiative system, these equations describe a time evolution with entropy increase, according to principle 2 of thermodynamics. However, for a superradiant system, the entropy may decrease, while the asymptotic solution of these equations, corresponding to constant entropy, describes a field radiation on the account of environmental heat absorption by Peltier effect.

PACS number(s): 03.65.Ca
Keywords: master equation; super radiance; quantum dot; photon, phonon; correlation

1. Introduction

Physics of open quantum systems has a long history [1, 2, 3, 4, 5]. A general method, for deriving a quantum master equation for a certain system with the density matrix $\rho(t)$, in a dissipative environment with the density matrix R, has been disclosed by Ford, Lewis, and O'Connell in [6]. In a previous paper [7], we derived a non-Markovian master equation for a system of Fermions in a complex dissipative environment of other Fermions, Bosons, and the free electromagnetic field. This equation includes a Markovian term, which describes correlated transitions of the system and environment particles, with energy conservation. For a system of electrons coupled to a coherent electromagnetic field, when the self-consistent field matrix elements are considered for the two-body potential between the system and environment particles, a thermal fluctuation Hamiltonian term, and a time non-local term, are also obtained in the second-order approximation [7].

We applied this equation to a super radiant semiconductor heterostructure converting environmental heat into coherent electromagnetic energy, we called quantum heat converter [8], working on the basis of a new kind of active quantum dots, we call quantum injection dots [9]. In this paper we discuss the agreement of the device operation with principle 2 of thermodynamics, which asserts that, in an isolated system, an ordering process is forbidden. However, there is a problem that this principle seems to be in contradiction with the highly ordered structure of the planet and, consequently, of the planetary system we live in, which is isolated from the rest of our galaxy [10]. In section 2, we describe the dissipative quantum dynamics of a superradiant semiconductor heterostructure by three quantum master equations, for the system of active quantum dots, for the electromagnetic field, and for the crystal lattice vibrations. In section 3, we describe the operation of a quantum heat converter. We derive an entropy equation, and show that for a superradiant system the entropy may decrease. In section 4, we give conclusions.

2. Dissipative dynamics of a superradiant semiconductor structure

We consider an electromagnetic field with two counter-propagating modes, with the densities $\rho_+^F(x,t), \rho_-^F(x,t)$, interacting with an active system of electrons with the density $\rho^S(t)$, a system of quasi-free electrons/holes in the conduction regions with the densities of states $g_\alpha^F(\varepsilon)$ and the occupation probabilities $f_\alpha^F(\varepsilon)$ of these states of energies ε, and a system of phonons generated by this field, with counter-propagating modes with densities $\rho_+^P(x,t), \rho_-^P(x,t)$, which decay in the environment of conduction electrons/holes. With the creation-annihilation operators a_+^+ - a_+ and a_-^+ - a_- for the two field waves, and the Fermi operators c_i^+ - c_i for the active electrons, we obtain the field quantum master equations

$$\begin{aligned}
&\frac{d}{dt}\rho_+^F(x,t) = -\mathrm{i}\omega[a_+^+a_+, \rho_+^F(x,t)] \qquad (1)\\
&+\sum_{j>i}\omega_{ji}\vec{K}_A\vec{r}_{ij}\left[\langle c_j^+c_i\rangle a_+e^{ikx} - \langle c_i^+c_j\rangle a_+^+e^{-ikx}, \rho_+^F(x,t)\right]\\
&+\mathrm{i}\frac{\omega}{2\omega_\nu}Kv\left[\langle a_{\nu+}^+\rangle a_+ + \langle a_{\nu+}\rangle a_+^+, \rho_+^F(x,t)\right]\\
&+\Lambda_x\int_0^x \left\{[a_+\rho_+^F(x',t'), a_+^+] + [a_+, \rho_+^F(x',t')a_+^+]\right\}e^{-\mathrm{i}k(x-x')}dx'\\
&\frac{d}{dt}\rho_-^F(x,t) = -\mathrm{i}\omega[a_-^+a_-, \rho_-^F(x,t)] \qquad (2)\\
&+\sum_{j>i}\omega_{ji}\vec{K}_A\vec{r}_{ij}\left[\langle c_j^+c_i\rangle a_-e^{ikx} - \langle c_i^+c_j\rangle a_-^+e^{-ikx}, \rho_+^F(x,t)\right]\\
&+\mathrm{i}\frac{\omega}{2\omega_\nu}Kv\left[\langle a_{\nu-}^+\rangle a_- + \langle a_{\nu-}\rangle a_-^+, \rho_-^F(x,t)\right]\\
&+\Lambda_x\int_x^{L_D} \left\{[a_-\rho_-^F(x',t'), a_-^+] + [a_-, \rho_-^F(x',t')a_-^+]\right\}e^{-\mathrm{i}k(x'-x)}dx',
\end{aligned}$$

where ω is the field frequency, ω_{ji} are transition quantum dot transition frequencies, $\vec{K}_A = \vec{1}_y\sqrt{\alpha\frac{\lambda}{\mathcal{V}_A}}$ is a vector in the field polarization direction $\vec{1}_y$, depending on the spectroscopic constant α, the field wavelength λ, and the quantization volume of a quantum dot $\mathcal{V}_A$, we call quantization vector, $K = \sqrt{\alpha\frac{\lambda}{\mathcal{V}}}$ is the quantization number of the electromagnetic field in a quantization volume $\mathcal{V}$, $a^+_{\nu+}$ - $a_{\nu+}$ are phonon creation-annihilation operators, ω_ν is the frequency of these phonons generated by the electromagnetic field with the same wave number $k_\nu = k$, v is the sound velocity in the crystal, Λ_x is the field dissipation constant, and L_D is the thickness of the active semiconductor structure, between the two mirrors including this structure. The field master equations (1)-(2) are space non-local, the field quantization volume $\mathcal{V}$ being much larger than the field wavelength λ. This electromagnetic field excites a vibration field described by similar master equations, with the phonon dissipation constant Λ^P_x. For a description of the whole system of interest, besides the master equations (1)-(2) of the electromagnetic field, and of the crystal vibrations induced by this field, we consider a master equation for the active system of electrons [8],

$$\frac{d}{dt}\rho^S(t) = -\frac{\mathrm{i}}{\hbar}[H,\rho^S(t)] + \sum_{ij}\lambda_{ij}\left\{[c_i^+c_j\rho^S(t),c_j^+c_i] + [c_i^+c_j,\rho^S(t)c_j^+c_i]\right\}, \qquad (3)$$

depending on the Hamiltonian $H = \sum_i \varepsilon_i c_i^+ c_i + V$ of these electrons with the electron-field potential V, while λ_{ij} are decay/excitation rates. In [8], explicit expressions of these rates are derived for a semiconductor heterostructure.

3. Entropy equation for a superadiant dissipative system

We consider a super radiant semiconductor device absorbing heat from the environment [8] (figure 1). When an electron current I is injected in the device, an electromagnetic field is generated by quantum transitions at points of maximum field, while the electrons are thermally excited on deep-level paths at the nodes of this field in a Fabry-Perot cavity. Really, the electron current I enhances the lower states of a deep-level path, while the higher states of this path are depleted. In this way, a thermal junction becomes colder, attracting environmental heat, which excites the electrons on the deep-level path of this junction. We define the entropy

$$S_{DLP}(t) = -\sum_k \rho^S_{kk}(t)\left[\ln\rho^S_{kk}(t) - \ln\rho^S_{kk}(\infty)\right], \qquad (4)$$

for the many-level system of the deep-level path of a thermal junction, and

$$S_{SQD}(t) = -\rho^S_{00}(t)\ln\frac{\rho^S_{00}(t)}{\rho^S_{00}(\infty)} - \rho^S_{11}(t)\ln\frac{\rho^S_{11}(t)}{\rho^S_{11}(\infty)}. \qquad (5)$$

for a two-level superradiant quantum dot of a superradiant junction (see figure 1b). From the quantum master equations (1)-(3), we derive field, population, and polarization equations. We obtain two entropy equations:

$$\frac{d}{dt}S_{DLP}(t) = \sum_{ki}\lambda_{ki}\rho^S_{ii}(\infty)\left[\frac{\rho^S_{kk}(t)}{\rho^S_{kk}(\infty)} - \frac{\rho^S_{ii}(t)}{\rho^S_{ii}(\infty)}\right]\left[\ln\frac{\rho^S_{kk}(t)}{\rho^S_{kk}(\infty)} - \ln\frac{\rho^S_{ii}(t)}{\rho^S_{ii}(\infty)}\right], \qquad (6)$$

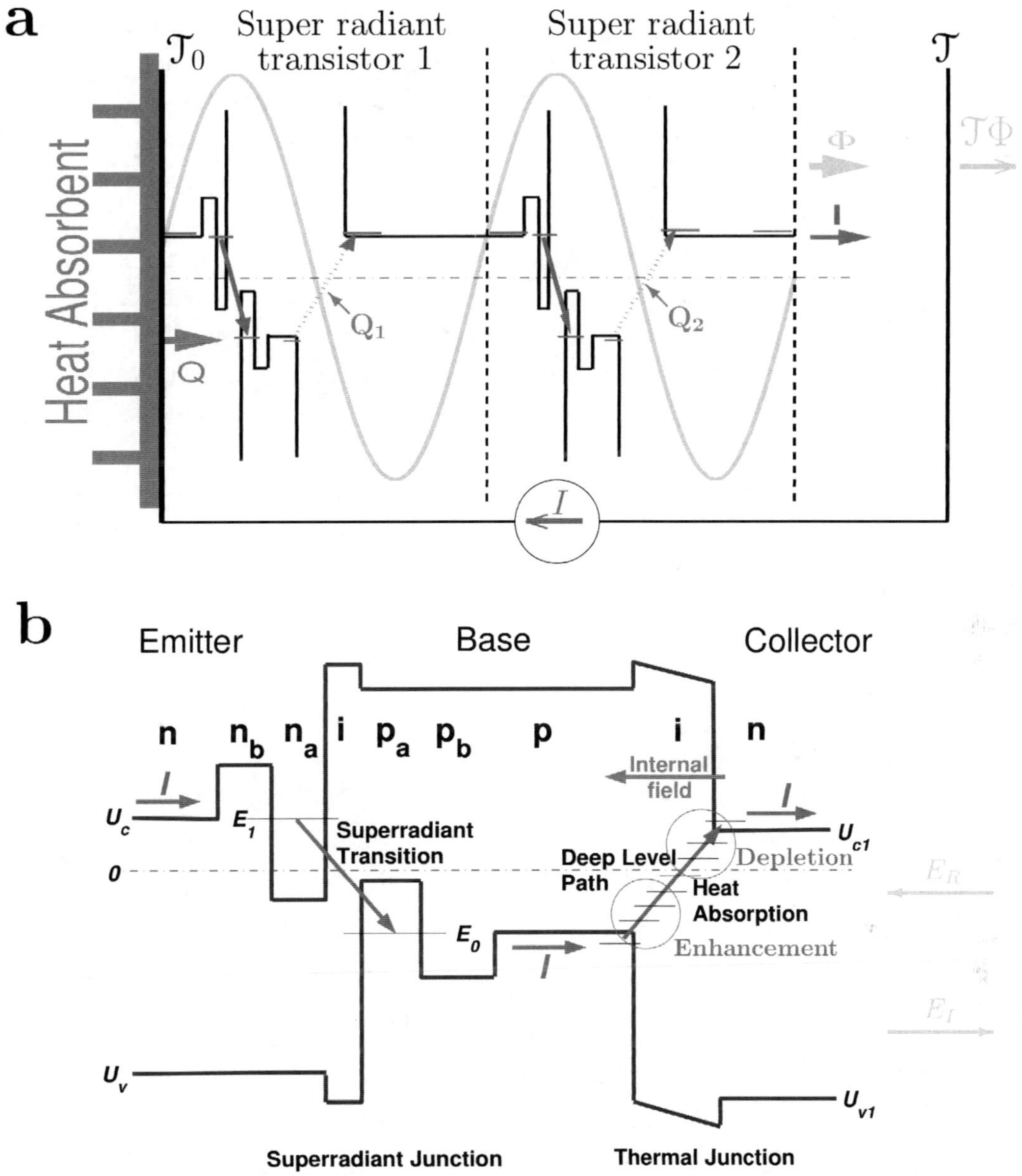

Figure 1. (a) Super radiant semiconductor device as a sequence of super radiant transistors, with super radiant quantum dots situated at points of maximum field, and deep-level paths at nodes of this field, absorbing heat from the environment $Q_1, Q_2, \ldots$; (b) Superradiant transistor: the electrons of a current I injected in the device by a potential difference $U_c - U_{c1}$ undergo a superradiant decay to a much lower potential $E_0 \ll E_1 \approx U_c$, in the superradiant junction, and a thermal excitation from E_0 to $U_{c1} \approx U_c$, in the thermal junction.

and

$$\frac{d}{dt}S_{SQD}(t) = \left\{ \frac{e^2}{4\hbar^2\omega^2 K_A^2} \left[\frac{d}{dt}|\mathcal{E}(t)|^2 + 2\gamma_F |\mathcal{E}(t)|^2 \right] \left[1 - \frac{\rho_{11}^S(t)}{\rho_{11}^S(\infty)} \right] \right.$$
$$\left. + 2\lambda_{10}\rho_{00}^S(\infty) \left[\frac{\rho_{11}^S(t)}{\rho_{11}^S(\infty)} - \frac{\rho_{00}^S(t)}{\rho_{00}^S(\infty)} \right] \right\} \left[\ln \frac{\rho_{11}^S(t)}{\rho_{11}^S(\infty)} - \ln \frac{\rho_{00}^S(t)}{\rho_{00}^S(\infty)} \right], \quad (7)$$

where $\mathcal{E}(t)$ is the electric field amplitude of the superradiant electromagnetic field, and $\gamma_F = \frac{\Lambda_x+\Lambda_x^P}{4k}$ is the decay rate of this field. Equation (6) for the time-evolution of a many-level system of Fermions is a positively defined form, which corresponds to the so-called principle 2 of thermodynamics. Equation (7) for a superradiant two-level system includes a similar positively defined form for the two-level system, and a field dependent form with the sign given by the logarithmic factor. Really, the field dependent term in the curly bracket is always positive: when the population $\rho_{11}(t)$ increases to a higher asymptotic value $\rho_{11}(\infty)$, the field amplitude $|\mathcal{E}(t)|$ also increases; when the population $\rho_{11}(t)$ decays to a lower asymptotic value $\rho_{11}(\infty)$, the field amplitude $|\mathcal{E}(t)|$ decreases. In other words, when the superradiant field increases, the population $\rho_{11}(t)$ increases to a higher asymptotic value $\rho_{11}(\infty)$, while the population $\rho_{00}(t)$ decreases to a lower asymptotic value $\rho_{00}(\infty)$, i.e. the logarithmic factor is negative, which means that the field entropy term decreases. We notice that the field term in equation (7) dominates the atomic term, which being proportional to the excitation rate λ_{10} from the lower state $|0\rangle$ to the higher state $|1\rangle$, is very small.

4. Conclusions

We described a superradiant semiconductor structure by a system of five quantum master equations, for the two counter-propagating modes of the electromagnetic field, the two counter propagating modes of the crystal vibrations excited by this field, and the active quantum dots. For a many level system of Fermions we obtained an entropy increase, in agreement with principle 2 of thermodynamics, while for a superradiant system of Fermions, the entropy may decrease. For the steady-state, while the entropy is constant, a conversion of environmental heat into coherent electromagnetic energy is possible. On the account of this energy, other physical systems may evolve to states with lower entropy. In more general terms, a self-organization of a sufficiently large system may emerge by superradint processes taking place in this system.

References

[1] Lindblad G 1976 *Commun. Math. Phys.* **48** 119-130
[2] Sandulescu A, Scutaru H 1987 *Ann. Phys.* **173** 277-317
[3] Alicki R and Lendi K 1987 *Quantum Dynamical Semigroups and Applications* in *Lecture Notes in Physics* **286** (Springer Verlag, Berlin)
[4] Weiss U 1999 *Quantum Dissipative Systems* (World Scientific, Singapore)
[5] Breuer H-P and Petruccione F 2002 *The Theory of Open Quantum Systems* (Oxford University Press)
[6] Ford G W, Lewis J T and O'Connell R F 1996 *Ann. Phys.* **252** 362-385
[7] Stefanescu E, Scheid W and Sandulescu A 2008 *Ann. Phys.* **323** 1168-1190
[8] Stefanescu E 2010 *Prog. Quant. Elec.* **34** 349-408
[9] Stefanescu E 2012 *Quantum Injection Dots* in *Fingerprints in the Optical and Transport Properties of Quantum Dots* Editor: Ameenah Al-Ahmadi, INTECH, Open Access
[10] Lieb E H and Yngvason J 1999 *Phys. Rept.* **310** 1-96

AUTHOR INDEX

AUTHOR INDEX